Statistical Modelling
in Hydrology

Statistical Modelling in Hydrology

Robin T. Clarke

Instituto de Pesquisas Hidráulicas
Universidade Federal do Rio Grande do Sul
Porto Alegre RS, Brazil

JOHN WILEY & SONS
Chichester • New York • Brisbane • Toronto • Singapore

Copyright © 1994 by John Wiley & Sons Ltd,
Baffins Lane, Chichester,
West Sussex PO19 1UD, England
National Chichester (0243) 779777
International (+44) 243 779777

Other Wiley Editorial Offices

John Wiley & Sons, Inc., 605 Third Avenue,
New York, NY 10158-0012, USA

Jacaranda Wiley Ltd, 33 Park Road, Milton,
Queensland 4064, Australia

John Wiley & Sons (Canada) Ltd, 22 Worcester Road,
Rexdale, Ontario M9W 1L1, Canada

John Wiley & Sons (SEA) Pte Ltd, 37 Jalan Pemimpin #05-04,
Block B, Union Industrial Building, Singapore 2057

Library of Congress Cataloging-in-Publication Data

Clarke, Robin T.
    Statistical modelling in hydrology / by Robin T. Clarke.
        p.   cm.
    Includes bibliographical references and index.
    ISBN 0-471-95016-5
    1. Hydrology — Mathematical models.   I. Title.
    GB656.2.M33C58   1994
    551.48′01′5118 — dc20   94-4944
                                                      CIP

British Library Cataloguing in Publication Data

A catalogue record for this book is available from the British Library

ISBN 0-471-95016-5

Typeset in 10/12pt Times by Laser Words, Madras, India
Printed and bound in Great Britain by Bookcraft (Bath) Ltd, Avon

For my mother and father

Contents

Preface

As with all human activity, fashions in hydrology come and go. In recent years there has been much emphasis on hydrology as a physical science: an emphasis to be welcomed in that it seeks to lessen the unpredictability of hydrological systems. However, physical hydrology is largely what hydrologists do when they are employed by universities and research institutions, and hydrologists who are not research workers (as well as some who are) must address difficult questions which necessarily require interpretation of past hydrological records of rainfall, climatological measurements, and river flow. These records are seldom complete, not always adequate for the job at hand, and have usually been collected by different individuals not all of whom were dedicated to accuracy.

Nevertheless, the development of a nation's water resources is critically dependent upon the interpretation of such records, and statistical methods are essential for this interpretation. Whilst questions of the 'What if ... ?' kind can best be answered using physics, other questions — such as 'How often ... ?' — are best answered by applying statistical methods.

This book is intended for hydrologists who interpret hydrological records. Much of the material is not new, but the advent of cheap and powerful computers has changed the way in which data are analysed, and the book aims to illustrate the changes which are taking place. For example, methods which, in the past, could rarely be used because of the effort needed to compute them, can now be used with ease, so that the use of 'quick-and-dirty' methods is no longer justified.

I acknowledge with deep gratitude the help from colleagues and students at the Instituto de Pesquisas Hidráulicas (IPH), Porto Alegre RS, Brazil. Without their continued support this book would not have been written. Some of the data used in Chapters 1 and 5 were collected as part of the ABRACOS Project and made available by the Institute of Hydrology (UK) and the Instituto Nacional de Pesquisas Espaciais (Brazil). ABRACOS is a collaboration between the Agência Brasileira de Cooperação and the UK Overseas Development Administration. I am indebted to my colleague and friend Dr John Roberts, of the Institute of Hydrology, for the upper of the two flood photographs on the book's cover.

The diskette available from The MathWorks is distributed by The MathWorks, Inc., 24 Prime Park Way, Natick, MA 01760-1520. Phone: (508)653-1415; Fax: (508)653-2997; E-mail: info@mathworks, com. Please fill in the reply card at the end of this book to order a copy. MATLAB and SIMULINK are registered trademarks of The Mathworks, Inc. GENSTAT is a registered trademark of the Lawes Agricultural Trust.

Finally, I have used the pronoun 'he' as an abbreviation for the more politically correct form 'he/she', but it is hoped that readers of both sexes will find the book useful.

Chapter 1

Some applications of statistical models in hydrology

1.1 HYDROLOGICAL VARIABILITY

In the study of the water cycle, both in its hydrological and water chemistry aspects, inferences must commonly be drawn from sequences of measurements that are recorded at discrete intervals of time, as illustrated by the following examples.

Figure 1.1 shows the sequences of annual floods (more precisely, the sequences of maximum values of mean daily discharge) occurring in each of the years 1934–84 for two sites in the

Ibirama	*	1342	625	619	797
Apiuna	1111	1914	1507	950	1111
Ibirama	1250	271	263	566	649
Apiuna	1742	1033	918	881	1960
Ibirama	236	474	763	592	981
Apiuna	495	566	1280	1100	2250
Ibirama	438	281	556	393	726
Apiuna	702	1680	1260	909	1620
Ibirama	897	969	566	1300	526
Apiuna	2630	1890	*	3090	1220
Ibirama	520	487	897	582	510
Apiuna	936	1240	2160	1550	1750
Ibirama	*	708	998	477	298
Apiuna	648	1460	1930	727	562
Ibirama	872	483	1040	1010	1240
Apiuna	1730	1020	2030	2210	2310
Ibirama	697	1406	801	741	1002
Apiuna	951	2760	1575	1640	2156
Ibirama	1090	*	589	490	2475
Apiuna	1847	3086	927	1539	4327
Ibirama	2125				
Apiuna	4314				

Figure 1.1   Annual floods (annual maximum mean daily discharges, cubic metres per second) for two basins on Rio Itajaí-Açú, Santa Catarina, Brazil: Rio Hercílio at Ibirama (area 3314 km^2) and Rio Itajaí-Açú at Apiuna (area 9242 km^2), 1934–84.

basin of the Rio Itajaí in the state of Santa Catarina, Brazil. The sequences are for the gauging stations Rio Hercílio at Ibirama (catchment area 3314 km^2) and Rio Itajaí-Açú at Apiuna (area 9242 km^2). Missing values—another common feature of hydrological sequences—are denoted by the symbol $*$. A noticeable feature of both sequences is the two extreme floods which occurred in the final two years of the record, 1983 and 1984, both of which are believed to be associated with *el Niño* events.

Figure 1.2 shows the mean monthly discharges, in cubic metres per second, for the first five years of record, and also for the last five years, at a gauging station on a tributary of the Rio Uruguay, Rio Lava Tudo at Fazenda Mineira (area 1147 km^2). Whilst there is some evidence of an annual cycle, with mean monthly discharge tending to be lower in the months November to January, there is much year-to-year variability. Water supply schemes using water extracted from the river would clearly need to take full account of this variability if serious shortfalls in supply are to be avoided.

Figure 1.3 shows hourly sequences of measured solar radiation, net radiation, wet and dry bulb temperatures, wind speed and direction for one 24-hour period in November 1990, as recorded by an automatic weather station in the city of Manaus, Amazonas. The sequence is part of a very much larger data set of hourly data extending over several years, collected for the

Month	Year 1	Year 2	Year 3	Year 4	Year 5
January	1.83	13.90	0.95	32.17	9.87
February	3.03	4.99	27.09	29.75	24.21
March	3.15	8.21	4.57	16.06	10.16
April	2.16	3.15	2.58	7.20	2.78
May	12.28	1.70	1.60	8.18	21.26
June	33.36	16.55	3.55	78.18	19.51
July	34.38	11.87	5.73	50.46	20.22
August	53.69	4.39	10.84	22.99	18.67
September	43.52	6.83	22.72	9.02	40.38
October	13.10	4.43	10.29	10.46	22.09
November	7.44	9.91	3.90	3.51	10.31
December	4.83	1.27	4.14	6.07	24.41

Month	Year 45	Year 46	Year 47	Year 48	Year 49
January	5.95	13.40	34.53	15.82	45.78
February	20.60	9.57	37.85	15.68	31.56
March	26.50	3.38	8.59	15.12	10.59
April	27.60	21.78	52.66	46.40	21.26
May	18.20	24.10	118.90	48.25	34.77
June	25.60	18.10	37.01	35.18	7.10
July	16.90	16.80	39.88	17.84	23.00
August	24.50	17.90	50.32	5.57	49.68
September	23.60	26.00	32.05	45.21	90.77
October	26.30	30.70	69.30	17.76	26.46
November	24.00	63.00	19.99	7.20	9.32
December	4.83	28.20	13.72	9.79	9.14

Figure 1.2   Mean monthly discharges (cubic metres per second) for the Rio Lava Tudo at Fazenda Mineira (area 1147 km^2).

SolR	NetR	Wet	Dry	Wspeed	Wdir
−2.0	−26.9	23.0	23.5	0.2	80.5
−1.7	−19.4	23.2	23.7	0.4	64.9
−1.8	−2.1	23.1	23.4	0.2	58.5
−1.2	−13.6	23.2	23.5	0.0	57.6
−1.4	−13.6	23.1	23.2	0.0	51.7
−0.8	−9.6	23.3	23.4	0.0	181.4
−0.4	−7.6	23.4	23.6	0.1	135.8
−0.9	−10.2	23.2	23.4	0.1	129.5
−1.1	−11.3	23.1	23.2	0.0	65.7
4.5	−4.0	23.3	23.4	0.0	80.2
58.0	30.3	23.7	23.9	0.0	134.2
128.4	76.8	24.0	24.7	0.1	74.2
289.7	186.4	24.4	26.0	0.3	87.8
639.9	424.5	25.1	28.2	0.7	97.7
542.0	355.7	25.5	29.1	0.7	93.2
508.8	333.6	25.5	29.7	0.9	72.8
695.1	455.2	25.6	30.8	1.0	56.3
682.8	442.2	25.7	31.7	0.8	91.9
520.6	330.5	25.6	31.7	0.9	59.5
394.3	235.5	25.7	31.8	0.5	78.0
207.0	92.5	25.3	31.2	0.2	94.7
29.8	−24.4	24.8	29.5	0.0	87.1
−2.4	−42.7	24.4	28.5	0.3	88.8
−2.4	−42.2	24.4	28.0	0.3	97.2

Figure 1.3   Meteorological data recorded by the Automatic Weather Station at Manaus City site for one day in November 1990. Data are hourly totals or means, calculated from measurements made every 10 seconds. Hours are Greenwich Mean Time.

Symbols are:

SolR   : Solar radiation (watts per square metre)
NetR   : Net radiation (watts per square metre)
Wet   : Wet bulb temperature (degrees Celsius)
Dry   : Dry bulb temperature (degrees Celsius)
Wspeed : Wind speed (metres per second)
Wdir   : Wind direction (degrees).

purpose of studying the energy balance within Amazonia and how it is affected by deforestation. The radiation data in particular show a clear diurnal pattern, with greater fluctuations around the trend in daylight hours.

Figure 1.4, used by kind permission of Dr Pierre Chevallier of ORSTOM, shows values for eight variables measured on 30 separate storms on a first-order basin of the River Booro-Borotou, area 1.36 km^2, in Ivory Coast. The storms occurred during the period from June 1984 to October 1987; unlike the data in Figures 1.1 to 1.3, the observations in each data sequence are unequally spaced in time, corresponding to the irregular occurrence of rain. The first variable, flood volume, has been ranked in decreasing order of magnitude; the other variables follow the same ranking.

All four of these examples show two characteristics: first, that the observations are ordered in time; and second, that the observations within each sequence commonly follow a general

Day	Month	Year	fv	fp	ft	fs	bf	pv	pt	pi
18	Aug	1985	11.10	4150	148	1	20.50	82.7	304	80.4
21	Aug	1987	4.03	1260	566	1	2.06	55.0	205	80.2
27	Sep	1987	2.60	1050	189	1	9.11	42.4	275	64.0
3	Aug	1985	1.89	466	394	1	6.75	54.8	431	41.9
30	Aug	1985	1.78	315	349	1	13.50	45.6	182	47.2
14	Sep	1984	1.73	183	492	2	4.40	57.0	254	35.1
5	Aug	1985	1.58	570	205	1	12.10	43.4	289	26.0
16	Jun	1984	1.56	682	201	1	0.19	56.4	200	95.7
13	Sep	1986	1.30	201	558	2	3.44	40.0	149	32.4
27	Jul	1985	1.25	466	218	1	1.81	54.4	222	38.3
31	Aug	1987	1.20	115	859	2	7.98	31.2	300	21.7
1	Sep	1984	1.13	62	836	2	2.75	41.8	489	21.0
31	Oct	1985	1.09	71	884	2	7.08	28.3	*	11.4
11	Jul	1985	1.04	265	343	1	2.40	33.7	148	35.0
15	Aug	1987	1.00	66	1041	2	0.45	52.5	327	52.6
3	Sep	1985	0.92	119	550	2	20.50	35.6	331	*
12	Sep	1985	0.88	173	363	1	19.60	24.8	54	43.0
13	Aug	1985	0.85	119	420	2	13.80	32.1	331	31.4
29	Sep	1987	0.84	201	328	1	20.80	19.9	50	28.1
17	Oct	1984	0.82	111	479	2	2.06	39.7	153	33.7
12	Oct	1987	0.79	61	160	2	10.20	22.5	215	19.3
2	Sep	1987	0.79	64	963	2	10.70	20.4	179	25.4
2	Oct	1984	0.76	131	445	2	3.13	34.8	279	47.7
8	Oct	1987	0.69	64	1009	2	11.90	18.9	161	*
30	Sep	1986	0.66	45	1281	2	2.00	29.4	172	46.6
4	Sep	1985	0.62	442	114	2	91.00	16.7	207	24.4
20	Jun	1987	0.60	136	363	2	0.00	70.1	252	73.7
17	Aug	1987	0.52	24	1273	2	2.32	24.2	256	25.6
3	Sep	1987	0.51	67	766	2	19.00	14.4	208	15.9
12	Sep	1986	0.50	18	1513	2	1.56	19.6	149	27.2

Figure 1.4   Values of eight variables measured on each of 30 different storms crossing the basin of the River Booro-Borotou, Ivory Coast, West Africa, during the period June 1984 to October 1987. Basin area: 1.36 km^2. Definition of symbols:

fv : Flood volume (millimetres)
fp : Peak discharge (litres per second)
ft : Flood duration (minutes)
fs : Indicator variable (time since last storm)
bf : Base flow (litres per second)
pv : Precipitation depth (millimetres)
pt : Precipitation duration (minutes)
pi : Maximum 30-minute rainfall intensity (millimetres per hour).

trend with considerable fluctuations around it. Unlike the sequences that occur in mathematics, where plotted points lie on smooth curves, sequences of hydrological observations often show wide scatter when plotted, although trends amongst the plotted points may be evident. Statistical methods are therefore required both to describe succinctly the behaviour of individual hydrological variables, and to elicit relations between them.

1.2 STATISTICAL MODELS

For many hydrological applications, it is sufficient to regard the pattern shown by a plotted variable (which we commonly denote by suffixed symbols such as y_t, q_t: the latter, when referring to variables derived from hydrographs) as consisting of the sum of two components: a 'systematic' part, and a 'random' part. The systematic part is commonly represented by a function, say $f(t)$, of time; this function will usually contain some fixed but unknown quantities which must be estimated from the available data. These quantities are 'parameters', and are commonly represented by the Greek letter θ which may have several components $\theta_1, \theta_2, \ldots, \theta_k$; thus the systematic part is written more fully as $f(t; \theta)$ where θ is a vector with k components.

The random part, representing fluctuations about the systematic component, is usually represented by the symbol ε_t or a_t, the suffix being a reminder that these components also have a particular time order. Thus with q_t being the variable of interest, we can write

$$q_t = f(t; \theta) + \varepsilon_t \tag{1.1}$$

Given appropriate assumptions about the ε_t (to be discussed later) equations like (1.1) represent a family of 'statistical models'. It is a 'family' because the precise form of the function $f(.)$ remains to be specified; and the numerical values of the parameters θ — together with other parameters describing properties of the random components ε_t — must also be specified. Typical problems encountered in the statistical modelling of hydrological series include how to determine which explicit form $f(t; \theta)$ is most appropriate for the particular hydrological variable being modelled; and how to obtain efficient estimates of the unknown parameters θ (and also the other parameters entering the model through the specification of ε_t).

As examples of the kinds of statistical model included in the family of models (1.1), we refer to the data shown in Figures 1.1 to 1.3. The 48 years of annual flood record shown in Figure 1.1 for the Rio Hercílio at Ibirama can be represented in the first instance by a simple model of the form $q_t = \mu + \varepsilon_t$, where q_t is the observed annual flood in year t. The systematic part of the model, μ, is independent of time, and is thus a parameter in the sense defined earlier; the random components ε_t — which can be either positive or negative — then represent the year-to-year fluctuations about the fixed value μ. Of course, if land development or other factors caused changes in the pattern of flooding throughout the period of record, it would not be appropriate simply to assume that the systematic part was constant; statistical analysis would reveal whether 'non-stationarity' of this kind was in evidence.

In the case of the record of mean monthly discharge (q_t) for the Rio Lava Tudo, shown in Figure 1.2, it would be reasonable in the first instance to explore the suitability of a model of the form

$$q_t = \alpha + \beta_1 \cos(2\pi t/12) + \beta_2 \sin(2\pi t/12) + \varepsilon_t \tag{1.2}$$

The systematic part of the model now is no longer constant but depends on t, and it contains three parameters α, β_1 and β_2. As before, if the pattern of mean monthly discharge yielded by the basin was changing through the period of record because of trends in land use or other factors, the model (1.2) would need to be modified to take this into account. Note that, in the absence of such trends, the model (1.2) still has a kind of 'stationarity', even though monthly discharge varies according to an annual cycle, in that q_t is varying harmonically about a fixed value α.

The data shown in Figure 1.3 for solar and net radiation show a more complicated systematic component. Within daylight hours, the pattern is approximately harmonic in form; in the hours of darkness, a systematic component of the type (1.1) is more appropriate. There is also some evidence that the fluctuations about the general trend are greatest when solar radiation is most intense.

1.3 STATISTICAL MODELS USING EXPLANATORY VARIABLES

The examples discussed above using the data from Figures 1.1 to 1.3 referred to statistical models with systematic components which relate q_t to time t. They were thus descriptive of the behaviour of a single variable and did not seek to establish relationships between two or more hydrological variables. Often we need to explore such relationships, and the systematic component of the model then itself includes hydrological measurements, termed *explanatory variables*. Thus, to explore the relation between mean monthly discharge q_t and monthly rainfall in months $t, t - 1, \ldots$ the general form (1.1) is modified to the form

$$q_t = f(p_t, p_{t-1}, \ldots; \; \boldsymbol{\theta}) + \varepsilon_t \qquad (1.3)$$

In equations like (1.3), the variable q_t is often termed the *response variable*, since the model expresses how the variable q_t responds to monthly rainfall as measured by the explanatory variables on the right-hand side. (In some texts, response variables and explanatory variables are termed 'dependent' and 'independent' variables, respectively). As before, an important part of the modelling process is the selection of the most appropriate form for the systematic component $f(.)$ on the right-hand side of equation (1.3). As an explicit example, suppose that q_t is storm runoff from a drainage basin during a short interval of time indexed by t, and that $p_t, p_{t-1}, \ldots$ are the depths of effective rainfall during the time intervals $t, t - 1, \ldots$. If it is assumed that the basin, when saturated, functions as a linear system; that the unit hydrograph (or, in the jargon of systems theory, the 'impulse response function') has k ordinates; and that q_t contains a random component of error ε_t associated with factors such as the inexact knowledge of mean areal precipitation, then the model relating q_t to the p_t takes the form

$$q_t = h_0 p_t + h_1 p_{t-1} + h_2 p_{t-2} + \cdots + h_{k-1} p_{t-k+1} + \varepsilon_t \qquad (1.4)$$

the systematic part of the model now containing the k parameters $\{h_i\}, \ i = 0, \ldots, k - 1$. Obviously we can write a general form for a statistical model with explanatory variables in the form

$$q_t = f(\{p_t\}, \{E_t\}, \{s_t\}, \ldots, \boldsymbol{\theta}) + \varepsilon_t \qquad (1.5)$$

having several sets of explanatory variables $\{p_t\} = \{p_t, p_{t-1}, \ldots\}$; $\{E_t\} = \{E_t, E_{t-1}, \ldots\}$; $\{s_t\} = \{s_t, s_{t-1}, \ldots\}$ which might represent, for example, rainfall, potential evaporation, soil moisture content, during the time intervals $t, t - 1, \ldots$.

In the family of statistical models containing explanatory variables, the simplest kind is that for which the function $f(.)$ is linear in the unknown parameters, as in equation (1.4) above. Whatever the form of the function $f(.)$, the explanatory variables may be 'causative' (in the sense that rainfall and soil moisture status determine surface runoff) or they may simply be variables that are useful for estimating q_t without necessarily implying a causative relation. Thus, during a long period of recession when the daily flow q_t is steadily growing smaller, the variable q_t may be a very good guide to the likely value of q_{t+1} in the absence of rainfall, using a model of the form $q_{t+1} = \theta q_t + \varepsilon_t$, although q_t is not causative of q_{t+1}.

1.4   THE RANDOM COMPONENT ε_t

Hitherto we have said very little about the properties of the quantities ε_t; by regarding them as random, however, we imply that they possess a probability distribution. The following assumptions will be made:

(i)   We shall assume that the mean value ('expected value' or 'expectation', denoted by the symbol E: see Section A.4 of the Appendix) of the random variable ε_t is zero: that is, $E[\varepsilon_t] = 0$.

(ii)   For much the greater part of this book, we shall also assume that the ε_t are statistically independent: that is, that the joint probability distribution $P[\varepsilon_{t_1}, \varepsilon_{t_2}, \varepsilon_{t_3}, \ldots]$, of each and every set of the ε_t, factorises to give the form

$$P[\varepsilon_{t_1}, \varepsilon_{t_2}, \varepsilon_{t_3}, \ldots] = P[\varepsilon_{t_1}]P[\varepsilon_{t_2}]P[\varepsilon_{t_3}]\ldots \qquad (1.6)$$

where, for example, $P[\varepsilon_{t_1}]$ is the marginal distribution of ε_{t_1}. Thus, using the model $q_t = \mu + \varepsilon_t$ for the annual flood data in Figure 1.1, the assumption of statistical independence implies that no information would be lost by rearranging the sequence in which the values q_t are written. The assumption of statistical independence is clearly a very strong one, and checking it should form a routine part of the process of selecting an appropriate statistical model, fitting it (that is, estimating its parameters) and verifying that the model assumptions are valid for the particular data set under study.

(iii)   We shall generally assume that the random variables $\{\varepsilon_t\}$ have the same probability distribution: that is, that the marginal distributions $P[\varepsilon_{t_1}], P[\varepsilon_{t_2}], \ldots$ introduced in (ii) above are all of identical mathematical form, with the same parameters. At the beginning of each modelling study we shall, of course, need to assume a particular form for this distribution, and the Appendix gives a summary, by no means exhaustive, of the principal characteristics of probability distributions commonly used in the statistical analysis of hydrological data. Just as analysis must verify whether the initial choice for the systematic component is appropriate, so it must also verify whether the initial choice of probability distribution is appropriate for the data under study.

Thus, every statistical model has parameters of two kinds: those associated with the systematic part of the model, and those associated with the particular probability distribution used for the random components ε_t.

1.5   PARSIMONY IN STATISTICAL MODEL BUILDING

We have seen that each statistical model contains some parameters associated with its systematic component, and others associated with its random component. If we were to be prodigal in our use of parameters, say by using a model with as many parameters in its systematic component as we had data values, we would find that the model fitted the data exactly: the systematic component would pass through every data point plotted (whether we are plotting one hydrological variable as a function of time, as in equation (1.1), or a response variable against an explanatory variable, as in equation (1.3)). In practice, however, such a model would perform very poorly, in the kind of applications discussed in Section 1.6.

A cardinal principle in the building of statistical models, in hydrology as in other sciences, is that we should use the smallest number of parameters necessary to describe adequately the characteristics of the data. Essentially, the model is like the summary of a technical document; if the summary is as long as the original document or longer, it is failing in its purpose. On the other hand, if a statistical model fails to describe important characteristics of the data being modelled, it is equally unsatisfactory — just as the summary of a technical report is unsatisfactory if it is so condensed that it omits mention of some of the report's most important conclusions. Fortunately, statistical theory provides quantitative procedures which enable us to ascertain, in many cases, whether the inclusion of additional parameters adds appreciably to the information already included within a particular model structure.

'Parsimony', in statistical model building, is the principle of limiting the parameters to a small number which adequately summarise the characteristics of the data, and in later chapters we shall discuss procedures which seek to establish parsimonious descriptions of data.

1.6   SOME HYDROLOGICAL USES OF STATISTICAL MODELS

We have used the term 'building' to describe the process of formulating statistical models of hydrological data. Like laying bricks, model formulation is an iterative process. We begin by advancing a candidate model which, from graphical inspection of the data or otherwise, appears likely to represent their principal characteristics. Next, we determine in which respects the candidate model fails in its intended purpose; this we do by plotting the 'fitted values' given by the model, and comparing the fitted values with data. The discrepancies between the fitted values and the observations will give clues as to how the candidate model should be modified. Having modified it, we again compare fitted values obtained from the model with observations, modifying the model a second time if necessary. We continue this iterative process until we can find no further improvement in model fit. Statistical theory provides us with methods by which to determine whether the current modification to the model structure has resulted in a substantial improvement, or whether the improvement is 'not significant'. In the latter case, the iterative process stops and we adopt, as the 'best-fitting' model, that for which no further significant improvement was found.

Clearly the fitting of statistical models to hydrological data is a process requiring constant interaction between modeller and computer. This interaction brings two dangers:

(i)   Modelling may become an end in itself, the modeller losing sight of the practical purpose for which the model was being originally developed.

(ii) The modeller may come to believe that his model is the hydrological reality, instead of being a greatly simplified description of something which is really very much more complicated. An extreme case of this danger would occur if the modeller chose to discard field observations of hydrological behaviour which did not agree with model 'predictions'. It is often these 'aberrant' observations which point to serious shortcomings in the statistical model and which, if studied, may lead ultimately to its improvement.

Another way of looking at the statistical modelling of hydrological data is to consider it as a procedure for setting up hypotheses about how hydrological systems behave. We may recognise that the 'current' model is not a particularly satisfactory description of system behaviour, and we shall not be too disturbed about replacing it by something better, as new data come along and as our insight into hydrological processes develops. The process of adopting a hypothesis until something better is available to replace it is the process by which hydrological science develops; but in some applications, particularly those where water resource systems are being designed, we are not in a position to wait until new data and new thinking come along. We must then pay particular care to make sure that the model that we use for design is as 'good' (in terms of representing the principal characteristics of hydrological data) as we can possibly make it. This we can do best by separating the available data into two parts, one of which is used to fit the model and the other to assess how well it performs. In some circumstances — such as where the data are a sequence of annual floods, the order of which bears no particular significance — we may divide the data several times over, repeating the fitting and testing of the model at each division. In other circumstances — as where the data are mean daily discharges, the order of which has significance — we must find ways of dividing the data into two parts which do not destroy this information.

In the remainder of this section, we give some examples of hydrological analyses for which statistical models are used. The list is by no means exhaustive, but serves to introduce applications which will be discussed at greater length during chapters which follow.

1.6.1 Estimating Floods with T-year Return Period

The data of Figure 1.1 are the maximum mean daily discharges observed at two sites on the Rio Itajaí system, for each year in the period 1934–84 (although a few years with incomplete records have been omitted). To avoid a clumsy phrase, we call the 'annual maximum mean daily discharge' the 'annual flood'. The annual flood which would recur, in the long run, with a frequency of once in T years, is a measure of flood susceptibility which is of interest to civil engineers designing dam spillways; to planning officials concerned with developing domestic housing schemes; to highway engineers building roads in river valleys; and to industrialists considering sites for new factories and power stations.

Using a model of the form $q_t = \mu + \varepsilon_t$ with an appropriate probability distribution $P[\varepsilon_t]$ for the random components $\varepsilon_t = q_t - \mu$, we see that μ is the mean annual flood, by virtue of the assumption $E[\varepsilon_t] = 0$. An estimate X_0 of the annual flood with return period T years is given by the following procedure. First, having selected a candidate distribution $P[\cdot]$, we obtain efficient estimates of the parameters of this distribution. Second, we determine whether the fitted distribution adequately represents the observed sequence of annual floods: if it does not, we modify it accordingly. Third, when a satisfactory model has been identified, we solve the integral

$$\int_{X_0}^{\infty} f(q; \theta)\,\mathrm{d}q = 1/T$$

where $f(q; \theta)$ is the probability density function of $q = \mu + \varepsilon$. Having thus obtained an estimate $\hat{X}_0$ of X_0, we shall probably require to calculate limits, enclosing $\hat{X}_0$, within which the 'true' value lies with a given 'confidence probability'.

1.6.2 Extending Streamflow Records

In developing countries, flood records as long as those for the Rio Hercílio at Ibirama and the Rio Itajaí at Apiuna are the exception rather than the rule; even where a record has been kept for many years, gaps may occur that destroy its continuity. An engineer designing a dam spillway, or a regional planner who finds he has only a short flood record at his disposal, must then explore whether longer records of annual floods exist for other gauging sites, possibly in neighbouring catchment areas. If longer flood records can be found for nearby sites, the possibility may exist of 'transferring information' to the site of interest: that is, of using the longer record to add information about the frequencies of extreme floods. Various procedures exist for effecting such a transfer; one approach is to use an appropriate bivariate distribution (see Appendix) for the flood flows at two sites, and to calculate the conditional distribution of the annual flood at the short-record site, given values of the annual flood at the long-record site (see Chapter 4). Information is thus 'transferred' from sites with long flow records to sites with shorter flow records; particular care may be necessary to avoid 'diluting' the information available at the site with short record. This can occur if the correlation between long-record and short-record sites is small.

1.6.3   Information Transfer to Sites without Flow Records

A problem that is yet more extreme than that described in Section 1.6.2 occurs where no flood records exist near a site for which a development is being proposed. A procedure then used is to assemble many flood records from the region, and to seek statistical relationships between flood characteristics (typically, the mean annual flood) and variables describing the catchment areas above the flow gauging sites. These explanatory variables may include measures of channel slope, percentage of area covered by open water surface, measures of rainfall regime, and so on. Having derived statistical relationships expressing the response variable (say, mean annual flood) in terms of the explanatory variables (channel slope, impermeable area, etc.), knowledge of the explanatory variables for the site without record can be used to estimate its flood characteristics (see Chapter 4). Problems will arise in practice because records of discharge are of variable length and of variable quality; and a major difficulty may arise where the majority of gauge sites are in the lower reaches of drainage basins, with few in the smaller, headwater catchment areas which are often of principal interest in flood studies.

1.6.4   Simulation of Flow Sequences for Water Resource Planning and Operation

Planners concerned with supplying water for domestic use, intensive agriculture, power generation, or industry, must ensure that the supply is sufficiently reliable. This does not mean, of course, that the supply should never fail, but that the frequency of failure is acceptably small, that the durations of failures when they occur are acceptably short, and that the magnitude of the shortfall is not too great. If water is abstracted from a river, a statistical model of its flow pattern through the year provides a way by which the planner can explore, by computer simulation, the frequency with which failures occur for any given abstraction rule. Several such rules will need to be explored, to build up a picture of how the frequency and severity of failures vary with abstraction rule.

Faced with such a problem, the planner may seek to build a statistical model of the type (1.2) above (although possibly more harmonic terms may be needed). Depending on circumstances, it may be necessary to use the mean discharges for periods shorter than a month; and if there are several possible abstraction sites on different rivers, it will be necessary to develop a 'multivariate' model which takes account both of the seasonal variability inflow at each abstraction site, and of any tendency for the flows in different rivers to rise and fall in phase. Note that models of the form (1.2) would be inappropriate if the data show significant tendencies for months of high flow to follow months of high flow, and conversely. It may then be more appropriate to look for a model expressing each month's flow in terms of flow in the preceding month (or months), of the type

$$q_t = f(q_{t-1}, q_{t-2}, \ldots; \ \theta) + \varepsilon_t$$

Although $q_{t-1}, q_{t_2}, \ldots$, on the right-hand side, are explanatory variables, there is not a causative relation between response variable and explanatory variables in this model — as would be the case if, say, monthly rainfall $p_t, p_{t-1}, \ldots$ entered the model as explanatory variables.

1.6.5 Modelling Rainfall Sequences for Soil Moisture Simulation

The previous examples have been concerned with statistical models of flow. In semi-arid regions, river flows are sparse and often intermittent; to assess the potential of a region for rain-fed agriculture, there is often no alternative but to use the records of daily rainfall which are more readily available and often of considerable length, even in developing countries. Statistical models of the occurrence of daily rainfall, and of the depth of rain when it occurs, have been developed from which it is possible to calculate probabilities of runs of dry and wet days; given reasonable assumptions about soil moisture capacities and evaporation rates, it has proved possible to estimate the probabilities that adequate soil water is available, for example at normal planting time, to support crop growth. Rainfall models of this type (see Chapter 6) belong to the important class of generalised linear models. Essentially the same statistical models may be used to model flow sequences for flow gauging sites in semi-arid regions, where periods of no flow alternate with periods when flow is measurable.

1.6.6 Short-term Forecasting of Flood Levels and Discharges

Models of the general form (1.5) may be used to obtain short-term forecasts of flood levels, which in turn can determine whether flood warnings are required. Forecasts are most easily found in response to rain that has already fallen, but for which the generated runoff has not yet reached the basin outfall; the 'lead time' for which such forecasts can be obtained is then limited by the response characteristics of the basin modelled. In practice, forecasts will also be needed of the basin's response to rain that has yet to fall. Statistical procedures may then provide useful estimates of the expected values of future rainfall; where weather-radar signals are available, statistical models may be used to derive forecasts of the direction of rain-cell movement and future rainfall intensity.

An important aspect of statistical models in flood forecasting is that it is possible, at least in theory, to obtain 'prediction intervals' for estimates of future flood levels: that is, intervals which have a specified probability of including the level reached by flood waters. Calculation of these intervals is theoretically possible where a model of the form (1.5) has been fitted with valid assumptions about the random components as specified in Section 1.4; a 'likelihood-based' confidence region (see Chapter 3) can then be calculated for the model parameters θ, and this confidence region can in theory be used to derive prediction intervals for forecasts of future discharge.

1.6.7 Exploration of Characteristics of Very Large Bodies of Data, to Define Small Subsets Useful as Explanatory Variables

The above examples all refer to the planning and operation of water resource systems in their quantitative aspects, discussed in terms of modelling the relationships between hydrological variables such as q_t and time t, or between q_t and explanatory variables. Where water quality is also to be considered, the number of variables yielded by chemical analysis is often large; indeed both the number of variables to be modelled and the number of explanatory variables commonly increase substantially. An important preliminary to modelling is then the preparation of summaries showing the essential characteristics of each variable; whilst the human eye can absorb the characteristics of ten numbers, it is less able to do so for 100 numbers, and still less able for 1000. Chapter 2 discusses techniques appropriate for exploring and summarising the main features of large data sets, using the facilities for graphical exploration of data, and subsets of data, which are now a common feature of good statistical packages.

1.6.8 Identification of Possible Instrument Errors, and Faulty Readings, Where Large Quantities of Hydrological Data Are Recorded Automatically

With modern hydrological instrumentation, large numbers of measurements are automatically recorded on solid-state memory devices which may remain unattended in the field for weeks or months at a time, particularly where access is difficult. Whilst the memory device may be extremely reliable, the sensors may be subject to error. For example, measurements of river level become unreliable where intake pipes of stilling wells become blocked with sediment or weed growth; net radiation measurements must be rejected where condensation occurs on the interior of the radiometer dome (caused, perhaps, by saturation of the silica gel, which should be keeping the inside of the radiometer dome dry). Errors of this kind may be sudden in onset and transitory in duration, or they may be gradual in onset and persistent in their effects.

The human eye is an excellent device for locating such errors. However, the quantities of data can become so large that it becomes impracticable for all records to be individually inspected. A computer, suitably programmed to identify errors that are known commonly to occur, is a useful aid in the 'quality controlling' of large bodies of hydrological data, if only to identify suspect values which can be confirmed or rejected by inspection. The tests built into quality control programs are usually based on knowledge of the physical processes being observed: for example, a wet bulb temperature cannot exceed the dry bulb temperature; latent heat flux cannot exceed the net radiation supporting it.

Methods of this kind are very good at identifying where variables fall outside ranges that are physically realistic. They perform less well where the onset of errors is gradual, or where transitory errors do not give rise to physically unrealistic data. Statistical procedures can sometimes assist in the identification of such errors; by studying relationships between hydrological variables and identifying where the relationships change, errors or suspect observations can be identified which are not detected by 'physical' tests. It is possible, for example, to identify data points which are particularly 'influential' when relationships are fitted, because they lie apart from the majority of data points. There is now a large body of applied statistical theory useful for the diagnosis of suspect observations or outliers.

1.7 COMPUTATIONAL FACILITIES FOR EFFICIENT FITTING OF STATISTICAL MODELS

Graphical procedures have long played an important part in the fitting of statistical models to hydrological data. Following the development of modern statistical computing packages for interactive use with desk-top computers, graphics have become a much more flexible tool for the data analyst. Use of graphical procedures play a central role in the methods described in this book, and it will be assumed that readers are familiar with a package such as one of those mentioned. Plotting by lineprinter is adequate for most purposes at the model exploration stage, serving to suggest which model appears appropriate initially and whether the assumptions required by the model are supported by the data.

Plotting hydrological data is essential as a means of understanding their principal characteristics, of exploring relationships between variables, and of comparing the fit of a model with observations. In addition to the use of graphical procedures, a common requirement is to reduce large data sets to summary forms that are more easily interpretable. We discuss these methods in the next chapter, using GENSTAT as one statistical package of several that are particularly appropriate for the purpose. Both GENSTAT (1992, 1993), and a related package GLIM, have been developed by statisticians at the Rothamsted Experimental Station, Harpenden, Hertfordshire, UK, with long experience in the analysis of data from many fields of applied science. GENSTAT, in particular, offers many options giving great flexibility in input and output, in graphical techniques, and in the kinds of analysis to be undertaken. Its disadvantage, however, is that the use of some of its library routines (of which there are very many) may be slow on a small desk-top computer, although this shortcoming has been substantially rectified in later versions of the package. It is also possible to run GENSTAT either in batch mode or in interactive mode; for long-running GENSTAT jobs, it is therefore convenient to leave the computer running in batch mode and to return later to look at output files. GLIM is a package written strictly for interactive analysis of data; it allows greater flexibility than most programs, but its output is less well labelled. Both packages have excellent facilities for the preparation of tables in which data are classified according to two or more factors, and tables so produced are of publication quality.

Where we need in the present text to use methods for numerical optimisation, we use the package MATLAB. This is extremely easy to use and is also very fast on a desk-top computer; it has the advantage of including many functions not available in GENSTAT (the gamma and incomplete gamma functions, for example). In the author's experience, MATLAB is particularly appropriate for the minimization of goodness-of-fit functions, and in the running water balance calculations used to obtain a fitted hydrograph, both of these calculations being discussed in Chapter 8.

GENSTAT and MATLAB, the packages used throughout this book, by no means exhaust the possibilities. BMDP is a library of integrated programs which will undertake the regression analyses described in Chapters 4, 5 and 6 of this book; it has a very wide range of diagnostic techniques based on model residuals. SAS (Statistical Analysis System) is a larger system of integrated programs.

1.8   GENSTAT AND MATLAB CONVENTIONS USED IN THIS BOOK

Most diagrams and tables in this book are derived from the output from GENSTAT or MATLAB programs. The book is therefore accompanied by a diskette, giving listings of these programs together with explanatory notes (please fill in the reply card at the end of this book to order a copy). Within GENSTAT, all comments between double quotes (" ") are ignored by the computer, and this facility has been used to annotate the programs which are in files with file extension .PRT; thus, FIG2-1.PRT is the program used to produce Figure 2.1, the first figure appearing in Chapter 2. The full output, including monitoring messages, from this program is given in a file with file-extension .OUT; thus the output from FIG2-1.PRT is given in FIG2-1.OUT. An edited form of this output appears in the text as Figure 2.1. To avoid the complication of tables and figures with the same number, all tables and figures derived from computer listings are termed figures for the purpose of this text.

Where programs are given in MATLAB, the computer ignores any line beginning with a percentage symbol, %. This facility has been used to annotate MATLAB programs.

Where, within the text or within program files, GENSTAT directives are referred to, these are always given in upper-case letters; thus CALCULATE, MODEL and CUMULATE are some of the GENSTAT directives appearing in this book. They are always written in full, although they may be abbreviated to four letters at most by the user; thus CALCULATE and CUMULATE can equally well be written CALC and CUM. Both GENSTAT and MATLAB have substantial 'help' facilities (enabling the user to get on-line assistance when uncertainties arise) and recent versions of GENSTAT also have a 'menu' structure to assist the user by listing the possible next steps in his analysis. Furthermore, both GENSTAT and MATLAB have facilities for the incorporation of existing subroutines written in FORTRAN and (in the case of MATLAB) the C language.

1.9 CUSTOMISED PACKAGES FOR HYDROLOGICAL ANALYSIS

A number of packages are available specifically for certain types of hydrological analysis. One such is HYDATA, a hydrological data processing and analysis system developed for use on personal computers by the Institute of Hydrology, Wallingford, England; similar packages are HYDROM and PLUVIOM, developed by ORSTOM and widely used in francophone countries particularly. Being menu-driven, they are very suitable for users who are uncertain about which analyses they wish to undertake; as examples of HYDATA capabilities, options are available for deriving rating curves, for comparing two data sequences on the same graph, for double mass plots, and for deriving flow duration curves. Options also exist for incorporating FORTRAN subroutines for other analyses.

Despite their appeal, we do not use such packages in this book, for two reasons. First, a working knowledge of GENSTAT and MATLAB can be acquired very easily, even though it may take much longer to acquire a full understanding of the many options available with, say, a GENSTAT directive. Second, both GENSTAT and MATLAB provide facilities not only for graphical exploration of data and for preliminary analyses, but also for fitting and verifying models.

1.10 EXPERT SYSTEMS

Section 1.7 mentioned GLIM, a package specifically designed for exploratory analysis of data. GLIMPSE (Generalised Interactive Modelling with PROLOG and Statistical Expertise) is a knowledge-based front-end for GLIM, designed to make the program easier to use by non-experts by incorporating statistical expertise into the system and allowing the user to give instructions to the system at a higher level than that required by most programs.

Cox and Snell (1989) give a valuable discussion of the role of expert systems in statistical analysis of data. They point out that expert systems, as generally understood, suppose that there is a definitively correct answer in each situation; the object of the expert system is to enable non-expert users to come close to that answer. They point out that for statistical analyses of realistic complexity, it is more a matter of providing general advice on how to approach analyses, especially those involving relatively complicated methods.

There has been considerable discussion about the role of expert systems in hydrology, and there can be little doubt that the use of expert systems for analysis of hydrological data will increase in the future. They have already appeared in the field of statistical analysis (Gale and

Pregibon, 1982). Cox and Snell (1989) point out that an expert system will normally have two components, a knowledge 'engine' and a statistical engine. GLIM is one example of the latter. In assessing the knowledge engine, Cox and Snell suggest that attention be paid to the following questions:

(i)   How authoritarian is it? Does it give advice and offer suggestions, for example on whether to omit suspect observations, or does it determine what to do in accordance with rigid rules?

(ii) How flexible is it, in for example allowing the user to proceed with straightforward parts of the analysis?

(iii) Is there a lexicon to explain unfamiliar technical terms and also information available on request to justify advice offered?

(iv) Does it allow easy passage in any order between various phases of analysis?

Expert systems in hydrological analysis are likely to be most useful where an inexperienced hydrologist requires to analyse sets of data which are not totally identical to the ideal cases and examples expounded in textbooks. The expert system can then lead him through procedures which more experienced hydrologists would follow, when confronted with data which do not fully satisfy the requirements normally needed for particular analyses. Such systems may well have an important role to play in the future of hydrological data analysis.

1.11   GEOGRAPHICAL INFORMATION SYSTEMS

These are another development made possible by the rapidly increasing power, and rapidly falling price, of computers. Topographic data, together with data on land use, channel networks, soil types and other variables which measure hydrological characteristics of a region, can now be brought together and accessed very quickly. To give one example, the Connecticut Natural Resources Center (Wallis, 1988, 1991) has a geographical information system (GIS) with map layers showing soil types, watershed boundaries, 100-year and 500-year flood-plain zones, surface and groundwater geology, and land use. Major difficulties, however, are the enormous effort often necessary to digitise historic records, and to ensure that, once digitised, they are free from error. Even in data bases prepared by agencies of the highest repute, the number of key-punch and other inconsistencies has been found to be very high. Nevertheless, the hydrological applications of GISs must be very considerable if the costs of establishing them can be reduced, and if the frequency of erroneous data entries can be made acceptably small.

1.12   FUTURE DEVELOPMENTS

With the rapid growth of networking facilities through Internet, much software ('freeware') is becoming available to anyone with a terminal able to access it. Data sets are also becoming available to network users. Chapter 2 in this book discusses the calculation of l-moments; an IBM Research Report (Hosking and Wallis, 1991) discusses statistical procedures, based on l-moments, which are useful in regional flood frequency analysis, and the FORTRAN software developed by these authors can be obtained, free and very rapidly, by sending an e-mail message to STATLIB, a component of Internet. The user simply sends the request:

send lmoments from general

to the address:

statlib @lib.stat.cmu.edu

Availability of both software and data sets, through the medium of networks of this kind, will increase dramatically in the future, giving hydrologists ready access to each other's working tools for the benefit of all.

With the explosion in the availability of freeware, the necessity for the purchase of packages such as GENSTAT and MATLAB can be questioned. The view of this author is that whilst freeware should not be ignored, there are very substantial benefits to be gained from the purchase and use of packages like GENSTAT which have been steadily developed over long periods, which have good support services, and into which library procedures are incorporated with due regard to how other parts of the package behave. Freeware may not have these advantages, and there is nothing the user can do if the freeware is found to be faulty. Furthermore, whilst freeware programs may have e-mail addresses to which users may send requests for help, the person offering help may move, or be away, at a time critical to the person needing assistance. There are therefore substantial advantages to be gained by using a fully supported computer package which is continually being developed and updated.

REFERENCES

Cox, D. R. and Snell, E. J. (1989). *Analysis of Binary Data* (2nd edn). Chapman and Hall, London.
Gale, W. A. and Pregibon, D. (1982). An expert system for regression analysis. *Proc. 14th Symp. on the Interface*. Springer-Verlag, New York, 110–17.
GENSTAT 5 Committee, Rothamsted Experimental Station (1992) *GENSTAT 5 Reference Manual*. Oxford University Press, Oxford
GENSTAT 5 Committee, Payne, R. W., Arnold, G. M. and Morgan, G. W. (eds) (1993) *GENSTAT 5 Procedure Library Manual*, Release 2[3]. NAG Ltd (Numerical Algorithms Group), Oxford.
Hosking, J. R. M. and Wallis, J. R. (1991). *Some Statistics Useful in Regional Frequency Analysis*. IBM Research Report RC 17096 (#75863).
MATLAB Reference Guide (August 1992) The Mathworks, Inc. Natick, MA.
Wallis, J. R. (1988). The GIS/hydrology interface: the present and the future. *Environmental Software*, **3**, 171–3.
Wallis, J. R. (1991). The interface between GIS and hydrology. In Loucks, D. P. and da Costa, J. R. (eds) *Decision Support Systems*. NATO ASI Series, Vol. G 26. Springer-Verlag, Berlin, 189–97

Chapter 2

Exploring and summarising hydrological data sets

2.1 MEASURES OF POSITION AND DISPERSION

We are concerned in this chapter with exploring the characteristics of hydrological data sequences by interactive use of a desk-top computer, with extensive use of computer graphics. Each data sequence has its own characteristics and the methods described do not constitute a recipe to be followed rigidly in every instance; they serve only to show the reader how the power and flexibility of desk-top computers can be brought to bear during the important stage of data exploration. The computer instructions used in this chapter are some of those available in GENSTAT, but since the chapter does not set out to be a GENSTAT user manual, the full set of options available in each instruction are not listed exhaustively. Our concern is to get the output necessary for the problem at hand, and leave the reader to follow more specialised reading if he wishes to explore refinements.

When data are given in time sequence, looking at the column of numbers on the printed page does not readily convey an idea of which values are more frequent and which are less frequent. We can easily get an idea of the largest and smallest values in the column, since when the READ instruction is used to put hydrological data into a GENSTAT calculation, the computer screen shows, for each variable read, the maximum and minimum values, together with the number of values read, and the number of missing values (denoted by '$*$' in Figure 1.1). From this information we find at once that, in the case of the flood data 1934–84 for the Rio Hercílio at Ibirama, the range of values is from 236 m^3 s^{-1} to 2475 m^3 s^{-1}; for the Rio Itajaí-Açú at Apiuna, the maximum and minimum annual floods are 495 m^3 s^{-1} and 4327 m^3 s^{-1}, respectively. A useful way of getting an idea of the 'spread' of values between these limits, however, is to tell GENSTAT to calculate the median and quartiles of the flood sequence. It will be recalled that the median and upper and lower quartiles are particular cases of 'quantiles', statistics characterising a sample of data. Formally, given a sample $\{x_i, i = 1, 2, \ldots, n\}$ a quantile q can be calculated for any proportion p lying in the range $[0, 1]$, such that q has the following properties:

(i)   at least the proportion p of the sample $\{x_i\}$ are less than or equal to q;

(ii) at least the proportion $1 - p$ of the $\{x_i\}$ are greater than or equal to q;

(iii) if $q = x_i$ and $q = x_{i+1}$ both satisfy (i) and (ii), then take $q = (x_i + x_{i+1})/2$.

The median is the quantile corresponding to $p = 0.5$, and the upper and lower quartiles correspond to $p = 0.25$ and $p = 0.75$, respectively. Having read the flood sequences for Ibirama and Apiuna into vectors with the same names, we calculate quantiles using the GENSTAT directive

> QUANTILE Ibirama, Apiuna

where the '>' is the computer prompt symbol, appearing only when GENSTAT is used interactively; and the words that follow are entered on the keyboard, after which the 'Enter' key is pressed. The directive shows that the median floods for Ibirama and Apiuna are 673 and 1545 m^3 s^{-1}, respectively. The lower and upper quartiles floods for Ibirama are 500 and 989.5 m^3 s^{-1}, and for Apiuna 985 and 1980 m^3 s^{-1}. (A simple modification to the quantile instruction produces quantiles for any other proportions p.) From these values for the medians and quartiles, we see at once that both data sets are skewed to the right, the difference between median and lower quartile being considerably less than the difference between upper quartile and median.

Thus the median, or 50% quantile, gives a measure of *position*, showing where the 'middle' observation of the sequence occurs, whilst differences between the quartiles gives a measure of the *dispersion* of data values. In later sections, where data are assumed to follow particular probability distributions, we shall generally use the arithmetic *mean* as a measure of position, and the *standard deviation* as a measure of data dispersion. Where we are not yet ready to specify the probability distribution from which data might be drawn, quantiles provide a useful summary of data characteristics.

2.2 GRAPHICAL PROCEDURES IN DATA REDUCTION
2.2.1 The Box-and-whisker Plot

Medians and quartiles are useful numerical summaries of the position, dispersion and skewness of a data set. The information in the numbers is conveyed with greater immediacy, however, if they are represented graphically using a 'box-and-whisker' plot. For the annual flood sequences for Ibirama and Apiuna, Figure 2.1 shows the diagrams produced by use of the GENSTAT library procedure WHISKER, having first read the flood sequences for Ibirama and Apiuna into vectors with the same names. The letter M marks the median position, asterisks the interval

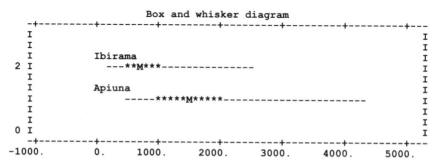

Figure 2.1   Box-and-whisker diagram of annual flood data from Rio Hercílio at Ibirama, and Rio Itajaí-Açú at Apiuna (see Figure 1.1). The symbol M marks the median; asterisks mark the interval between the quartiles; dashes mark the range of each variable. Units on the horizontal scale are cubic metres per second.

between quartiles, and dashed lines mark the intervals between maximum and minimum values. Thus the simplest form of the WHISKER instruction, used when GENSTAT is used interactively to explore the Ibirama and Apiuna annual flood records, is

> WHISKER Ibirama
> WHISKER Apiuna

although options (not discussed here) are available which permit the printing of diagram titles, of titles for the horizontal axes, and which specify what kind of graphics to use (the options being 'lineprinter', as in the present case, or 'highquality', where a graph plotter is to be used to produce a prettier format).

2.2.2 Histograms

An alternative to the use of quantiles and box-and-whisker plots as a means of displaying data characteristics in easily assimilable form is by the use of histograms. The GENSTAT directive

>HISTOGRAM Ibirama: & Apiuna

produces the displays shown in Figure 2.2; class intervals are determined automatically but, as an option, can be specified as part of the command. If the frequencies in the class intervals are stored in the variates *Ibfreq* and *Apfreq*, with the class interval midpoints stored in *Ibx* and *Apx*, the GENSTAT instructions

>CALCULATE a=CUMULATE(Ibfreq): & b=CUMULATE(Apfreq)

```
        Histogram of Ibirama
                         -      400      6    *******
                400  -    800     24    ***************************
                800  -   1200     11    ************
               1200  -   1600      5    *****
               1600  -   2000      0
               2000  -   2400      1    *
               2400  -              1    *
        Missing values:     3

        Scale: 1 asterisk represents 1 unit.

        Histogram of Apiuna
                         -      800      6    ******
                800  -   1600     21    *********************
               1600  -   2400     17    *****************
               2400  -   3200      4    ****
               3200  -   4000      0
               4000  -   4800      2    **
               4800  -              0
        Missing values:     1

        Scale: 1 asterisk represents 1 unit.
```

Figure 2.2   Histograms of annual flood data for Rio Hercílio at Ibirama and Rio Itajaí-Açú at Apiuna (see Figure. 1.1).

stores the cumulative frequencies in the vectors a and b, and the directives

>BARCHART a; lbx : & b; Apx

produce barchart diagrams of the cumulative frequency distribution as shown in Figure 2.3. Here, the plots shown are those produced by lineprinter. By including an option of the form

>BARCHART [highquality] a; lbx; PEN=1

GENSTAT produces graph-plotter output. Use of different parameters for PEN results in plots of different colours.

A more sophisticated procedure, AKAIKEHISTOGRAM, which provides better defini-tion between the groups, is available as a GENSTAT library procedure. In the procedure HISTOGRAM illustrated above, the 'default' number of groups is taken as the square root of the number of observations; so that if the number of observations is very large, the number of

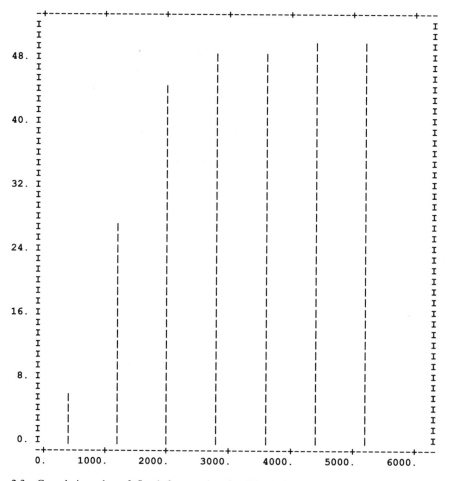

Figure 2.3 Cumulative plot of flood frequencies for Rio Itajaí-Açú at Apiuna (for raw data, see Figure 1.1).

groups may be unnecessarily large for displaying the characteristics of the data. Again, if the shape of the histogram is very complex, the square-root rule can give groups that are too few in number to display adequately the features of the data. The procedure AKAIKEHISTOGRAM uses a method which optimises the 'Akaike's information criterion' (AIC) to produce more informative groupings. When the procedure AKAIKEHISTOGRAM is applied to the annual flood sequences of Ibirama and Apiuna, the result is as shown in Figure 2.4.

Akaikehistogram of Ibirama flood sequence

```
         -     470     7    ------
 470 -    890    25    ----------------------
 890 -   1310    12    -----------
1310 -   1730     2    --
1730 -   2150     1    -
2150 -            1    -
```

Scale: 1 — represents 1 unit(s).

Akaikehistogram of Apiuna flood sequence

```
         -    1390    22    -------------------
1390 -   2390    22    -------------------
2390 -   3390     4    ----
3390 -            2    --
```

Scale: 1 — represents 1 unit(s).

Akaikehistogram of logged Ibirama flood sequence

```
         -    5.70     5    +++++
5.70 -   6.06     1    +
6.06 -   6.42    15    +++++++++++++++
6.42 -   6.78    11    +++++++++++
6.78 -   7.14    11    +++++++++++
7.14 -   7.50     3    +++
7.50 -            2    ++
```

Scale: 1 + represents 1 unit(s).

Akaikehistogram of logged Apiuna flood sequence

```
         -    6.32     1    +
6.32 -   6.56     4    ++++
6.56 -   6.80     2    ++
6.80 -   7.04    11    +++++++++++
7.04 -   7.28     4    ++++
7.28 -   7.52    11    +++++++++++
7.52 -   7.76    11    +++++++++++
7.76 -   8.00     2    ++
8.00 -   8.24     2    ++
8.24 -            2    ++
```

Scale: 1 + represents 1 unit(s).

Figure 2.4   Histograms of the Ibirama and Apiuna annual flood sequences without transformation; and after log transformation, using the GENSTAT AKAIKEHISTOGRAM procedure (see Sakamoto, Ishiguro and Kitagawa, 1986).

2.2.3 Cumulative Frequency Diagrams

Diagrams of cumulative frequencies in which each observation can be identified, instead of being grouped into class intervals as in Section 2.2.2, are also plotted very simply. The GENSTAT instructions

```
>CALCULATE Ibsort=SORT(Ibirama): & Apsort=SORT(Apiuna)
```

put the ordered sequences (in increasing order of magnitude) into the variates *Ibsort* and *Apsort*. Asterisks, representing the three missing values at Ibirama, are written into the first three elements of *Ibsort*; similarly, an asterisk appears as the first element of *Apsort*, corresponding to the one missing value at this site. The following instructions calculate the proportion i/N, corresponding to the ith value in the ordered sequence of N non-missing values (recall also that GENSTAT ignores all text included between double quotes):

```
>VARIATE [VALUES=1 . . . 51] I,A
>
>"This instruction defines 2 vectors I and A, each with values
> 1 to 51."
>
>
>CALCULATE Iind=(I.GT.3): & Aind=(A.GT.1)
>& Icum = CUMULATE(Iind): & Icum=Icum/48
>& Acum = CUMULATE(Aind): & Acum=Acum/50
>
>"The first line of instructions defines vectors Iind
>and Aind, the elements of each being 1 where values
>are in the ordered flood sequences are not missing
>and 0 otherwise. The second line of instructions calculates
>running totals of the elements of Iind, from i=1 to i=48,
>and then the values of i/48. The last line does the same
>for the Apiuna flood sequence."
```

Here, $n = 48$ for Ibirama and $n = 50$ for Apiuna (see Figure 1.1). Then the instructions

```
>GRAPH Icum;Ibsort: & Acum; Apsort
```

result in the two cumulative frequency diagrams for the Ibirama and Apiuna flood sequences shown in Figure 2.5. Individual points are plotted as asterisks (other choices of symbol are possible); the numbers 2, 3 and 4 in the diagrams indicate where two, three or four points overlap, this being caused by the limited discrimination with which points can be plotted using a lineprinter. The symbols '3' and '*' occurring at the ends of the horizontal axes denote the number of missing values in each plot. In both diagrams, the two extreme floods that occurred in 1983 and 1984 are clearly identifiable; at Apiuna, the floods in these two years were 4327 and 4314 m³ s⁻¹, respectively, and are too close to be distinguished in the relatively coarse lineprinter plot, their occurrence being denoted by the symbol '2'.

    So far we have used only the data from Figure 1.1 on flood sequences to illustrate aspects of data exploration. The data from Figure 1.2 are for first five years of mean monthly discharge for the River Lava Tudo at Fazenda Mineira; we use them to illustrate some GENSTAT

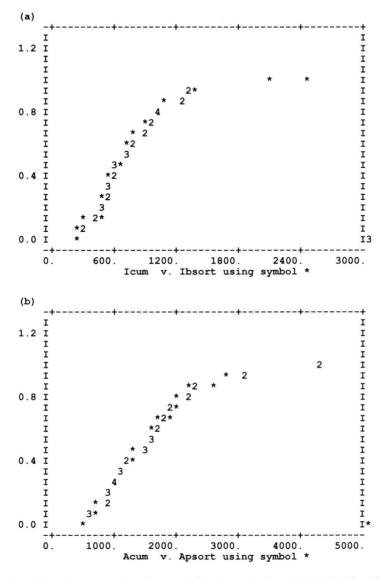

Figure 2.5   Cumulative frequency plots for annual flood records from (a) the Rio Hercílio at Ibirama, and (b) the Rio Itajaí-Açú at Apiuna. (For data, see Figure 1.1.)

techniques for obtaining plots of variables against time. The data, which are part of a 47-year (= 564 month) sequence of mean monthly discharges, were read from a computer file named SIMDAT.MAT by the following instructions.

```
>VARIATE [NVALUES=564] q
>OPEN 'SIMDAT.MAT';CHANNEL=2;FILETYPE=input
>READ[CHANNEL=2] q
```

To abstract the first five years(= 60 months) of this sequence, we define a vector t of length 564, with elements having the values 1, 2, ..., 564, and then restrict the length of q to include only those values for which the elements of t are less than or equal to 60 (that is, five years of monthly flows). Hence:

```
>VARIATE [VALUES=1...564] t
>RESTRICT q,t; CONDITION= t.LE.60
```

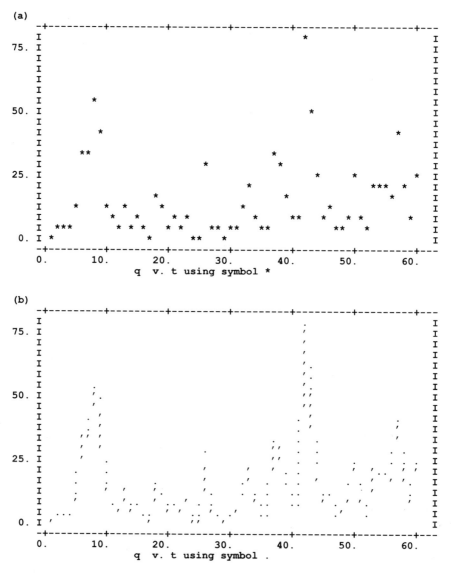

Figure 2.6   Plots of mean monthly flows for Rio Lava Tudo at Fazenda Mineira: first five years of record. The use of a different plotting symbol in (b) is for purposes of readability.

The GENSTAT instruction:

>GRAPH q;t

then produces the rather unsatisfactory plot shown in Figure 2.6(a). We can distinguish a little more detail if we amend the last instruction to be the following:

>GRAPH q;t; METHOD=line

which yields the slightly less inferior plot shown in Figure 2.6(b). In this case, the quality of the plots would be much improved by using a graph plotter instead of a lineprinter; this can be easily achieved using the 'highquality' option for GENSTAT graphical output. Nevertheless, we are concerned in this chapter with techniques for data exploration, for which plots on the computer screen are generally adequate.

2.2.4 Flow Duration Curves

Directives for putting hydrological data in ascending order are particularly useful where it is necessary to calculate flow duration curves, showing proportions, plotted on the vertical axis, of values in a sequence of flows which exceed particular flow values, plotted on the horizontal axis. Figure 2.7 shows how a flow duration curve can be simply obtained, using as illustration the 365 mean daily flows for 1945, from the Rio Hercílio at Ibirama. Using GENSTAT's RESTRICT command, it is also straightforward to abstract, from a long record of mean daily flows, the flows for any particular year or set of years, and to plot the flow duration curve for this reduced data set. For example, given a four-year record of mean daily flow, denoted by MeanDF, with the fourth year a leap year, the GENSTAT instructions to extract data for the third year and plot the corresponding flow duration curve in the manner described above are as follows:

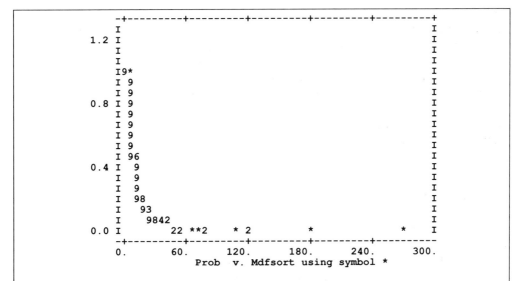

Figure 2.7   Flow duration curve for mean daily flows, Rio Hercílio at Ibirama: mean daily flows for year 1945.

```
>VARIATE [NVALUES=1461] MeanDF, MDFSorted, Proportions
>     &      [VALUES=1 ... 1461] DayNumber
> CALCULATE MDFSorted = 0    "Initialize; now select data:-"
>RESTRICT MeanDF, DayNumber, MDFSorted, Proportions;\
>             CONDITION=(DayNumber.GT.730).AND.(DayNumber.LT.1096)
> CALCULATE MDFSorted=SORT(MeanDF) : & DayNumber=DayNumber-730
>     &      Proportion=1-DayNumber/365
> GRAPH [NROWS=20; NCOLUMNS=70] Proportion; MDFSorted
> CALCULATE DayNumber=DayNumber+730
> RESTRICT MeanDF, DayNumber,MDFSorted, Proportions
```

Refinements in the use of the GRAPH directive include options for the insertion of a general title for the graph, and of titles for its axes; and for plotting several graphs on the same figure, with different symbols. The last RESTRICT directive in the above piece of program restores the data sequences to their original form, ready for subsequent analyses.

2.2.5   Double Mass Curves

It is sometimes suspected that hydrological records may lack internal consistency. For example, the site of one rain gauge in a network of gauges may have been changed at some time during the period of record; or the pattern of flow at a flow gauging structure may have been altered by works on the channel or by agricultural or urban development. To explore whether such inconsistencies occur, it is often useful to plot the cumulative rainfall (or flow) recorded at the suspect site against the cumulative rainfall (flow) recorded at other sites within the network.

Use of the GENSTAT directive:

```
> CALCULATE y=CUMULATE(x)
```

gives the running totals $x(1), x(1) + x(2), x(1) + x(2) + x(3), \ldots$ of a hydrological sequence x, storing the running totals in the variable y. Applying the directive to two sequences of data, say *RAIN1* and *RAIN2* (for example, sequences of monthly rainfall totals recorded by two rain gauges, one of which may be suspect) the two sets of running totals can be plotted against each other to give a double mass curve, useful for identifying possible inhomogeneities in records. As an example, the file FIG2-8.PRT on the accompanying diskette contains monthly rainfall data for the headwater catchments of the rivers Wye and Severn in Wales; GENSTAT instructions given in the file, for the calculation of a double mass curve, give the output summarised in Figure 2.8.

2.2.6   Contour Plots

For the calculation of maximum-likelihood estimates of model parameters, as set out in later chapters, we frequently need to explore the nature of the likelihood surface in the region of a point at which we believe the surface to have a maximum value; the GENSTAT directive CONTOUR (illustrated later) gives a plot of contours of the likelihood surface. More generally, the CONTOUR directive can be used to plot contours of any two-way array of numbers specifying values on a rectangular grid; applications could therefore include (after interpolation, if necessary) the representation of terrain, groundwater contours, reservoir siltation, and soil moisture volume fraction.

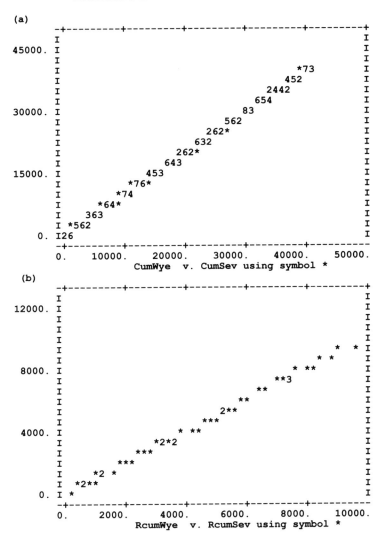

Figure 2.8   Double mass curve of monthly rainfall (a) for Wye and Severn catchments, 1971–87; (b) for the first four years of data only.

2.3   MORE GENERAL NUMERICAL SUMMARIES OF DATA SETS

We have given examples of the use of the GENSTAT procedure QUANTILE, from which numerical descriptors were obtained of the position and dispersion of the values in a data set for a hydrological variable. A more general routine yielding numerical summaries is the following (given in its most basic form):

```
>DESCRIBE Ibirama, Apiuna
```

This procedure prints, for the Ibirama and Apiuna annual flood sequences, the following statistics: mean, minimum, maximum, number of non-missing values, number of missing values,

Summary statistics for Ibirama
Number of observations = 48
Number of missing values = 3
Mean = 783.708
Median = 673.000
Minimum = 236.000
Maximum = 2475.000
Lower quartile = 500.000
Upper quartile = 989.500

Summary statistics for Apiuna
Number of observations = 50
Number of missing values = 1
Mean = 1624.080
Median = 1544.500
Minimum = 495.000
Maximum = 4327.000
Lower quartile = 951.000
Upper quartile = 1960.000

Figure 2.9   Summary statistics for annual flood series from Rio Hercílio at Ibirama, and Rio Itajaí-Açú at Apiuna. Summaries obtained using GENSTAT directive DESCRIBE.

the median, and the upper and lower quartiles, as shown in Figure 2.9. These are printed by default; however, further options may be introduced by the following:

```
>DESCRIBE [SELECTION=nval,nobs,nmv,mean,median,min,max,\
>range,q1,q3,var,sd,sem,%cv,sum,uss,skew,seskew] Ibirama, Apiuna
```

where, with $\{x_i\}$ denoting either annual flood sequence,

nval   is number of values
nobs   is number of non-missing values
nmv   is number of missing values
mean   is the arithmetic mean ($\Sigma x_i/nobs; = \bar{x}$)
median   is the median of the sequence $\{x_i\}$
min   is the minimum of the sequence $\{x_i\}$
max   is the maximum of the sequence $\{x_i\}$
range   is the difference (*max−min*)
q1   is the lower quartile of the $\{x_i\}$
q3   is the upper quartile of the $\{x_i\}$
var   is the variance, $\Sigma(x_i - \bar{x})^2/(nobs - 1)$
sd   is the standard deviation ($= \sqrt{var}$)
sem   is the standard error of the mean ($= sd/\sqrt{nobs}$)
%cv   is the coefficient of variation ($= (100sd/mean)\%$)
sum   is the total, Σx_i
ss   is the corrected sum of squares, $\Sigma(x_i - \bar{x})^2$
uss   is the uncorrected sum of squares, Σx_i^2
skew   is the skewness
seskew   is the standard error of the skewness coefficient.

Summary statistics for fp

Number of values =	30
Number of observations =	30
Mean =	389.823
Median =	133.500
Minimum =	18.400
Maximum =	4150.000
Range =	4131.600
Lower quartile =	65.500
Upper quartile =	442.000
Variance =	593410.062
Standard deviation =	770.331
Standard error of mean =	140.643
Coefficient of variation =	36.079
Skewness =	4.113
Standard Error of Skewness =	0.427

Summary statistics for pi

Number of values =	30
Number of observations =	28
Mean =	40.175
Median =	34.350
Minimum =	11.400
Maximum =	95.700
Range =	84.300
Lower quartile =	25.500
Upper quartile =	47.450
Variance =	455.532
Standard deviation =	21.343
Standard error of mean =	4.033
Coefficient of variation =	10.040
Skewness =	1.068
Standard Error of Skewness =	0.441

Figure 2.10   Summary statistics for variables *fp* (peak discharge, litres per second) and *pi* (maximum 30-minute rainfall intensity, millimetres per hour): data from 30 events on basin of the River Booro-Borotou, Ivory Coast (see Figure 1.4).

The 'skewness' is here calculated as

$$(M_3 - 3M_1M_2 + 2M_1^3)/(M_2 - M_1^2)^{3/2}$$

where, writing N for *nobs*, $M_i = \Sigma x^i / N$, and 'seskew' is

$$6N(N - 1)/[(N - 2)(N + 1)(N + 3)]$$

There is a large number of options available with this procedure; Figure 2.10 shows some typical output, for which some of the above options were excluded, for the variables fp and pi of Figure 1.4.

2.4   EXPLORING RELATIONS BETWEEN VARIABLES

The above sections have discussed the use of GENSTAT for getting to know the statistical characteristics of a sequence of hydrological data. Having obtained a feel for each variable

Identifier	Minimum	Mean	Maximum	Values	Missing	
Day	1.00	14.90	31.00	30	0	
Year	1984	1986	1987	30	0	
fv	0.500	1.501	11.100	30	0	Skew
fp	18.4	389.8	4150.0	30	0	Skew
ft	114.0	577.0	1513.0	30	0	
fs	1.000	1.633	2.000	30	0	
bf	0.00	10.77	91.00	30	0	Skew
pv	14.40	38.08	82.70	30	0	
pt	50.0	233.5	489.0	30	1	
pi	11.40	40.18	95.70	30	2	

Figure 2.11   Summary of variables given in GENSTAT output, after reading data from 30 storms on the basin of the River Booro-Borotou (see Figure 1.4).

separately, it will then be necessary to explore the relationships between them. We illustrate some of the available techniques using data from Figure 1.4, on the 30 storms on the basin of the River Booro-Borotou in the Ivory Coast.

We begin by reading the data from the computer file PIERRE.DAT. We note that, in this data file, the months are given in text form, so that we must specify a variate called *Month* in which to read text rather than numbers. Variates are also specified for the day (*Day*); year (*Year*); flood volume in millimetres (*fv*); peak discharge in litres per second (*fp*); flood duration in minutes (*ft*); flood shape (*fs*); base flow in litres per second (*bf*); precipitation depth in millimetres (*pv*); precipitation duration in minutes (*pt*); and precipitation intensity in millimetres per hour (*pi*). To reserve space for these variables, we enter the following instructions:

```
>TEXT [NVALUES=30] Month
>VARIATE [NVALUES=30] Day,Year,fv,fp,ft,fs,bf,pv,pt,pi
>OPEN 'PIERRE.DAT'; CHANNEL=2; FILETYPE=input
```

We then read the data from the file PIERRE.DAT by the instruction:

```
>READ [CHANNEL=2] Day,Month,Year,fv,fp,ft,fs,bf,pv,pt,pi
```

GENSTAT replies with the output shown in Figure 2.11, showing for each non-textual variable the minimum, mean, maximum, number of values read, and number of asterisks encountered, indicating missing values.

Now suppose we wish to explore the relation between peak discharge (*fp*) and baseflow (*bf*); the GENSTAT instruction:

```
>GRAPH fp;bf
```

produces the graph shown in Figure 2.12. The point corresponding to the event of 4 September 1985, when baseflow was at the high level of 91 litres per second, shows up on the extreme right of the graph; to see what happens if we remove this point, we enter the command:

```
>RESTRICT fp,bf; CONDITION=bf.LT.90
```

which in the present case removes the values of *bf* and *bf* for just the one event (all others having baseflow less than 90 l s^{-1}). Graphing the remaining points by means of

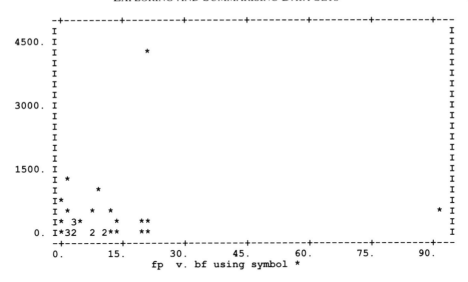

Figure 2.12   Data from River Booro-Borotou, Ivory Coast: plot of peak discharge, *fp*, against baseflow, *bf*. For data, see Figure 1.4.

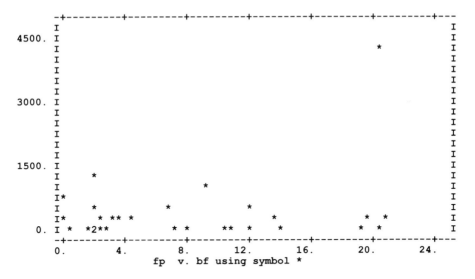

Figure 2.13   Data as for Figure 2.12, but with event of 4 September 1985 omitted by means of GENSTAT's RESTRICT directive.

```
>GRAPH fp;bf
```

produces the graph shown in Figure 2.13. The outstanding point is that for the flood of 18 August 1985, with peak discharge of over 4000 l s^{-1} . To restore the point that was removed, we can enter the command:

```
>RESTRICT fp, bf
```

and the full data set is once more available. To remove both points — the one with base flow greater than 90 l s^{-1}, and the other with peak discharge greater than 4000 l s^{-1} — we could enter:

```
>RESTRICT fp, bf; CONDITION=(fp.GT.4000).OR.(bf.GT.90)
```

and again the full data set can be restored by means of the command

```
>RESTRICT fp, bf
```

We move now to another example. In Section 2.2.3 above we read the 47 years of mean monthly flows for the Rio Lava Tudo at Fazenda Mineira (Figure 1.2) into a single vector, q, of length 564. We could have read the data into an array with 47 rows, representing years, and 12 columns, representing months, by using the following commands:

```
>VARIATE [NVALUES=47]jan,feb,mar,apr,may,jun,jul,aug,sep,\
>oct,nov,dec
>OPEN 'SIMDAT.MAT'; CHANNEL=2;FILETYPE=input
>READ [CHANNEL=2]jan,feb,mar,apr,may,jun,jul,aug,sep,oct,nov,dec
```

GENSTAT responds by printing out the data summary shown in Figure 2.14. If we were preparing to fit a Thomas–Fiering model (see Chapter 4) we should need to look at the data graphically by plotting each month's mean discharge against that of the preceding month; for example,

```
>GRAPH feb;jan
```

would produce the plot shown in Figure 2.15. We can see by the clustering of points around the horizontal axis that the assumption of equal dispersion around a fitted linear relationship, necessary for the Thomas–Fiering model, is not satisfied. A Thomas–Fiering model fitted without having looked at the data on a graph would result in the generation of negative flows; initial plotting of the data identifies the problem and saves the time that would have been spent in fitting an inappropriate model.

Identifier	Minimum	Mean	Maximum	Values	Missing	
jan	0.95	21.99	61.93	47	0	
feb	2.87	25.40	72.33	47	0	
mar	3.15	19.19	72.21	47	0	Skew
apr	2.16	18.68	63.52	47	0	
may	1.60	20.85	118.90	47	0	Skew
jun	2.87	25.23	78.18	47	0	
jul	5.16	33.84	239.15	47	0	Skew
aug	1.14	43.31	247.02	47	0	Skew
sep	2.14	42.47	129.72	47	0	
oct	4.43	35.14	211.27	47	0	Skew
nov	2.45	23.25	81.68	47	0	
dec	1.27	19.66	76.23	47	0	Skew

Figure 2.14   Summary statistics given in GENSTAT output: 47 years of monthly data from Rio Lava Tudo at Fazenda Mineira (see Figure 1.2), read month by month.

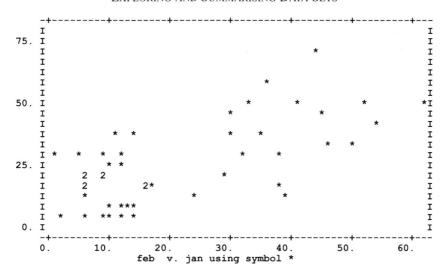

Figure 2.15   Data from Rio Lava Tudo at Fazenda Mineira (see Figure 1.2): 47 years of mean monthly flows in February plotted against mean monthly flows in January.

2.4.1   Plotting Using Lagged Variables

In preliminary data analyses, it is often necessary to plot variables against their 'lagged' values: that is, to plot x_t against x_{t-1} . This kind of plot is particularly useful when checking whether the assumption of statistical independence is supported by the data. Where this assumption is invalid, there is commonly a positive correlation between x_t and x_{t-1} which is detected when the two variables are plotted. In GENSTAT, lagged variables are easily constructed by use of the directive SHIFT; given a sequence of data x_1, x_2, x_3, x_4, x_5 stored in the variate x, the instruction

```
>CALCULATE y=SHIFT(x;-1)
```

leaves the variate y containing x_2, x_3, x_4, x_5, $*$, where the asterisk denotes a missing value. The values in the sequence have therefore been shifted one place to the left, with the first value transferred to the end. If the -1 in SHIFT is replaced by $+1$, the sequence is shifted one place to the right; the variate y then contains $*$, x_1, x_2, x_3, x_4. Generalisation to

```
>CALCULATE y=SHIFT(x;s)
```

is obvious. A related GENSTAT directive has the form

```
> CALCULATE y=CIRCULATE(x;s)
```

for which, with $s = -1$, the variate y would contain the sequence x_2, x_3, x_4, x_5, x_1, with obvious generalisations. Also useful is another GENSTAT function, DIFFERENCE$(x;s)$; after the instruction

```
>CALCULATE y=DIFFERENCE(x;s)
```

the variate y contains $x_t - x_{t-s}$. If the s is omitted, the first differences $x_t - x_{t-1}$ are formed. If $t - s < 1$ or $t - s > N$, with N the number of data values in the sequence, the tth element of y is set to 'missing', normally the symbol $*$. GENSTAT has a comprehensive set of procedures for identifying, fitting and using linear time-series models (not included in this book) of the kind described by Box and Jenkins (1970) and termed ARIMA models; working with differences between variate values is a characteristic of one important class of models of the ARIMA family.

2.5　ALTERNATIVE DESCRIPTORS OF A DATA SAMPLE: *l*-MOMENTS

In recent years, a powerful alternative to the use of moments has been developed, as described by Hosking (1986, 1990; Hosking, Wallis and Wood, 1985, and related papers). These alternative descriptors are called *l-moments*; they are, in fact, weighted sums of the data values after placing them in rank order. After ranking the data in ascending order to give the sequence $x_1, x_2, \ldots, x_n$ where $x_1 \le x_2 \le x_3 \le \cdots \le x_n$, two sets of quantities are required, given by

$$a_r = n^{-1} \sum_{i=1}^{n} \binom{n-i}{r} x_i \bigg/ \binom{n-1}{r} \qquad r = 0, 1, 2, \ldots, n-1 \qquad (2.1)$$

$$b_r = n^{-1} \sum_{i=1}^{n} \binom{i-1}{r} x_i \bigg/ \binom{n-1}{r} \qquad r = 0, 1, 2, \ldots, n-1 \qquad (2.2)$$

where $\binom{k}{j}$ is zero if $k < j$.

From these quantities a_r and b_r the *l*-moments can be calculated. Two expressions are given for each *l*-moment, which serve as a check on the calculation:

$$
\begin{aligned}
l_1 &= a_0 & l_1 &= b_0 \\
l_2 &= a_0 - 2a_1 & l_2 &= 2b_0 - b_1 \\
l_3 &= a_0 - 6a_1 + 6a_2 & l_3 &= 6b_2 - 6b_1 + b_0 \\
l_4 &= a_0 - 12a_1 + 30a_2 - 20a_3 & l_4 &= 20b_3 - 30b_2 + 12b_1 - b_0
\end{aligned}
\qquad (2.3)
$$

The *l*-moments l_1, l_2, l_3, l_4 (higher-order *l*-moments also exist) are important for several reasons: first, because as descriptors of a sample of data they are less affected by extreme observations than conventional moments; second, because they serve to provide good estimates of model parameters for use in iterative calculations leading to more efficient maximum-likelihood estimates (see Chapter 3); and third, because in the case of certain distributions such as the Wakeby distribution (see Chapter 3) there is no straightforward iterative procedure using maximum-likelihood estimates, so that estimation of model parameters by the use of *l*-moments is the only practical alternative.

Calculation of the expressions a_r, b_r of equations (2.1) and (2.2) is easily programmed; see, for example, Figure 2.16, which gives a listing in MATLAB code. For the data from the Rio Hercílio at Ibirama, for example, it is found that

$$
\begin{aligned}
a_0 &= 783.7083 & b_0 &= 783.7083 \\
a_1 &= 279.6170 & b_1 &= 504.0913 \\
a_2 &= 159.2735 & b_2 &= 383.7478 \\
a_3 &= 107.9338 & b_3 &= 314.7439 \\
a_4 &= 80.1442 & b_4 &= 269.2901
\end{aligned}
$$

```
x = [236.0
     263.0
     271.0
     281.0
     . . . . . .
     1342.0
     1406.0
     2125.0
     2475.0];

Nx=length(x);
for j=1:5,
    r=j-1;
    const1=fac(r);
    const2=fac(Nx-1);
    const3=fac(Nx-1-r);
    den=const2/(const1*const3);
    sigma=0;
    for i=1:Nx,
    U=Nx-i-r;
    if U>=0,
    coeff=fac(Nx-i)/(const1*fac(U));
    sigma=sigma+coeff*x(i);
    end;
    end;
a(j)=sigma/(Nx*den);
end
'Elements of a'
a'
'Values of first 4 l-moments are'
l1=a(1)
l2=a(1)-2*a(2)
l3=a(1)-6*a(2)+6*a(3)
l4=a(1)-12*a(2)+30*a(3)-20*a(4)
```

Figure 2.16   MATLAB code for calculating the first four l-moments. The data in the vector x are annual floods for Rio Hercílio at Ibirama (see Figure 1.1). Nx is the number of data values, 48 in this case. Only the first and last four data values are printed in this edited listing.

from which the expressions (2.3) giving the l-moments in terms of a_r and b_r yield

$$l_1 = 783.7083$$

$$l_2 = 224.4743$$

$$l_3 = 61.6473$$

$$l_4 = 47.8333$$

Hosking and others have shown how these four l-moments can be used to describe the position, dispersion, skewness and kurtosis of a data set. We return to the use of l-moments in Chapter 3, where we use them to estimate the parameters of a Wakeby distribution. The reader is referred to Section 1.12, where an e-mail address is given from which much free FORTRAN software on l-moments may be obtained.

2.6   THE IMPORTANCE OF GRAPHICAL STUDIES OF RESIDUALS

We have demonstrated in this chapter a very few of the wide range of procedures available for plotting, exploring and summarising hydrological data sets; we have used GENSTAT for our purposes, although other computer programs have similar facilities. The procedures used have been used, in most cases, in their simplest forms, using the default options; much more sophisticated work can be done when the user becomes familiar with the many other options available. The essential point, however, is the following: no statistical modelling study should be undertaken without first looking at the data by extensive use of computer plotting facilities. Plotting long sequences of data year by year (or period by period) will help to show whether there are temporal changes throughout the period of record; plotting one variable against another will help to identify 'outliers' such as those present in Figure 1.4. Finally, whilst this chapter has been concerned with the use of data-exploratory procedures prior to undertaking statistical modelling, we shall continue to make extensive use of them in later chapters, for the purpose of assessing the goodness of fit between model predictions and data. It is by plotting observations against the fitted values given by alternative models, that the most suitable model can be identified — a model parsimonious in its parameters, and describing adequately the principal features of the data.

REFERENCES

Box, G. E. P. and Jenkins, G. M. (1974). *Time Series Analysis: Forecasting and Control.* Holden Day, San Francisco.

Hosking, J. R. M. (1986). *The Theory of Probability-weighted Moments.* Res. Rep. RC13412. IBM Research, Yorktown Heights, NY.

Hosking, J. R. M. (1990). *l*-moments: analysis and estimation of distributions using linear combinations of order statistics. *J. Royal Statist. Soc. B,* **52** (2), 105–24.

Hosking, J. R. M., Wallis, J. R. and Wood, E. F. (1985). Estimation of the generalised extreme value distribution by the method of probability-weighted moments. *Technometrics,* **27** (3), 251–61.

Sakamoto, Y., Ishiguro, M. and Kitagawa, G. (1986). *Akaike Information Statistics.* Reidel, Dordrecht.

EXERCISES AND EXTENSIONS

2.1. The following data, reproduced by kind permission of G. Kite, show lake levels (in metres) at four limnimetric sites on Lake Victoria, precipitation over the lake, and levels of Lakes Mobuto, Malawi and Tanganyika for the period 1950–80. The data are given in computer file X2-1.DAT on the diskette accompanying this book. Assume that the Lake Victoria levels were measured at the end of each year. Using the computer package of your choice:

(a) Plot each lake level against year.

(b) Calculate the differences Δh in level of Lake Victoria at the four sites, and plot them against precipitation.

(c) Plot Lake Victoria levels against levels in Lakes Mobuto, Malawi, and Tanganyika.

(d) Plot a histogram of Lake Victoria precipitation.

(e) Put the precipitation data in increasing order, and hence estimate quantiles.

(f) Plot a curve of cumulative rainfall, and comment on its appearance.

Gauge datums are as follows (metres above mean sea level):

Entebbe	1123.46
Jinja	1122.95
Kisumu	1122.34
Mwanza	1125.83
Butiaba (L. Mobuto)	609.82

Year	L. Victoria levels (m)				L. Victoria Precipitation (mm)
	Entebbe	Jinja	Kisumu	Mwanza	
1950	10.25	10.69	10.59	7.59	1640
1951	10.13	10.65	10.83	7.52	2199
1952	10.15	11.18	11.39	8.05	1457
1953	10.80	11.31	11.34	8.20	1523
1954	10.53	11.06	11.32	7.88	1514
1955	10.36	10.86	11.19	7.74	1793
1956	10.34	10.84	11.21	7.70	1550
1957	10.40	10.91	11.25	7.75	1483
1958	10.53	11.02	11.27	7.70	1469
1959	10.44	10.94	11.26	7.74	1646
1960	10.34	10.84	11.36	7.57	1781
1961	10.35	10.87	11.14	7.70	2201
1962	11.43	11.95	12.16	8.72	1596
1963	11.88	12.40	12.64	9.23	2019
1964	12.39	12.91	13.16	9.77	1871
1965	12.39	12.89	13.16	9.75	1615
1966	11.98	12.49	12.76	9.35	1720
1967	11.83	12.33	12.63	9.20	1726
1968	11.78	12.32	12.45	9.13	1989
1969	12.06	12.58	12.76	9.39	1745
1970	11.83	12.36	12.53	9.22	2023
1971	11.95	12.46	12.63	9.23	1329
1972	11.68	12.18	12.35	9.09	1715
1973	11.83	12.28	12.51	9.15	1433
1974	11.54	12.06	12.23	8.89	1763
1975	11.46	11.97	12.17	8.78	1963
1976	11.47	12.04	12.24	8.87	2068
1977	11.34	11.70	12.05	8.69	1712
1978	11.61	12.03	12.32	9.00	1412
1979	12.06	12.56	12.75	9.38	1355
1980	12.00				

Year	Levels (m) in		
	L. Mobuto	L. Malawi	L. Tanganyika
1950	10.32	473.05	773.32
1951	9.94	472.90	773.14
1952	10.57	473.29	773.71
1953	10.83	472.74	774.11
1954	10.25	472.38	773.95
1955	10.43	472.44	773.63
1956	10.46	473.29	773.67
1957	10.69	473.35	773.85
1958	10.76	473.90	774.12
1959	10.76	473.42	774.02
1960	10.56	472.93	773.87
1961	10.98	472.74	773.89
1962	12.57	472.93	774.51
1963	13.85	473.72	775.55

(continued)

Year	Levels (m) in		
	L. Mobuto	L. Malawi	L. Tanganyika
1964	13.88	474.27	776.35
1965	13.61	474.06	776.49
1966	12.59	473.81	776.10
1967	12.30	473.60	775.70
1968	11.94	473.42	775.83
1969	12.28	473.60	776.21
1970	12.02	473.48	775.78
1971	12.14	473.51	775.63
1972	11.60	473.29	775.35
1973	11.59	473.35	775.12
1974	11.42	474.02	774.67
1975	11.30	473.96	774.63
1976	11.70	474.27	774.37
1977	11.32	474.02	774.35
1978	11.87	474.63	774.60
1979	12.27	475.73	774.65
1980			

2.2. The data in the diskette file X2-2.DAT show the monthly water balance for Lake Chad for the period from May 1968 to April 1973, taken from G. Vuillaume, 'Bilan hydrologique mensuel et modelisation sommaire du régime hydrologique du lac Tchad', *Cah. ORSTOM*, ser. Hydrol., Vol. XVIII, No 1, 1981. The data are reproduced by kind permission of ORSTOM.

(a) Plot water level, lake area, and lake volume against time, and against each other.

(b) Plot, on the same graph, input to the lake from precipitation, and input to the lake from inflow. What do you conclude?

2.3. The following data (given also in diskette file X2-3.DAT) show monthly totals of potential evapo-transpiration *ET* (in millimetres, calculated by Penman formula, at Bol-Matafo, on the north-eastern shore of Lake Chad. (a) Using the program of your choice, plot *ET* against time. (b) Abstract from the total record of *ET*, the values for the period from May 1968 to April 1973; plot these against the total losses given in column (12) of the data in Exercise 2.2 above. What do you conclude about losses from Lake Chad relative to *ET*?

Year 19:	May	Jun	Jul	Aug	Sep	Oct	Nov	Dec	Jan	Feb	Mar	Apr
65–66	*	*	*	*	137	176	171	120	153	124	216	215
66–67	209	*	*	*	142	208	181	144	118	129	186	219
67–68	244	196	188	141	143	188	166	140	130	138	176	215
68–69	219	178	150	160	170	204	175	157	151	163	201	222
69–70	206	200	187	145	154	194	188	156	157	173	230	253
70–71	237	223	180	139	140	224	194	150	132	154	232	244
71–72	232	210	188	140	144	212	172	143	132	157	233	220
72–73	218	193	190	181	184	178	177	148	151	159	229	247
73–74	247	222	205	149	175	217	176	155	127	154	204	217
74–75	222	203	124	126	143	184	176	151	137	156	219	228
75–76	227	205	175	178	148	210	170	145	125	167	213	227
76–77	212	180	160	145	185	178	186	149	134	160	217	258
77–78	221	198	186	106	163	193	154	146	*	*	*	*

Chapter 3

Fitting distributions

3.1 THE NULL MODEL

In this chapter we consider the simple model in which the variable y_t is expressed in the form:

$$y_t = \mu + \varepsilon_t \tag{3.1}$$

We assume that the random components of the model satisfy the conditions $E[\varepsilon_t] = 0$; $E[\varepsilon_t^2] = \sigma_\varepsilon^2$; and $E[\varepsilon_t \varepsilon_s] = 0$ for $t \neq s$. Indeed, we shall make an even stronger assumption, namely that the observations y_t (and therefore the ε_t also) are statistically independent, all with the same probability distribution. The emphasis of the chapter is upon the estimation of the model parameters for different distributions for the y_t that are commonly used where statistical procedures are used to answer questions about the frequencies of hydrologically extreme events. The simple model (3.1), which includes no 'explanatory' variables — that is, variables which explain, to a lesser or greater degree, the way in which y_t varies — is called the 'null' model. In the null model, each item of data y_t has the same expected value μ; where explanatory variables are included, the expected value of each y_t will be different, depending upon the values of the explanatory variables x_t. Later chapters in this book discuss more complex models in which the systematic component μ on the right-hand side of (3.1) is replaced by forms involving explanatory variables; thus in Chapter 4, we consider the case where μ is replaced by a linear combination of explanatory variables. Models with the above assumptions (statistical independence of the y_t; y_t all with the same probability distribution) are commonly appropriate for a wide range of hydrological problems, particularly those where the y_t are the largest or smallest values occurring in a hydrological year. Thus we are concerned in this chapter with statistical models for variables of which a very few examples are the following: annual maximum discharge; annual minimum discharge; minimum value during the hydrological year of a k-day running mean of mean daily flows; annual rainfall; and annual maximum 30-minute rainfall intensity. We note that total annual runoff may or may not be safely assumed to be statistically independent from year to year; in a first-order basin with rapid response, total annual runoff will closely follow the annual regime and the assumption of statistical independence will probably be justified. However, total annual runoff from a large drainage basin with plentiful storage will probably be correlated from one year to the next; the models discussed in the present chapter would then not be appropriate. On the other hand, we can usually be fairly confident that measurements of annual maximum discharge, and of annual maximum mean daily flow, are statistically independent, although we must not lose sight of the fact that a degree of statistical dependence may be introduced because discharges are estimated using a stage–discharge relationship. It is probably safe to assume that such dependence does not greatly affect conclusions drawn from a statistical analysis of annual floods, although no attempts seem to have been made to quantify it.

3.2 THE LIKELIHOOD FUNCTION

To fix ideas, suppose that $y_1, y_2, \ldots, y_N$ are the annual maximum instantaneous discharges (or 'annual floods') observed at a gauging station in a sequence of years; the data for the Rio Itajaí at Apiuna, shown in Figure 1.1, are typical of this kind of data. If the assumption of statistical independence is valid, there is no need for the sequence to be unbroken; indeed, the order of the observations is then also unimportant. If, however, we needed at some stage to study whether the flood response of the basin had tended to change throughout the period of record — perhaps as a consequence of land-use change — then the order in which the observations had occurred would be extremely important, and the statistical analysis would need to take account of it. The large body of statistical methods, known as time-series analysis, utilises information contained in the order in which non-independent observations occur.

Let us suppose that the probability distribution, common to all the y_t, is $f(y; \theta)$, where θ is a vector containing a small number of parameters, commonly two or three. We note in passing that the quantity μ of equation (3.1) will be a function of the θ, since by the definition of the expected value operator $E[.]$ given in Section A.4 in the Appendix:

$$E[y_t] = \int_{-\infty}^{\infty} yf(y; \theta)\mathrm{d}y = \mu \tag{3.2}$$

and the integral will be a function of the parameters θ.

We come now to an important definition which underlies much of the material in this book. Since the y_t are statistically independent, the probability of obtaining the observations y_t is

$$f(y_1; \theta) \cdot f(y_2; \theta) \cdot \ldots \cdot f(y_N; \theta)\mathrm{d}y_1 \cdot \mathrm{d}y_2 \ldots \mathrm{d}y_N \tag{3.3}$$

assuming that $y_1, y_2, \ldots, y_N$ is a random sample from an infinite population. Given the observations y_t, the *likelihood function* $L(\theta; y_1, \ldots, y_N)$ is defined by

$$L(\theta; y_1, \ldots, y_N) = \prod_{t=1}^{N} f(y_t; \theta) \tag{3.4}$$

that is, the likelihood function is proportional to the probability of obtaining the given observations, regarded as a function of the unknown parameters θ. The likelihood function is particularly important because, when the statistical model is 'correct', the function $L(\theta)$ contains all the information in the data. However, different models will give rise to different likelihood functions, and we have seen that one of the tasks confronting us is to find which model is the most appropriate for representing characteristics of the data at hand. Situations will arise in which we are unable to conclude that one model is more appropriate than another (or, indeed, than several others); in such cases each different, but equally plausible, model will give rise to its own likelihood function.

Hence, whilst it is true to say that the likelihood contains all the available information when the statistical model is 'correct', this is not particularly helpful if we are unable to say conclusively which model from several alternatives is the more appropriate.

Having defined the likelihood function, we discuss its use in estimating the parameters θ. To do this, we consider the surface defined by $L(\theta)$ (where, to simplify notation, we have dropped the y_t from the expression for the likelihood function) as θ varies. Clearly different values for θ will give rise to different values for $L(\theta)$. From all the possible θ values, we would expect intuitively that the value of θ would be most appropriate which maximises the probability of the values y_t actually observed: that is, which maximises the likelihood function $L(\theta)$. More formally, for a likelihood function $L(\theta)$ of a function of k parameters θ, the *maximum-likelihood estimate* $\hat{\theta}$ of θ is defined as the value of $\hat{\theta}$ for which

$$L(\hat{\theta}) \geq L(\theta)$$

for all θ. Now we have seen (equation (3.4)) that $L(\theta)$ is a product of terms like $f(y_t; \theta)$; if we work with the logarithm of $L(\theta)$,

$$l(\theta) = \sum_{t=1}^{N} \ln f(y_t; \theta) \tag{3.5}$$

then the values of θ which maximise $l(\theta)$ will also maximise $L(\theta)$. Frequently we can find the value of θ which maximises $L(\theta)$ and $l(\theta)$ by differentiation, that is by solving the equations

$$\partial l/\partial\theta = \partial \ln L/\partial\theta = L^{-1}\partial L/\partial\theta = 0 \tag{3.6}$$

These equations in general will not have explicit solutions for θ (that is, a solution cannot be found of the form 'θ equals an expression containing just the observations y_t', without θ appearing on both sides of the equality sign). It will then be necessary to solve the equations iteratively, and some examples are given below. When the vector of parameters θ has only two or three elements, it is always advisable to plot the surfaces $L(\theta)$ or $l(\theta)$ by methods which are also demonstrated below using GENSTAT. The surface is not, of course, plotted just to indicate where its maximum values occur; it is the whole course of the function which contains the totality of information provided by the data. To quote Box and Jenkins (1970):

> situations can occur where the likelihood function has two or more peaks ... [or] ... sharp ridges and spikes. All of these situations have logical interpretations. In each case, the likelihood function is telling us something which we ought to know. Thus, the existence of two peaks of approximately equal heights implies that there are two sets of values of the parameters which might explain the data. The existence of obliquely-oriented ridges means that a value of one parameter, considerably different from its maximum-likelihood value, could explain the data if accompanied by a value of the other which deviated appropriately ... to understand the estimation situation, we must examine the likelihood function both analytically and graphically.

Box and Jenkins proceed to warn against a too simplistic usage of equations (3.6) above to obtain maximum-likelihood estimates:

> The treatment afforded the likelihood method has, in the past, often left much to be desired and ineptness in the practitioner has sometimes been mistaken for deficiency in the method. The treatment has often consisted of
>
> (i) differentiating the log-likelihood and setting first derivatives to zero to obtain the maximum-likelihood (ML) estimates;
>
> (ii) deriving approximate variances and covariances of these estimates from the second derivatives of the log-likelihood or from the expected values of the second derivatives.

Mechanical application of the above can, of course, produce nonsensical answers. This is so, first, because of the elementary fact that setting derivatives to zero does not necessarily produce maxima, and second, because the information which the likelihood function contains is only fully expressed by the ML estimates and by the second derivatives of the log-likelihood, if the quadratic approximation is adequate over the region of interest. To know whether this is so for a new estimation problem, a careful analytical and graphical investigation is usually required.

Since these words were written, the ability of computers to assist in graphical aspects of statistical estimation has developed immensely, and it behoves hydrologists to exploit this development to the full.

To reinforce still further the strictures of Box and Jenkins, quoted above at length, it is sometimes stated in the hydrological literature that 'maximum-likelihood estimates gave nonsensical values'. Typically, the practitioner has attempted to solve equations (3.6) for, say, two parameters θ_1, θ_2, by a search over the entire (θ_1, θ_2) space, when the search should more properly have been conducted over a subset of this space. In such circumstances it may be more appropriate to look for the maximum of the log-likelihood surface by graphical searches along the boundaries of the permissible region, to identify the $\hat{\theta}$ for which $L(\hat{\theta}) > L(\theta)$, instead of solving equations (3.6). In other cases, the reported 'failure' of the ML procedure to produce estimates of which the practitioner approved, may be simply an indication that he was forcing the data into the wrong statistical model.

So far, we have assumed that the data $y_1, \ldots, y_N$ are a sample drawn at random from the population with probability density $f(y; \theta)$, and the subsequent development requires the notion of repetitive sampling from the same population. It can be argued that, for hydrological applications, the concept of repetitive sampling is inappropriate, the data being uniquely associated with the particular times at which measurements were made. Therefore, the argument runs, the basis for the application of statistical methods based upon repetitive sampling does not exist. Whilst this contains an element of truth, it does appear that statistical methods are helpful in the analysis of hydrological data, not least because they provide a means of separating the 'haphazard' variability in a data set (associated with the random component of the model) from variability associated with regularities such as those represented by the systematic component of a model. Whilst acknowledging that the idea of repetitive sampling may sometimes give rise to conceptual difficulties, we put these concerns aside and proceed under the assumption that the data $y_1, \ldots, y_N$ constitute one of a population of such samples, which for the course of this chapter are drawn independently from the probability density $f(y; \theta)$.

3.3 FITTING LOG-NORMAL DISTRIBUTIONS BY MAXIMUM LIKELIHOOD

When a sequence of annual floods is plotted as a histogram, its shape is very commonly positively skewed. A family of probability distributions which exhibits this shape is the *log-normal* family, which we now discuss. The characteristic skewness of log-normal distributions, and the fact that they avoid the non-zero probability (however small) associated with some other distributions, combine to make this family of distributions one that is very widely used in hydrological analysis. Besides its use, demonstrated below, for annual maximum mean daily discharges, it has also been used to model flood peak discharges, and annual, monthly and daily rainfall. Its use extends far beyond hydrology; for example, it has been used to describe earthquake magnitudes, lengths of quiescent periods between earthquakes, distribution of sediment particle sizes, and many other variables occurring in nature. The paper by

Stedinger (1980) contains an excellent description of hydrological applications of the log-normal distribution.

3.3.1 The Two-parameter Log-normal Distribution

We illustrate maximum-likelihood estimation procedures for the case when the y_t have the log-normal distribution with two parameters:

$$f(y; \mu, \sigma^2) = \{1/[\sigma y \sqrt{(2\pi)}]\} \exp[-(\ln y - \mu)^2/(2\sigma^2)] \qquad (0 \le y < \infty) \qquad (3.7)$$

The logarithm of the likelihood function, $l(\mu, \sigma^2)$ in which the two parameters previously denoted by $\theta = (\theta_1, \theta_2)$ are now denoted by $\theta_1 = \mu$ and $\theta_2 = \sigma^2$, is

$$l(\mu, \sigma^2) = -N \ln \sigma - (1/2) \ln(2\pi) - \sum \ln y_t - \sum_t (\ln y_t - \mu)^2/(2\sigma^2)$$

from which $\partial l/\partial \mu = 0$ gives

$$\sum (\ln y_t - \mu) = 0$$

and $\partial l/\partial \sigma = 0$ gives

$$-(N/\sigma) + (1/\sigma^3) \sum (\ln y_t - \mu)^2 = 0$$

These two equations have the explicit solution

$$\hat{\mu} = \sum (\ln y_t)/N \qquad \text{and} \qquad \hat{\sigma}^2 = \sum (\ln y_t - \hat{\mu})^2/N$$

We take for an example the data for the Rio Itajaí at Apiuna, shown in Figure 1.1. Having read the data by GENSTAT into a variate *Apiuna*, the annual floods y_t are converted to (Napierian) logs and the mean (denoted by *muhat*) and variance (*sighat2*) calculated by:

```
>CALCULATE logAp=LOG(Apiuna)   :   &   muhat=MEAN(logAp)
>    &   sighat2=VARIANCE(logAp)
>PRINT muhat, sighat2
```

giving the values $\hat{\mu} = 7.2697$ and $\hat{\sigma}^2 = 0.253\,316$ ($\hat{\sigma} = 0.5033$). We plot the log-likelihood surface using the GENSTAT statistical function LLNORMAL(x; m; v) or LLN(x; m; v) which calculates the values of the likelihood of a sample x from a normal distribution with mean m and variance v; in the present example, x is replaced by the variate *logAp*, and a grid of points (m, v) is selected in the neighbourhood of the likelihood maximum. For example, with a 7×7 grid of values of the parameters μ and σ, stored in the 7×7 matrices named **mu** and **sig**, we can calculate the 7×7 grid of log-likelihood values by the commands:

```
>SCALAR rmu,rsig
>FOR i=1...7
>FOR j=1...7
>CALCULATE rmu=mu$[i;j] : & rsig=sig$[i;j]
>& logL$[i;j]=LLNORMAL(logAp;rmu;rsig)
>ENDFOR
>ENDFOR
```

logL 1	2	3	4	5	6	7
1 −4698.0898	−3123.1177	−2337.7556	−1867.8019	−1555.3383	−1332.7480	−1166.2533
2 −2090.2703	−1384.5713	−1033.8459	−824.6742	−686.0651	−587.6567	−514.2984
3 −525.5781	−341.4436	−251.5001	−198.7973	−164.5012	−140.6019	−123.1255
4 −4.0148	6.2652	9.2815	9.8279	9.3532	8.4161	7.2654
5 −525.5794	−341.4445	−251.5008	−198.7979	−164.5016	−140.6023	−123.1258
6 −2090.2727	−1384.5734	−1033.8473	−824.6751	−686.0659	−587.6575	−514.2991
7 −4698.0942	−3123.1208	−2337.7576	−1867.8037	−1555.3397	−1332.7493	−1166.2543

Figure 3.1   Values of log-likelihood in the region of the maximum: two-parameter log-normal distribution fitted to Apiuna flood record after log transformation. logL is evaluated over a 7×7 grid, centred on the maximum, with increments equal to 0.2 times the maximum-likelihood estimates of μ and σ^2. Values shown are values of log-likelihood, plus N times $\ln \sqrt{(2\pi)}$ (this accounts for the positive values shown)

Contour plot of logL at intervals of 470.792

** Scaled values at grid points **

−9.9791	−6.6338	−4.9656	−3.9674	−3.3037	−2.8309	−2.4772
−4.4399	−2.9409	−2.1960	−1.7517	−1.4573	−1.2482	−1.0924
−1.1164	−0.7253	−0.5342	−0.4223	−0.3494	−0.2987	−0.2615
−0.0085	0.0133	0.0197	0.0209	0.0199	0.0179	0.0154
−1.1164	−0.7253	−0.5342	−0.4223	−0.3494	−0.2986	−0.2615
−4.4399	−2.9409	−2.1960	−1.7517	−1.4573	−1.2482	−1.0924
−9.9791	−6.6337	−4.9656	−3.9674	−3.3037	−2.8309	−2.4772

```
             0.0000   0.1667   0.3333   0.5000   0.6667   0.8333   1.0000
           '        '        '        '        '        '        '
   1.000-000   222    444444           66666666666666666            -
            222    444444           66666666666666666
        222    44444           6666666666666               88
          4444     66666666666                    888888888888
        4444    666666666          8888888888888888888888
        4       66666666               888888888888888888888888888888
   0.833-   6666666              888888888888888888888888888888888888888-
        666666             88888888888888888888888888888888888888888
        6            888888888888888888888888888888888
            888888888888888888888888888
      8888888888888888888888
      8888888888
   0.667-888                                                          -
                  00000000000000000000000000000000000000000000000000
   0.500-   0000000000000000000000000000000000000000000000000000000
            00000000000000000000000000000000000000000000000000000000-
   0.333-888                                                          -
      8888888888
      888888888888888888888
            888888888888888888888888888
        6            88888888888888888888888888888888888
        666666             888888888888888888888888888888888888888
   0.167-   6666666              88888888888888888888888888888888888888-
        4            66666666               888888888888888888888888888
          4444     666666666               8888888888888888888888
        4444    66666666666                    888888888888
        222    44444           6666666666666               88
            222    444444           66666666666666666
   0.000-000   222    444444           66666666666666666            -
           '        '        '        '        '        '        '
```

Figure 3.2   Contour plot of log-likelihood function values, given in Figure 3.1: log-likelihood obtained by fitting a two-parameter log-normal distribution to 51 years of flood record from Rio Itajaí-Açú at Apiuna, after log transformation.

Figure 3.1 shows the values of the log-likelihood calculated by these commands, and Figure 3.2 shows the values plotted by lineprinter using the GENSTAT command

>CONTOUR logL

It is seen that the log-likelihood surface is symmetric in the direction of μ (vertically downwards, in the Figure 3.2) but is very asymmetric in the perpendicular direction, corresponding to the direction of σ^2. The log-likelihood surface has its maximum at the centre of Figure 3.2, corresponding to the point in the fourth row and fourth column of the array for which contours are drawn. Clearly, we should be unjustified in assuming that the log-likelihood surface can be adequately represented by a quadratric surface near its maximum.

We also wish to know whether the log-normal distribution is a good representation of the annual flood sequence for Apiuna. A graphical procedure appropriate for this purpose is the *normal plot*, in which the ordered sequence of the values in *logAp* is plotted against the expected values of the order statistics from a normal distribution. In fact we plot the ordered sequence of values in *logAp* against the *normal equivalent deviates* corresponding to $(t - 0.375)/(N + 0.25)$, having first ordered the logarithms of the annual floods in the variate *logAp* using:

>CALCULATE logAp=SORT(logAp)

If the record of annual floods at Apiuna is well represented by a normal distribution, the ordered sequence of log floods in *logAp* should show a strong linear relation with the values of NED$((t - 0.375)/(N + 0.25))$. Figure 3.3 shows that this is indeed the case.

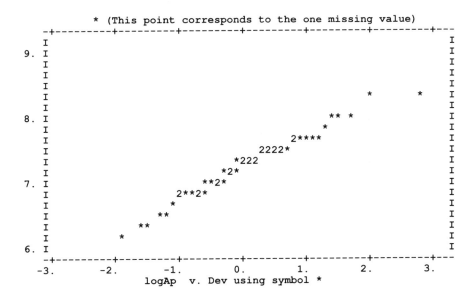

Figure 3.3   Normal plot of 51 years' record of annual flood (maximum mean daily discharge) for Rio Itajaí-Açú at Apiuna. Data as in Figure 1.1, but transformed to logarithms.

Having estimated the parameters of the log-normal model, we can now estimate the annual flood with return period T years, by calculating the value Y_0 which satisfies the following equation:

$$\frac{1}{\hat{\sigma}\sqrt{(2\pi)}} \int_{-\infty}^{Y_0} y^{-1} \exp\left\{-\frac{1}{2\hat{\sigma}^2}[\ln y - \hat{\mu}]^2\right\} dy = 1 - T^{-1} \tag{3.8}$$

If, for example, we require the flood with return period 100 years, so that $T^{-1} = 0.01$, we can calculate it easily using the GENSTAT function NED(p), which calculates the $p\%$ quantile for the standard normal distribution $N(0, 1)$. Thus, for the annual flood data for Apiuna, we use the commands

```
>CALCULATE Y0=muhat + sighat*NED(0.99) : & Y0=EXP(Y0)
>PRINT Y0 : "(This value of Y0 is then the 100-year flood)"
```

in which *muhat* and *sighat* are the maximum-likelihood estimates of μ and σ, the parameters of the two-parameter log-normal distribution.

A characteristic of the two-parameter log-normal distribution is that $\ln y$ has a normal distribution:

$$f(z; \mu, \sigma^2) = \{1/[\sigma\sqrt{(2\pi)}]\} \exp[-(z - \mu)^2/(2\sigma^2)]$$

Transformation to normality by taking logarithms is a property of the log-normal distribution; however, in practice, hydrological data follow a probability distribution that is known to be skew but not necessarily log-normal. A more general transformation, of which the log transform is a particular case, is the family of Box–Cox transformations, defined by

$$z = \begin{cases} (y^\lambda - 1)/\lambda & \text{when } \lambda \text{ is non-zero} \\ \log y & \text{when } \lambda \text{ is zero} \end{cases}$$

where λ is to be estimated from the data. The estimation procedure consists of the following:

(i)   Assume a range of values for λ: usually values in the range between 2 and -2 are appropriate.

(ii) For each value of λ within this range, transform the y-sequence of data into the z-sequence given by the above transformation.

(iii) For each value of λ, calculate the maximum of the log-likelihood function, say $\max[\ln L(\lambda)]$.

(iv) Select the value of λ for which this quantity is maximised.

It may be shown that the value of λ which transforms the data sequence from the Rio Hercílio at Ibirama is -0.055; clearly this value does not look very different from zero, and we can in fact test whether it differs significantly from that quantity. To do this, we take the plot of $\ln L(\lambda)$ against λ, and do the following:

(i)   Draw the (horizontal) tangent to the curve $\ln L(\lambda)$ at the value of λ at which the curve reaches its maximum.

(ii)   Draw a second horizontal line below this tangent, the perpendicular distance between
        these two horizontal parallel lines being the $100(1 - \alpha)\%$ quantile of the χ^2 distribution
        with one degree of freedom (the GENSTAT instructions, for $\alpha = 0.05$, are

> CALCULATE y=CED(0.95;1) : PRINT y

        from which the 95% quantile of the χ^2 distribution is 3.841, the horizontal distance
        between the two lines).

(iii)   Mark the two points where the lower horizontal line cuts the curve $\ln L(\lambda)$; these points
        define two values of λ, say λ_L, λ_U, which define a $100(1 - \alpha)\%$ confidence interval for
        the true value of λ, the parameter of the transformation. If the interval defined by λ_L,
        λ_U includes zero, corresponding to a log transformation, then the value of λ maximising
        $\ln L(\lambda)$ (in this case -0.055) does not differ significantly from this quantity, at the
        $100\alpha\%$ significance level.

We conclude that a log transformation gives a good approximation to normality. We shall
return to Box–Cox transformations at various points during the remainder of this book.
    When should we use a tranformation — such as a Box–Cox transformation — to put
skewed data into a scale which is approximately normal, and when should we use a
skewed distribution — such as the Gumbel distribution discussed below — without resorting
to transformation of data? To answer this question, we need to consider the log-likelihood
surfaces given by the two options. If, for one of the two options, the maximum of the log-
likelihood function was much greater than for the other, this would suggest that one of the
two options is more consistent with the data than its alternative. We return to the question of
choosing between distributions in Section 3.6 below, where we use likelihood methods in an
attempt to establish whether a log-normal or gamma distribution is more appropriate.

3.3.2   The Three-parameter Log-normal Distribution

The log-normal distribution fitted to the flood record at Apiuna had two parameters, μ and σ
(or μ and σ^2). A modified form of this distribution, having greater flexibility for the modelling
of flood records, contains three parameters, denoted by μ, σ and a:

$$f(y; \mu, \sigma, a) = \{1/[\sigma (y-a)\sqrt{(2\pi)}]\} \exp\{-(\ln[y-a]-\mu)^2/(2\sigma^2)\} \qquad (a \le y < \infty) \quad (3.9)$$

The two-parameter log-normal is, of course, a special case of this distribution, having $a = 0$.
    Estimates of the parameters of the three-parameter distribution can be calculated using the
GENSTAT function LLNORMAL$(x; m; v)$ introduced earlier. For suppose that we assume that
the parameter a is fixed and known, say a_0; then use of the variable $y_t^* = y_t - a_0$ enables us
to calculate the estimates $\hat{\mu}_0$ and $\hat{\sigma}_0^2$, given by

$$\hat{\mu}_0 = \sum(\ln y_t^*)/N$$
$$\hat{\sigma}_0^2 = \sum(\ln y_t^* - \hat{\mu}_0)^2/N$$

The corresponding value of the log-likelihood $l(\hat{\mu}_0, \hat{\sigma}_0^2, a_0)$ for this value of a_0 is

$$l(\hat{\mu}_0, \hat{\sigma}_0^2, a_0) = -N \ln \hat{\sigma}_0 - N \ln(2\pi)/2 - \sum \ln y_t^* - \sum(\ln y_t^* - \hat{\mu}_0)^2/(2\hat{\sigma}_0^2) \qquad (3.10)$$

Apart from the constant term, $-N \ln(2\pi)/2$, this is the quantity calculated by the GENSTAT function LLNORMAL. Hence, by choosing a succession of values $a_0, a_1, \ldots$, we can build up a *conditional likelihood function* showing how $l(\hat{\mu}, \hat{\sigma}^2, a)$ varies with a. The value of a for which this conditional likelihood has its maximum defines the maximum-likelihood estimate of the parameter a, which in turn defines the maximum-likelihood estimates of μ and σ. We illustrate the procedure briefly with the annual flood sequence from the Rio Hercílio at Ibirama. Having read the annual flood record into the variate *Ibirama*, we set up a loop like the following:

```
>FOR a=0,50...200
>CALCULATE loglba=LOG(Ibirama-a)   :   & muhat=MEAN(loglba)
>& sig2hat=VARIANCE(loglba)
>& logL = LLNORMAL(loglba;muhat;sig2hat)
>PRINT a, logL
>ENDFOR
```

with the results shown in Figure 3.4. For the values of a included in the loop, the log-likelihood is a maximum at the point $a = 0$ (although it would be advisable to confirm that this is the maximum by using, as the first line in the loop definition, the opening command

```
>FOR a=−50,10...50
```

which would explore the log-likelihood in the neighbourhood of $a = 0$ more thoroughly). Physical considerations suggest, however, that negative values of the parameter a are unwarranted, as giving rise to the possibility, although small, of negative flows.

    The above method for estimating the parameter a requires us to explore the likelihood surface, always a desirable thing to do. However, alternative methods are also available; it is possible, for example, to estimate all three parameters simultaneously by solving the non-linear equations

$$\partial l/\partial \mu = 0$$

$$\partial l/\partial \sigma = 0$$

$$\partial l/\partial a = 0$$

by an iterative procedure, subject, of course, to the constraint that the solution for a must be less than y_1. Kite (1977) shows that algebraic manipulation of the three equations for μ, σ

a	logL
0.0	8.594
50.0	4.437
100.0	−0.4701
150.0	−6.628
200.0	−15.56

Figure 3.4   Fitting a three-parameter log-normal distribution to sequence of annual floods (annual maximum mean daily discharges) from Rio Hercílio at Ibirama: 51 years of record, three missing values. Log-likelihood calculated for a series of values of the parameter a, defining the lower bound of the distribution. Value of $\log L$ is log-likelihood plus N times $\ln \sqrt{(2\pi)}$, which accounts for the positive values.

and a results in a single equation for the parameter a, roots of which can be calculated by numerical methods.

3.4 ESTIMATION BY THE METHOD OF MOMENTS

The above discussion has illustrated the estimation of model parameters using the method of maximum likelihood. Other estimation procedures also exist, and hydrologists have long been accustomed to estimating parameters by the *method of moments*. Because this method gave estimates which were often easily calculated, it was widely used before the advent of general computer programs for statistical analysis, even when it was known that the method of moments gives estimates that are inferior to those given by maximum likelihood. Now that excellent computer programs are widely available and can be used on inexpensive desk-top computers, the use of the method of moments to obtain definitive estimates ('moment estimates') of parameters can rarely be justified.

This is not to say, however, that moment estimates do not have uses in intermediate stages of parameter estimation. It was mentioned earlier that maximum likelihood frequently led to equations that must be solved by iterative procedures which require initial estimates of the parameters, subsequently improved by the iterative procedures of maximum likelihood; moment estimates are useful for initiating such procedures.

Earlier hydrological texts placed considerable emphasis on estimation by moments; in this book, however, we emphasise estimation procedures based upon the likelihood function. We note that the method of moments consists in equating moments calculated from data with moments calculated from the probability distribution from which the data are assumed to be a random sample. Thus if $\bar{y}$, s^2 are the mean and variance of a data sequence $y_1, y_2, \ldots, y_N$, estimates of parameters θ_1 and θ_2 of a two-parameter probability distribution $f(y; \theta_1, \theta_2)$ are estimated, using the method of moments, by solving the following two equations for $\hat{\theta}_1$ and $\hat{\theta}_2$:

$$\bar{y} = \int_{-\infty}^{\infty} yf(y; \theta_1, \theta_2)\,dy$$

$$\vdots$$

$$= \mu$$

say, which will be a function of θ_1 and θ_2; and

$$s^2 = \int_{-\infty}^{\infty} (y - \mu)^2 f(y; \theta_1, \theta_2)\,dy$$

$$\vdots$$

$$= \sigma^2$$

say, which will also be a function of θ_1 and θ_2.

Where the probability distribution has three parameters, the third moment calculated from the data $(= \Sigma(y_t - \bar{y})^3/N)$ would be equated to the third moment calculated as

$$\int_{-\infty}^{\infty} (y - \mu)^3 f(y; \theta_1, \theta_2, \theta_3)\,dy$$

The essential point is that one equation is obtained, by equating moments of the sample with moments of the distribution, for each parameter to be estimated. In particular cases, such as the normal distribution (see Appendix), moment estimates and maximum-likelihood estimates of some of the parameters are equivalent; that they are not equal in general is demonstrated by the two-parameter log-normal distribution, for which the moment estimates are obtained by solving the following equations for μ and σ:

$$\bar{y} = \exp(\mu + \sigma^2/2)$$

and

$$s^2 = \exp(2\mu + \sigma^2) \cdot [\exp(\sigma^2) - 1]$$

These two equations for μ and σ can be solved to give

$$\tilde{\mu} = \ln \bar{y} - (1/2) \ln(1 + s^2/\bar{y}^2)$$
$$\tilde{\sigma}^2 = \ln(1 + s^2/\bar{y}^2)$$

which are clearly different from the maximum-likelihood estimates. Kite (1977) lists a number of methods other than the method of moments by which estimates of distribution parameters may be obtained for the probability distributions discussed in this chapter.

3.5 FITTING GAMMA DISTRIBUTIONS BY MAXIMUM LIKELIHOOD

The two- and three-parameter gamma distributions have been widely used in hydrology, not least for the purpose of modelling the frequencies of annual floods. Like observed data from many phenomena, gamma distributions are skewed to the right. Amongst other hydrological applications, they have been used to describe rainfall depths; a particularly effective model of daily rainfall in semi-arid regions (Stern and Coe, 1984) uses two-parameter gamma distributions for which the parameters (μ and κ in the description below) vary harmonically throughout the year. Applications of their model include: calculation of probabilities of runs of wet or dry days, beginning at any day in the year (from which, for example, probabilities of dry periods can be calculated at critical periods in crop development); and probabilities that sufficient water is available in the top metre of soil to sustain a planted crop, day by day throughout the year.

3.5.1 The Two-parameter Gamma Distribution

Using the symbols μ and κ (instead of θ_1 and θ_2), the two-parameter distribution can be written

$$f(y; \mu, \kappa) = (\kappa/\mu)^\kappa y^{\kappa-1} \exp[-\kappa y/\mu]/\Gamma(\kappa) \qquad 0 \leq y < \infty; \mu > 0; \ \kappa > 0 \qquad (3.11)$$

for which the mean is μ and the variance μ^2/κ (the coefficient of variation being $1/\sqrt{\kappa}$). The parameter κ determines the shape of the distribution. If $0 < \kappa < 1$, the distribution increases without limit as y approaches zero and decreases monotonically as $y \to \infty$. The special case $\kappa = 1$ gives the exponential distribution. If $\kappa > 1$ the density is zero at the origin, and there is a single mode at $y = \mu - \mu/\kappa$. The distribution is positively skewed for all κ, and approaches the normal distribution as $\kappa \to \infty$.

Given a sample (say of annual floods) that are statistically independent, the log-likelihood may be written:

$$l(\mu, \kappa) = N\kappa \ln(\kappa/\mu) + (\kappa - 1) \sum \ln y_t - (\kappa/\mu) \sum y_t - N \ln \Gamma(\kappa) \qquad (3.12)$$

The equation $\partial l/\partial\mu = 0$ yields immediately the estimate of μ:

$$\hat{\mu} = \bar{y} \tag{3.13}$$

but the equation $\partial l/\partial\kappa = 0$ has no explicit solution:

$$\ln\kappa - \partial\ln\Gamma(\kappa)/\partial\kappa = \ln\bar{y} - \overline{\ln y} \tag{3.14}$$

The quantity $\partial\ln\Gamma(\kappa)/\partial\kappa$ is the *digamma function*. This, together with the trigamma function (the derivative of the digamma function) is easily programmed both in GENSTAT and MATLAB. The two GENSTAT procedures given below (a) calculate the digamma function, which is used by means of calls such as

> DIGAMMA [PRINT = yes] X

which prints the value of digamma (x); and (b) find the root of equation (3.14), by means of calls such as

> GREENDUR [PRINT = yes] d; ROOT = kappa

where d contains the expression on the right-hand side of equation (3.14). After execution of the call on GREENDUR, 'kappa' contains the desired root.

```
PROCEDURE 'DIGAMMA'
OPTION NAME = \
'PRINT' ; ''(I: string (no, yes) default no) controls whether
the digamma function is printed''               \
MODE = t; DEFAULT = 'no'
PARAMETER NAME = \
'X', ''(I: scalar) argument of digamma function''          \
'DIGAMAX'; ''(O: scalar) calculated digamma function''    \
MODE = p
''Check that the procedure arguments are valid using
Library procedure CHECKARGUMENTS''
CALCULATE E = 0
CHECKARGUMENTS [ERROR = E] PRINT, X, DIGAMAX; ''Names of the arguments'' \
SET = 2 ('yes'), 'no';                           \
TYPE = 'text', 2('scalar');                      \
DECLARED = 2 ('yes'), 'no';                      \
VALUES = ! T (no, yes), 2(*);                    \
PRESENT = ('yes'), 'no'
EXIT [CONTROL=procedure] E
SCALAR s, c, s3, s4, s5, d1; VALUE = 0.00001, 8.5, 0.08333333333, \
0.008333333333, 0.003968253968, −0.5772156649
SCALAR y, r
''Now check that the argument is positive:- ''
IF X < 0
    PRINT 'X < 0: invalid argument'
    EXIT [CONTROL = procedure]
```

```
ENDIF
      IF UNSET (DIGAMAX)
      SCALAR Digamma
      ASSIGN Digamma; POINTER = DIGAMAX
ENDIF
CALCULATE DIGAMAX = 0
IF X < s
      CALCULATE DIGAMAX = d1−1/X
      EXIT [CONTROL = procedure]
ENDIF
CALCULATE y=X
IF y < c
      CALCULATE DIGAMAX = DIGAMAX−1/y
      &             y=y+1
ENDIF
CALCULATE      r=1/y
& DIGAMAX = DIGAMAX + LOG(y) − 0.5∗r
&            r = r ∗ r
& DIGAMAX = DIGAMAX − r∗ (s3 − r∗ (s4 − r∗s5))
IF 'yes' · IN · PRINT
      PRINT [IPRINT = ∗] 'X=', X, 'Digamma = ', DIGAMAX; DECIMALS = ∗, 4, ∗, 8
ENDIF
ENDPROCEDURE
''
''

PROCEDURE 'GREENDUR'
OPTION NAME = \
'PRINT'; "(I: string {no, yes} default no) controls whether
or not the root is printed"
MODE = t; DEFAULT ='no'
PARAMETER NAME = \
'D', "(I: scalar) the RHS of the gamma log-likelihood equation" \
'ROOT'; "(O: scalar) the root of the gamma log-likelihood equation" \
MODE = p
"Check that the procedure arguments are valid using
Library procedure CHECKARGUMENTS "
CALCULATE E = 0
CHECKARGUMENTS [ERROR = E] PRINT, D, ROOT; "Names of the arguments" \
SET = 2 ('yes'), 'no';                         \
TYPE = 'text', 2('scalar');                    \
DECLARED = 2 ('yes'), 'no';                    \
VALUES = !T (no, yes), 2(∗);                   \
PRESENT = ('yes'), 'no'
EXIT [CONTROL = procedure] E
IF UNSET (ROOT)
      SCALAR Root
```

```
    ASSIGN Root; POINTER = ROOT
ENDIF
SCALAR rop, ro, dig, test
CALCULATE rop = 1
FOR I = 1...50
    CALCULATE ro = rop
    DIGAMMA X = ro; DIGAMAX = dig
    CALCULATE rop = ro * (LOG (ro) − dig)/D
    &              test = ABS (rop − ro)
    IF 'yes' .IN.PRINT
            PRINT I, 'iterations', 'ROOT is', ro; DECIMALS = 0, *, *, 8
    ENDIF
    IF TEST < 0.000001
            CALCULATE ROOT = ro
            EXIT [NTIMES = 2]
    ENDIF
ENDFOR
IF 'yes' .IN.PRINT
        PRINT 'After', I, 'iterations'; DECIMALS = *, 0, *
        PRINT [IPRINT = *] 'Root of gamma log-likelihood equation is', ROOT;\
        DECIMALS = *, 8
ENDIF
ENDPROCEDURE
```

However, we can explore the log-likelihood surface directly, using the statistical function LLGAMMA($x;m;d$) of GENSTAT, which provides the log-likelihood function for the gamma distribution with mean m ($= \mu$, in our notation) and index d ($= \kappa$). Putting m equal to the maximum-likelihood estimate $\hat{\mu}$ for this parameter, we can choose a sequence of values for the index d, for each of which we can use LLGAMMA($x;m;d$) to calculate the log-likelihood function, as a function of the index d. Thus, for the data from the Rio Hercílio at Ibirama, we have $\hat{\mu} = m = 783.708$. The variance of the Ibirama data is 193 060; equating this quantity to the second moment about the mean for the distribution, μ^2/κ, we find the moment estimate of κ to be $\tilde{\kappa} = 3.1814$. Using LLGAMMA($x;m;d$) to plot the log-likelihood between 3 and 5, we find that for $d = 3.9$, 4.0 and 4.1 we have log-likelihood values of $-36.903\,16$, $-36.896\,51$ and $-36.906\,18$, respectively. The following GENSTAT instructions calculate the log-likelihood between 3.90 and 4.10:

```
>FOR i=1...21
>CALCULATE d=(i−11)*0.01+4 : & logL=LLGAMMA(Ibirama;783.708;d)
>PRINT d,logL; DECIMALS=3,6
>ENDFOR
```

from which we find that $\ln L$ has values $-36.896\,526$, $-36.896\,454$ and $-36.896\,515$ where d has values 3.98, 3.99 and 4.00, showing a maximum near 3.99. Replacing the second line in the above instructions by

```
>CALCULATE d=(i−11)*0.001+3.99
```

we find that 3.993 is an improved estimate, although numerical errors are beginning to influence the least significant digits, as the following table shows:

d	$\ln L$	d	$\ln L$
3.980	−36.896 526	3.991	−36.896 481
3.981	−36.896 549	3.992	−36.896 477
3.982	−36.896 492	3.993	−36.896 422
3.983	−36.896 431	3.994	−36.896 477
3.984	−36.896 496	3.995	−36.896 488
3.985	−36.896 465	3.996	−36.896 454
3.986	−36.896 496	3.997	−36.896 567
3.987	−36.896 473	3.998	−36.896 492
3.988	−36.896 481	3.999	−36.896 557
3.989	−36.896 477		
3.990	−36.896 454		

The method described in the last paragraph obliges the user to explore the log-likelihood surface in some detail. This is always a good thing to do, although if procedures are available for the calculation of $\Gamma(\kappa)$ or $\partial \ln \Gamma(\kappa)/\partial\kappa$, equation (3.14) can then be solved numerically to obtain an estimate of κ. Alternatively, the log-likelihood function (3.12) can be evaluated for a series of values of κ, and the maximum of the log-likelihood identified, conditional on κ; the value of κ corresponding to the maximum of these conditional maxima will give the maximum-likelihood estimate of κ, the estimate of μ already being given by $\hat{\mu} = \bar{y}$.

Where the digamma function is available, it is possible to obtain the maximum-likelihood estimate of κ by using the following iterative procedure:

$$\kappa_{i+1} = \kappa_i [\ln \kappa_i - \partial \ln \Gamma(\kappa)/\partial\kappa_{\kappa=\kappa_i}]/(\ln A - \ln G) \qquad (3.15)$$

where A and G are the arithmetic and geometric means, respectively, of the observations $y_1, y_2, y_3, \ldots, y_n$. Starting, say, with the estimate of κ given by the method of moments, this is substituted in the right-hand side of equation (3.15) to calculate an improved estimate of κ, given by the left-hand side. The method can be shown to converge rapidly for the Ibirama and Apiuna data of Figure 1.1. Thus, for data from the Rio Hercílio at Ibirama used above, the procedure (3.15) gives $\hat{\kappa} = 3.9897\,294$ after five iterations. The file DIG.PRT, on the diskette which accompanies this book, gives MATLAB code for iterative use of equation (3.15) to calculate the maximum likelihood estimate of κ.

3.5.2 The Three-parameter Gamma Distribution

The above example has considered the maximum-likelihood estimation of parameters μ and κ in the two-parameter gamma distribution. As with the three-parameter log-normal model, a three-parameter form of the gamma distribution can be obtained by measuring the variable y from the point $(a, 0)$ as origin. The distribution then becomes

$$f(y; \mu, \kappa) = (\kappa/\mu)^\kappa (y-a)^{\kappa-1} \exp[-\kappa(y-a)/\mu]/\Gamma(\kappa) \qquad a \leq y < \infty \qquad (3.16)$$

a form which is also known (see Kite, 1977) as the Pearson type III distribution; important references are Bobee (1975) and Bobee and Robitaille (1977). A straightforward but tedious method for fitting the three parameters μ, κ and a is that used for the three-parameter log-normal: we assume a value, say a_0, for the parameter a, and explore the log-likelihood function using the GENSTAT function LLGAMMA(x;m;d). Having identified the maximum-likelihood estimates $\hat{\mu}$ and $\hat{\kappa}$ conditional on the chosen value of a_0, we calculate the value of the log-likelihood (see equation (3.12) above, with $y - a_0$ used instead of y) with these estimates substituted in place of μ and κ; call this log-likelihood $\ln L(\hat{\mu}, \hat{\kappa} \,|\, a_0)$. This calculation is repeated for a series of values a_0 so that $\ln L(\hat{\mu}, \hat{\kappa} \,|\, a_0)$ can be plotted against the value of a_0. By interpolation, we then calculate the value of a_0 for which the conditional log-likelihood is a maximum; this determines the maximum-likelihood estimate of the parameter a, and hence of μ and κ.

As with the three-parameter log-normal distribution, another method of calculating maximum-likelihood estimates of the parameters μ, κ, a is to search for the solution to the three non-linear equations

$$\partial l / \partial \mu = 0$$

$$\partial l / \partial \kappa = 0 \qquad\qquad (3.17)$$

$$\partial l / \partial a = 0$$

subject to the constraint that $a < y_1$. However, for some data sets, the equations (3.17) may have no solution lying within the space of permissible parameter values; particular care is necessary if the iterative search crosses the boundaries of the space within which feasible estimates must lie. Where a search procedure crosses a boundary delineating the region of feasible values, this may be demonstrating that the three-parameter gamma distribution is inappropriate for the data at hand.

When a three-parameter gamma distribution is fitted to the variable $\ln y$, instead of to the variable y itself, the resulting distribution is the 'log-Pearson type III' distribution, recommended in 1967 by the US Federal Water Resources Council for adoption as the standard flood frequency distribution by all US government agencies. In describing the investigations which resulted in this recommendation, Benson (1968) explained that no rigorous statistical criteria exist by which probability distributions of flood flows can be compared, so that the choice of the log-Pearson type III distribution was, to some degree, subjective. The following section shows how distributions can be compared by means of their respective likelihood functions, removing to some degree the subjectivity of choice (although subjectivity in the choice of candidate distributions remains); and, contrary to recommendations such as those of the US Federal Water Resources Council, the theme of this book is that each series of annual floods should 'speak for itself' concerning which probability distribution, or distributions, afford the 'best' description of its characteristics.

3.6   USING THE LIKELIHOOD FUNCTION TO CHOOSE BETWEEN THE GAMMA AND LOG-NORMAL DISTRIBUTIONS

Both the log-normal and gamma distributions are of broadly similar form in that both are skewed with a longer upper tail; indeed, it is this characteristic which makes them suitable for representing annual floods and other hydrological variables, where skewness is invariably present. The natural question arises: which distribution is the more appropriate?

The question is discussed at some length by Atkinson (1987). Given N statistically independent data values, he recommends the following procedure:

 (i)   Fit the log-normal model to the data.

 (ii) Using the fitted log-normal model, generate 100 samples of size N.

 (iii) For each sample generated, calculate the maximum of the log-likelihood function (a) assuming the log-normal model; (b) assuming the gamma model. Call the values of these maxima LN and G.

 (iv) Plot LN against G, showing on the graph the values of LN and G for the historic data sequence.

 (v)   Repeat steps (i) to (iv), fitting the gamma distribution instead of the log-normal and generating the 100 sequences of N from the fitted gamma distribution.

To explain the interpretation of these plots, assume first that the point (LN,G), derived from the N data values — let us call it the 'observed point' — lies within the cloud of points given by both sets of 100 generated samples of size N; the interpretation is that the log-normal and gamma models are indistinguishable. Suppose, however, that, when the 100 samples from the log-normal distribution are generated, the maximised gamma log-likelihood G of the observed point is appreciably greater than its value LN, causing the observed point to lie remote from the cluster; this indicates that the gamma model is the more appropriate. Finally, suppose that, when the 100 simulated samples are generated from the gamma distribution, the value of the maximised log-likelihood LN is appreciably greater than the maximised log-likelihood G, causing the observed point to lie removed from the cluster of 100 points; this indicates that the log-normal model is the more appropriate. We illustrate using the data from the Rio Hercílio at Ibirama.

First, we note the necessity for generating 100 samples of size 48 (the number of non-missing observations for Ibirama) (a) following a log-normal distribution, and (b) following a gamma distribution. Since, for Ibirama, the maximum-likelihood estimate of the shape parameter of the gamma distribution is $\hat{\kappa} = 3.9897$, we adopt the simplification of putting $\kappa = 4$, since integral values of κ simplify the generation of random samples from the gamma distribution. Generation of random samples is easily effected using the GENSTAT functions GRLOGNORMAL and GRGAMMA.

Figure 3.5 shows the plot of the maximized log-likelihoods for 100 samples generated from the two-parameter log-normal distribution fitted to the Ibirama data, with $\hat{\mu} = 6.5335$, $\hat{\sigma}^2 = 0.262\,567$. The log-normal maximised log-likelihoods are plotted along the horizontal axis; the gamma maximised log-likelihoods along the vertical. For the 'observed' point, the maximised log-likelihood when the log-normal distribution is fitted, is $\hat{\mu} = 6.5335$ and $\hat{\sigma}^2 = 0.262\,567$, is

$$-N \ln \hat{\sigma} - (N/2) \ln(2\pi) - N\hat{\mu} - N/2 = -349.6231$$

When the two-parameter gamma distribution is fitted, the maximised log-likelihood is

$$N\hat{\kappa} \ln \hat{\kappa} - N\hat{\kappa} \ln \hat{\mu} + N(\hat{\kappa} - 1) \ln G - N\hat{\kappa} - N \ln \Gamma(\hat{\kappa}) = -350.5050$$

where $\kappa = 3.9897$, $\hat{\mu}$ is now 783.7, and G is the geometric mean of the Ibirama data. The point $LN = -349.62$, $G = -350.50$ is shown as a cross in Figure 3.5. The two maximised log-likelihoods are very close. The 'observed' point falls well within the cluster of maximised log-likelihoods calculated from the 100 generated samples. A similar picture is obtained when the hundred samples are generated from the corresponding gamma distribution; We conclude that, for the annual flood sequence for the Rio Hercílio at Ibirama, the two-parameter log-normal and the two-parameter gamma distributions are indistinguishable.

LN	−357.6320	−356.6304	−342.7636	−348.6733	−353.9288
GL	−359.2558	−356.4149	−344.6691	−348.5476	−355.6705
LN	−347.6913	−342.6220	−342.5874	−349.0608	−351.7328
GL	−347.7819	−343.4935	−344.0645	−354.6300	−352.4300
LN	−354.3585	−354.7365	−343.5129	−354.7606	−351.8121
GL	−354.5826	−355.0871	−342.4035	−356.5163	−352.2529
LN	−349.3663	−343.4201	−358.1784	−338.7454	−351.5044
GL	−350.6412	−343.1481	−358.3579	−338.6291	−352.6496
LN	−351.4032	−357.7685	−356.4671	−355.7259	−343.1000
GL	−349.4488	−356.5664	−356.6399	−357.5356	−343.6684
LN	−345.3958	−347.1989	−358.3473	−362.2802	−350.1895
GL	−347.8105	−350.5627	−357.0914	−363.9075	−352.0540
LN	−344.4122	−356.6723	−343.9934	−336.9668	−356.0018
GL	−343.6694	−357.4263	−346.0195	−336.5645	−357.5746
LN	−354.0929	−355.3354	−349.1431	−351.8242	−355.7144
GL	−353.8250	−354.8418	−352.4936	−352.4561	−357.3138
LN	−349.7734	−345.9212	−357.0583	−354.6562	−350.5491
GL	−353.2923	−346.8025	−358.1295	−353.8510	−350.9088
LN	−346.1417	−346.0460	−343.3867	−343.2334	−343.1985
GL	−347.2663	−348.0346	−344.2180	−344.3104	−345.9515
LN	−346.5591	−357.6745	−343.9098	−350.8196	−340.7997
GL	−346.3300	−358.2516	−344.7217	−351.8604	−339.8780
LN	−354.0800	−336.3315	−353.6376	−353.3702	−346.8191
GL	−353.7950	−336.5577	−354.9416	−353.1728	−350.8825
LN	−358.0169	−342.2326	−346.4273	−343.9388	−348.3978
GL	−358.2283	−339.7429	−350.0931	−343.6223	−348.3484
LN	−351.7054	−348.2852	−348.5285	−351.7033	−341.6389
GL	−353.0110	−350.8375	−348.4786	−351.7484	−341.9385
LN	−359.2220	−350.5015	−339.8371	−346.2367	−344.9320
GL	−359.5772	−352.4779	−339.0571	−348.7006	−342.7549
LN	−353.6910	−349.4449	−354.7814	−346.6239	−342.1453
GL	−353.8176	−350.4109	−356.0056	−350.7303	−343.3626
LN	−347.7149	−340.7674	−353.2047	−350.3677	−351.6318
GL	−347.6572	−341.0983	−353.2291	−349.8553	−355.5763
LN	−348.5739	−350.0955	−348.7946	−348.2190	−346.9024
GL	−350.4172	−349.6769	−348.8883	−348.6518	−350.0808
LN	−351.1482	−340.1623	−338.6663	−356.2952	−355.3136
GL	−351.8894	−339.6830	−339.3673	−355.6662	−356.2730
LN	−347.8789	−354.4089	−350.7775	−352.4182	−350.2206
GL	−346.6353	−355.6029	−352.3076	−354.3072	−349.4890

Figure 3.5   GENSTAT output of 100 maximised log-likelihoods, corresponding to 100 samples of size 48 generated from a log-normal distribution, with parameters estimated as described in the text (LN: log-normal distribution fitted; GL: gamma distribution fitted, both to samples generated from the log-normal distribution).

Figure 3.5 (*continued*)

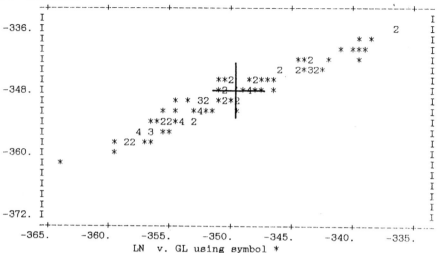

LN   v. GL using symbol *

3.7 FITTING THE GUMBEL DISTRIBUTION BY MAXIMUM LIKELIHOOD

The Gumbel distribution is one of a general class of distributions obtained by considering the probability distribution of the largest or smallest of a number, say n, of random variables. When these random variables, say X_t $(t = 1, 2, \ldots, n)$, are statistically independent and have the same probability distribution, limiting forms can be derived for the distributions of largest and smallest values, as the number n of variables grows to infinity. These limiting forms can be expected to describe the behaviour of $\max(X_1, X_2, \ldots, X_n)$ and $\min(X_1, X_2, \ldots X_n)$, even when the underlying distribution of the X_t is not known exactly.

Because the annual maximum mean daily discharge is the largest of the 365 mean daily discharges occurring throughout the year, one might expect that that annual maximum would follow one of the limiting forms: in particular, the Gumbel, or type I, limiting distribution discussed below. Whilst this may be the case, we must recognise that the justification for using the Gumbel distribution, like that for using the log-normal or gamma, is pragmatic rather than theoretical. This is because the conditions of statistical independence and common distribution for the X_t do not hold when the X_t are mean daily flows.

The Gumbel distribution is defined by

$$f(y; \alpha, u) = \alpha \exp[-\alpha(y - u) - \exp\{-\alpha(y - u)\}] \qquad -\infty < y < \infty \qquad (3.18)$$

with parameters α and u; its mean and variance are $u + \gamma/\alpha$ and $\pi^2/6\alpha^2$ respectively, where γ is Euler's constant, $0.577\ldots$. Like the log-normal and gamma distributions, it has positive skew. The log-likelihood function is given by

$$l(u, \alpha) = -N \ln \alpha - \alpha \sum(y_t - u) - \sum \exp[-\alpha(y_t - u)] \qquad (3.19)$$

from which the equations $\partial l/\partial u = 0$, $\partial l/\partial \alpha = 0$ give

$$\sum \exp[-\alpha(y_t - u)] = N \qquad (3.20)$$

and

$$N/\alpha - \sum(y_t - u) + \sum(y_t - u) \exp[-\alpha(y_t - u)] = 0 \qquad (3.21)$$

These equations have no explicit solution. However, some manipulation gives

$$1/\alpha = \bar{y} - \left\{ \sum y \exp(-\alpha y) \right\} / \left\{ \sum \exp(-\alpha y) \right\} \tag{3.22}$$

$$\exp(-\alpha\mu) = \sum \exp(-\alpha y)/N \tag{3.23}$$

Equation (3.22) involves only α, and can be solved by Newton–Raphson iteration; substitution of $\hat{\alpha}$ in equation (3.23) yields an estimate $\hat{\mu}$.

An alternative method of solution illustrates the procedure to be used when it is not possible to eliminate parameters to leave one equation with one unknown, as in equation (3.22). A more general Newton–Raphson solution must then be used. Using the moment estimates u_0, α_0 to start the iteration, improved estimates are given by $u_0 + \Delta u$, $\alpha_0 + \Delta\alpha$, where Δu and $\Delta\alpha$ are the solutions of the following two linear equations:

$$\begin{aligned} a_{11}\Delta u + a_{12}\Delta\alpha = b_1 \\ a_{21}\Delta u + a_{22}\Delta\alpha = b_2 \end{aligned} \tag{3.24}$$

where

$$\begin{aligned} a_{11} &= -\alpha^2 Z \\ a_{12} &= a_{21} = N - Z + \alpha W \\ a_{22} &= -N/\alpha^2 - V \\ b_1 &= -N\alpha + \alpha Z \\ b_2 &= -N/\alpha + \sum(y_t - u) - W \end{aligned} \tag{3.25}$$

and

$$\begin{aligned} Z &= \sum \exp[-\alpha(y_t - u)] \\ W &= \sum(y_t - u)\exp[-\alpha(y_t - u)] \\ V &= \sum(y_t - u)^2 \exp[-\alpha(y_t - u)] \end{aligned}$$

To illustrate the maximum-likelihood estimation of u and α, we use the simpler procedure, solving equation (3.23) for α and substituting the solution in (3.24) to obtain the estimate of u. Using data from the Rio Hercílio at Ibirama, moment estimates of u and α are $\tilde{u} = 586$ and $\tilde{\alpha} = 0.002\,919$, respectively (found by solving the equations $783.7 = u + \gamma/\alpha$ and $193\,060 = \pi^2/(6\alpha^2)$, where 783.7 and 193 060 are the sample mean and variance). Figure 3.6 gives a short series of GENSTAT instructions which perform the iterative calculation described above; Figure 3.7 shows that after five iterations, the calculation had converged to give the maximum-likelihood estimates

$$\begin{aligned} \hat{u} &= 602.98 \\ \hat{\alpha} &= 0.003\,418\,48 \end{aligned} \tag{3.26}$$

Now for the Gumbel distribution, the cumulative distribution function is

$$F(y; u, \alpha) = \exp[-\exp(-\alpha(y - u)]$$

so that the T-year flood is given by calculating the value of y for which $F(y; \hat{u}, \hat{\alpha}) = 1 - 1/T$. Thus the T-year flood is given by

$$y_T = \hat{u} - (1/\hat{\alpha})\ln[-\ln(1 - 1/T)] \tag{3.27}$$

```
VARIATE [VALUES= 1342,625, 619, 797,1250,271, 263,566,649, 236, 474,\
                 763,592, 981, 438, 281,556, 393,726,897, 969, 566,\
                 1300,526, 520, 487, 897,582, 510,708,998, 477, 298,\
                 872,483,1040,1010,1240,697,1406,801,741,1002,1090,\
                 589,490,2475,2125] y
SCALAR u,alf; VALUE=586, 0.002919
"
"
"
"
```

"This is a GENSTAT program for calculating the parameters u, alf of a Gumbel distribution. The above data are the sequence of annual floods (i.e., annual maximum mean daily discharges) for the Rio Hercílio at Ibirama. The method runs through 20 iterations of the Newton–Raphson calculation for alf, and then calculates u. For these data, convergence is achieved after five iterations."

```
"
"
CALCULATE ybar=MEAN(y)
FOR i=1...20
    CALCULATE e = EXP(−alf*y)
          &      ye = y*e
          &     y2e = y*y*e
          &    sume = SUM(e)
          &   sumye = SUM(ye)
          & sumy2e = SUM(y2e)
          &       f = 1/alf−ybar+sumye/sume
          &   fdash = −1/(alf*alf)+(sumye*sumye−sume*sumy2e)/(sume*sume)
          &    dalf =−f/fdash
          &     alf =alf+dalf
          & Absdalf =ABS(dalf)
    PRINT 'Iteration =',i,'alfa=',alf;DECIMALS=*,0,*,8
    EXIT Absdalf>0.000001
  ENDFOR
CALCULATE se = SUM(EXP(−alf*y))
        &       se = se/NOBSERVATIONS(y)
        &        u = −LOG(se)/alf
PRINT u; DECIMALS=6
```

Figure 3.6   GENSTAT code for fitting a Gumbel distribution by maximum likelihood, using the sequence of annual floods for the Rio Hercílio at Ibirama (see Figure 1.1).

We can assess the goodness of fit visually by plotting the ordered sample, say $y_{(t)}$, against the quantiles of the standardised Gumbel distribution corresponding to probabilities $(t-1/2)/N$ for $t = 1,\ldots, N$. If the annual flood data for the Rio Hercílio at Ibirama are well fitted by a Gumbel distribution, the plotted points should lie near to a straight line. Figure 3.8 shows that this is approximately the case, the two outlying points corresponding to the two extreme floods of 1983 and 1984.

As always, we plot of the log-likelihood in the vicinity of its maximum. To obtain this plot, we first set up an array of values (say of dimension 7×7) for each of the parameters u and α. This can be done by means of:

```
                        i=1          alfa=0.00335866767887
                        i=2          alfa=0.00341764894901
                        i=3          alfa=0.00341848080000
                        i=4          alfa=0.00341848096037
                        i=5          alfa=0.00341848096037
                        i=6          alfa=0.00341848096037
                        i=7          alfa=0.00341848096037
                        i=8          alfa=0.00341848096037

                     u = 6.029803183202101e+002
```

Figure 3.7   Fitting Gumbel parameters u, *alfa* to annual flood series for Rio Hercílio at Ibirama (data as in Figure 1.1). MATLAB was used to solve equation (3.22) by Newton–Raphson; equation (3.23) then gave the estimate of u.

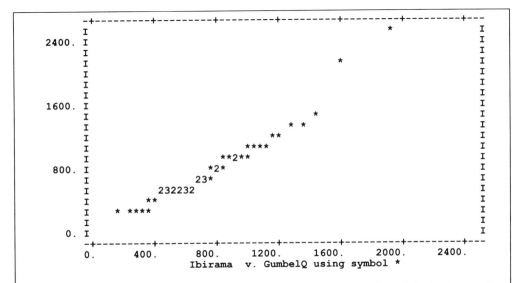

Figure 3.8   Annual flood sequence for Rio Hercílio at Ibirama (see data in Figure 1.1): data in ascending order are plotted against quantiles of the Gumbel distribution.

	logL 1	2	3	4	5	6	7
1	−359.712921	−358.168518	−357.584717	−357.771515	−358.591370	−359.941711	−361.744080
2	−357.208496	−355.168427	−354.092468	−353.797699	−354.152924	−355.061310	−356.449951
3	−355.390381	−353.049500	−351.703491	−351.179108	−351.354736	−352.143188	−353.481506
4	−354.365051	−351.969788	−350.642670	−350.225983	−350.614349	−351.738159	−353.553528
5	−354.255585	−352.115479	−351.180725	−351.319519	−352.455444	−354.551361	−357.600342
6	−355.204163	−353.706390	−353.643585	−354.928253	−357.534973	−361.485931	−366.843384*
7	−357.375000	−357.001495	−358.423553	−361.627899	−366.677155	−373.698486	−382.879395

Figure 3.9   Ordinates of log-likelihood surface: log-likelihood surface for a Gumbel distribution fitted to series of annual floods for Rio Hercílio at Ibirama (see Figure 1.1). Grid for log-likelihood plot centred on maximum-likelihood estimates $\hat{u} = 602.98$, $\hat{\alpha} = 0.003\,418\,48$ (giving the log-likelihood in the fourth row, fourth column of the matrix log$L$).

```
Contour plot of logL at intervals of 3.265
** Scaled values at grid points **
-109.4449   -109.3305   -109.7660   -110.7474   -112.2937   -114.4439   -117.2556
-108.7801   -108.3214   -108.3022   -108.6956   -109.4939   -110.7039   -112.3446
-108.4896   -107.8342   -107.5479   -107.5904   -107.9383   -108.5802   -109.5139
-108.5231   -107.7896   -107.3832   -107.2556   -107.3745   -107.7187   -108.2746
-108.8371   -108.1202   -107.7080   -107.5474   -107.6012   -107.8427   -108.2525
-109.3939   -108.7692   -108.4397   -108.3494   -108.4582   -108.7364   -109.1616
-110.1609   -109.6879   -109.5091   -109.5663   -109.8174   -110.2310   -110.7829


          0.0000      0.1667      0.3333      0.5000      0.6667      0.8333      1.0000
          ,           ,           ,           ,           ,           ,           ,
   1.000-000000000000000000000000          8888888    66666      444      2-
         00000000000000000000000000000         888888      66666     444
         00000             0000000000000000       888888     66666     44
         00                  0000000000000         888888      6666
                             00000000000         888888      6666
                            00000000000       888888       6666
   0.833-                   0000000000       888888      -
                            0000000000        88888
                  22222222                 0000000000         88
                 22222222222222222         000000000
                 222222222222222222222222       000000000
                 2222222222222222222222222222        000000000
   0.667-        2222222222222222222222222222222222       000000-
                 222222222222222222222222222222222222222       000
                 22222222222222222222222222222222222222222222
                 2222222222222222222222222222222222222222222222222
                 2222222222222222222222222222222222222222222222222222
                 222222222222222222222222222222222222222222222222222222
   0.500-        22222222222222222222222222222222222222222222222222222222      -
                 2222222222222222222222222222222222222222222222222222222222
                 22222222222222222222222222222222222222222222222222222222222
                 222222222222222222222222222222222222222222222222222222222
                 2222222222222222222222222222222222222222222222222222222
                 222222222222222222222222222222222222222222222222222
   0.333-        222222222222222222222222222222222222222222222222      -
                 2222222222222222222222222222222222222222222
                 2222222222222222222222222222222222222
         00          222222222222222222222222
         000
         0000
   0.167-000000                                                  0000-
         000000000                                         000000000
         000000000000                                000000000000000
         00000000000000000              000000000000000000000000000
         000000000000000000000000000000000000000000000000000000000000000
         0000000000000000000000000000000000000000000000000000000000
   0.000-    00000000000000000000000000000000000000000000000          -
         ,           ,           ,           ,           ,           ,           ,
```

Figure 3.10   Contour plot of log-likelihood surface shown in Figure 3.8: Gumbel distribution fitted to series of annual floods for Rio Hercílio at Ibirama (see Figure 1.1).

```
>SCALAR a,u;VALUES=602.98,0.00341848
>MATRIX[ROWS=7;COLUMNS=7;VALUES=7(-3,-2,-1,0,1,2,3)] uval
>  &   [ROWS=7;COLUMNS=7;VALUES=(-3,-2,-1,0,1,2,3)7] aval
>  &   [ROWS=7;COLUMNS=7] logL
>CALCULATE uval=u+0.1*u*uval:    &    aval=a+0.1*a*aval
```

To calculate the values of the log-likelihood over the grid of u- and a-values stored in **uval** and **aval**, we use

```
>FOR i=1...7
>FOR j=1...7
>CALCULATE a=aval$[i;j]  :   &   u=uval$[i;j]
> &            x=Ibirama−u   :   &   y=EXP(−a∗x)
> &      p=48∗LOG(a)−a∗SUM(x)−SUM(y)   :   &   logL$[i;j]=p
>ENDFOR
>ENDFOR
```

The instructions

```
>PRINT logL   :   CONTOUR logL
```

produce the numerical values of the log-likelihood at the 7×7 grid-points shown in Figure 3.9, and the lineprinter contour diagram shown in Figure 3.10. It is seen that the contours are approximately elliptical in the neighbourhood of the maximum; we should not go too far wrong, in this case, by summarising the log-likelihood surface in terms of the values of u and α at which the surface has its maximum, and the second derivatives $-\partial^2 \ln L/\partial u^2$, $-\partial^2 \ln L/\partial \alpha^2$, $-\partial^2 \ln L/\partial u \partial \alpha$, at this point.

3.8   THE WEIBULL DISTRIBUTION

The Weibull distribution belongs to the same family of limiting extreme value distributions as the Gumbel distribution of the previous section; an alternative name is the extreme value type III distribution, the Gumbel distribution being extreme value type I. It is useful for modelling the distribution of annual *minimum* flows, and has cumulative distribution function

$$F(y) = 1 - \exp[-(y/\alpha)^k] \qquad (y > 0) \tag{3.28}$$

and probability density

$$f(y) = (k/\alpha)(y/\alpha)^{k-1} \exp[-(y/\alpha)^k] \tag{3.29}$$

where the quantities k and α are parameters of the distribution. As with the Gumbel distribution, these can be estimated by maximising the log-likelihood function, either by direct search or by using a Newton–Raphson iterative method to solve the equations $\partial l/\partial k = 0$, $\partial l/\partial \alpha = 0$. With initial estimates k and α (which, in the example given below, are the moment estimates) corrections Δk and $\Delta \alpha$ are calculated by solving the two linear equations

$$a_{11}\Delta k + a_{12}\Delta\alpha = b_1$$
$$a_{21}\Delta k + a_{22}\Delta\alpha = b_2 \tag{3.30}$$

where

$$a_{11} = -N/k^2 - \sum[\ln(y_t/\alpha)^2](y_t/\alpha)^k$$
$$a_{12} = -N/\alpha + (1/\alpha)\sum(y_t/\alpha)^k + (k/\alpha)\sum[\ln(y_t/\alpha)](y_t/\alpha)^k$$
$$a_{22} = Nk/\alpha^2 - k(k+1)/\alpha^2 \sum(y_t/\alpha)^k \tag{3.31}$$
$$b_1 = -N/k - \sum[\ln(y_t/\alpha)][1 - (y_t/\alpha)^k]$$
$$b_2 = (k/\alpha)\sum[1 - (y_t/\alpha)^k]$$

summation in each case being over the observations y_t.

3.81	7.27	7.74	8.68	6.14
9.15	9.62	10.60	3.34	3.34
1.86	4.48	8.68	7.27	3.81
9.62	3.96	1.20	5.80	7.65
3.58	5.48	3.96	8.30	6.00
3.58	7.65	5.86	3.58	10.20
10.90	6.24	2.82	11.50	5.24
10.60	10.60	12.20	6.24	8.40
15.70	7.70	6.60	7.46	12.50
16.20	31.80	17.70		

Figure 3.11   Annual minimum mean daily flows for Rio Hercílio at Ibirama (units: cubic metres per second)

```
VARIATE [NVALUES=48] Ibirama
SCALAR N,nits; VALUE=48,10
OPEN 'ita.min'; CHANNEL=2; FILETYPE=input
READ [CHANNEL=2] Ibirama
CLOSE 'ita.min'; CHANNEL=2; FILETYPE=input
PRINT [ORIENTATION=across] Ibirama;DECIMALS=2
''
''

"Now set starting values for Weibull parameters.In the case of minimum annual flows for the
Rio Hercílio at Ibirama, these are"
''
''

SCALAR k,alf;VALUE=1.6,5.6
''
''

FOR i=1...nits
CALCULATE u=LOG(Ibirama/alf)
    &        v=(Ibirama/alf)**k
    &        a11=-N/(k*k)-SUM(u*u*v)
    &        a12=-N/alf+(1/alf)*SUM(v)+(k/alf)*SUM(u*v)
    &        a22=N*k/(alf*alf)-k*(k+1)*SUM(v)/(alf*alf)
    &        a21=a12
    &        b1=-N/k-SUM(u*(1-v))
    &        b2=(k/alf)*SUM(1-v)
    &        den=a11*a22-a12*a12
    &        dk=(a22*b1-a12*b2)/den
    &        dalf=(-a21*b1+a11*b2)/den
    &        k=k+dk
    &        alf=alf+dalf
PRINT i,k,alf; DECIMALS=0,6,6
ENDFOR
```

Figure 3.12   GENSTAT code for fitting a Weibull distribution by maximum likelihood, using the minimum flows for the Rio Hercílio at Ibirama

Calculation of the starting values for k and α requires a little more calculation than in the Gumbel case. For the Weibull distribution, the mean, variance and coefficient of variation $cv(= \sigma/\mu)$ are given by

$$E[Y] = \alpha\Gamma(1 + 1/k)$$

$$\mathrm{var}(Y) = \alpha^2[\Gamma(1 + 2/k) - \Gamma^2(1 + 1/k)]$$

$$\frac{1}{cv^2} = \Gamma(1 + 2/k)/\Gamma^2(1 + 1/k) - 1$$

respectively; the last equation, which involves k alone, can by solved approximately to give the starting value for this parameter, and the solution substituted in the first equation to provide a starting value for α.

We illustrate the calculation using annual minimum flows ('annual minimum mean daily discharge') for the Rio Hercílio at Ibirama; these data are shown in Figure 3.11. The mean, variance and $1/cv^2$ are easily found to be 7.971, 25.790 562 and 0.405 915; therefore solving the equation

$$\Gamma(1 + 2/k)/\Gamma^2(1 + 1/k) = 1.405 915$$

gives a starting value for the parameter k of about 1.6. Substitution then gives, as an approximation to α, 5.6.

	k	alf
1	1.932809	7.883084
2	1.749239	8.448105
3	1.732590	8.914838
4	1.726078	8.991042
5	1.725910	8.993291
6	1.725910	8.993293
7	1.725910	8.993293
8	1.725910	8.993293
9	1.725910	8.993293
10	1.725910	8.993293

Figure 3.13   Fitting a Weibull distribution to annual minimum mean daily flows for the Rio Hercílio at Ibirama: maximum-likelihood fitting, by Newton–Raphson iterative solution. Left-hand column shows the iteration number, k and alf are the Weibull parameters.

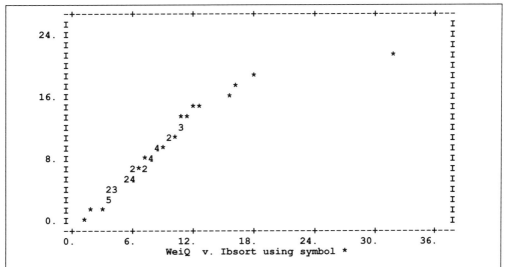

Figure 3.14   Plot of ordered minimum mean daily flows for Rio Hercílio at Ibirama against quantiles after fitting a Weibull distribution by maximum likelihood.

Contour plot of logL at intervals of 5.000

** Scaled values at grid points **

−33.0678	−30.7606	−29.9026	−29.9049	−30.4969	−31.5380	−32.9545
−32.7782	−30.2100	−29.0704	−28.7980	−29.1472	−30.0045	−31.3297
−32.5156	−29.7188	−28.3482	−27.8810	−28.1142	−28.9883	−30.5383
−32.3158	−29.3728	−27.9041	−27.4519	−27.8998	−29.3151	−31.9255
−32.2572	−29.3703	−28.1446	−28.2648	−29.8344	−33.2762	−39.4062
−32.5362	−30.2374	−30.2134	−32.5708	−38.1372	−48.6133	−67.1090
−33.7684	−33.7550	−38.3037	−49.3447	−71.1794	−112.3009	−189.3776

Figure 3.15   Contour plot (see also Figure 3.16 of log-likelihood surface after fitting a Weibull distribution to annual minimum mean daily flows for Rio Hercílio at Ibirama. The log-likelihood surface maximum occurs at the grid-point in the fourth row, fourth column.

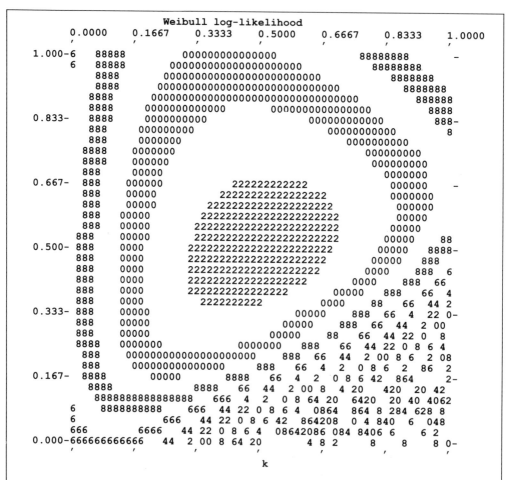

Figure 3.16   Contour plot of log-likelihood surface obtained by fitting a Weibull distribution to the annual miminum flows for Rio Hercílio at Ibirama (see data in Figure 3.11, and log-likelihoods in Figure 3.15).

Figure 3.12 shows a series of GENSTAT commands which calculate the maximum-likelihood estimates of k and α, by iterative solution of the above equations (3.30). The number of iterations, labelled *nits*, was set at 10; however, as shown in Figure 3.13, the calculation had effectively converged after six iterations.

We can get an idea of the goodness of fit by plotting the ordered observations, $y_{(t)}$ say, against the quantiles of the fitted distribution corresponding to probabilities $(t - 1/2)/N$ where, for the Ibirama data, $N = 48$ and $t = 1, \ldots, 48$. If the Weibull distribution provides a good fit, we should expect a linear relation between these two quantities. Figure 3.14 shows this plot; it is clear that the data contain one aberrant point, corresponding to the year 1983 when the minimum daily flow was 31.8 m^3 s^{-1}.

As argued above, it is good practice to plot the log-likelihood function in the neighbourhood of its maximum $\hat{k} = 1.725\,91$, $\hat{\alpha} = 8.993\,29$. The CONTOUR instruction can be used for this purpose; Figure 3.15 shows the values of the log-likelihood function at each of a grid of points centred upon the maximum-likelihood estimates of k and α, at intervals of 0.2 times k and α. The point in the centre of the grid has the maximum log-likelihood, as it should have; Figure 3.16 shows the contours of this plot. It will be observed that, in the immediate vicinity of the maximum, the contours are approximately elliptical; further away from the maximum, contours are more distorted.

3.9   THE GENERALISED EXTREME VALUE DISTRIBUTION

The penultimate distribution that we consider in this chapter is the generalised extreme value (GEV) distribution, having three parameters, with cumulative density function

$$F(x; u, \alpha, k) = \exp[-\{1 - k(x - u)/\alpha\}^{1/k}] \tag{3.32}$$

and probability density

$$f(x; u, \alpha, k) = (1/\alpha)\{1 - k(x - u)/\alpha\}^{1/k-1} \cdot \exp[-\{1 - k(x - u)/\alpha\}^{1/k}] \tag{3.33}$$

where

$$\xi + \alpha/k \leq x < \infty \qquad k < 0;$$
$$-\infty < x < \infty \qquad k = 0$$
$$-\infty < x \leq \xi + \alpha/k \qquad k > 0$$

It is assumed in what follows that k is non-zero. For if $k = 0$, the distribution becomes identical with the Gumbel; it can also be shown that the GEV distribution is related to the Weibull.

Once again, we obtain estimates of the model parameters u, α, k by maximising the log-likelihood. Denoting by Z_t the variable $1 - k(x - u)/\alpha$, the log-likelihood is given by

$$\ln L = -N \ln \alpha + (1/k - 1) \sum \ln Z_t - \sum Z_t^{1/k}$$

or

$$\ln L = -N \ln \alpha - (1 - k) \sum y_t - \sum \exp(-yt) \tag{3.34}$$

where $y_t = -(1/k) \ln[1 - k(x_t - u)/\alpha]$ and summation is over the suffix t. The maximum-likelihood estimates of u, α and k are obtained by iterative calculation of three linear equations with solutions Δu, $\Delta \alpha$, Δk. These equations are

$$a_{11}\Delta u + a_{12}\Delta\alpha + a_{13}\Delta k = b_1$$
$$a_{21}\Delta u + a_{22}\Delta\alpha + a_{23}\Delta k = b_2 \qquad (3.35)$$
$$a_{31}\Delta u + a_{32}\Delta\alpha + a_{33}\Delta k = b_3$$

where the matrix $A = [a_{ij}]$ is symmetric. With the definition of y_t given above, we find by differentiating with respect to u, α, k:

$$\partial y/\partial u = -\exp(ky)/\alpha$$
$$\partial y/\partial \alpha = [1 - \exp(ky)]/(\alpha k)$$
$$\partial y/\partial k = -[1 - \exp(ky)]/k^2$$

whence the vector $[b_1, b_2, b_3]$ on the right-hand side of equations (3.35) is given by

$$-b_1 = -(1 - k)\sum(\partial y/\partial u) + \sum(\partial y/\partial u)\exp(-y)$$
$$-b_2 = -(N/\alpha) - (1 - k)\sum(\partial y/\partial \alpha) + \sum(\partial y/\partial \alpha)\exp(-y)$$
$$-b_3 = \sum y - (1 - k)\sum(\partial y/\partial k) + \sum(\partial y/\partial k)\exp(-y)$$

and where, for ease of notation, we have written y instead of y_t and have omitted the range of summation from 1 to N. To calculate the a_{ij}, we need the second derivatives of y_t with respect to u, α and k, and it is easily found that

$$\partial^2 y/\partial u^2 = k(\partial y/\partial u)^2$$
$$\partial^2 y/\partial u\partial\alpha = -(\partial y/\partial u)/\alpha + k(\partial y/\partial u)(\partial y/\partial \alpha)$$
$$\partial^2 y/\partial \alpha^2 = -(\partial y/\partial \alpha)/\alpha + (\partial y/\partial u)(\partial y/\partial \alpha)$$
$$\partial^2 y/\partial u\partial k = -\alpha(\partial y/\partial u)^2/k$$
$$\partial^2 y/\partial \alpha\partial k = -\alpha(\partial y/\partial u)(\partial y/\partial \alpha)/k$$
$$\partial^2 y/\partial k^2 = -2(\partial y/\partial k)/k + \alpha(\partial y/\partial u)(\partial y/\partial k)/k$$

from which

$$a_{11} = -(1 - k)\sum \partial^2 y/\partial u^2 - \sum \exp(-y)(\partial y/\partial u)^2 + \sum \exp(-y)(\partial^2 y/\partial u^2)$$
$$a_{12} = -(1 - k)\partial^2 y/\partial u\partial\alpha - \sum \exp(-y)(\partial y/\partial u)(\partial y/\partial \alpha) + \sum \exp(-y)(\partial^2 y/\partial u\partial\alpha)$$
$$a_{22} = (N/\alpha^2) - (1 - k)\sum(\partial^2 y/\partial \alpha^2) - \sum \exp(-y)(\partial y/\partial \alpha)^2 + \sum \exp(-y)(\partial^2 y/\partial \alpha^2)$$
$$a_{13} = \sum(\partial y/\partial u) - (1 - k)\sum(\partial^2 y/\partial u\partial k) - \sum \exp(-y)(\partial y/\partial u)(\partial y/\partial k)$$
$$\qquad + \sum \exp(-y)(\partial^2 y/\partial u\partial k)$$
$$a_{23} = \sum(\partial y/\partial \alpha) - (1 - k)\sum(\partial^2 y/\partial \alpha\partial k) - \sum \exp(-y)(\partial y/\partial \alpha)(\partial y/\partial k)$$
$$\qquad + \sum \exp(-y)(\partial^2 y/\partial \alpha\partial k)$$
$$a_{33} = 2\sum(\partial y/\partial k) - (1 - k)\sum(\partial^2 y/\partial k^2) - \sum \exp(-y)(\partial y/\partial k)^2 + \sum \exp(-y)(\partial^2 y/\partial k^2)$$

The iterative calculation proceeds essentially as in the case of the Gumbel and Weibull distributions, although with considerably greater complexities. Basically in equations (3.35) we are iteratively solving, by Newton–Raphson, the equations $\partial l/\partial u = 0$, $\partial l/\partial \alpha = 0$, $\partial l/\partial k = 0$. We illustrate with the data from the Rio Hercílio at Ibirama shown in Figure 1.1.

The first step is to obtain estimates of the three parameters that may be used to start the iterative calculation leading to the maximum-likelihood estimates. One approach (Hosking, 1986) is to calculate a quantity c from the equation

$$c = (2b_1 - b_0)/(3b_2 - b_0) - \ln 2/\ln 3$$

where b_0, b_1 and b_2 are the b-coefficients defined in Section 2.7. Having calculated c, an estimate of k is given by $k = 7.8590c + 2.9554c^2$. Using the values $b = 783.7083$, $b_1 = 504.0913$ and $b_2 = 383.7478$ calculated in Section 2.7 for the Rio Hercílio at Ibirama, we find that $c = -0.0202$ and that the estimate of k is -0.1575. The other two parameters α and u are given by

$$\tilde{\alpha} = l_2\tilde{k}/((1 - 2^{-\tilde{k}})\Gamma(1 + \tilde{k}))$$

$$\tilde{u} = l_1 - \tilde{\alpha}\{1 - \Gamma(1 + \tilde{k})\}/\tilde{k}$$

where l_1 and l_2 are the first and second l-moments defined in Section 2.6. Using the numerical values $l_1 = 783.7083$ and $l_2 = 224.4743$ found for the Ibirama data, we obtain the approximate solution

$$\tilde{u} = 575.7452; \quad \tilde{\alpha} = 273.6357; \quad \text{and } \tilde{k} = -0.1575$$

where the 'tilde' shows that these estimates differ from the maximum-likelihood estimates (for which a 'hat' is used).

Using these initial estimates, we are now able to begin the iterative solution of equations (3.35) above. GENSTAT and MATLAB codes for the calculation are available which are simplified transcriptions of the published FORTRAN Algorithm AS 215 (Hosking, 1985) as amended by MacLeod (1988). However, it is preferable to use the published FORTRAN code directly, with MacLeod's amendments, for the calculation, because estimation of the parameter k involves subtractions of numbers which are very nearly equal in magnitude, and when k is small a substantial loss of precision may result. This can be avoided in FORTRAN by using double-precision arithmetic. Both GENSTAT and MATLAB have facilities for calling subprograms in FORTRAN; in the case of MATLAB, there are routines in the External Interface Library which sit upon the user's FORTRAN code, and through which this code can be run and the results communicated to MATLAB. In the case of GENSTAT, there is a facility PASS, which does work specified in FORTRAN subprograms supplied by the user, but not linked into GENSTAT. This would be particularly useful where, for example, a GEV distribution is to be fitted to many sets of data, and where GENSTAT is to be used subsequently to plot estimates of the GEV parameters, or to explore relationships between them and catchment characteristics by means of regression analysis.

The calculation requires considerable care, as the following discussion illustrates. For the Ibirama data, a much-simplified GENSTAT program, working in single precision arithmetic only, was used to solve the equations (3.35) repeatedly for Δu, $\Delta \alpha$ and Δk, using as starting values the l-moment estimates $\tilde{u} = 575.7452$, $\tilde{\alpha} = 273.6357$, $\tilde{k} = -0.1575$. Each solution was used to obtain new values $u = u + \Delta u$, $\alpha = \alpha + \Delta \alpha$, $k = k + \Delta k$; convergence was assumed when the largest of Δu, $\Delta \alpha$, Δk was less than 0.0001 in absolute magnitude. With this criterion, the calculation converged after five iterations. The estimates of u, α and k after each iteration were:

Iteration:	$\hat{u}$:	$\hat{\alpha}$:	$\hat{k}$:
1	570.168 701	259.133 209	−0.156 525
2	570.602 234	260.642 242	−0.156 430
3	570.606 689	260.663 727	−0.156 418
4	570.606 873	260.663 818	−0.156 417
5	570.606 873	260.663 788	−0.156 418

Although 'convergence' was obtained, it should be emphasised that convergence of the iterative calculation is not guaranteed, and whether or not 'convergence' occurs is dependent upon (a) the choice of convergence criteria, (b) the starting values for the iteration, and (c) the maximum number of iterations and/or likelihood function evaluations permitted in the program. Hosking (1985) says 'The use of good initial parameter estimates greatly increases the speed of the algorithm. The choice $k = 0$, $\alpha = s\sqrt{(6/\pi)}$, $u = \bar{x} - \gamma\alpha$, where $\bar{x}$ and s are the sample mean and standard deviation and $\gamma = 0.577\,215\,66$ is Euler's constant, usually give rapid convergence. More accurate initial estimates, obtained for example by the method of probability-weighted moments (Hosking et al., 1985), may be preferred in some applications.' It is instructive, therefore, to see what happens when we use as starting values the moment estimates $\tilde{u} = \bar{x} - \gamma\alpha = 433.2$, $\tilde{\alpha} = s\sqrt{(6/\pi)} = 607.2$ calculated from the Ibirama data, with $\tilde{k} = 0$, as suggested by Hosking; he uses a different accuracy criterion for each parameter, namely a relative accuracy of 10^{-4} in u and α, and an absolute accuracy of 10^{-4} for k. Applying these to the Ibirama data, convergence was not achieved after 50 iterations and 100 evaluations of the log-likelihood, the final values of u, α and k being 433.2179, 607.1944, and 0.0487: rather different from those shown above using l-moments to start the calculation. But what happens if we start Algorithm AS 215 from the l-moment estimates as starting values, as we had done with the single-precision GENSTAT program giving the results tabulated above? The following output is now obtained from AS 215 (where 'XI' is 'u' in our notation):

MAXIMUM-LIKELIHOOD ESTIMATION OF GENERALIZED EXTREME-VALUE DISTRIBUTION

ITER	EVAL	XI	ALPHA	K	ACTION	LOG-L	GNORM
1	1	575.7452	273.6357	−.1575	NEWTON	−368.258	2.27D+01
2	2	674.0687	351.2219	−.1369	NEWTON	−362.962	7.40D+00
3	3	712.7181	397.3403	−.1521	NEWTON	−362.048	1.91D+00
4	4	720.5558	411.2930	−.1698	NEWTON	−361.982	2.35D−01
5	5	721.1177	412.6025	−.1733	NEWTON	−361.981	3.94D−03
6	6	721.1227	412.6183	−.1734	NEWTON	−361.981	1.00D−06
	6	721.1227	412.6183	−.1734	CONVGD		

We see that the estimates are now $\hat{u} = 721.1227$, $\hat{\alpha} = 412.6183$, $\hat{k} = -0.1734$, very different from the estimates obtained earlier from the same starting point, using single-precision arithmetic and a much more basic program for the Newton–Raphson calculation.

We can draw two conclusions from this. First, that the use of good starting values is particularly important, and some preliminary plotting of the log-likelihood function (e.g. using the GENSTAT directive CONTOUR, holding one parameter constant and allowing the other two to vary) is desirable to give some idea of its behaviour. Second, that careful attention must be paid to the numerical accuracy of quantities at intermediate stages of the calculation, if misleading results are to be avoided.

Some data sets will arise, however, for which convergence cannot be obtained, even with the most careful programming. Of these, Hosking writes 'If no choice of initial values leads to convergence, a likely explanation is that the sample configuration is grossly atypical of the generalised extreme value distribution with $|k| < 0.5$; in particular, samples with large negative skewness often fail to give convergence.' However, the chances of convergence are increased substantially by the use of good computer code for the estimation of GEV parameters, which must make allowance for the possible occurrence of infeasible estimates of u, α, k during the course of the iterations (that is, estimates for which $u + \alpha/k$ becomes greater than some of the data values when k is negative, or less than some of the data values when k is positive). Algorithm AS 215 tests for such occurrences, and its use is recommended; the version incorporating Macleod's amendments is given in file GEV.FOR accompanying this book.

3.10   THE WAKEBY DISTRIBUTION

As one further example, we discuss briefly a probability distribution less widely used than those generally employed in hydrological frequency studies. The Wakeby distribution is of a rather different nature in that neither $f(x; \theta)$ nor its integral $F(x; \theta)$ is explicitly defined. Instead, the distribution is defined implicitly by the equation

$$x = \xi + \alpha\{1 - (1 - F)^\beta\}/\beta - \gamma\{1 - (1 - F)^{-\delta}\}/\delta \qquad (3.36)$$

where there are five parameters α, β, γ, δ and ξ; it is assumed that $\beta + \delta \geq 0$. It is seen that x varies between $x_{\min} = \xi$, corresponding to the cumulative probability density $F = 0$, and $x_{\max} = \xi + \alpha/\beta - \gamma/\delta$, corresponding to $F = 1$.

This distribution is more complex than those discussed earlier, first because it is not possible to write down the likelihood function corresponding to a set of observations, and second because it has five parameters $(\alpha, \beta, \gamma, \delta, \xi)$ instead of the two or three of the distributions previously discussed. Since the likelihood function cannot be written explicitly, it is not straightforward to calculate maximum-likelihood estimates. The Wakeby distribution has not yet entered into common hydrological usage, being still the subject of research; whilst it is entirely appropriate, and indeed necessary, for new distributions to be explored, it is not yet clear whether use of the Wakeby distribution brings practical benefits that outweigh its greater complexity.

We note that, in theory, it is possible to obtain the Wakeby probability density function in tabulated form, for given values of the five parameters $(\xi, \alpha, \beta, \gamma, \delta)$, by the following computing procedure. First, for each of a (dense) sequence of values x, solve equation (3.36) iteratively, to give the cumulative distribution function $F(x; \xi, \alpha, \beta, \gamma, \delta)$. Then, differentiate F numerically to give $f(x; \xi, \alpha, \beta, \gamma, \delta)$. For a given set of data values x, therefore, it is theoretically possible to construct and maximise a likelihood function in the five-dimensional parameter space. However, the computational work would be very considerable and well beyond the power of desk-top computers which are the tool for other methods described in this book.

Hosking (1986; see also Hosking 1990, 1991) has shown how the five parameters may be estimated by the use of probability-weighted moments (estimated by the a_i of Section 2.7), although the efficiency of his estimation procedures is not stated. Using data for the Rio Hercílio at Ibirama, and the definitions of a_i given in Section 2.7, the procedure is as follows. First, calculate

$$N_1 = a_0 - 24a_1 + 81a_2 - 64a_3$$

$$N_2 = a_0 - 12a_1 + 27a_2 - 16a_3$$

$$N_3 = a_0 - 6a_1 + 9a_2 - 4a_3$$

$$C_1 = 8a_1 - 81a_2 + 192a_3 - 125a_4$$

$$C_2 = 4a_1 - 27a_2 + 48a_3 - 25a_4$$

$$C_3 = 2a_1 - 9a_2 + 12a_3 - 5a_4$$

Then estimates of β and $-\delta$ are given as the roots of the quadratic equation

$$(N_2C_3 - N_3C_2)z^2 + (N_1C_3 - N_3C_1)z + (N_1C_2 - N_2) = 0$$

the larger root giving β. Estimates of the other parameters are given by

$$\tilde{\gamma} = -(1 - \tilde{\delta})(2 - \tilde{\delta})(3 - \tilde{\delta})[(4N_2 - N_1) + (4N_3 - N_1)\tilde{\beta}]/[6(\tilde{\beta} + \tilde{\delta})]$$

$$\tilde{\alpha} = (1 + \tilde{\beta})(2 + \tilde{\beta})(3 + \tilde{\beta})[(4N_2 - N_1) - (4N_3 - N_2)\tilde{\delta}]/[6(\tilde{\beta} + \tilde{\delta})]$$

$$\tilde{\xi} = a_0 - \tilde{\alpha}/(1 + \tilde{\beta}) - \tilde{\gamma}/(1 - \tilde{\delta})$$

To illustrate the calculation, we fit a Wakeby distribution to the data from the Rio Hercílio at Ibirama. It was found in Section 2.6 that the values of the a_i were

$$a_0 = 783.7083$$

$$a_1 = 279.6170$$

$$a_2 = 159.2735$$

$$a_3 = 107.9338$$

We find

$N_1 = 66.290\,600$	$C_1 = 41.047\,100$
$N_2 = 1.748\,000$	$C_2 = -4.699\,100$
$N_3 = 107.732\,600$	$C_3 = 20.257\,100$

The quadratic equation in z thus becomes

$$541.655\,671z^2 - 3079.255\,492z - 383.256\,489 = 0$$

with roots $5.806\,747$ and $-0.121\,852$. Hence

$$\tilde{\beta} = 5.806\,747$$

$$\tilde{\delta} = 0.121\,852$$

The equations giving γ, α and ξ give

$$\tilde{\gamma} = -324.655\,957$$

$$\tilde{\alpha} = -1468.143\,577$$

$$\tilde{\xi} = 1369.102\,977.$$

Hence the fitted distribution is of form

$$x(F) = 1369.1 - 1468.1[1 - (1 - F)^{5.807}]/5.807$$

$$+ 324.66\{1 - (1 - F)^{-0.1218}\}/0.1218$$

On putting $F = 0$, $F = 1$, we find that the upper and lower limits of x are 1369 and 3782 m^3 s^{-1} respectively; however, the observed sequence of annual floods at Ibirama ranges from 236 to 2475 m^3 s^{-1}, the discrepancy at the lower limit suggesting that the Wakeby distribution is inappropriate for this flood record.

3.11   THE PRECISION OF MAXIMUM-LIKELIHOOD ESTIMATES

One aspect of the maximum-likelihood procedures described above was the plotting of the likelihood surface in the neighbourhood of its maximum. This is because the curvature of the surface at this point is closely related to the precision of the maximum-likelihood estimates; intuitively, if the likelihood surface is 'flat' in this region, with very gentle curvature and shallow gradient, the maximum point will be ill determined and the maximum-likelihood estimates will have low precision. Conversely, if the curvature is very pronounced and the gradients very steep, the maximum of the likelihood surface will be clearly defined with high precision. Thus, curvature of the likelihood surface provides a measure of precision; furthermore, curvature of the likelihood surface is related to the rate of change of gradient in the vicinity of the maximum.

Statistical theory shows that, when the log-likelihood function is approximately quadratic in the neighbourhood of its maximum, the precision of maximum-likelihood estimates is given by the inverse of the matrix of second derivatives $-\partial^2 l/\partial\theta_i\partial\theta_j$, calculated at the likelihood maximum. More specifically, if $\hat{\theta}$ is the vector of maximum-likelihood estimates of the model parameters θ, the matrix $[\mathrm{var}\,\hat{\theta}]$ of variances and covariances is given by any of the three equivalent relations

$$E[-\partial^2 l/\partial\theta_i\partial\theta_j] = E[-\partial^2 l/\partial\theta_i\partial\theta_j]_{\theta=\hat{\theta}} = -[\partial^2 l/\partial\theta_i\partial\theta_j]_{\theta=\hat{\theta}}$$

Statistical theory also shows that, when data are plentiful (when N is 'large'), the estimates $\hat{\theta}$ are approximately normally distributed, the means of the (approximate) p-dimensional normal distribution being the 'true' parameter values θ, and the variance-covariance matrix being

$$[\mathrm{var}\,\theta] = E[-\partial^2 l/\partial^2 l\partial\theta_i\partial\theta_j]^{-1} \tag{3.37}$$

and the larger the value of N, the better is the normal approximation. We write 'true' in quotes because the statistical models that we use to describe the characteristics of, for example, annual floods, are themselves no more than approximations to whatever procedure nature uses when generating flood sequences; and we are deluding ourselves if we believe that floods in nature necessarily follow a log-normal, gamma, Gumbel, or any other distribution. The hydrological literature is replete with papers which, by means of extensive computer simulations, have studied the biases and root-mean-square errors in all kinds of estimates, when the real world has been chosen to be log-Pearson type III, or Gumbel, or three-parameter log-normal. But the real world is never so simple and, in this writer's opinion, the conclusions of such papers have greater importance as statistical exercises than for what they have to say about the practice of estimating the frequencies of flooding.

With the help of careful statistical analysis, however, we may be able to reject those statistical models which are not consistent with the data and to conclude that one, or perhaps several, remaining models serve equally well. If several models remain, the choice between them must be made on grounds that are other than statistical. Whichever of these models is adopted, it will be desirable to obtain confidence limits for its parameters. Approximate confidence limits can be given using the diagonal terms from the matrix given by the right-hand side of equation (3.37). Using the fact that each component of the vector $\hat{\theta}$ is approximately normally distributed, then approximate 95% confidence limits for the component (say) θ_1 of θ are given by $\hat{\theta}_1 \pm 1.96 \sqrt{(\text{var } \hat{\theta}_1)}$, var $\hat{\theta}_1$ being given by the first element in the leading diagonal of the matrix var $\hat{\theta}$. Limits for 99% and 99.9% confidence probabilities are calculated using normal deviates 2.58 and 3.29 instead of 1.96.

Approximate confidence limits thus calculated are appropriate when one parameter is considered individually; if confidence limits are calculated for all of the p parameters in the vector θ, the p confidence statements will not all be simultaneously true. Where confidence statements are required for all parameters simultaneously, we must use a joint confidence region which, again with N large, is shown by statistical theory to be

$$2l(\hat{\theta}; x) - 2l(\theta; x) \le \chi^2_{p,\alpha} \tag{3.38}$$

where $100(1 - \alpha)\%$ is the confidence level; χ^2 is the tabulated value of the χ^2 distribution for p degrees of freedom having probability α in its upper tail; and, as usual, p is the number of elements in the vector θ. Essentially, equation (3.38) defines those θ-values which are sufficiently close to the value $\hat{\theta}$ at which the log-likelihood $l(\theta; x)$ has its maximum.

We give now some examples of the calculation of var $\hat{\theta}$ for the distributions presented earlier. First, for the *two-parameter log-normal distribution*, the matrix $-E[-\partial^2 l/\partial \theta_i \partial \theta_j]$ becomes

$$\begin{bmatrix} -E[\partial^2 l/\partial \mu^2] & -E[\partial^2 l/\partial \mu \partial \sigma] \\ -E[\partial^2 l/\partial \mu \partial \sigma] & -E[\partial^2 l/\partial \sigma^2] \end{bmatrix} = (N/\sigma^2) \begin{bmatrix} 1 & 0 \\ 0 & 1/(2\sigma^2) \end{bmatrix} \tag{3.39}$$

Inversion of the matrix on the right-hand side, as is required by equation (3.37), gives

$$\begin{bmatrix} \text{var } \hat{\mu} & \text{cov}(\hat{\mu}, \hat{\sigma}) \\ \text{cov}(\hat{\mu}, \hat{\sigma}) & \text{var } \hat{\sigma} \end{bmatrix} = \begin{bmatrix} \sigma^2/N & 0 \\ 0 & 2\sigma^4/N \end{bmatrix} \tag{3.40}$$

For the *three-parameter log-normal distribution*, the matrix $-E[-\partial^2 l/\partial \theta_i \partial \theta_j]$ becomes

$$\begin{bmatrix} -E[\partial^2 l/\partial \mu^2] & -E[\partial^2 l/\partial \mu \partial \sigma] & -E[\partial^2 l/\partial \mu \partial a] \\ -E[\partial^2 l/\partial \mu \partial \sigma] & -E[\partial^2 l/\partial \sigma^2] & -E[\partial^2 l/\partial a \partial \sigma] \\ -E[\partial^2 l/\partial \mu \partial \sigma] & -E[\partial^2 l/\partial a \partial \sigma] & -E[\partial^2 l/\partial a^2] \end{bmatrix} = (N/\sigma^2)\mathbf{J} \tag{3.41}$$

where

$$\mathbf{J} = \begin{bmatrix} (\sigma^2 + 1)\exp[2(\sigma^2 - \mu)] & \exp[\sigma^2/2 - \mu] & -\exp[\sigma^2/2 - \mu] \\ \exp[\sigma^2/2 - \mu] & 1 & 0 \\ \exp[\sigma^2/2 - \mu] & 0 & 1/(2\sigma^2) \end{bmatrix} \quad (3.42)$$

Inversion of the matrix $(N/\sigma^2)\mathbf{J}$ then gives expressions for var $\hat{\mu}$ (the element in the first row, first column of the inverse), var $\hat{\sigma}$ (second row, second column), var $\hat{a}$ (third row, third column), together with the covariances $\text{cov}(\overline{\mu}, \overline{\sigma})$, $\text{cov}(\overline{\mu}, \overline{a})$ and $\text{cov}(\overline{\sigma}, \overline{a})$.

For the *two-parameter gamma distribution* given by equation (3.11), the 2×2 matrix $-E[-\partial^2 l/\partial\theta_i\partial\theta_j]$ becomes

$$\begin{bmatrix} -E[\partial^2 l/\partial\mu^2] & -E[\partial^2 l/\partial\mu\partial\kappa] \\ -E[\partial^2 l/\partial\mu\partial\kappa] & -E[\partial^2 l/\partial\kappa^2] \end{bmatrix} = N \begin{bmatrix} \kappa/\mu^2 & 0 \\ 0 & \partial\psi/\partial\kappa - 1/\kappa \end{bmatrix} \quad (3.43)$$

where $\psi = \partial \ln \Gamma(\kappa)/\partial\kappa$, the digamma function. Inversion of the matrix on the right-hand side gives var $\overline{\mu}$, var $\hat{\kappa}$; $\text{cov}(\hat{\mu}, \hat{\kappa})$ is zero.

For the *Gumbel distribution*, given by equation (3.18), inversion of the matrix $-E[-\partial^2 l/\partial\theta_i\partial\theta_j]$ leads to

$$\text{var}\,\hat{\alpha} = \alpha^2/(\pi^2/6)$$
$$\text{var}\,\hat{u} = \alpha^{-2}[1 + (1 - \gamma)^2/(\pi^2/6)] \quad (3.44)$$
$$\text{cov}(\hat{\alpha}, \hat{u}) = -(1 - \gamma)/(\pi^2/6)$$

where, as before, γ is Euler's constant.

It is straightforward, although tedious, to find expressions for $-E[-\partial^2 l/\partial\theta_i\partial\theta_j]^{-1}$ for the other distributions discussed earlier; a good text for this topic is the book by Kite (1977).

3.12 CONFIDENCE INTERVALS FOR *T*-YEAR FLOODS: OR MORE GENERALLY, QUANTILES

3.12.1 Confidence Intervals Using the Central Limit Theorem

The preceding section has discussed ways of obtaining confidence regions for the parameters θ of a probability distribution. In most hydrological applications, however, the model parameters are of less interest than the quantiles of the distribution which they define. This section therefore considers the problem of making confidence statements about quantiles derived from the fitted distribution $f(x; \hat{\theta})$, where we assume that the estimates $\hat{\theta}$ have been obtained by maximum likelihood. We recall also that the symbol used earlier in this work to represent a quantile was X_0 where X_0 is the solution of the equation

$$G = \int_{-\infty}^{X_0} f(x; \hat{\theta})\mathrm{d}x - P = 0 \quad (3.45)$$

where, for a *T*-year flood, $P = 1 - 1/T$.

One straightforward method of calculating approximate 95% confidence intervals for X_0 is by the use of the normal approximation $X_0 \pm 2\sqrt{(\text{var } X_0)}$, where var X_0 can be calculated with the help of the variances and covariances in the equation var $\hat{\theta} = [-E(\partial^2 l/\partial\theta_i\partial\theta_j)]^{-1}$. Since X_0 is a function of $\theta_1, \theta_2, \ldots$, we can write, when N is large,

$$\text{var } X_0 = [(\partial G/\partial \theta_1)^2 \text{ var } \hat{\theta}_1 + (\partial G/\partial \theta_2)^2 \text{ var } \hat{\theta}^2$$

$$+ 2(\partial G/\partial \theta_1)(\partial G/\partial \theta_2) \text{ cov}(\hat{\theta}_1, \hat{\theta}_2)]/[f(X_0, \hat{\theta}_1, \hat{\theta}_2)] \qquad (3.46)$$

where additional terms of an obvious kind must be included when the model contains additional parameters $\theta_3, \theta_4, \dots$. The variances and covariances are read directly from the matrix expression for var $\hat{\theta}$; whilst $\partial G/\partial \theta_1$, for example, is found by differentiating equation (3.45) to give

$$\partial G/\partial \theta_1 = \int_{-\infty}^{X_0} \partial f/\partial \theta_1 dx \qquad (3.47)$$

Kite (1977) gives a very good treatment of the topic. For the normal and two-parameter log-normal distributions, Stedinger (1983) has shown how the non-central t-distribution may be used to derive exact confidence intervals for quantiles of these distributions. If $\xi_\alpha(p)$ and $\xi_{1-\alpha}(p)$ are the 100α and $100(1 - \alpha)$ percentiles of the non-central t-distribution with non-centrality parameter $\delta = z_p \sqrt{n}$ (with z_p the $100p$ percentile of the standard normal distribution, $N(0, 1)$) and degrees of freedom $n - 1$ (where n is the number of observations in the sample), then $100(1 - 2\alpha)\%$ exact confidence intervals for x_p, the pth quantile of the distribution $N(\mu, \sigma^2)$, are given by

$$(\bar{x} - s_x \xi_\alpha(p), \bar{x} + s_x \xi_{1-\alpha}(p))$$

where $\bar{x} = \Sigma x_t / n$ and $s_x = \sqrt{[\Sigma(x_t - \bar{x})^2/(n - 1)]}$ are the mean and standard deviation of the sample. However, procedures for calculating percentiles of the non-central t-distribution are not yet available in GENSTAT or MATLAB, although Stedinger reports that a routine for calculating $\xi(p)$ is available from the International Mathematical Subroutine Library (7500 Bellaire Boulevard, Houston, Texas 77036, USA).

Stedinger adapts the above method (Stedinger, 1983; Chowdhury and Stedinger, 1991), which gives exact confidence intervals for quantiles of the normal and two-parameter log-normal distributions, to obtain approximate confidence intervals for quantiles from the Pearson type III and log-Pearson type III distributions. Ashkar and Bobee (1988) also present results leading to approximate confidence intervals for these distributions.

3.12.2   Likelihood-based Confidence Intervals

An alternative procedure, and one of great generality for the calculation of confidence intervals for the quantiles of any probability distribution, is to go back to the likelihood surface, as follows.

We have stated that the region defined by inequality (3.38) gives a $100(1 - \alpha)\%$ confidence region for the parameters θ of the model. Now corresponding to each point θ lying within this region, or on its boundary, we can calculate a value X_0 using equation (3.45); by choosing many points θ we generate many X_0 values, the largest and smallest of which will define a $100(1 - \alpha)\%$ confidence region for the quantile X_0. It can be shown, in fact, that it is only necessary to compute X_0 for values of θ which lie on the boundary of the confidence region defined by inequality (3.38).

Figure 3.17 shows 95% confidence limits that were calculated by this procedure, for a series of flood records from the basin of the Rio Itajaí-Açú, assuming (a) a Gumbel distribution; (b) a gamma two-parameter distribution; (c) Box–Cox transformation to a normal scale. All three

(a) Gumbel distribution.

River, Site	XL	X0	XU
Itajaí-Açú, Rio do Sul	1697	1864	9161
Hercílio, Ibirama	2023	2192	4289
Itajaí, Apiuna	3925	4279	7075
Itajaí, Indaial	5589	6062	14781

(b) Two-parameter gamma distribution.

River, Site	XL	X0	XU
Itajaí-Açú, Rio do Sul	1423	1770	2400
Hercílio, Ibirama	1566	1936	2571
Itajaí, Apiuna	3227	3971	5234
Itajaí, Indaial	4155	5124	6771

(c) Normal distribution, after Box–Cox transformation to normal scale.

River, Site	XL	X0	XU
Itajaí-Açú, Rio do Sul	1297	1507	1833
Hercílio, Ibirama	1684	2293	3641
Itajaí, Apiuna	3430	4530	6767
Itajaí, Indaial	4451	5943	9053

Figure 3.17 Values of 100-year flood (maximum mean daily flow, X_0) together with 95% lower and upper confidence limits denoted by X_L, X_U. Confidence limits are likelihood-based confidence limits. They are calculated for three cases: (a) Gumbel distribution; (b) two-parameter gamma distribution; (c) normal distribution, after Box–Cox transformation to normal scale.

calculations are straightforward in MATLAB (this author has found MATLAB preferable to GENSTAT for this purpose, since the relevant GENSTAT optimisation procedure fails if more than 16 variables are used when defining the function to be optimised), and confidence intervals calculated using the likelihood surface in this way have greater accuracy than confidence intervals calculated using a normal approximation.

Several computational methods can be used to obtain the upper and lower confidence limits X_U and X_L for the T-year flood X_0. The simplest method is to search along transects, starting at the point $(\hat{\theta}_1, \hat{\theta}_2)$ on the likelihood surface where the maximum occurs. Thus, starting at this point, small steps are taken in each of a number of directions, and at each step the log-likelihood is calculated to determine whether the current point lies inside the region

$$2[l(\hat{\theta}; x) - l(\theta; x)] \le \chi_\alpha^2$$

or outside it, where $\alpha = 0.05$ or 0.01 determines the confidence level. As soon as the current point lies outside this region, the corresponding value of X_0 is recorded, and the process is reinitiated, moving along a transect in another direction. Having repeated the calculation along, say, C transects, the largest and smallest of the C values X_0 are taken as X_U, X_L. Other procedures for calculating X_U and X_L require the computation of

$$\min_\theta [1 + (1 + g^2(\theta))/X_0(\theta)]$$

and

$$\min_\theta [X_0(\theta)\{1 + g^2(\theta)\}]$$

where

$$g(\theta) = -2\{l(\hat{\theta}; x) - l(\theta; x)\} - \chi_\alpha^2$$

This calculation may take several hours on a 386 desk-top computer, using the Nelder–Mead algorithm for minimisation; the transect method, described above, requires two minutes or so per transect.

3.13 ASSESSING GOODNESS OF FIT

3.13.1 Cumulative Plots

Earlier parts of this chapter have been concerned with procedures for estimating parameters of statistical models, estimating quantiles (functions of the model parameters), and obtaining measures of their reliability. On occasions, we have referred to procedures for testing the agreement between a fitted probability distribution and the sample of data used in fitting it. We now return to the problem of assessing goodness of fit, and present a general procedure appropriate for the purpose.

Suppose that, given the sample $y_1, y_2, \ldots, y_N$, we have fitted the probability density function $f(y; \hat{\theta})$. The cumulative density function is denoted by $F(y; \hat{\theta})$, so that

$$F(y; \hat{\theta}) = \int_{-\infty}^{y} f(u; \hat{\theta}) du$$

Let the sample $y_1, y_2, \ldots, y_N$, ranked in increasing order, be denoted by $y_{(1)}, y_{(2)}, \ldots, y_{(N)}$. If the density function $f(y; \hat{\theta})$ is a good representation of the data, we should expect intuitively that there would be a close agreement between (a) the ranked observations $y_{(i)}, i = 1, \ldots, N$, and (b) the expected values of the N quantiles of the distribution $f(y; \hat{\theta})$. If the agreement is good, a plot of $y_{(i)}$ for $i = 1, \ldots, N$ against the expected values of the N quantiles would give a straight line.

The procedure is illustrated in Figure 3.18. The cumulative distribution is denoted by the curve $F(y; \theta)$, which is mapped onto a uniform distribution on the vertical axis, along which F is measured, lying between $F = 0$ and $F = 1$. This interval is now divided into N equal divisions, corresponding to the $N + 1$ points $F = 0, F = 1/N, F = 2/N, \ldots, F = 1$. The expected values of the N quantiles of $f(y; \hat{\theta})$ correspond on the F-axis to the midpoints of these N intervals: that is, to $F = 1/(2N), F = 3/(2N), \ldots, F = (2N - 1)/(2N)$. Points $U_1, U_2, \ldots, U_N$ on the y-axis are found by making the inverse mapping: formally, we calculate U_i by solving

$$(i - \tfrac{1}{2})/N = \int_{-\infty}^{U_i} f(y; \hat{\theta}) dy$$

or $U_i = F^{-1}((i - \tfrac{1}{2})/N)$. Calculation of the N values U_i is followed by plotting the pairs $\{y_{(i)}, U_i\}$; if the plotted points lie close to a straight line, we conclude that the distribution $f(y; \hat{\theta})$ constitutes a good fit to the data.

The above procedure for testing goodness of fit is quite general. In the particular case where the distribution $f(y; \hat{\theta})$ is normal, a formal test of goodness of fit can be based upon the correlation coefficient between $y_{(i)}$ and U_i. We calculate the familiar quantity

$$r = \sum U_i(y_{(i)} - \bar{y})/[\sum U_i^2 \sum (y_{(i)} - \bar{y})^2]$$

and compare the calculated value of this correlation coefficient with tabulated values prepared by Filliben (1975); the null hypothesis is $H_0 : \rho = 1$ against the alternative $H_1 : \rho < 1$.

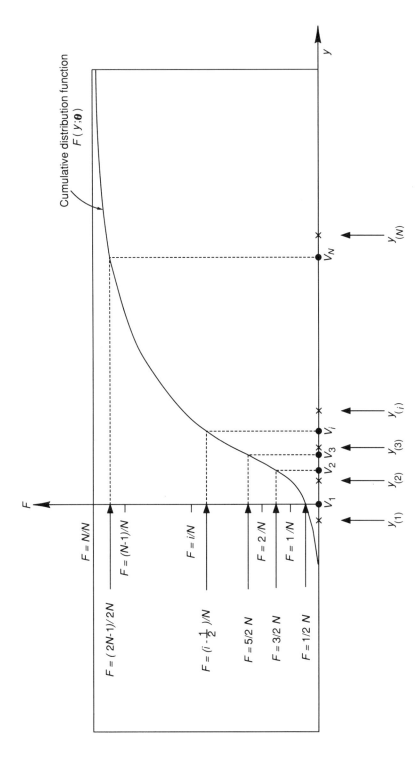

Figure 3.18 General scheme for assessing goodness of fit by comparing an ordered sample $y_{(1)}$, $y_{(2)}$, $y_{(i)}$, $y_{(N)}$ with $F^{-1}((1 - \frac{1}{2})/N)$, $F^{-1}((2 - \frac{1}{2})/N), \ldots, F^{-1}((i - \frac{1}{2})/N), \ldots, F^{-1}((N - \frac{1}{2})/N)$.

3.13.2   Envelope Curves

We have used cumulative plots at various points throughout this chapter, the plots being linear when the agreement between the data and the postulated distribution is good. But a difficulty with all such plots is knowing whether they are sufficiently linear, or whether the inevitable irregularities amount to more than just random fluctuations. A particular difficulty with such plots is that the ordering of the observations introduces a dependence between the values; and one or two random departures may show up as distortions to the curve, rather than as local irregularities. In consequence, features may be read into the data which are not present in reality.

One way of overcoming the risk of adopting spurious features as reality is by the use of simulation envelopes, as described by Atkinson (1987). We deal here only with the case where the underlying distribution is normal, to which the log-normal distribution can be transformed. Suppose we have a sequence of N annual floods which may follow a log-normal distribution. We have transformed the data to logarithms, have estimated the parameters by $\hat{\mu}$ and $\hat{\sigma}$, and have plotted the ordered sample of N values against the expected quantiles from a standard normal distribution. Let us suppose that the cumulative plot shows some bends or kinks, which cause us to have some doubt about the log-normal hypothesis. The simulation envelopes, calculated as described below, give us an idea of how far such kinks are to be expected in random samples of size N from a standard normal distribution.

To find the envelopes, we proceed as follows:

(i)   We generate a number of samples of size N, from the standard normal distribution, and transform them to a distribution $N(\hat{\mu}, \hat{\sigma}^2)$. The number of samples is commonly 19, for reasons set out below.

(ii) For each sample, we put the N values of the 'pseudo-sample' in ascending order. We can regard the 19 samples as being the 19 rows of a table, having N columns, the 'pseudo-observations' in each row being in ascending order.

(iii) Using the notional $19 \times N$ table, we calculate the maximum and minimum of each column. The maxima will determine the upper envelope, the minima the lower envelope.

(iv) We plot the two envelopes together with the ordered sample of N data values. If the latter plot lies within the two envelopes, then we have reason to believe that the sample of N values, $\ln y_t$, $(t = 1, \ldots, N)$ conforms to a normal distribution, and that the N data values y_t $(t = 1, \ldots, N)$ conform to a log-normal distribution.

We illustrate the procedure in Figure 3.19, using the sequence (Figure 1.1) of annual maximum mean daily flows for the Rio Hercílio at Ibirama. The file FIG3-19.PRT on the diskette accompanying this book shows the GENSTAT instructions yielding the two envelopes shown in the figure; the calculation took approximately two minutes on a 386 desk-top computer. We see that even the two extreme floods, for the years 1983 and 1984, lie within the band formed by the two envelopes and bounded by horizontal dashes (colons mark points at which data points and envelope points are so close as to be indistinguishable, given the coarse resolution of the printer). However, the plot given by the sample (shown as asterisks in the figure) wanders outside the envelope for two of the smaller data values, although the departure is not great. We should not be too much in error if, on the evidence of this plot, we were to assume that a two-parameter log-normal distribution gives a satisfactory fit to the flood sequence from the Rio Hercílio at Ibirama.

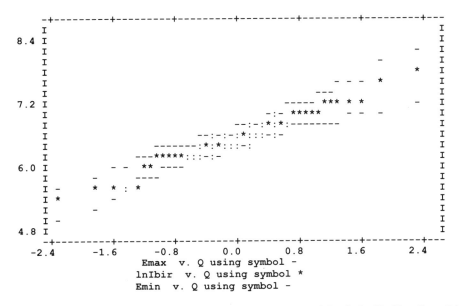

Figure 3.19 Upper and lower envelopes for normal plot of log annual floods for Rio Hercílio at Ibirama:
envelopes obtained by simulating 19 samples of size 48 from normal distribution with parameters $\hat{\mu}$, $\hat{\sigma}$
equal to mean and standard deviation of logs of annual floods.

We can easily generate pseudo-samples from other probability distributions relevant to flood
frequency studies. Figure 3.20 shows the plot of envelopes for fitting a Gumbel distribution
to the same flood sequence (Rio Hercílio at Ibirama); the plot given by the observed flood
sequence lies entirely within the band defined by the two envelopes, and once again the two
extreme floods of 1983 and 1984 are seen not to be so extreme as to cause rejection of the
hypothesis that Hercílio floods follow a Gumbel distribution.

Finally, Figure 3.21 shows the envelopes obtained when the annual minimum mean daily
flows are for the Rio Hercílio at Ibirama are fitted by a Weibull distribution (the GENSTAT
library has a procedure GRWEIBULL for generating pseudo-random variables from a Weibull
distribution). The aberrant value of 31 m^3 s^{-1}, already suspect from the plot of Figure 3.14,
lies well outside the band defined by the two envelopes. This minimum flow occurred in 1983;
thus whilst the maximum mean daily flow for that year was not far removed from the general
pattern, the low flow appears to be highly unusual.

Why do we generate 19 samples, in order to define the upper and lower envelopes? Together
with the observed data sample, we have 20 samples altogether; by generating 19 samples, there
is a chance of 1 in 20 that a particular data value lies outside the envelope. We note, however,
that because the ordering of the observed value introduces correlation between the plotted
values, the exact probability that the envelope contains the sample is not easily calculated. The
envelopes should not be interpreted, therefore, in any rigid probability sense, but rather as a
diagnostic tool for identifying whether data values which appear extremely different from the
rest of the sample really are very different. Furthermore, if several data values — perhaps those
in an upper or lower end of the ordered sample — all lie outside the band formed by the two
envelopes, we would doubt very strongly whether the fitted distribution is really appropriate
for the data.

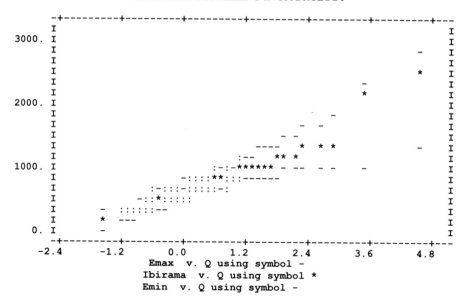

Figure 3.20   Upper and lower envelopes for cumulative plot (ordered data plotted against corresponding quantiles): Gumbel distribution fitted to flood sequence from Rio Hercílio at Ibirama.

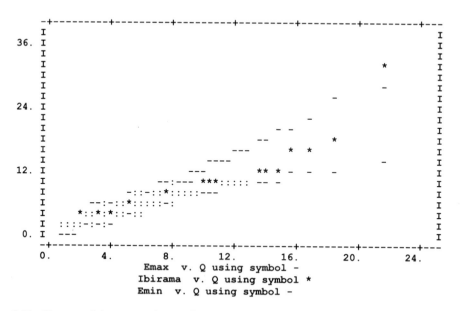

Figure 3.21   Upper and lower envelopes of cumulative plot (of ordered data against corresponding quantiles): Weibull distribution fitted to annual minimum mean daily flows, Rio Hercílio at Ibirama (48 years of record).

We have discussed envelope plots for the log-normal, Gumbel and Weibull distributions; we have said nothing so far about envelope plots for the two-parameter gamma distribution. For the three distributions discussed, the generation of random samples was straightforward; we generated samples of appropriate size from a uniform distribution (using the GENSTAT function URAND) and transformed them to the distribution required. In the case of the (log-)-normal, we used the GENSTAT function NED (Normal Equivalent Deviate); simple formulae give the transformations from the uniform distribution to the Gumbel and Weibull distributions. For the gamma distribution, transformation from the uniform distribution to a gamma scale is straightforward so long as the shape parameter κ can be considered a whole number, but is less straightforward when κ is not a whole number. We return to the problem of generating samples from a gamma distribution in Chapter 7.

3.14 CONCLUDING REMARKS

The enormous literature on flood frequency studies gives many other probability distributions, as well as procedures for estimating their parameters, which have not been mentioned in this chapter. In addition to the distributions mentioned earlier in this chapter, Cunnane (1989), in a work that is already a classic, identifies other 'candidate' distributions not discussed here — the two-component extreme value, the log-logistic, and the generalised logistic distributions — and gives many references to their origin, development and use. Corresponding to the plethora of probability distributions, there is a multitude of papers describing estimation procedures for parameters.

In this chapter, and throughout this book, we have emphasised the role of the likelihood function for estimation (whether for obtaining point estimates of parameters, or interval estimates of quantiles), and for selecting the most appropriate from a set of candidate probability distributions. The reason we have done this is that, if the probability distribution is 'correct', the likelihood function contains all the available information on the model parameters and it is sensible to base estimation procedures on this function. Furthermore, likelihood-based confidence regions have a firm theoretical basis and are efficient when long or moderately long data sequences are available.

An important word in the last paragraph is the word 'correct'. As mentioned earlier in this chapter, no statistical model can be entirely 'correct' (in the sense that it describes the exact physical law used by nature when producing any sequence of annual maximum or minimum flows), but is at best a good representation of the statistical characteristics of the data sequence. Furthermore, several models may be indistinguishable in the sense that any one of them could serve as a representation of these characteristics — just as we have seen that there is very little to choose between the two-parameter gamma and log-normal distributions for the annual flood sequence on the Rio Hercílio at Ibirama. Pragmatism must be the essence of any study of hydrological extremes; as stated by one of the most influential writers on flood frequency analysis (Cunnane, 1989) 'theoretical arguments cannot, *per se*, identify a best choice of distribution for floods'.

Lastly, statistical software for hydrological application continues to develop rapidly. The latest (June 1994) Release 3 of the GENSTAT package contains a new directive, DISTRIBU-TION, which enables the user to fit any one of a very considerable number of distributions, both discrete and continuous, by maximum likelihood. If, for example, we have read the records of annual floods for Ibirama and Apiuna into two variates with these names, three-parameter log-normal distributions can be fitted to both data sets by means of the directive

DISTRIBUTION [DISTRIBUTION=logNormal; CONSTANT=estimate] DATA=\
    Ibirama, Apuina; PARAMETERS=IbTheta, ApTheta; SE=IbSE,\
    ApSE; VCOVARIANCE=IbVar, ApVar

which stores the maximum-likelihood estimates of the three log-normal parameters in *IbTheta*, *ApThera* for Ibirama and Apiuna, respectively. The standard errors of these estimates are stored in *IbSE* and *ApSE*, and the variance-covariance matrices in *IbVar* and *ApVAr*. Omission of the CONSTANT option in the square brackets fits a two-parameter log-normal distribution. By putting DISTRIBUTION=gamma or DISTRIBUTION=Weibull instead of DISTRIBUTION=logNormal, two- and three-parameter versions of these distributions can be fitted. Other continuous distributions fitted by the DISTRIBUTION directive are: the sum of two normal distributions, with four or five parameters, depending on whether or not the two normal distributions have common variance); the exponential distribution; the Pareto distribution (two and three parameters); and beta distributions Types I and II. Any one of nine discrete distributions can also be fitted by the same directive.

REFERENCES

Ashkar, F. and Bobee B. (1988). Confidence intervals for flood events under a Pearson 3 or log Pearson 3 distribution. *Water Resources Bulletin*, **24**, 639–49.

Atkinson, A. C. (1987). *Plots, Transformations and Regression: An Introduction to Graphical Methods of Diagnostic Regression Analysis*. Clarendon Press, Oxford.

Benson, M. A. (1968). Uniform flood frequency estimating methods for federal agencies. *Water Resour. Res.*, **4**, 891–908.

Bobee, B. (1975). The log Pearson type 3 distribution and its applications in hydrology. *Water Resour. Res.*, **14**, 365–9.

Bobee, B. and Robitaille, R. (1977). The use of the Pearson Type 3 distribution and log Pearson type 3 distribution revisited. *Water Resour. Res.*, **13**, 427–43.

Box, G. E. P. and Jenkins, G. M. (1970). *Time Series Analysis: Forecasting and Control*. Holden Day, San Francisco.

Chowdhury, J. U. and Stedinger, J. R. (1991). Confidence interval for design floods with estimated skew coefficient. *ASCE Jour. Hydraulic Eng.*, **117**, 811–31.

Cunnane, C. (1989). *Statistical Distributions for Flood Frequency Analysis*. Operational Hydrology Report No 33. World Meteorological Organisation, Geneva.

Filliben, J. J. (1975). The probability plot correlation test for normality. *Technometrics*, **17**, 111–17.

Hosking, J. R. M. (1985) Algorithm AS 215: maximum-likelihood estimation of the parameters of the generalized extreme value distribution. *Appl. Statist.* **34**, 301–310.

Hosking, J. R. M. (1986). *The Theory of Probability-weighted Moments*. Res. Rep. RC13412. IBM Research, Yorktown Heights, NY.

Hosking, J. R. M. (1990). *l*-moments: analysis and estimation of distributions using linear combinations of order statistics. *J. R. Statist. Soc. B.*, **52**, 105–24.

Hosking, J. R. M. (1991). *Fortran Routines for Use with the Method of l-moments. Version 2*. Res. Rep. RC17097 IBM Research, Yorktown Heights, NY.

Hosking, J. R. M., Wallis J. R. and Wood, E. F. (1985) Estimation of the generalized extreme-value distribution by the method of probability-weighted moments. *Technometrics*, **27**, 251–61.

Kite, G. W. (1977). *Frequency and Risk analyses in Hydrology*. Water Resources Publications, Fort Collins, CO.

Macleod, A. J. (1988). A remark on Algorithm AS 215: maximum-likelihood estimation of the parameters of the generalized extreme-value distribution. *Appl. Statist.* **37**, 198–199.

Stedinger, J. R. (1980). Fitting log-normal distributions to hydrologic data. *Water Resour. Res.*, **16**, 481–90.

Stedinger, J. R. (1983). Confidence intervals for design events. *J. Hydraul. Eng. ASCE*, **109**, 13–27.
Stern, R. D. and Coe, R. (1984). A model fitting analysis of daily rainfall data. *J. R. Statist. Soc. A*, **147**, 1–34.

EXERCISES AND EXTENSIONS

3.1. The log-logistic cumulative distribution function is given by

$$F(y; a, b, c) = \{1 + [(y - a)/b]^{-1/c}\}^{-1} \qquad y > a; a, b, c > 0$$

Derive the three equations which must be solved to give the maximum-likelihood estimates of the three parameters a, b and c.

3.2. The two-component Gumbel distribution is given by

$$F(y; \alpha_1, \alpha_2, u_1, u_2, p) = p \exp[-\exp(-\alpha_1(y - u_1))]$$

$$+ (1 - p) \exp[-\exp(-\alpha_2(y - u_2))] \qquad -\infty < y < \infty$$

and has been used to describe annual flood sequences which may arise from two distinct physical causes, such as summer convective storms and spring snowmelt. It is an example of a *mixture* of distributions, of the general form

$$f(y; \theta, \mathbf{p}) = p_1 f(y; \theta_1) + p_2 f(y; \theta_2) + \cdots + p_k f(y; \theta_k)$$

where $\Sigma p_i = 1$; the components $f(y; \theta_1)$, $f(y; \theta_2)$, ... are probability densities having the same mathematical form, but in general different parameters. Derive the five equations which must be solved to give the maximum-likelihood equations. If the parameter $\mathbf{p}$ is known, how do the equations simplify?

3.3. The generalized logistic distribution is defined by

$$F(y) \begin{cases} \{1 + [1 - \gamma(y - \alpha)/\beta]^{1/\gamma}\}^{-1} & \gamma \neq 0 \\ \{1 + \exp[-(y - \alpha)/\beta]\}^{-1} & \gamma = 0 \end{cases}$$

where, if $\gamma = 0$ then y can take any value, and if $\gamma \neq 0$ we require:

$$\alpha + \beta/\gamma \leq y < \infty \qquad \gamma < 0$$

$$-\infty < y \leq \alpha + \beta/\gamma \qquad \gamma > 0$$

Explore the characteristics of this distribution.

Chapter 4

Linear Relationships with Explanatory Variables

4.1 PRINCIPLES OF REGRESSION ANALYSIS

Regression analysis seeks to establish a relationship between a variable y and p other variables $x_1, x_2, \ldots, x_p$ which may explain how y varies. In linear regression, the explanatory variables combine to form a systematic component which is linear in unknown parameters, of the form $\beta_0 x_0 + \beta_1 x_1 + \beta_2 x_2 + \cdots + \beta_p x_p$; any observation of the random variable Y, say y_t, deviates from the systematic component of the model by an amount denoted by ε_t, so that the model equation can be written

$$y_t = \beta_0 x_{0t} + \beta_1 x_{1t} + \beta_2 x_{2t} + \cdots + \beta_p x_{pt} + \varepsilon_t \tag{4.1}$$

where $x_{0t}, x_{1t}, x_{2t}, \ldots, x_{pt}$ are the observations of the variables $x_0, x_1, x_2, \ldots, x_p$ corresponding to the tth observation of y, namely y_t. Equation (4.1) can be written

$$y_t = \mathbf{x}_t^T \boldsymbol{\beta} + \varepsilon_t$$

where $\mathbf{x}_t$ is the vector $[x_{0t}, x_{1t}, x_{2t}, \ldots, x_{pt}]^T$ and $\boldsymbol{\beta}$ is the vector of parameters $[\beta_0, \beta_1, \beta_2, \ldots, \beta_p]^T$. In many applications, the values of x_{0t} — the variable multiplying the first regression coefficient β_0 — are all equal to one, so that the component $\beta_0 x_{0t}$ becomes a constant term, β_0; for the time being we will retain the full expression.

Given N observations of the variable Y, say $y_1, y_2, \ldots, y_t, \ldots, y_N$, together with the corresponding observations on each of the x-variables, we can write the N relations (4.1) in the form

$$\mathbf{y} = \mathbf{X}\boldsymbol{\beta} + \boldsymbol{\varepsilon} \tag{4.2}$$

where $\mathbf{y}$ is now an $N \times 1$ vector of the observations $\{y_t\}$ on y, $\boldsymbol{\beta}$ is a $(p+1) \times 1$ vector of parameters, $\boldsymbol{\varepsilon}$ is an $N \times 1$ vector of random components, and $\mathbf{X}$ is an $N \times (p+1)$ matrix with the observations on the variables $x_0, x_1, x_2, \ldots, x_p$ in the first, second, third,..., pth columns. In the particular case when the x_{0t} are all equal to one, the first column of the matrix $\mathbf{X}$ will consist entirely of ones. If the ε_t are random variables with expected value zero, we can write, for the vector random variable $\mathbf{Y}$ of which $\mathbf{y}$ is the realisation,

$$E[\mathbf{Y}] = \mathbf{X}\boldsymbol{\beta}$$

Except where otherwise indicated, it will be assumed that the deviations ε_t are statistically independent, normally distributed, and with unknown variance σ_ε^2. When $x_{0t} = 1$ for all t, the parameter β_0 represents an intercept on the axis along which y is measured. If the systematic component of the model contains only one other x-variable, so that, with a slight change of notation, this systematic component can be written $\beta_0 + \beta_1 x$, the statistical model is a simple linear regression; if the systematic component contains more than one explanatory variable, it is a multiple (linear) regression. In the hydrological applications discussed below, we shall sometimes use the symbol q, instead of y, when the variable being modelled is derived from a flow record.

4.2 HYDROLOGICAL APPLICATIONS OF LINEAR REGRESSION ANALYSIS

Common statistical problems in linear regression analysis include the following: efficient estimation of the parameters β and σ_ε^2; estimation of $E[Y_{t+1}]$ and of y_{t+1}, given observations on the variables $x_1, x_2, \ldots, x_p$ at time $t + 1$, including the calculation of prediction intervals for these estimates; the testing of hypotheses about the parameters β, including the selection, from possibly many alternatives, of the 'best' set of explanatory variables for predicting y; the combination of estimates of the parameters β and σ_ε^2 given by different data sets.

However, the purpose of this book is not to give an account of the statistical theory appropriate for solution of the above problems, since excellent texts already exist for that purpose. We are concerned, rather, with presenting an account of some of the many applications of linear regression analysis in the water sciences; with illustrating, by example, some of the more modern techniques which further extend the usefulness of linear regression analysis in this field; and in particular, with showing that the wise use of regression analysis requires more than a mechanical application of an appropriate computer routine.

The following examples show a little of the breadth of application of linear regression analysis in hydrological science.

4.2.1 The Thomas–Fiering Model for the Simulation of Monthly Flow Sequences

The Thomas–Fiering model is a procedure for deriving artificial sequences, commonly of mean monthly discharge ('flow'), which have statistical characteristics similar in important respects to those of the record of monthly flow observed at a river flow gauging station. The artificial sequences generated by the Thomas–Fiering model are used to simulate the behaviour of a water resource system which, for example, may abstract water from the river near the gauging site for the purpose of supplying domestic, industrial and irrigation requirements. The model, as commonly formulated, consists of twelve simple linear regressions in which February flow (y) is regressed on January flow (explanatory variable x); March flow (y) is regressed upon February flow (explanatory variable x) . . . .

The twelve regressions $q_{feb} = \alpha_{feb} + \beta_{feb} q_{jan} + \varepsilon_{feb}$, $q_{mar} = \alpha_{mar} + \beta_{mar} q_{feb} + \varepsilon_{mar}, \ldots$ are used to generate the artificial flow sequences by selecting the $\varepsilon_{feb}, \varepsilon_{mar}, \ldots$ commonly from normal distributions with appropriate variances, and adding these random components to the systematic components $\alpha_{feb} + \beta_{feb} q_{jan}, \alpha_{mar} + \beta_{mar} q_{mar}, \ldots$. Clearly, scope exists for including explanatory variables other than flow in the preceding month; for example, flows in several of the preceding months may be included, and monthly precipitation also. There is scope also for using distributions other than the normal for generating the random components ε, whilst modification to the basic model will also be necessary if the flow gauge is sited on a river with intermittent flow.

4.2.2 Representation of a Flow Rating Curve by a Polynomial

Yevjevich (1972) discusses an example in which a parabolic rating curve $Q = \beta_0 + \beta_1 H + \beta_2 H^2$ is fitted to measured discharges Q and stages H, using 13 pairs of observations (Q, H). Since the parameters $\beta_0, \beta_1, \beta_2$ occur linearly in the equation for the rating curve, the systematic component of the statistical model for Q is of the multiple regression form. We discuss this example more fully in Chapter 5.

4.2.3 Fitting of Sediment Rating Curves, Giving Sediment Concentration in Terms of Discharge

Concentration of suspended sediment, c, is commonly related to river discharge q by an equation of the form $c = \alpha q^\beta$. Observations on c will deviate from the expression on the right-hand side of this equation; if the deviations combine additively with the systematic component to give the form

$$c = \alpha q^\beta + \varepsilon$$

then the systematic component of the model is clearly non-linear in the parameters α and β. However, in certain cases a linear relation may be appropriate between the logarithms of c and q, of the form

$$\ln c = \alpha^* + \beta^* \ln q + \varepsilon^*$$

so that the statistical model has now been expressed in linear regression form.

4.2.4 Estimating Missing Values in Flow Records

Some of the Figures 1.1 to 1.4 contained 'missing values', there denoted by asterisks. Denoting by y a sequence of, say, annual floods at a gauging site with incomplete record, and by $x_1, x_2, \ldots, x_p$ the annual floods at nearby gauging stations with records that are free from missing values, multiple regression can be used to estimate missing values in the sequence y. It will, however, be necessary to apply regression analysis with particular care to ensure that basic assumptions are satisfied. We return to this example in Chapter 7.

4.2.5 Regionalisation of Mean Annual Floods

Where flood characteristics must be estimated for a site without flow records, it is common to use flood records from other stations within the region to establish relationships between (for example) mean annual flood q and basin characteristics (area A, channel slope S, percentage P of basin area that is lake or reservoir, etc.) as explanatory variables. For this purpose it is common to fit multiple regression models of the form

$$\ln q = \beta_0 + \beta_1 \ln A + \beta_2 \ln S + \beta_3 \ln P + \cdots + \varepsilon \tag{4.3}$$

Care must also be taken in this instance because the records available from the regional stations will be of different lengths, so that the mean annual floods q will require different weightings in the analysis. Clearly in this example there is scope for examining whether Box–Cox transformations with parameters λ, other than the log transformation for which $\lambda = 0$, are more appropriate for some or all of the variables in the regression.

4.2.6 Calculation of Unit Hydrograph Ordinates from Storm Runoff (y) and Effective Rainfall (x) Sequences

One of the examples in Chapter 1 concerned the relationship between storm runoff q_t and effective rainfall $p_t, p_{t-1}, p_{t-2}, \ldots$. When expressed in the form

$$q_t = h_0 p_t + h_1 p_{t-1} + h_2 p_{t-2} + \cdots + h_k p_{t-k} + \varepsilon \tag{4.4}$$

this is essentially a linear regression, the parameters $h_0, h_1, \ldots, h_k$ being the ordinates of the 1-hour unit hydrograph (assuming q_t, p_t are given at hourly intervals). Particular care is needed in the model fitting to ensure that the ordinates are physically meaningful: some data sets may yield negative ordinates of the unit hydrograph. One procedure which avoids this problem is the 'constrained linear system' (CLS) model, in which the model (4.4) is fitted subject to the constraints that all ordinates h_i must be non-negative. However, where such constraints are introduced, the model (4.4) is no longer a simple linear regression.

4.2.7 Models Representing Certain Non-linear Systems

The linear relation (4.4) can be derived from linear systems considerations, where the response $q(t)$ by a physically realistic system is related to an input variable $p(t)$ by

$$q(t) = \int_{t=0}^{\infty} p(\tau) h(t - \tau) d\tau \tag{4.5}$$

In discrete time, equation (4.5) becomes

$$q_t = \sum_{t=0}^{\infty} p_\tau h_{t-\tau}$$

and, substituting a 'large' integer k instead of the infinity in the summation sign, we obtain exactly equation (4.4).

A natural generalisation, in the case of non-linear system behaviour, is to replace the integral in equation (4.5) by an expansion of the form

$$q(t) = \int_{\tau=0}^{\infty} h(\tau) p(t - \tau) d\tau + \int_{\tau_1=0}^{\infty} \int_{\tau_2=0}^{\infty} p(t - \tau_1) p(t - \tau_2) h(\tau_1, \tau_2) d\tau_1 d\tau_2 + \cdots \tag{4.6}$$

which becomes, when put in discrete form,

$$q_t = \sum_{\tau=0}^{\infty} h_\tau p_{t-\tau} + \sum_{\tau_1=0}^{\infty} \sum_{\tau_2=0}^{\infty} h_{\tau_1, \tau_2} p_{t-\tau_1} p_{t-\tau_2} + \cdots \tag{4.7}$$

Although the generalisation (4.7) is non-linear in the input variables p_t, it remains linear in the parameters $h_\tau, h_{t_1, t_2}, \ldots$. In principle, therefore, it is possible to estimate them by linear regression (the infinite upper limits to the summations being replaced by finite values, and the number of terms in the expansion of integrals also being truncated). In practice, computational considerations are likely to render such a straightforward solution difficult.

4.2.8 Fitting Surfaces

It is sometimes appropriate to represent the spatial variability of a hydrological variable (say annual rainfall, denoted by R) in terms of spatial coordinates (x_1, x_2) in the form

$$R = \beta_0 + (\beta_1 x_1 + \beta_2 x_2) + (\beta_{11} x_1^2 + \beta_{12} x_1 x_2 + \beta_{22} x_2^2) + \cdots + \varepsilon$$

where the parameters β are estimated using data on annual rainfall from a network of gauges. Given data $\{R^{(i)}, x_1^{(i)}, x_2^{(i)}\}$, $i = 1, 2, \ldots, N$, where N is the number of gauges in the network, the estimation problem remains one of multiple regression, since the parameters β to be estimated occur linearly in the model. This remains true where x_1, x_2 are replaced by harmonic functions (giving a Fourier expansion with unknown coefficients); and where x_1, x_2 are replaced by orthogonal polynomials, whether in one or more dimensions. Orthogonal polynomials, a specialised technique for fitting curves and surfaces, are simply transformations from the variables x_1, x_2, x_1^2, $x_1 x_2$, x_2^2 to new variables $P_1(x_1, x_2)$, $P_2(x_1, x_2)$, ... which are polynomials in x_1, x_2, subject to the constraints that $\Sigma P_1(x_1, x_2) = \Sigma P_2(x_1, x_2) = \cdots = 0$; $\Sigma P_1(x_1, x_2) P_2(x_1, x_2) = \Sigma P_1(x_1, x_2) P_3(x_1, x_2) = \cdots = 0$ over the values of the explanatory variable, thereby simplifying the calculation of the parameters β. Draper and Smith (1981) give an excellent discussion of the principles involved for the one-dimensional case where the polynomials $P_1(x)$, $P_2(x)$, ... are linear, quadratic, ... polynomials in just one variable x.

The above polynomial model for R expresses how a response variable varies over a space of two dimensions; clearly a hydrological variable such as hydraulic conductivity $H(x, y, z)$ could be similarly represented in a space of three dimensions, such as that within a soil mass defined within the region $-\infty < x < \infty$; $-\infty < y < \infty$; $z > 0$, with z measured vertically downwards.

Where the number of explanatory variables is large and the correlations amongst them appreciable, it is sometimes appropriate to calculate the *principal components* of the explanatory variables (that is, linear combinations of the explanatory variables which are mutually uncorrelated), and to model the response variable in terms of a reduced number of principal components as new explanatory variables. These 'new' explanatory variables are those principal components which have largest variances, these being given by the latent roots of the variance-covariance matrix of the explanatory variables. For full details, the reader is referred to Draper and Smith (1981).

4.2.9 Quality Control of Hydrometeorological Data

Climatological data are increasingly being collected by automatic weather stations sited in remote areas, the data being stored for subsequent collection on field visits, or transmitted by satellite link to the laboratory. The volumes of data thus collected are often extremely large; nevertheless, the field sensors are liable to failure, or to give false readings, for any one of a number of reasons. For example, the Perspex domes of radiometers become dirty, or the silica gels which should keep them dry become saturated, so that condensation obscures the interior of the dome. Data collected and transmitted from remote locations therefore require particularly close scrutiny. However, the volumes of data are such that the 'quality control' of data, at least in its initial stages, must be effected by computer programs which scan the data sequences and look for discrepancies of types which are known to occur with particular field sensors.

Tests made by the quality control programs are commonly based on knowledge of the permissible ranges within which the measured variables should lie, these limits being determined by experience. In temperate latitudes, hourly solar radiation data sequences are marked as suspect if more than eleven hourly solar radiation values exceed 1400 W m^{-2}; net radiation is marked as suspect if all hourly measurements within the day are less than 9 W m^{-2} or if net radiation is greater than 80% of solar radiation over grass.

The diagnostic tests of multiple regression analysis, used to identify data which are particularly influential, also have a part to play in the identification of suspect data. The 'physical' methods described in the last paragraph can identify where variables exceed certain limits, but cannot identify where gross errors have occurred whilst leaving variables within these limits. It is in such cases that regression diagnostics can complement (not replace) physical methods. Pozzi (1993) explored what happened when errors of known magnitude (typically 5% or 10%, and of random sign) were introduced randomly into sequences of data recorded by automatic weather stations; regression diagnostic procedures identified a large proportion of such errors, which were not detected by the physical tests.

4.3 THE BASICS OF LINEAR REGRESSION

We take as the model

$$y_t = \mathbf{x}_t^{\mathrm{T}} \boldsymbol{\beta} + \varepsilon_t \tag{4.8}$$

with the assumptions $E[\varepsilon_t] = 0$; $E[\varepsilon_t^2] = \sigma_\varepsilon^2$; and $E[\varepsilon_t \varepsilon_s] = 0$ for $t \neq s$. We also assume that the 'errors' or 'residuals' ε_t are normally distributed. To calculate estimates $\hat{\boldsymbol{\beta}}$ of the $p + 1$ parameters $\boldsymbol{\beta}$, we calculate the values $\hat{\boldsymbol{\beta}}$ which minimise the sum of squared errors, $\varepsilon^{\mathrm{T}} \varepsilon$, where ε is the $N \times 1$ vector with typical element ε_t; that is, the estimates of $\boldsymbol{\beta}$ minimise

$$R(\boldsymbol{\beta}) = (\mathbf{y} - \mathbf{X}\boldsymbol{\beta})^{\mathrm{T}}(\mathbf{y} - \mathbf{X}\boldsymbol{\beta}) \tag{4.9}$$

By differentiating equation (4.9) with respect to the elements of the vector $\boldsymbol{\beta}$, and setting the resulting derivatives equal to zero, it can be shown that the estimates $\hat{\boldsymbol{\beta}}$ satisfy the set of $p + 1$ linear equations

$$\mathbf{X}^{\mathrm{T}}(\mathbf{y} - \mathbf{X}\hat{\boldsymbol{\beta}})^{\mathrm{T}} = \mathbf{0}$$

giving

$$(\mathbf{X}^{\mathrm{T}}\mathbf{X})\hat{\boldsymbol{\beta}} = \mathbf{X}^{\mathrm{T}}\mathbf{y} \tag{4.10}$$

or, assuming that the inverse $(\mathbf{X}^{\mathrm{T}}\mathbf{X})^{-1}$ exists,

$$\hat{\boldsymbol{\beta}} = (\mathbf{X}^{\mathrm{T}}\mathbf{X})^{-1}\mathbf{X}^{\mathrm{T}}\mathbf{y} \tag{4.11}$$

Substituting for $\boldsymbol{\beta}$ in expression (4.9), we obtain the minimum value of the sum of squares

$$R(\hat{\boldsymbol{\beta}}) = (\mathbf{y} - \mathbf{X}\hat{\boldsymbol{\beta}})^{\mathrm{T}}(\mathbf{y} - \mathbf{X}\hat{\boldsymbol{\beta}})$$
$$= \mathbf{y}^{\mathrm{T}}[\mathbf{I} - \mathbf{X}(\mathbf{X}^{\mathrm{T}}\mathbf{X})^{-1}\mathbf{X}^{\mathrm{T}}]\mathbf{y} \tag{4.12}$$

We shall use this relation frequently, together with the following alternative form:

$$R(\hat{\boldsymbol{\beta}}) = \mathbf{y}^{\mathrm{T}}\mathbf{y} - \hat{\boldsymbol{\beta}}^{\mathrm{T}}(\mathbf{X}^{\mathrm{T}}\mathbf{y}) \tag{4.13}$$

Equation (4.13) is the basis for the analysis of variance (ANOVA) in which the total sum of squares $y_1^2 + y_2^2 + \cdots + y_N^2 = \mathbf{y}^{\mathrm{T}}\mathbf{y}$ for the observations $\{y_t\}$ is divided into two components, one of which is $\hat{\boldsymbol{\beta}}^{\mathrm{T}}\mathbf{X}^{\mathrm{T}}\mathbf{y}$, and the other is the quantity $R(\hat{\boldsymbol{\beta}})$. Statistical texts (see, for example, Draper and Smith, 1981) show the following:

(i) If the vector $\boldsymbol{\beta}$ has $p + 1$ elements, and the vector $\mathbf{y}$ has N components, the 'degrees of freedom' associated with the components $\mathbf{y}^T\mathbf{y}$, $\hat{\boldsymbol{\beta}}\mathbf{X}^T\mathbf{y}$ and $R(\hat{\boldsymbol{\beta}})$ are N, $p + 1$, and $N - p - 1$, respectively.

(ii) Given the above assumptions concerning the ε_t, the quantities $\hat{\boldsymbol{\beta}}\mathbf{X}^T\mathbf{y}/(p + 1)$ and $R(\hat{\boldsymbol{\beta}})/(N - p - 1)$ are both distributed as χ^2 with $p + 1$ and $N - p - 1$ degrees of freedom, when the null hypothesis that the vector $\boldsymbol{\beta} = \mathbf{0}$ is true. In which case,

(iii) The ratio of these two quantities, $\hat{\boldsymbol{\beta}}\mathbf{X}^T\mathbf{y}/(p + 1)$ divided by $R(\hat{\boldsymbol{\beta}})/(N - p - 1)$, is distributed as a 'variance ratio', or F-statistic, with $p + 1$ and $N - p - 1$ degrees of freedom. Tables exist for this statistic, but its quantiles can be easily calculated by GENSTAT, as shown below. The importance of this result is that we can use it to test whether the inclusion of an additional variable in the regression leads to a significant reduction in the residual sum of squares; or, to put it more formally, we can test the null hypothesis that the regression coefficient for the newly introduced variable is zero. By comparing the calculated ratio

$$F_{calc} = \frac{\hat{\boldsymbol{\beta}}\mathbf{X}^T\mathbf{y}/(p + 1)}{R(\hat{\boldsymbol{\beta}})/(N - p - 1)}$$

with the quantiles of the distribution which this ratio would have, if the null hypothesis were true, we can weigh the evidence for and against the null hypothesis. For if F_{calc}, lies far into the upper tail of the tabulated distribution, so that the probability of obtaining by chance a value as large as, or larger than, F_{calc}, is very small, we can conclude either (a) that the null hypothesis is true, and that purely by chance nature has given us a set of data which is extremely odd; or (b) that the null hypothesis is false. The weight of evidence is therefore in favour of (b); and the farther that F_{calc} lies into the upper tail of the tabulated F-distribution, the greater is the weight of evidence against the null hypothesis.

Now in practice, we are interested in testing which of a series of linear regression models, involving an increasing number of explanatory variables, is best suited to explaining how the variable y_t varies. So we commonly begin with a model very similar to those of the preceding chapter (except that now the variable Y is distributed normally):

$$y_t = \beta_0 + \varepsilon_t$$

in which y_t is a constant, β_0, apart from the random error ε_t. We then fit a model with one explanatory variable x_1 added, namely

$$y_t = \beta_0 + \beta_1 x_1 + \varepsilon_t$$

and then we may need, possibly, to introduce a second explanatory variable x_2, so that the model then becomes:

$$y_t = \beta_0 + \beta_1 x_1 + \beta_2 x_2 + \varepsilon_t$$

and so on. Using equation (4.13), each model fitted will give a new value of $R(\hat{\boldsymbol{\beta}})$ which will become smaller with each new parameter fitted. Indeed, if we were to fit N parameters, $R(\hat{\boldsymbol{\beta}})$ would be zero, since we have only N data values y_t to fit N parameters $\beta_0, \beta_1, \beta_2, \ldots, \beta_{N-1}$. If we denote the sequence of values of $R(\hat{\beta})$, corresponding to the models of increasing numbers of explanatory variables, by $R(\hat{\beta}_0)$; $R(\hat{\beta}_0, \hat{\beta}_1)$; $R(\hat{\beta}_0, \hat{\beta}_1, \hat{\beta}_2)$, ... then we can summarise the whole procedure as shown in Table 4.1, in which we have assumed that k explanatory variables is the maximum number that we would wish to include.

It should be noticed that the expression $\hat{\beta}^T X^T y$ which throughout Table 4.1, differs depending upon the number of parameters fitted. Thus, where β_0 alone is fitted, the matrix X consists of a single column of ones, being of dimension $N \times 1$; where β_0 and β_1 are fitted, the matrix X is of dimension $N \times 2$, having ones in its first column and the N values of the explanatory variable x_1 in its second column; and so on. In particular, the value of $\hat{\beta} X^T y$ corresponding to the fitting of β_0 is therefore just $N\bar{y}^2$, and it is convenient to subtract this quantity, together with its corresponding degree of freedom, from the last line in Table 4.1. Further simplification also results if we rearrange the table to give differences between the $R(\hat{\beta})$ which result from successively fitting additional parameters, giving Table 4.2.

Two points must be made about the Tables 4.1 and 4.2. First, as shown in Table 4.2, each new parameter (and its associated explanatory variable) included in the model results in a reduction in the residual sum of squares $R(\hat{\beta})$ which, as has been noted, would become zero if we were to fit N parameters. Frequently, the reductions in the residual sum of squares (RSS) associated with the first few parameters are substantial, the reductions becoming successively smaller with each additional parameter included. It is one of the objectives of model fitting to determine at what point the reduction in RSS becomes too small to warrant the inclusion of additional parameters in the model; and for this purpose the theoretical results (i), (ii) and (iii), given above, are used. Suppose, for example, that we have fitted first β_1, for which we obtained a very substantial reduction $R(\hat{\beta}_0) - R(\hat{\beta}_0, \hat{\beta}_1)$ in Table 4.2; if we then fit successively $\beta_2, \beta_3, \ldots$, and get reductions $R(\hat{\beta}_0, \hat{\beta}_1) - R(\hat{\beta}_0, \hat{\beta}_1, \hat{\beta}_2)$, $R(\hat{\beta}_0, \hat{\beta}_1, \hat{\beta}_2) - R(\hat{\beta}_0, \hat{\beta}_1, \hat{\beta}_2, \hat{\beta}_3)$, ..., which are roughly equal to $R(\hat{\beta}_0, \ldots, \hat{\beta}_k)$ — the 'unexplained' error left over when all available explanatory variables have been included — this would suggest that fitting $\beta_2, \beta_3, \ldots$ brings

Table 4.1 Summary of procedure for fitting linear models with increasing numbers of parameters β_0; β_0, β_1; $\beta_0, \beta_1, \beta_2$; ...; $\beta_1, \ldots, \beta_k$, corresponding to $0, 1, 2, \ldots, k$ explanatory variables $x_1, x_2, \ldots, x_k$.

Parameters fitted	Degrees of freedom	Sum of squares	Notes
β_0	1	$\hat{\beta}^T X^T y$	Here, β is the scalar $\hat{\beta}_0$ and
	$N - 1$	$R(\hat{\beta}_0)$	X is $1 = [1, 1, \ldots]^T$.
β_0, β_1	2	$\hat{\beta}^T X^T y$	Here, $\hat{\beta} = [\hat{\beta}_0, \hat{\beta}_1]^T$ and
	$N - 2$	$R(\hat{\beta}_0, \hat{\beta}_1)$	$X = [1, x_1]$.
$\beta_0, \beta_1, \beta_2$	3	$\hat{\beta}^T X^T y$	Here, $\hat{\beta} = [\hat{\beta}_0, \hat{\beta}_1, \hat{\beta}_2]^T$ and
	$N - 3$	$R(\hat{\beta}_0, \hat{\beta}_1, \hat{\beta}_2)$	$X = [1, x_1, x_2]$.
...	...	...	
	...	...	
$\beta_0, \ldots, \beta_k$	$k + 1$	$\hat{\beta}^T X^T y$	
	$N - k - 1$	$R(\hat{\beta}, \ldots, \hat{\beta}_k)$	
	N	$y^T y$	In each case, total degrees of freedom and total sum of squares are N and $y^T y$, respectively.

Table 4.2   Analysis of variance (ANOVA) table, derived from Table 4.1, and showing the reduction in the sum of squares function $R(\hat{\beta})$ which results from fitting additional parameters (i.e, including additional explanatory variables).

	Degrees of freedom	Sum of squares
Fitting β_1	1	$R(\hat{\beta}_0) - R(\hat{\beta}_0, \hat{\beta}_1)$
Fitting β_2 in addition to β_1	1	$R(\hat{\beta}_0, \hat{\beta}_1) - R(\hat{\beta}_0, \hat{\beta}_1, \hat{\beta}_2)$
Fitting β_3 in addition to β_1, β_2	1	$R(\hat{\beta}_0, \hat{\beta}_1, \hat{\beta}_2) - R(\hat{\beta}_0, \hat{\beta}_1, \hat{\beta}_2, \hat{\beta}_3)$
$\cdots$	$\cdots$	$\cdots$
Fitting β_k in addition to $\beta_1, \ldots, \beta_{k-1}$	1	$R(\hat{\beta}_0, \hat{\beta}_1, \ldots, \hat{\beta}_{k-1}) - R(\hat{\beta}_0, \hat{\beta}_1, \ldots, \hat{\beta}_k)$
Error	$N - k - 1$	$R(\hat{\beta}_0, \ldots, \hat{\beta}_k)$
Total	$N - 1$	$\mathbf{y}^T\mathbf{y} - N\bar{y}^2$

about very little reduction in error sum of squares, beyond that resulting when β_1 is fitted. The F-test described in (iii) above formalises this heuristic argument.

Second, the presentation in Tables 4.1 and 4.2 has assumed that only one additional parameter is added at each step, so that the degrees of freedom for the RSS is reduced by one at each step. At times, however, it is more appropriate to add groups of parameters, instead of adding them one at a time. For example, when calculating the regression of mean annual floods from a set of drainage basins on explanatory variables representing their topographical, geological and climatic characteristics, it would be appropriate to start, perhaps, by regressing mean annual flood on the climatic variables; and then to include, as additional explanatory variables, those describing basin geology; then those describing topography. A second example occurs where, say, mean monthly flow q_t is to be represented by a harmonic expansion of the form

$$q_t = \alpha_0 + \beta_1 \cos(2\pi t/12) + \gamma_1 \sin(2\pi t/12)$$
$$+ \beta_2 \cos(4\pi t/12) + \gamma_2 \sin(4\pi t/12) + \varepsilon_t$$

In such a case, we should usually be interested in fitting successively the parameters α_0; $\alpha_0, \beta_1, \gamma_1$; $\alpha_0, \beta_1, \gamma_1, \beta_2, \gamma_2$ — thereby adding parameters two at a time.

Table 4.2 can easily be modified to include such cases. If we fitted β_0, followed by the group $\beta_1, \beta_2, \ldots, \beta_p$, followed again by the group $\beta_{p+1}, \beta_{p+2}, \ldots, \beta_{p+q}$, then Table 4.2 would be modified to the form shown in Table 4.3.

Table 4.3   Partitioning of sums of squares and degrees of freedom.

	Degrees of freedom	Sum of squares
Fitting $\beta_1, \beta_2, \ldots, \beta_p$	p	$R(\hat{\beta}_0) - R(\hat{\beta}_0, \hat{\beta}_1, \ldots, \hat{\beta}_p)$
Fitting $\beta_{p+1}, \ldots, \beta_{p+q}$ after fitting $\beta_1, \ldots, \beta_p$	q	$R(\hat{\beta}_0, \hat{\beta}_1, \ldots, \hat{\beta}_p) - R(\hat{\beta}_0, \hat{\beta}_1, \ldots, \hat{\beta}_{p+q})$
Error	$N - p - q$	$R(\hat{\beta}_0, \hat{\beta}_1, \ldots, \hat{\beta}_{p+q})$
Total	$N - 1$	$\mathbf{y}^T\mathbf{y} - N\bar{y}^2$

Table 4.4   ANOVA for the case of simple linear regression.

	Degrees of freedom	Sum of squares	Mean square
Regression on x (i.e. fitting β_1)	1	S_{xy}^2/S_{xx} (i.e. $R(\hat{\beta}_0) - R(\hat{\beta}_0, \hat{\beta}_1)$)	S_{xy}^2/S_{xx}
Error	$N - 2$	$S_{yy} - S_{xy}^2/S_{xx}$ (i.e. $R(\hat{\beta}_0, \hat{\beta}_1)$)	$(S_{yy} - S_{xy}^2/S_{xx})/(N - 2)$
Total	$N - 1$	S_{yy}	

4.4   SPECIAL CASE: SIMPLE LINEAR REGRESSION

In the case of one explanatory variable, say x, the ANOVA becomes as shown in Table 4.4 (where we have added an extra column showing 'mean squares': these are the 'sums of squares' divided by the corresponding degrees of freedom).

In Table 4.4 we have used the notation $S_{xx} = \Sigma(x_t - \bar{x})^2$, $S_{xy} = \Sigma(x_t - \bar{x})(y_t - \bar{y})$, and $S_{yy} = \Sigma(y_t - \bar{y})^2 = \mathbf{y}^T\mathbf{y} - N\bar{y}^2$. If the null hypothesis $\beta_1 = 0$ is correct, so that the variable x is of no value for explaining how y varies, the two quantities in the mean squares column are distributed as χ^2 with 1 and $N - 2$ degrees of freedom, respectively; and their ratio is then distributed as the F-statistic with 1 and $N - 2$ degrees of freedom. As a test of the value of including x as an explanatory variable in the model of y_t, we compare the numerical value of the F-statistic calculated from the data with the tabulated value of the F-statistic for 1 and $N - 2$ degrees of freedom.

Although the above algebra appears daunting on first encounter, the reader is advised to spend a little time mastering the principles leading to it, although the formulae themselves are not important for the practitioner. The calculations are easily undertaken by any good statistical computing program. We illustrate the procedure using GENSTAT, but other programs (SAS, GLIM) are also appropriate.

4.4.1   Example: Linear Regression between Annual Floods at Two Sites, Records of Different Lengths

Since there are fewer missing annual floods for the sequence for the Rio Hercílio at Apiuna than for the Rio Itajaí-Açú at Apiuna, we can explore whether the annual floods at Apiuna (x) can be used to fill in the missing values for the annual floods (y) at Ibirama. So we consider the linear regression of y on x, first plotting the points to get an idea of how x and y vary jointly. Thus, having read the data as variates *Ibirama* and *Apiuna*, we can use the command

```
>RESTRICT Ibirama,Apiuna; CONDITION=(Ibirama.GT.0).AND.(Apiuna.\
>GT.0)
>GRAPH Ibirama;Apiuna
```

to produce the plot shown in Figure 4.1. (The RESTRICT directive removes the missing values; without this, missing values are shown on the periphery of the diagram, which we prefer to avoid.) The two extreme floods for the years 1983 and 1984 both stand out clearly, and they both have a substantial effect on the analysis, as will be seen later.

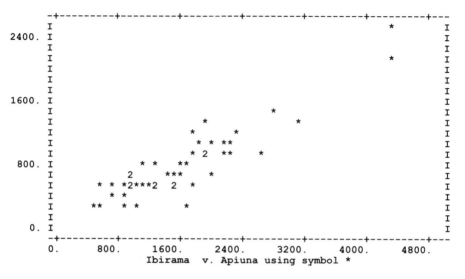

Figure 4.1   Plot of annual maximum mean daily flow for Rio Hercílio at Ibirama (vertical axis) against annual maximum mean daily flow for Rio Itajaí-Açú at Apiuna (horizontal axis). Both variables are measured in units of cubic metres per second.

The GENSTAT commands

>MODEL Ibirama : FIT Apiuna

result in the printing of the analysis shown in Figure 4.2. The 'Summary of analysis' shows the total sum of squares $\mathbf{y}^T\mathbf{y} - N\bar{y}^2 = 9\,025\,393$ divided into two components: one ('Regression') for the fitting of β_1, with value 7 265 785, and the other ('Residual') for $R(\hat{\beta}_0, \hat{\beta}_1)$, with value 1 759 608. The table shows the mean squares in the column headed 'm.s.', equal to the sums of squares ('s.s.') divided by degrees of freedom ('d.f.'). From this column we calculate the variance ratio, or F-statistic, as 7 265 785/39 102 = 185.82. To test whether the variable *Apiuna* is useful for predicting values of the variable *Ibirama* (or, more formally, to test the null hypothesis $\beta_1 = 0$) we compare this variance ratio (or F-statistic) with tabulated values of the variance ratio statistic with 1 and 45 degrees of freedom. Or we can use GENSTAT to calculate the quantiles of this statistic, by means of the command

>CALCULATE F=FED(0.95;1;45) : PRINT F

which gives the 95% quantile of the variance ratio distribution as $F = 4.057$. The 99% and 99.9% quantiles are found to be 7.234 and 12.39, respectively. Since the calculated variance ratio of 185.81 lies far into the upper tail of the distribution of the variance ratio, calculated on the supposition that the null hypothesis $\beta_1 = 0$ is true, we reject this hypothesis, and conclude (as is obvious from the plot in Figure 4.1) that the Apiuna flood sequence is extremely useful for predicting missing annual floods at Ibirama. The output showing 'Percentage variance accounted for' shows that the simple linear regression $E[Y] = \beta_0 + \beta_1 x$ had $R(\hat{\beta}_0, \hat{\beta}_1)$ equal to $(100 - 80.1) = 19.9\%$ of the total sum of squares 9 025 393, so that the regression has accounted for a considerable part of the variation in Y, the annual flood at Ibirama.

***** Regression Analysis *****

Response variate: Ibirama
    Fitted terms: Constant, Apiuna

*** Summary of analysis ***

	d.f.	s.s.	m.s.
Regression	1	7265785.	7265785.
Residual	45	1759608.	39102.
Total	46	9025393.	196204.

Percentage variance accounted for 80.1

* MESSAGE: The following units have large residuals:

17	−2.73
50	2.31

* MESSAGE: The following units have high leverage:

50	0.250
51	0.248

*** Estimates of regression coefficients ***

	estimate	s.e.	t
Constant	12.5	63.8	0.20
Apiuna	0.4775	0.0350	13.63

Figure 4.2 Output following GENSTAT computation of linear regression of annual maximum mean daily flow for Rio Hercílio at Ibirama (response variable) on annual maximum mean daily flow for Rio Itajaí-Açú at Apiuna (explanatory variable).

Lower down in the output, we see a section headed 'Estimates of regression coefficients'. The column headed 'estimate' in Figure 4.2 shows that the numerical estimates of β_0 and β_1 are 12.5 and 0.4775, giving a regression equation of $\hat{y} = 12.5 + 0.4775x$ (that is, 'estimated annual flood at Ibirama' $= 12.5 + 0.4775$ times 'observed annual flood at Apiuna'). The standard errors of the $\hat{\beta}_0$ and β_1 are also given, from which, if required, confidence intervals can be calculated for β_0 and β_1 as $12.5 \pm 63.8t$ and $0.4775 \pm 0.035t$, where t is the value of the t-statistic with 45 degrees of freedom, read from tables. Or, we can use GENSTAT to avoid using statistical tables by noting that the tabulated value of t for 45 degrees of freedom is the square root of the F-statistic, used above, for 1 and 45 degrees of freedom; hence the GENSTAT command

```
>CALCULATE t=SQRT(FED(0.95;1;46))   :   PRINT t
```

gives the value of the t-statistic that we require (namely that which defines upper and lower tails of the t-distribution with combined probability 5%). The above PRINT instruction shows that the value of t, with 45 degrees of freedom, defining a probability of 0.05 in the two upper tails of its distribution, is 2.014; hence the 95% confidence limits for β_0 and β_1 are $12.5 \pm 2.014 \times 63.8$ and $0.4775 \pm 2.014 \times 0.035$, or $(-116.0, 141.0)$ and $(0.407, 0.548)$, respectively. By substituting the probabilities 0.99 and 0.999 in place of 0.95 in the above GENSTAT command for the calculation of F, we find that the t-values necessary for the calculation of 99% and 99.9% confidence limits for β_0 and β_1 are 2.687 and 3.515, respectively. We note that the confidence limits for β_0 include the value of zero, which is to be expected.

The output shown in Figure 4.2 is the minimum produced by the GENSTAT directives MODEL and FIT. We can very easily extend this basic output to give, for the linear models discussed in this chapter, the fitted values $\hat{y}_t = 12.5 + 0.4775x$; the *standardised residuals* (essentially, the residuals $y_t - \hat{y}_t$ standardised to have the same variance, a desirable characteristic if we wish to make normal probability plots); the *leverages*, already mentioned in the output of Figure 4.2, and discussed more fully later in this chapter; and the variance-covariance matrix of the estimates $\hat{\beta}$. In addition to printing these various quantities, they can also be stored for future use, and this is particularly useful where, for example, we wish subsequently to plot the residuals in various ways. We illustrate the possibilities, continuing the example of the simple linear regression of Ibirama annual floods on Apiuna annual floods. Figure 4.3 shows an extension to the output of Figure 4.2, obtained using the GENSTAT directive RKEEP, which prints (and stores) the standardised residuals and fitted values. We noted in Chapter 2 that, when exploring whether a regression model is appropriate, we need to pay particular attention to the residuals; if we obtain a curvilinear plot when the residuals are plotted against the explanatory variable (*Apiuna* here), this would indicate that a term like (*Apiuna*)^2 should be included in the regression. If we find a pattern when the residuals are plotted in the order in which the data were recorded, this would suggest a non-stationarity in the relation between the Ibirama and Apiuna annual floods, with one or other — possibly both — varying with time. Again, if we plot the residuals against the fitted values $\hat{y}_t$, we may perhaps observe that the residuals tend to be larger, in absolute magnitude, for larger values of $\hat{y}_t$; this would suggest that the variance of residuals σ_ε^2 is not constant, as the model requires. Thus all of these plots (of residuals against explanatory variable, of residuals against time order, and of residuals against fitted values) serve the important purpose of diagnosing the correctness of the model. Figure 4.3 shows how these plots can be obtained using GENSTAT instructions of the kind

```
>GRAPH res;Apiuna   :   & res; Ordr   :   & res;fitval.
```

None of the three plots suggests any relationship between the residuals and *Apiuna*, *Ordr* or $\hat{y}_t$. To this extent, therefore, the structure of the linear regression model appears to be appropriate. But we recall the warning messages given in the basic output of Figure 4.2 to the effect that units 17 and 50 have large residuals, and that units 50 and 51 have high leverage; these are discussed in Section 4.7 below.

We recall that the purpose of fitting the statistical model was to estimate the missing annual floods at Ibirama, using the more complete record at Apiuna. The annual floods at Apiuna for which the Ibirama floods are missing, are 1111, 648 and 3086 m^3 s^{-1}; using the GENSTAT command

```
>PREDICT Apiuna;LEVELS=!(1111,648,3086)
```

output is produced which shows that estimates of the missing Ibirama floods are 543, 322 and 1486 m^3 s^{-1}, respectively. Using the values of the *t*-statistic calculated earlier, we can, if required, calculate limits for our predictions.

This example has considered the estimation of missing values, by linear regression, at one site only. Frequently, we have sequences with missing values at several sites, and it is necessary to estimate the missing values in each of the sequences. GENSTAT has a useful library procedure, MULTMISS, for this problem, in which the missing values are estimated iteratively until the differences between estimates in successive iterations are all acceptably small. We return to this topic in Chapter 7.

```
 22   RKEEP Ibirama;RESIDUALS=res;FITTEDVALUES=fitval
 23   ''
-24   ''
 25   PRINT [ORIENTATION=across] res,fitval
      res        2.1268     -0.5479      0.7869      1.3036      2.0739
      fitval     926.5       732.2       466.2       543.1       844.4
      res       -1.2071     -0.9683      0.6848     -1.5336     -0.0674
      fitval     505.8       450.9       433.2       948.5       248.9
      res        0.9954      0.7131      0.2783     -0.5450      0.4677
      fitval     282.8       623.8       537.8      1087.0       347.8
      res       -2.7286     -0.2982     -0.2763     -0.3073     -1.9301
      fitval     814.8       614.2       446.6       786.1      1268.4
      res        0.2761     -0.9963     -0.3542      0.3117     -0.6029
      fitval     915.0      1488.1       595.1       459.5       604.7
      res       -0.7548     -0.8726     -1.7292     -0.0088      0.3269
      fitval    1044.0       752.7       848.2       709.7       934.1
      res        0.6075      0.0890      0.1705     -0.0854      0.2978
      fitval     359.7       280.9       838.6       499.6       981.9
      res       -0.2974      0.6407      1.1861      0.3942      0.1859
      fitval    1067.9      1115.6       466.7      1330.5       764.6
      res       -0.2794     -0.2057      1.0001      0.6894     -1.3161
      fitval     795.7      1042.1       894.5       455.2       747.4
      res        2.3145      0.3059
      fitval    2078.8      2072.5
 26   GRAPH[NROWS=20;NCOLUMNS=70]res;Apiuna
  1
```

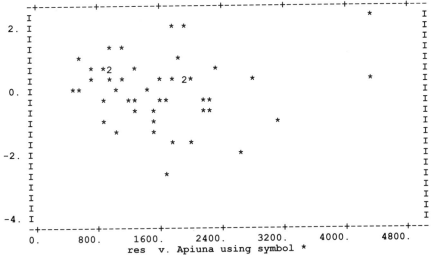

```
 27              &          res;Ordr
  1
```

Figure 4.3   Regression of annual flood for Rio Hercílio at Ibirama on annual flood for Rio Itajaí-Açú at Apiuna: listing of residuals *(res)* and fitted values *(fitval)*; and plots (a) of residuals against explanatory variable *(Apiuna)*; (b) of residuals against data order *(Ordr)*; (c) of residuals against fitted values *(fitval)*.

Figure 4.3   (*continued*)

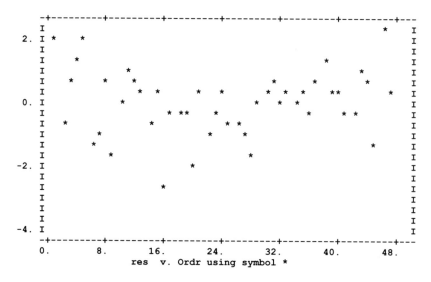

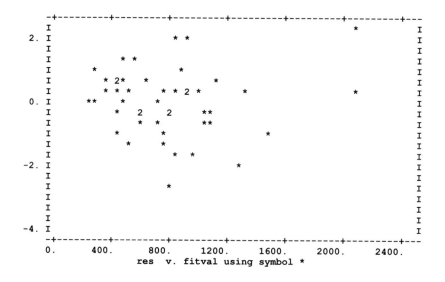

4.4.2   Example: Regression of Annual Runoff on Annual Rainfall: Plynlimon Experimental Catchments

The data in Figure 4.4 are annual totals, in millimetres, of runoff, rainfall and Penman's potential evapotranspiration (ET), for the headwater basins of the rivers Wye and Severn (areas 10.55 km^2 and 8.70 km^2, respectively) in the Welsh uplands. The Wye is upland pasture, and a large part (about 67.5%, including roads and channels within the forest) of the Severn basin is planted to coniferous forest; the purpose of the study, which has been widely disseminated,

Year	WyeQ	WyeP	WyeET	SevQ	SevP	SevET
1971	1509	1993	583	*	1948	583
1972	1758	2131	534	1567	2201	534
1973	2100	2606	551	1822	2504	551
1974	2222	2793	513	2075	2847	513
1975	1416	2101	550	1430	2122	550
1976	1344	1722	555	1158	1731	555
1977	2118	2522	506	1956	2720	506
1978	2017	2354	425	1932	2479	432
1979	2378	2748	472	2220	2797	491
1980	2259	2618	488	2081	2636	511
1981	2314	2822	498	2215	2763	493
1982	1961	2301	564	1914	2338	582
1983	2273	2693	530	2101	2650	500
1984	1768	2087	568	1602	2135	563
1985	2038	2295	492	1897	2355	461
1986	2371	2655	512	2194	2720	470
1987	2162	2358	464	1910	2355	461

Figure 4.4   Annual totals for water balance of the Institute of Hydrology's Wye and Severn experimental catchments. Reproduced by permission of the Institute of Hydrology. Symbols (all in millimetres):
$WyeQ$:   annual runoff, Wye catchment
$WyeP$:   annual precipitation, Wye catchment
$WyeET$:   annual Penman evapotranspiration, Wye catchment.
$SevQ$, $SevP$ and $SevET$ defined similarly.

is to determine whether runoff is reduced when British uplands are planted to forest. The 'Plynlimon study' (see Kirby, Newson and Gilman, 1991) has involved intensive research particularly of the hydrological processes which determine runoff in such areas; hence the data shown in Figure 4.4 are a very small part of a much larger data archive.

Before making a regression analysis, we begin by making some preliminary observations on the data. We note first that the annual total Penman ET is the same for both basins in each of the years 1971 to 1977, which is perhaps surprising in view of the different albedos (0.1 and 0.2) taken for the Severn and Wye, respectively; inspection of the monthly totals shows that these, too, are equal over this period. The mean values over the whole period of record, are (in millimetres):

	Wye	Severn	Difference
Rainfall	2399.9(17)	2429.5(17)	−29.6(17)
	±74.85	±75.82	±17.19
Runoff	2000.5(17)	1879.6(16)	+151.6(16)
	±77.78	±73.07	±17.87
ET	517.9	515.1	+2.8(17)
	±9.94	±10.53	±4.20
Difference $(\overline{P} - \overline{Q})$,	399.4(17)	549.9(16)	
rainfall minus annual runoff	±29.35	±26.05	

Over the period of a year, the water balance equation $P = Q + AE + \Delta S$ (where AE is actual evaporation and ΔS is change in storage) might be expected to reduce to $AE = P - Q$

apart from random errors, since ΔS should return to somewhere near its initial value. The annual 'loss' through evaporation and transpiration is therefore 150.5 ± 39.24 mm greater for the forested Severn than for the Wye pasture (more sophisticated analyses would take into account the fact that the higher parts of the Severn basin, approximately 23% of the total basin area, are also upland pasture).

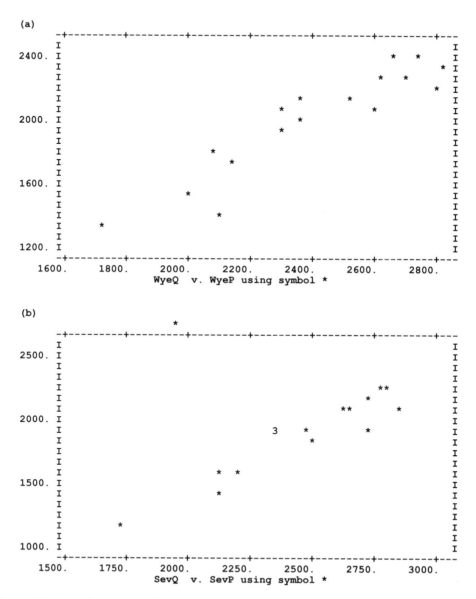

Figure 4.5   Plots of annual runoff against annual precipitation (a) for the Wye experimental catchment, 1971–87; (b) for the Severn experimental catchment, 1971–87. The asterisk above the border of the second plot is for the year 1971, when there were problems with flow measurement on the Severn (hence, a missing data point: see Figure 4.4).

So how does linear regression come into this? Suppose we plot annual runoff against annual rainfall for each basin; we should expect the fitted line $Q = \hat{\beta}_0 + \hat{\beta}_1 P$ to be comparable to the annual water balance relation $Q = P - AE$, so that $\hat{\beta}_1$ should be close to one, with $-\hat{\beta}_0$ giving an estimate of annual actual evaporation. Plotting Q against P for the Wye and Severn, we obtain the plots shown in Figure 4.5. Calculating the regressions by the GENSTAT instructions

> MODEL WyeQ : FIT WyeP

with similar instructions for the Severn, we find (see output in Figure 4.6) that the fitted lines for the two basins are:

$$WyeQ = -322 + 0.9675WyeP$$

$$\pm\, 237 \pm 0.0975$$

and

$$SevQ = -397 + 0.9254SevP$$

$$\pm\, 219 \pm 0.0885$$

The output of Figure 4.6 also shows two warning messages about the Wye regression, and one about the Severn regression. For both regressions, unit 6 (that is, the sixth line of data in the list shown in Figure 4.4) has large 'leverage'; at this stage, we have not yet discussed the meaning of leverage, but the nature of the message suggests that something is unusual about the data for this year. In fact, the year 1976 was a year of very low rainfall (relatively, since mid-Wales is wet) such that 1976 is known as a drought year, despite having an annual rainfall of about 1700 mm. Data from this year give the point shown in the bottom left-hand corner of Figure 4.5. So the regression analysis, by identifying data from 1976 as giving rise to high leverage, has pointed out that there is something unusual about the data for this year. This is one of the important uses of leverage in regression analysis: namely to identify data which contribute in some special way to the calculated regression, and which may require closer inspection to determine why they are so unusual.

But what about the message concerning unit 5, in the Wye regression analysis? We are told that this point has a large standardised residual of -2.52, and the unit 5 gives data for the year 1975. We note that the residual is negative in sign, so that the annual runoff y_t lies substantially below the fitted value $\hat{y}_t$; do the data provide any clue as to why this should be so? The answer is yes; if we go back to the data of *monthly* precipitation and runoff for the Wye, we find that in May 1975 problems were encountered with the measurement of flow, probably because of siltation within the flume by which discharge is measured. In consequence, flow for the month of May 1975 is not known with certainty, and was set equal to zero in the monthly data sequence (with an appropriate note stating why). Hence the annual runoff from the Wye for 1975 is underestimated. So the regression analysis identified that the standardised residual for 1975 was unusually large in the Wye regression analysis, and reference to the original data suggest why this is so. The point illustrated by the analysis is that careful study of the residuals and leverages will often lead us to identify data which, for whatever reason, contribute in some unusual way to the regression: possibly, but not necessarily, because of errors in the data.

We return now to the interpretation of the estimates $\hat{\beta}_0$, $\hat{\beta}_1$ in the two regression analyses. We note that both slopes $\hat{\beta}_1$ are less than one, although t-tests with 15 and 14 degrees of

***** Regression Analysis *****

Response variate: WyeQ
        Fitted terms: Constant, WyeP

*** Summary of analysis ***

	d.f.	s.s.	m.s.
Regression	1	1515722.	1515722.
Residual	15	232572.	15505.
Total	16	1748294.	109268.

Percentage variance accounted for 85.8

* MESSAGE: The following units have large residuals:
              5   −2.52

* MESSAGE: The following units have high leverage:
              6   0.34

*** Estimates of regression coefficients ***

	estimate	s.e.	t
Constant	−322.	237.	−1.36
WyeP	0.9675	0.0979	9.89

WyeQ	WyeFv	WyeRes	WyeLev
1509	1607	−0.8570	0.1611
1758	1740	0.1505	0.1035
2100	2200	−0.8383	0.0850
2222	2381	−1.3865	0.1542
1416	1711	−2.5189	0.1140
1344	1345	−0.0053	0.3427
2118	2119	−0.0047	0.0680
2017	1956	0.5051	0.0601
2378	2337	0.3518	0.1336
2259	2211	0.3999	0.0882
2314	2409	−0.8354	0.1688
1961	1905	0.4672	0.0649
2273	2284	−0.0939	0.1119
1768	1698	0.6017	0.1193
2038	1899	1.1554	0.0656
2371	2247	1.0470	0.0990
2162	1960	1.6740	0.0599

***** Regression Analysis *****

Response variate: SevQ
        Fitted terms: Constant, SevP

*** Summary of analysis ***

	d.f.	s.s.	m.s.
Regression	1	1211870.	1211870.
Residual	14	155042.	11074.
Total	15	1366912.	91127.

Percentage variance accounted for 87.8

Figure 4.6   Regression analyses of annual runoff on annual precipitation, Institute of Hydrology's Wye and Severn experimental catchments. See data of Figure 4.4.

Continued

_____ Figure 4.6 *(continued)* _____

```
    * MESSAGE: The following units have high leverage:
                        6              0.44

    *** Estimates of regression coefficients ***

                   estimate          s.e.              t
    Constant       −397.             219.            −1.81
    SevP             0.9254            0.0885         10.46
```

SevQ	SevFv	SevRes	SevLev
*	1406	*	0.0000
1567	1640	−0.7387	0.1097
1822	1921	−0.9699	0.0639
2075	2238	−1.7005	0.1686
1430	1567	−1.4087	0.1430
1158	1205	−0.6005	0.4376
1956	2121	−1.6588	0.1104
1932	1898	0.3375	0.0628
2220	2192	0.2884	0.1430
2081	2043	0.3783	0.0845
2215	2160	0.5551	0.1276
1914	1767	1.4495	0.0729
2101	2056	0.4492	0.0881
1602	1579	0.2325	0.1369
1897	1783	1.1248	0.0702
2194	2121	0.7391	0.1104
1910	1783	1.2529	0.0702

freedom for the Wye and Severn respectively (remembering that one year's runoff from the Severn is missing) shows that they are not significantly so. The differences from one — namely $1 - 0.9675 = 0.0325 \pm 0.0975$, and 0.0746 ± 0.0885 — are interesting for the following reason. In linear regression analysis, the explanatory variables x_i are not random variables and are taken to be fixed quantities free from measurement error. (In 'time-series' models of the autoregressive type $q_t - \mu = \phi(q_{t-1} - \mu) + \varepsilon_t$, on the other hand, all three quantities q_t, q_{t-1}, and ε_t are regarded as random variables). In the present example, the explanatory variable is total annual rainfall, measured using networks of rain gauges (in the Wye, 21 ground-level gauges read monthly, and three recording rain gauges; in the Severn, 11 gauges at canopy level in the forest, with seven in the unforested part of the upper Severn). Annual rainfall for both basins may therefore be expected to be subject to random errors of measurement. Furthermore, these errors would be expected to be larger for the Severn than for the Wye, due to the greater difficulty of measuring rainfall above the forest canopy.

The effect of random errors in the explanatory variable is to reduce the estimated regression coefficient, which becomes negatively biased (see, for example, discussions in Draper and Smith, 1981, on this effect). The regression coefficients calculated for the Wye and Severn are both slightly less than their expected value of unity, the differences being measures of the biases introduced by errors in rainfall measurement; and, as might be expected, the bias is larger for the Severn than for the Wye although neither difference is large enough to cause concern. It is interesting that, in another experimental basin, Coal Burn in the North of England, the data for which are also reported in the Institute of Hydrology Report 106 (Roberts, 1989), the regression of annual runoff on annual rainfall has a regression coefficient $\hat{\beta}_1$ of 0.796 ± 0.138, which is less than unity, although not significantly so; in Coal Burn, annual rainfall is estimated from four gauges in a total gauged area of 152 ha; these four gauges are in the valley bottom and near the basin outfall. One would expect that the errors in estimating mean areal rainfall from such a network would be larger, and this is confirmed by the value of the regression coefficient β_1 when annual runoff Q for Coal Burn is regressed upon annual rainfall P, $E[Q] = \hat{\beta}_0 + \hat{\beta}_1 P$.

Returning to the regression output, why are the standard errors of the estimates $\hat{\beta}_0$ ($\hat{\beta}_0 = -397 \pm 219$ mm) for the Wye and Severn so different from the values of $\overline{P} - \overline{Q}$ (399.4 $\pm$ 29.35 mm), shown in the table on p. 101? We recall that the regression of annual runoff on annual rainfall should be very like the water balance equation calculated over intervals of a year, namely $Q = P - E$, so we should expect the estimates of β_0, and their standard errors, to be very like the values of the 'losses' shown in the table above. However, the parameters β_0 are intercepts on the axis of Q, calculated for the value of $P = 0$; and the standard error of $\hat{\beta}_0$ can be shown to be

$$\pm\sqrt{s_\varepsilon^2[1/n + \overline{x}^2/\sum(x_t - \overline{x})^2]}$$

where s_ε^2 is the variance of residuals about the calculated regression line (so that it estimates σ_ε^2). The difference $\overline{P} - \overline{Q}$ shown in the above table (see p. 101 is calculated not near $P = 0$ but near $P = \overline{P}$, and its standard error also includes a contribution from the year-to-year variation in annual rainfall; we recall also that the linear regression assumes that the annual rainfalls P_t are fixed quantities and not subject to random variation.

4.4.3   Example: Comparison of Linear Regressions

Where regressions have been fitted to several sets of data, it is commonly required to test whether the fitted regressions differ significantly amongst themselves; and, if it is found that they do not, to combine them into a single regression equation with estimates of the regression coefficients β_i having better precision. The calculation is essentially an application of the general technique described in Table 4.1. Suppose we have k groups of data, the groups consisting of $n_1, n_2, \ldots, n_k$ observations on both the response variable y and the explanatory variables $x_1, x_2, \ldots, x_p$. We begin by fitting the model $y = \beta_0 + \varepsilon$ (a model with the parameters β_0 and σ_ε^2 common to all groups). Then we introduce more parameters by fitting the model $y = \beta_0 + \beta_1 x_1 + \cdots + \beta_p x_p + \varepsilon$ (having the $p + 1$ parameters $\beta_0, \beta_1, \ldots, \beta_p$ and the variance σ_ε^2 common to all k groups). Then we introduce still more parameters by fitting the model $y = \beta_0 + \beta_1^i x_1 + \beta_2^i x_2 + \cdots + \beta_p^i x_p + \varepsilon$ for $i = 1, 2, \ldots, k$ (in which the 'slope' parameters $\beta_1, \ldots, \beta_p$ differ in the k data sets). And finally, we fit the model $y = \beta_0^i + \beta_1^i x_1 + \cdots + \beta_p^i x_p + \varepsilon$ for $i = 1, 2, \ldots, k$ (in which both the intercepts β_0 and the slope parameters β_j, $j = 1, \ldots, p$, differ from data set to data set). At each stage in this sequence of model fittings, the residual sum of squares given by $R(\hat{\beta})$ in equation (4.13) is reduced; and we determine at which point the reduction becomes sufficiently small that it is 'not significant'. In practice, the calculation proceeds in two steps: in the first, we test whether the 'slopes' $\beta_1^i, \beta_2^i, \ldots, \beta_p^i$ (with $i = 1, 2, \ldots, k$) differ significantly from data set to data set, and, if they do not, we pool them to give common estimates $\tilde{\beta}_1, \tilde{\beta}_2, \ldots, \tilde{\beta}_p$. If, however, the slope coefficients do differ significantly from data set to data set, the analysis proceeds no further, since it would be inappropriate to use pooled estimates of the slope parameters.

Where it is valid to calculate the pooled estimates $\tilde{\beta}_1, \tilde{\beta}_2, \ldots, \tilde{\beta}_p$, we then proceed to the second step, in which we test whether the intercepts $\hat{\beta}_0^1, \hat{\beta}_0^2, \ldots, \hat{\beta}_0^k$ differ significantly from data set to data set. If they do not, we calculate a pooled intercept $\tilde{\beta}_0$; in this case, we have now derived estimates of all the parameters $\beta_0, \beta_1, \ldots, \beta_p$, pooled over all k data sets. If, however, we have found that the intercepts $\hat{\beta}_0^1, \hat{\beta}_0^2, \ldots, \hat{\beta}_0^k$ do differ significantly we have k parallel regressions, with slope parameters pooled over the k data sets.

We illustrate the calculation using annual runoff Q and annual rainfall P for three subcatchments within the Wye and Severn basins. The data are shown in Figure 4.7. For the four subcatchments of the Severn, the four linear regressions (see output in Figure 4.8) of annual flow on annual rainfall are

$$SevQ = -409 + 0.9425 SevP$$
$$\pm 181 \pm 0.0739 \quad (r^2 = 0.942)$$
$$HoreQ = -246 + 0.8485 HoreP$$
$$\pm 227 \pm 0.0982 \quad (r^2 = 0.897)$$
$$TanQ = -54 + 0.8060 TanP$$
$$\pm 278 \pm 0.1100 \quad (r^2 = 0.841)$$
$$HafQ = -39 + 0.8141 HafP$$
$$\pm 220 \pm 0.0891 \quad (r^2 = 0.902)$$

Year	WyeP	CyffP	IagoP	WyeQ	CyffQ	IagoQ		
1975	2101	2160	2134	1687	1693	1740		
1976	1722	1744	1750	1344	1294	1345		
1977	2531	2539	2593	2118	2069	2172		
1978	2349	2358	2402	2008	2044	2086		
1979	2745	2752	2824	2378	2409	2481		
1980	2559	2557	2637	2259	2270	2356		
1981	2768	2799	2810	2418	2427	2534		
1982	2286	2239	2376	1961	1961	2050		
1983	2671	2698	2705	2273	2262	2327		
1984	2087	2069	2136	1767	1745	1765		
1985	2295	2323	2304	2037	2015	2060		
Year	SevP	HoreP	TanP	HafP	SevQ	HoreQ	TanQ	HafQ
1975	2122	2217	2233	*	1454	1549	1619	*
1976	1731	1785	1745	1732	1217	1220	1262	1265
1977	2720	2838	2735	2712	2040	2035	2093	2103
1978	2452	2555	2560	2442	1931	1863	1900	1906
1979	2797	2841	2851	2841	2219	2201	2185	2210
1980	2635	2692	2763	2641	2081	2047	2114	2097
1981	2776	2826	2938	2775	2215	2172	2266	2183
1982	2338	2333	2461	2335	1914	1881	2045	2010
1983	2650	2765	2751	2563	2101	2073	2277	2119
1984	2135	2178	2147	2102	1602	1566	1782	1655
1985	2355	2424	2451	2319	1897	1987	2149	1975

Figure 4.7 Annual precipitation and annual runoff, 1975–85, for three subcatchments of the Wye (Wye, area 3.98 km²; Cyff, area 3.13 km²; Iago, area 1.02 km²); and four subcatchments of the Severn (Severn, area 1.16 km²; Hore, area 3.08 km²; Tanllwyth, area 0.89 km²; Hafren, area 3.67 km²).

```
***** Regression Analysis *****

Response variate:    SevQ
    Fitted terms:    Constant, SevP

*** Summary of analysis ***

                 d.f.        s.s.          m.s.
Regression        1        989536.       989536.
Residual          9         54820.         6091.
Total            10       1044356.       104436.

Percentage variance accounted for 94.2

* MESSAGE: The following units have high leverage:
                        2            0.53

*** Estimates of regression coefficients ***

                 estimate          s.e.            t
Constant          -409.            181.          -2.26
SevP                0.9425           0.0739       12.75

***** Regression Analysis *****

Response variate:    HoreQ
    Fitted terms:    Constant, HoreP

*** Summary of analysis ***

                 d.f.        s.s.          m.s.
Regression        1        846263.       846263.
Residual          9         86028.         9559.
Total            10        932292.        93229.

Percentage variance accounted for 89.7

* MESSAGE: The following units have high leverage:
                        2            0.52

*** Estimates of regression coefficients ***

                 estimate          s.e.            t
Constant          -246.            227.          -1.08
HoreP               0.8485           0.0902        9.41

***** Regression Analysis *****

Response variate:    TanQ
    Fitted terms:    Constant, TanP

*** Summary of analysis ***

                 d.f.        s.s.          m.s.
Regression        1        828388.       828388.
Residual          9        137898.        15322.
Total            10        966286.        96629.

Percentage variance accounted for 84.1
```

Figure 4.8   Regression analyses of annual runoff on annual precipitation for four subcatchments of the
Severn (see data in Figure 4.7) and for three subcatchments of the Wye.

_____ Continued _____|

_____ Figure 4.8 (*continued*) _____

∗ MESSAGE: The following units have high leverage:
                         2   0.55

∗∗∗ Estimates of regression coefficients ∗∗∗

	estimate	s.e.	t
Constant	−54.	278.	−0.19
TanP	0.806	0.110	7.35

∗∗∗∗∗ Regression Analysis ∗∗∗∗∗

Response variate: HafQ
   Fitted terms: Constant, HafP

∗∗∗ Summary of analysis ∗∗∗

	d.f.	s.s.	m.s.
Regression	1	691514.	691514.
Residual	8	66313.	8289.
Total	9	757826.	84203.

Percentage variance accounted for 90.2

∗ MESSAGE: The following units have high leverage:
                         2   0.59

∗∗∗ Estimates of regression coefficients ∗∗∗

	estimate	s.e.	t
Constant	−39.	220.	−0.18
HafP	0.8141	0.0891	9.13

∗∗∗∗∗ Regression Analysis ∗∗∗∗∗

Response variate: WyeQ
   Fitted terms: Constant, WyeP

∗∗∗ Summary of analysis ∗∗∗

	d.f.	s.s.	m.s.
Regression	1	1029028.	1029028.
Residual	9	23980.	2664.
Total	10	1053008.	105301.

Percentage variance accounted for 97.5

∗ MESSAGE: The following units have high leverage:
                         2   0.50

∗∗∗ Estimates of regression coefficients ∗∗∗

	estimate	s.e.	t
Constant	−343.	121.	−2.82
WyeP	0.9964	0.0507	19.65

∗∗∗∗∗ Regression Analysis ∗∗∗∗∗

Response variate: CyffQ
   Fitted terms: Constant, CyffP

∗∗∗ Summary of analysis ∗∗∗

	d.f.	s.s.	m.s.
Regression	1	1099454.	1099454.
Residual	9	54605.	6067.
Total	10	1154060.	115406.

Percentage variance accounted for 94.7

Figure 4.8 *(continued)*

```
* MESSAGE: The following units have high leverage:
                    2              0.48
* * * Estimates of regression coefficients * * *

                estimate            s.e.              t
Constant        -430.               183.            -2.35
CyffP              1.0259             0.0762         13.46

* * * * * Regression Analysis * * * * *

Response variate:   lagoQ
    Fitted terms:   Constant, lagoP

* * * Summary of analysis * * *

                d.f.        s.s.              m.s.
Regression       1       1239301.          1239301.
Residual         9         29533.             3281.
Total           10       1268834.           126883.

Percentage variance accounted for 97.4

* MESSAGE: The following units have high leverage:
                    2              0.50

* * * Estimates of regression coefficients * * *

                estimate            s.e.              t
Constant        -490.               134.            -3.67
lagoP              1.0612             0.0546         19.43
```

with all regressions calculated from 11 years of annual totals except for the Hafren (ten years). For all four regressions, we receive warning messages about the high leverages of unit 2, corresponding to the data (see Figure 4.7) for 1976 which, as we have seen, was a very atypical year for rainfall.

To formalise the calculation, denote sums of squares and products for x and y by S_{xx}, S_{xy} and S_{yy}: thus $S_{xx} = \Sigma(x_t - \bar{x})^2$, $S_{xy} = \Sigma(x_t - \bar{x})(y_t - \bar{y})$. We then have values for these quantities from each of the k data sets, where $k = 3$ for the Wye since it has three subcatchments, and $k = 4$ for the Severn. Denote the sums of squares and products for the k data sets by

$$
\begin{array}{ccc}
S^1_{xx} & S^1_{xy} & S^1_{yy} \\
S^2_{xx} & S^2_{xy} & S^2_{yy} \\
\cdots & \cdots & \cdots \\
S^k_{xx} & S^k_{xy} & S^k_{yy}
\end{array}
$$

with totals over the three columns

$$
\sum S^i_{xx}, \quad \sum S^i_{xy}, \sum S^i_{yy}.
$$

We shall also require to calculate the quantities

$$
A = \sum S_{yy},
$$

$$
B = \left(\sum S_{xy}\right)^2 / \left(\sum S_{xx}\right),
$$

$$
C = \sum \left(S^2_{xy}/S_{xx}\right)
$$

For the four Severn subcatchments, we have:

	S_{xx}	S_{xy}	S_{yy}
Severn	101 269.11	95 446.04	94 941.42
Hore	106 853.24	90 667.21	84 753.78
Tanllwyth	115 838.92	93 400.18	87 844.18
Hafren	104 333.36	84 939.94	75 782.61
Sum	428 294.63	364 453.37	343 321.99

To test whether the slope coefficients $\hat{\beta}_1^1, \hat{\beta}_1^2, \ldots, \hat{\beta}_1^k$ differ significantly amongst k data sets, calculate the following analysis of variance:

	df	SS	MS	F
Pooled slope	1	B	B	
Differences	$k - 1$	$C - B$	$(C - B)/(k - 1)$	
between slopes				
Pooled error	$\Sigma n_i - 2k$	$A - C$	$(A - C)/(\Sigma n_i - 2k)$	

where A, B and C are as defined above. If, at the appropriate significance level, the F-test of the ratio of mean squares (differences between slopes)/(pooled error) is *less* than the tabulated value of F for $k - 1$ and $\Sigma n_i - 2k$ degrees of freedom, we pool the k slope coefficients $\hat{\beta}_1^1, \hat{\beta}_1^2, \ldots, \hat{\beta}_1^k$ to give $\tilde{\beta}_1 = \Sigma S_{xy}/\Sigma S_{xx}$. If, on the other hand, the ratio of mean squares is *greater* than the tabulated value of F, we conclude that the k regressions differ and that no pooling is possible.

For the data from the four Severn subcatchments, the above ANOVA becomes:

	df	SS	MS	F
Pooled slope	1	310 128.238	310 128.238	339 ∗ ∗ ∗
Differences	3	1 221.914	407.305	<1, NS
between slopes				
Pooled error	35	31 971.838	913.481	
Total	39	343 321.99		

The ANOVA shows no significant evidence that the regression slopes for the four Severn subcatchments differ significantly, the pooled slope being therefore $\Sigma S_{xy}/\Sigma S_{xx} =$ 364 453.37/428 294.63 = 0.8509, with standard error $\sqrt{(913.481/428\,294.63)} = \pm 0.0462$. The pooled slope is significantly less than one, the value expected from water balance considerations. Whilst errors in rainfall measurement appear to influence annual totals, there is no evidence that these errors are greater in some subcatchments than in others. It is of interest to compare the Severn regressions with those from the three Wye subcatchments, for which the fitted lines are

$$WyeQ = -343 + 0.9964WyeP$$

$$\pm 121 \quad \pm 0.0507 \quad (r^2 = 0.975)$$

$$CyffQ = -430 + 1.0259CyffP$$

$$\pm 183 \quad \pm 0.0762 \quad (r^2 = 0.947)$$

$$IagoQ = -490 + 1.0612IagoP$$

$$\pm 134 \quad \pm 0.0546 \quad (r^2 = 0.974)$$

for which the slopes appear consistent with absence of errors in rainfall measurement, and the intercepts, with signs changed, are estimates of annual evaporation losses; the pooled slope is, in fact, 1.0285 ± 0.0355.

Supposing that the above analysis has led to the pooling of slope coefficients to give $\tilde{\beta}_1$, we are now in a position to test whether the intercepts $\hat{\beta}_0^1, \hat{\beta}_0^2, \ldots, \hat{\beta}_0^k$ differ significantly. For this we require the sums of squares and products calculated over all Σn_i data values. Denote these by T_{xx}, T_{xy} and T_{yy}, so that, for example, $T_{xx} = \Sigma(x_{ij} - \bar{x})^2$ where $\bar{x} = \Sigma x_{ij}/\Sigma n_i$, $(i = 1, 2, \ldots, k; \; j = 1, 2, \ldots, n_i)$, and the summation in T_{xx} is over both i and j. We form the analysis of variance:

	df	SS	MS	F
Overall regression	1	T_{xy}^2/T_{xx}		
Differences between slopes	$k - 1$	$C - B$		
Differences between intercepts	$k - 1$	by difference $(= D)$		
Pooled error	$\Sigma n_i - 2k$	$A - C$		
Total	$\Sigma n_i - 1$	T_{yy}		

In this analysis, it is most convenient to find the sum of squares for 'Differences between intercepts' as

$$D = T_{yy} - (A - C) - (C - B) - T_{xy}^2/T_{xx}$$

If, at the appropriate level of significance, the ratio (mean square for differences between intercepts/mean square for pooled error) is less than the tabulated F-statistic for $k - 1$ and $\Sigma n_i - 2k$ degrees of freedom, a pooled intercept, $\tilde{\beta}_0 = \bar{y} - \tilde{\beta}_1\bar{x}$, can be calculated from all the k data sets. If the ratio is greater than the tabulated F, we must use the k distinct regressions

$$y = \hat{\beta}_0^1 + \tilde{\beta}_1 x$$

$$y = \hat{\beta}_0^2 + \tilde{\beta}_1 x$$

$$\ldots \qquad \ldots$$

$$y = \hat{\beta}_0^k + \tilde{\beta}_1 x$$

having common slope $\tilde{\beta}_1$ but different intercepts $\hat{\beta}_0^1, \hat{\beta}_0^2, \ldots, \hat{\beta}_0^k$.

The ANOVAs given above are, in fact, special cases of the more general forms given in Tables 4.1 and 4.2. The ANOVA for testing whether the k slope parameters β_1^i differ significantly can be cast in the following form, analogous to Table 4.1:

	Degrees of freedom
Fitting β_0^i	k
Fitting β_0^i and β_1	$k + 1$
Fitting β_0^i and β_1^i	$2k$
$R(\hat{\beta}_0^i, \hat{\beta}_1^i)$	$\Sigma n_i - 2k$
Total	Σn_i

which is equivalent to:

	Degrees of freedom
Fitting β_1, after fitting β_0^i	1
Fitting β_1^i, after fitting β_1 and β_0^i	$k - 1$
$R(\hat{\beta}_0^i, \hat{\beta}_1^i)$	$\Sigma n_i - 2k$
Total	$\Sigma n_i - k$

A similar formulation can be constructed for testing the differences between the intercepts.

The above example has dealt with the case in which there is just one explanatory variable x. More generally, we may need to consider the case of p explanatory variables $x_1, x_2, \ldots, x_p$. The ANOVA for testing whether the p 'slope' coefficients differ significantly amongst the k data sets becomes:

	df	SS	MS	F
Pooled $\beta_1, \beta_2, \ldots, \beta_p$	p	A		
Differences between $\beta_1, \beta_2, \ldots, \beta_p$	$p(k - 1)$	B		
Pooled error	$\Sigma n_i - kp - k$	C		
Total	$\Sigma n_i - k$	D		

where A, B, C and D are calculated as follows.

(i) Calculate the $(p + 1) \times (p + 1)$ symmetric matrix of sums of squares (SS) and products (SP) for each of the k data sets, that for the ith data set being

$$\begin{bmatrix} S_{00}^i & S_{01}^i & S_{02}^i & \cdots & S_{0p}^i \\ S_{10}^i & S_{11}^i & S_{12}^i & \cdots & S_{1p}^i \\ \cdots & \cdots & \cdots & \cdots & \cdots \\ S_{p0}^i & S_{p1}^i & S_{p2}^i & \cdots & S_{pp}^i \end{bmatrix}$$

where S_{00}^i is the SS for y; S_{01}^i is the SP for y with x_1; and so on. Denote the $p \times p$ matrix, obtained by omitting the first row and column, by $\mathbf{U}^i$. Denote the $p \times 1$ vector $[S_{10}^i, S_{20}^i, \ldots, S_{p0}^i]^{\mathrm{T}}$ by $\mathbf{Z}^i$.

(ii) Calculate the $p \times p$ matrix $\mathbf{W} = \Sigma \mathbf{U}^i$ and the $p \times 1$ vector $\mathbf{V} = \Sigma \mathbf{Z}^i$, so that

$$\mathbf{W} = \begin{bmatrix} \sum S_{11}^i & \sum S_{12}^i & \sum S_{13}^i & \cdots & \sum S_{1p}^i \\ \sum S_{21}^i & \sum S_{22}^i & \sum S_{23}^i & \cdots & \sum S_{2p}^i \\ \cdots & & \cdots & \cdots & \\ \sum S_{p1}^i & \sum S_{p2}^i & \sum S_{p3}^i & \cdots & \sum S_{pp}^i \end{bmatrix} \qquad \mathbf{V} = \begin{bmatrix} \sum S_{01}^i \\ \sum S_{02}^i \\ \cdots \\ \sum S_{0p}^i \end{bmatrix}$$

(iii) Calculate $A = \mathbf{V}^{\mathrm{T}} \mathbf{W}^{-1} \mathbf{V}$, and enter it in the above table.

(iv) Calculate $B = \Sigma (\mathbf{Z}^i)^{\mathrm{T}} (\mathbf{U}^i)^{-1} \mathbf{Z}^i - A$, and enter it in the above table.

(v) Calculate $D = \Sigma S_{00}^i$, and enter it in the above table; hence obtain C by subtraction.

(vi) Compare the ratio of mean squares, (differences between $\beta_1, \beta_2, \ldots, \beta_p$)/(pooled error), with the tabulated F-statistic, to determine whether a pooling of the $\hat{\beta}_1^i, \hat{\beta}_2^i, \ldots, \hat{\beta}_p^i$ is permissible; if it is, the pooled estimates are $[\tilde{\beta}_1, \tilde{\beta}_2, \ldots, \tilde{\beta}_p]^{\mathrm{T}} = \mathbf{W}^{-1} \mathbf{V}$. If it is not, the k multiple regressions must remain distinct.

(vii) If the pooling leading to $\tilde{\beta}_j (j = 1, 2 \ldots, p)$ is permissible, proceed to determine whether the k intercepts $\hat{\beta}_0^1, \hat{\beta}_0^2, \ldots, \hat{\beta}_0^k$ differ significantly. Calculate the $p \times p$ matrix $\mathbf{W}^*$ and the $p \times 1$ vector $\mathbf{V}^*$, analogous to $\mathbf{W}$ and $\mathbf{V}$, but calculated using all the Σn_i data, from all the k data sets combined. Calculate also the total SS for all observations of the response variable y, over all k data sets; denote this by T_{00}.

(viii) Construct the ANOVA table:

	df	SS	MS	F
Overall regression	p	$(\mathbf{V}^*)^{\mathrm{T}} (\mathbf{W}^*)^{-1} \mathbf{V}^*$		
Differences between β_0^i	$k - 1$	(obtained by subtraction)		
Differences between 'slope' coefficients	$p(k - 1)$	B		
Pooled error	$\Sigma n_i - pk - k$	C		
Total	$\Sigma n_i - 1$	T_{00}		

(ix) Use the ratio of mean squares, (differences between β_0^i)/(pooled error), to determine whether a pooled estimate of a common intercept is permissible.

An important point concerning the above analyses is that they assume a constant error variance σ_ε^2 in all k data sets. The lines for 'Pooled error' in the ANOVA tables give mean squares which estimate this common variance; the analyses may be misleading where the k individual error variances are highly heterogeneous.

4.4.4 Comparison of Regressions Using the Dummy-variable Procedure

The calculations set out above appear complicated. However, statistical packages such as GENSTAT provide a simple means of testing for parallelism of regressions, and for differences between intercepts, by use of the *dummy-variable* procedure. To illustrate, suppose we wish to compare the regressions of annual runoff Q on annual rainfall P for the three Severn subcatchments Hore, Tanllwyth and Hafren: first testing whether the regressions are parallel, and then testing for differences between intercepts. With three subcatchments we form two dummy variables, z_1 and z_2, say, having values 0 or 1. The values of z_1 and z_2 are both 0 for the Hore; they are 1 and 0, respectively, for the Tanllwyth; and have values 0 and 1 for the Hafren. We also form two other variables, which we will denote Pz_1, calculated as the product of P with z_1 where P is annual precipitation, and $Pz_2 = P \times z_2$. We consider the multiple regression

$$Q = \beta_0 + \beta_1 P + \gamma_1 z_1 + \gamma_2 z_2 + \delta_1 P z_1 + \delta_2 P z_2 + \varepsilon \qquad (4.14)$$

Consider the form of this regression for the three subcatchments when the δs are both non-zero. For the Hore, Tanllwyth and Hafren we have, respectively,

$$Q = \beta_0 + \beta_1 P + \varepsilon$$
$$Q = (\beta_0 + \gamma_1) + (\beta_1 + \delta_1)P + \varepsilon \qquad (4.15)$$
$$Q = (\beta_0 + \gamma_2) + (\beta_1 + \delta_2)P + \varepsilon$$

so that all three slopes and all three intercepts are different. We therefore see that fitting the regression equation (4.14) and testing the hypothesis that $\delta_1 = \delta_2 = 0$ is equivalent to testing whether estimates of the three slopes β_1, $\beta_1 + \delta_1$ and $\beta_1 + \delta_2$ differ significantly. If, having accepted this hypothesis, we then test the more restrictive hypothesis $\gamma_1 = \gamma_2 = \delta_1 = \delta_2 = 0$, we see from equations (4.15) that we are testing for the coincidence of the three regressions. Having read the 32 pairs (Q, P) for the three Severn subcatchments, and formed the variables z_1, z_2, Pz_1 and Pz_2, the relevant GENSTAT instructions for this calculation are as follows:

```
>MODEL Q
>TERMS P,z1,z2,Pz1,Pz2
>FIT [PRINT=model,summary,accumulated,estimates, correlations,\
>monitoring] P,z1,z2,Pz1,Pz2
>DROP [PRINT=model,summary,accumulated,estimates,correlations,\
>monitoring] Pz1,Pz2
>DROP [PRINT=model,summary,accumulated,estimates,correlations,\
>monitoring] z1,z2
```

The resulting GENSTAT output is given in Figure 4.9. We note that, once again, there are warning messages of high leverages for units 2, 13 and 24 which are the units with 1976 data from the Hore, Tannllwyth and Hafren subcatchments. Abstracting the essential part from the full listing, we obtain:

	df	SS	MS	F
Fitting β_1	1	2 343 625	2 343 625	
Additional for fitting γ_1, γ_2	2	82 411	41 205	3.69*
Fitting β_1, γ_1, γ_2	3	2 426 036		
Additional for fitting δ_1, δ_2	2	1 203	602	NS
Fitting β_1, γ_1, γ_2, δ_1, δ_2	5	2 427 239		
Residual	26	290 239	11 163	
Total 31		27 174 478		

```
   1    "Comparison of regressions of annual runoff on annual precipitation,
  -2    Severn subcatchment data, using the method of dummy variables."
   3    ''
  -4    ''
   5    VARIATE[NVALUES=33]Year,Q,P,z1,z2,Pz1,Pz2
   6         & [VALUES=1...33]Order
   7    OPEN 'fig4-7.sev';CHANNEL=2;FILETYPE=input
   8    READ[CHANNEL=2]Year,P,Q

       Identifier    Minimum    Mean    Maximum    Values    Missing
          Year         1975     1980      1985        33         0
             P         1732     2486      2938        33         1
             Q         1220     1932      2277        33         1
   9    CLOSE 'fig4-7.sev';CHANNEL=2;FILETYPE=input
  10    ''
 -11    ''
  12    CALCULATE Hore=Order.LE.11
  13       &   Tanllwyth=(Order.GT.11).AND.(Order.LE.22)
  14       &     Hafren=Order.GE.23
  15    ''
 -16    ''
  17    "Now calculate the dummy variables z1 and z2."
  18    ''
 -19    ''
  20    CALCULATE z1=0   :  & z2=0
  21         &        z1=z1+Tanllwyth
  22         &        z2=z2+Hafren
  23    ''
 -24    ''
  25    "Now calculate the variables Pz1 and Pz2."
  26    ''
 -27    ''
  28    CALCULATE Pz1=P*z1    :  &  Pz2=P*z2
  29    ''
 -30    ''
  31    "Now fit the models for testing equality of slopes and parallelism."
  32    ''
 -33    ''
  34    MODEL Q          : TERMS P,z1,z2,Pz1,Pz2
  35    FIT[PRINT=model,summary, accumulated,estimates,correlations,\
  36         monitoring]P,z1,z2,Pz1,Pz2
   1
  36 ................................................................
```

***** Regression Analysis *****

Response variate: Q
    Fitted terms: Constant, P, z1, z2, Pz1, Pz2

*** Summary of analysis ***

	d.f.	s.s.	m.s.
Regression	5	2427239.	485448.
Residual 5	26	290239.	11163.
Total	31	2717478.	87661.
Change	−5	−2427239.	485448.

Figure 4.9   GENSTAT output listing for analysis comparing regressions in terms of slopes, and then of intercepts, using data from Severn subcatchments for the Hore, Tanllwyth and Hafren. Data 1975–85, with missing values. Regressions are of annual runoff on annual precipitation.

_Continued _

Figure 4.9 *(continued)*

Percentage variance accounted for 87.3

* MESSAGE: The following units have large residuals:
22	2.25

* MESSAGE: The following units have high leverage:
2	0.52
13	0.55
24	0.59

* * * Accumulated analysis of variance * * *

Change	d.f.	s.s.	m.s.	v.r.
+ P	1	2343625.	2343625.	209.94
+ z1	1	6070.	6070.	0.54
+ z2	1	76341.	76341.	6.84
+ Pz1	1	549.	549.	0.05
+ Pz2	1	654.	654.	0.06
Residual	26	290239.	11163.	
Total	31	2717478.	87661.	

* * * Estimates of regression coefficients * * *

	estimate	s.e.	t
Constant	−246.	245.	−1.00
P	0.8485	0.0975	8.71
z1	192.	341.	0.56
z2	206.	354.	0.58
Pz1	−0.042	0.135	−0.31
Pz2	−0.034	0.142	−0.24

* * * Correlations * * *

Constant	1	1.000					
P	2	−0.992	1.000				
z1	3	−0.719	0.713	1.000			
z2	4	−0.693	0.687	0.498	1.000		
Pz1	5	0.715	−0.721	−0.991	−0.496	1.000	
Pz2	6	0.680	−0.686	−0.489	−0.991	0.495	1.000
		1	2	3	4	5	6

```
  37     DROP[PRINT=model,summary, accumulated,estimates,correlations,\
  38          monitoring]Pz1,Pz2
1
38 ......................................................................
```

* * * * * Regression Analysis * * * * *

Response variate: Q
     Fitted terms: Constant, P, z1, z2

* * * Summary of analysis * * *

	d.f.	s.s.	m.s.
Regression	3	2426036.	808679.
Residual	28	291443.	10409.
Total	31	2717478.	87661.
Change	2	1203.	602.

Percentage variance accounted for 88.1

Figure 4.9 *(continued)*

* MESSAGE: The following units have large residuals:
$$22 \qquad 2.34$$

*** Accumulated analysis of variance ***

Change	d.f.	s.s.	m.s.	v.r.
+ P	1	2343625.	2343625.	209.94
+ z1	1	6070.	6070.	0.54
+ z2	1	76341.	76341.	6.84
+ Pz1	1	549.	549.	0.05
+ Pz2	1	654.	654.	0.06
Residual	26	290239.	11163.	
− Pz2	−1	−654.	654.	0.06
− Pz1	−1	−549.	549.	0.05
Total	31	2717478.	87661.	

*** Estimates of regression coefficients ***

	estimate	s.e.	t
Constant	−181.	140.	−1.30
P	0.8228	0.0546	15.07
z1	86.3	43.5	1.98
z2	120.9	44.7	2.71

*** Correlations ***

Constant	1	1.000			
P	2	−0.975	1.000		
z1	3	−0.136	−0.021	1.000	
z2	4	−0.211	0.061	0.486	1.000
		1	2	3	4

```
  39    DROP[PRINT=model,summary, accumulated,estimates,correlations,\
  40          monitoring]z1,z2
1
  40 ....................................................................
```

***** Regression Analysis *****

Response variate:　Q
　　Fitted terms:　Constant, P

*** Summary of analysis ***

	d.f.	s.s.	m.s.
Regression	1	2343625.	2343625.
Residual	30	373853.	12462.
Total	31	2717478.	87661.
Change	2	82410.	41205.

Percentage variance accounted for 85.8

* MESSAGE: The following units have large residuals:
$$22 \qquad 2.24$$

* MESSAGE: The following units have high leverage:
$$2 \qquad 0.171$$
$$13 \qquad 0.187$$
$$24 \qquad 0.193$$

___ Figure 4.9 *(continued)* ___

```
            * * * Accumulated analysis of variance * * *

      Change      d.f.        s.s.           m.s.           v.r.
      + P          1       2343625.       2343625.        209.94
      + z1         1          6070.          6070.          0.54
      + z2         1         76341.         76341.          6.84
      + Pz1        1           549.           549.          0.05
      + Pz2        1           654.           654.          0.06
      Residual    26        290239.         11163.
      − Pz2       −1          −654.           654.          0.06
      − Pz1       −1          −549.           549.          0.05
      − z2        −1        −76341.         76341.          6.84
      − z1        −1         −6070.          6070.          0.54

      Total       31       2717478.         87661.

            * * * Estimates of regression coefficients * * *

                      estimate          s.e.            t
      Constant         −98.            149.           −0.65
      P                  0.8163          0.0595        13.71

            * * * Correlations * * *

      Constant      1       1.000
      P             2      −0.991         1.000
                            1              2
```

Thus we see no evidence to reject the hypothesis that the three regression slopes are equal, the value of the F-statistic being $602/11\,163$, less than 1. However, there appears to be some evidence that the intercepts differ significantly, the F-statistic being $41\,205/11\,163 = 3.69$, greater than the tabulated value of 3.37 with 2 and 26 degrees of freedom.

The use of dummy variables is a technique with very wide application beyond the testing of hypotheses concerning several regressions. Draper and Smith (1981, pp. 250–57) show how dummy variables can be used to test for the existence of one or more trends in data, and to estimate the parameters of two intersecting linear regressions, in the cases when it is known which points lie on each trend, and when it is not.

4.5 WEIGHTED LINEAR REGRESSION

The above example discussed the case in which all N observations on the response variable y_t are given equal weight in the analysis. The data of Figure 4.10 are the mean annual floods (MAFs) for 23 flow gauging stations in the basin of the Rio Itajaí; the figure also shows the catchment area above each gauging station, and the number of years of record from which each MAF was calculated. It is then necessary to give MAFs calculated from long records, such as those for the Rio Itajaí-Açú at Apiuna, more weight in the analysis than short records, like that for the Itajaí-Açú at Warnow, with only three years of record. The weight to be given to each y-value is clearly the number of years of record; in fact, assuming that $y_t - \mathbf{x}_t^{\mathrm{T}}\boldsymbol{\beta}$ is normally distributed with zero mean and variance σ_ε^2/n_t, where n_t is the number of years of record used to calculate y_t, the tth MAF, then the log-likelihood function is the sum of terms like

$$n_t(y_t - \mathbf{x}_t^{\mathrm{T}}\boldsymbol{\beta})^2/\sigma_\varepsilon^2 \tag{4.16}$$

River	Station	Years	Area	MAF
R. Itajaí do Oeste	Taió	51	1575	294
R. das Pombas	Pouso Redondo	32	130	49
R. Trombudo	Trombudo Central	20	432	87
R. Adago	Barracão	19	163	175
R. Itajaí do Sul	Barracão	118	364	257
R. Itajaí do Sul	Saltinho	5	483	135
R. Itajaí do Sul	Jacaraca	24	720	265
R. Itajaí do Sul	Rio do Sul	35	5100	784
R. Itajaí do Sul	Rio do Sul Novo	6	5100	951
R. Hercílio	Barra da Prata	6	1420	629
R. Hercílio	Ibirama	48	3314	784
R. Neisse Central	Neisse Central	24	195	77
R. Itajaí-Açú	Apiuna	50	9242	1624
R. Itajaí-Açú	Warnow	3	9714	3194
R. Benedito Novo	Benedito Novo	49	692	188
R. Benedito	Timbó	44	1342	431
R. Itajaí-Açú	Indaial	46	11151	2112
R. do Testo	Rio do Testo	33	105	31
R. Itajaí-Açú	Itoupava Seco	14	11719	1632
R. Garcia	Garcia	32	127	58
R. Luiz Alves	Luiz Alves	36	204	54
R. Itajaí-Mirim	Botuvera	5	859	187
R. Itajaí-Mirim	Brusque	44	1240	225

Figure 4.10   Data on mean annual floods, *MAF* (mean of annual maximum mean daily flow) for 23 flow gauging sites on the Rio Itajaí-Açú river system, Brazil. Variate *Years* gives number of years of record; *Area* is basin area in square kilometres; *MAF* is in cubic metres per second.

and the least squares function $R(\beta)$ becomes

$$(\mathbf{y} - \mathbf{X}\beta)^{\mathrm{T}}\mathbf{W}^{-1}(\mathbf{y} - \mathbf{X}\beta) \tag{4.17}$$

where $\mathbf{W}$ is an $N \times N$ matrix with diagonal terms $1/n_1, 1/n_2, \ldots, 1/n_t, \ldots, 1/n_N$ and zeros elsewhere. The equations giving estimates of the parameters β are then

$$(\mathbf{X}^{\mathrm{T}}\mathbf{W}^{-1}\mathbf{X})\hat{\beta} = \mathbf{X}^{\mathrm{T}}\mathbf{W}^{-1}\mathbf{y} \tag{4.18}$$

and the residual sum of squares becomes

$$R(\hat{\beta}) = \mathbf{y}^{\mathrm{T}}\mathbf{W}^{-1}\mathbf{y} - \beta^{\mathrm{T}}\mathbf{X}^{\mathrm{T}}\mathbf{W}^{-1}\mathbf{y} \tag{4.19}$$

which is a simple generalisation of equation (4.13). Generalisations of Tables 4.1, 4.2 and 4.3 follow immediately.

Returning to the data on MAFs shown in Figure 4.10, let us calculate a weighted regression of MAF on catchment area. Having first read the data into variates *MAF*, *Area* and *Years*, we follow the standard practice of plotting the data points with the result shown in Figure 4.11; then the GENSTAT commands

>MODEL [WEIGHT=Years] MAF : FIT Area

effects the calculation required. First, however, we compute the regression without taking account of differences in record length; Figure 4.12 shows the listing obtained when no account

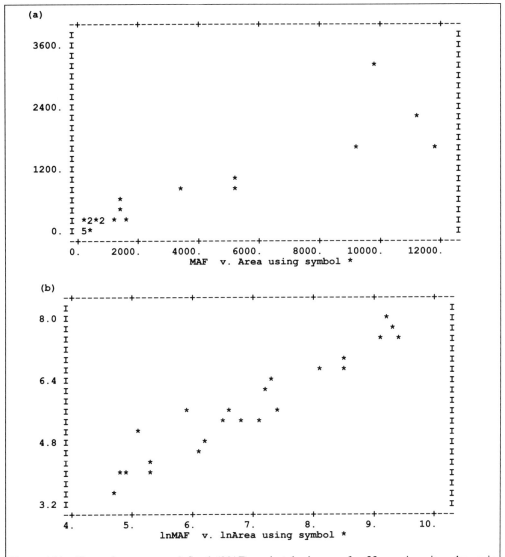

Figure 4.11 Plots of mean annual flood (*MAF*) against basin area for 23 gauging sites shown in
Figure 4.10. (a) Plot of untransformed data; (b) plot of data after transformation to logs.

is taken of the differing lengths of record. Regressing *MAF* on *Years* gives the regression

$$MAF = 76.2 + 0.1907Area$$

$$(\pm 87.2) \quad (\pm 0.0184)$$

with $R^2 = 82.8\%$. If both *MAF* and *Area* are logged, the regression becomes

$$\ln MAF = 0.101 + 0.8029 \ln Area$$

$$(\pm 0.383) \quad (\pm 0.0540)$$

```
   1    "Data for weighted and weighted regression of mean annual flood on basin area."
   2    ''
  -3    ''
   4    VARIATE[VALUES=293.5,48.5,86.7,174.5,256.7,134.8,264.9,\
   5    783.5,951.3,628.7,783.7,77.0,1624.0,3194.0,187.5,430.8,\
   6    2112.0,31.3,1632.0,58.3,53.6,187.0,225.2]MAF
   7      & [VALUES=51,32,20,19,118,5,24,35,6,6,48,24,50,3,49,44,46,33,\
   8    14,32,36,5,44]Years
   9      & [VALUES=1575,130,432,163,364,483,720,5100,5100,1420,3314,195,\
  10    9242,9714,692,1342,11151,105,11719,127,204,859,1240]Area
  11    VARIATE[NVALUES=23]Res,Fitvals,Levs,lnRes,lnFitvals,lnLevs
  12    ''
 -13    ''
  14    ''
 -15    ''
  16    "First, unweighted regression using raw data, and after taking logs."
  17    MODEL MAF    :  FIT Area
   1
  17 ..........................................................................................
```

***** Regression Analysis *****

Response variate: MAF
      Fitted terms: Constant, Area

*** Summary of analysis ***

	d.f.	s.s.	m.s.
Regression	1	11951994.	11951994.
Residual	21	2348538.	111835.
Total	22	14300531.	650024.

Percentage variance accounted for 82.8

* MESSAGE: The following units have large residuals:

14	4.20
19	−2.40

* MESSAGE: The following units have high leverage:

14	0.187
17	0.253
19	0.283

*** Estimates of regression coefficients ***

	estimate	s.e.	t
Constant	76.2	87.2	0.87
Area	0.1907	0.0184	10.34

```
  18    RKEEP MAF;RESIDUALS=Res;FITTEDVALUES=Fitvals;LEVERAGES=Levs
  19    PRINT MAF,Fitvals,Res,Levs
```

MAF	Fitvals	Res	Levs
293.5	376.5	−0.2543	0.04837
48.5	101.0	−0.1624	0.06586
86.7	158.6	−0.2218	0.06116
174.5	107.3	0.2079	0.06532
256.7	145.6	0.3431	0.06217
134.8	168.3	−0.1033	0.06042
264.9	213.5	0.1584	0.05719
783.5	1048.5	−0.8169	0.05897
951.3	1048.5	−0.2997	0.05897
628.7	346.9	0.8643	0.04964

Figure 4.12   GENSTAT output given by regression of mean annual flood (*MAF*) on basin area without transformation; and after taking logs. Data as shown in Figure 4.10.

_____ Continued _____

Figure 4.12 *(continued)*

MAF	Fitvals	Res	Levs
783.7	708.0	0.2315	0.04415
77.0	113.4	−0.1125	0.06480
1624.0	1838.2	−0.7022	0.16800
3194.0	1928.2	4.1980	0.18705
187.5	208.1	−0.0635	0.05755
430.8	332.1	0.3030	0.05033
2112.0	2202.2	−0.3120	0.25338
31.3	96.2	−0.2009	0.06628
1632.0	2310.4	−2.3960	0.28307
58.3	100.4	−0.1303	0.06591
53.6	115.1	−0.1901	0.06466
187.0	240.0	−0.1630	0.05545
225.2	312.6	−0.2683	0.05129

```
20    CALCULATE lnMAF=LOG(MAF)  :  & lnArea=LOG(Area)
21    MODEL lnMAF   :   FIT lnArea
 1
21..........................................................
```

***** Regression Analysis *****

Response variate: lnMAF
     Fitted terms: Constant, lnArea

*** Summary of analysis ***

	d.f.	s.s.	m.s.
Regression	1	34.616	34.6156
Residual	21	3.285	0.1564
Total	22	37.900	1.7227

Percentage variance accounted for 90.9

* MESSAGE: The following units have large residuals:

                          4   2.60

*** Estimates of regression coefficients ***

	estimate	s.e.	t
Constant	0.101	0.383	0.26
lnArea	0.8029	0.0540	14.88

```
22    RKEEP lnMAF;RESIDUALS=lnRes;FITTEDVALUES=lnFitvals;LEVERAGES=lnLevs
23    PRINT lnMAF,lnFitvals,lnRes,lnLevs
```

lnMAF	lnFitval	lnRes	lnLevs
5.682	6.012	−0.8552	0.04709
3.882	4.009	−0.3448	0.12205
4.462	4.973	−1.3306	0.05703
5.162	4.191	2.5961	0.10569
5.548	4.836	1.8597	0.06302
4.904	5.063	−0.4140	0.05372
5.579	5.384	0.5066	0.04566
6.664	6.955	−0.7739	0.09207
6.858	6.955	−0.2589	0.09207
6.444	5.929	1.3323	0.04559
6.664	6.609	0.1434	0.06960
4.344	4.335	0.0237	0.09409
7.393	7.433	−0.1089	0.13442
8.069	7.473	1.6245	0.13856
5.234	5.352	−0.3055	0.04620
6.066	5.884	0.4712	0.04494
7.655	7.583	0.1973	0.15053
3.444	3.838	−1.0744	0.13923

__ Figure 4.12 *(continued)* _____

lnMAF	lnFitval	lnRes	lnLevs
7.398	7.623	−0.6211	0.15501
4.066	3.991	0.2026	0.12384
3.982	4.371	−1.0333	0.09136
5.231	5.525	−0.7609	0.04399
5.417	5.820	−1.0425	0.04423

with $R^2 = 90.9\%$ (where, following accepted usage, we denote by R^2 the coefficient of determination, equal to the regression sum of squares divided by the total sum of squares). Various messages are printed in Figure 4.12 showing points (i.e. basins) having large residuals, or large leverages; we note particularly that unit 14 was reported as having a large residual and a large leverage, in the regression before log transformation of the data: this unit corresponds to the Rio Itajaí-Açú at Warnow, a site with only three years of record, the shortest of all 23 sites.

Figure 4.13 shows the corresponding linear regressions obtained by taking account of the different record lengths. Without log transformation, we have a regression

$$MAF = 72.9 + 0.17252 Area$$
$$(\pm 41.2) \quad (\pm 0.00879)$$

with $R^2 = 94.6\%$, rather greater than that obtained by ignoring differences in record length. After log transformation, the regression is now

$$\ln MAF = 0.046 + 0.8014 \ln Area$$
$$(\pm 0.320) \quad (\pm 0.0449)$$

with $R^2 = 93.5\%$. The graphs shown in the output show plots of the standardised residuals, *RES*, against ln*MAF* and against ln*Area*; the large residual associated with item 4, the basin on the Rio Adago above Barracão, can clearly be identified. The large leverages associated with certain points are difficult to explain with the information available; one possible explanation, for example, may be that the rating curves at sites with high leverages are in error, so that discharges — and hence mean daily flows — are consistently over- or underestimated. Checking such matters would require a return to the raw data available for the sites in question.

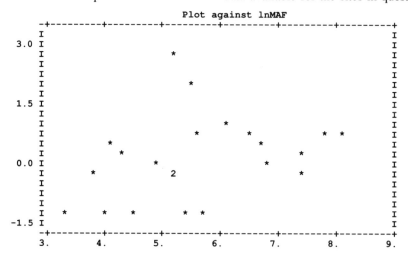

Plot against lnMAF

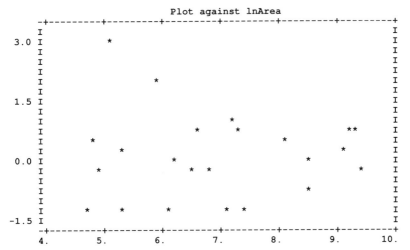

Graphs showing plots of **lnMAF** standardised residuals against **lnMAF** and **lnArea**

```
    1     "Data for weighted and weighted regression of mean annual flood on basin area."
    2     ''
   -3     ''
    4     VARIATE[VALUES=293.5,48.5,86.7,174.5,256.7,134.8,264.9,\
    5     783.5,951.3,628.7,783.7,77.0,1624.0,3194.0,187.5,430.8,\
    6     2112.0,31.3,1632.0,58.3,53.6,187.0,225.2]MAF
    7        & [VALUES=51,32,20,19,18,5,24,35,6,6,48,24,50,3,49,44,46,33,\
    8     14,32,36,5,44]Years
    9        & [VALUES=1575,130,432,163,364,483,720,5100,5100,1420,3314,195,\
   10     9242,9714,692,1342,11151,105,11719,127,204,859,1240]Area
   11     VARIATE[NVALUES=23]Res,Fitvals,Levs,lnRes,lnFitvals,lnLevs
   12     ''
  -13     ''
   14     ''
  -15     ''
   16     "Now weighted regression using raw data, and after taking logs."
   17     MODEL[WEIGHT=Years] MAF   :   FIT Area
  1
 17 ............................................................................
```

***** Regression Analysis *****

Response variate: MAF
  Weight variate: Years
    Fitted terms: Constant, Area

*** Summary of analysis ***

	d.f.	s.s.	m.s.
Regression	1	2.642E+08	2.642E+08
Residual	21	1.440E+07	6.857E+05
Total	22	2.786E+08	1.266E+07

Percentage variance accounted for 94.6

Figure 4.13   GENSTAT output given by regression of mean annual flood (*MAF*) on basin area without transformation; and after taking logs. Regression now calculated using number of years of record as weights: data as shown in Figure 4.10.

Continued

Figure 4.13 *(continued)*

* MESSAGE: The following units have large residuals:

14	3.05
19	−2.26

* MESSAGE: The following units have high leverage:

13	0.307
17	0.428

∗ ∗ ∗ Estimates of regression coefficients ∗ ∗ ∗

	estimate	s.e.	t
Constant	72.9	41.2	1.77
Area	0.17252	0.00879	19.63

18	RKEEP MAF;RESIDUALS=Res;FITTEDVALUES=Fitvals;LEVERAGES=Levs
19	PRINT MAF,Fitvals,Res,Levs

MAF	Fitvals	Res	Levs
293.5	344.6	−0.4619	0.08861
48.5	95.3	−0.3329	0.07646
86.7	147.4	−0.3355	0.04429
174.5	101.0	0.3957	0.04502
256.7	135.7	0.6329	0.04054
134.8	156.2	−0.0582	0.01093
264.9	197.1	0.4113	0.04959
783.5	952.8	−1.2569	0.07422
951.3	952.8	−0.0044	0.01272
628.7	317.9	0.9243	0.01071
783.7	644.7	1.2100	0.07567
77.0	106.6	−0.1800	0.05640
1624.0	1667.4	−0.4450	0.30741
3194.0	1748.8	3.0544	0.02056
187.5	192.3	−0.0428	0.10192
430.8	304.4	1.0551	0.07967
2112.0	1996.7	1.2485	0.42807
31.3	91.0	−0.4318	0.07936
1632.0	2094.7	−2.2620	0.14565
58.3	94.8	−0.2596	0.07652
53.6	108.1	−0.4127	0.08441
187.0	221.1	−0.0926	0.01001
225.2	286.8	−0.5151	0.08126

20	CALCULATE lnMAF=LOG(MAF): & lnArea=LOG(Area)
21	MODEL[WEIGHT=Years] lnMAF: FIT lnArea
1	

21 ....

∗ ∗ ∗ ∗ ∗ Regression Analysis ∗ ∗ ∗ ∗ ∗

Response variate:	lnMAF
Weight variate:	Years
Fitted terms:	Constant, lnArea

∗ ∗ ∗ Summary of analysis ∗ ∗ ∗

	d.f.	s.s.	m.s.
Regression	1	968.04	968.043
Residual	21	63.94	3.045
Total	22	1031.98	46.908

Percentage variance accounted for 93.5

_____ Figure 4.13 *(continued)* _____

* MESSAGE: The following units have large residuals:
                        4   2.68
* MESSAGE: The following units have high leverage:
                              13   0.234
                              17   0.241

*** Estimates of regression coefficients ***

	estimate	s.e.	t
Constant	0.046	0.320	0.14
lnArea	0.8014	0.0449	17.83

```
22    RKEEP lnMAF;RESIDUALS=lnRes;FITTEDVALUES=lnFitvals;LEVERAGES=lnLevs
23    PRINT lnMAF,lnFitvals,lnRes,lnLevs
```

lnMAF	lnFitval	lnRes	lnLevs
5.682	5.945	−1.1274	0.08467
3.882	3.946	−0.2271	0.14261
4.462	4.909	−1.1685	0.04160
5.162	4.128	2.6839	0.07339
5.548	4.772	1.9281	0.04143
4.904	4.998	−0.1216	0.00978
5.579	5.318	0.7483	0.03957
6.664	6.887	−0.8033	0.11211
6.858	6.887	−0.0414	0.01922
6.444	5.862	0.8199	0.00967
6.664	6.542	0.5172	0.11637
4.344	4.271	0.2124	0.08256
7.393	7.363	0.1351	0.23409
8.069	7.403	0.6656	0.01448
5.234	5.286	−0.2201	0.08182
6.066	5.817	0.9796	0.07004
7.655	7.514	0.6312	0.24133
3.444	3.775	−1.1967	0.16763
7.398	7.554	−0.3484	0.07565
4.066	3.928	0.4834	0.14470
3.982	4.308	−1.1950	0.12026
5.231	5.460	−0.2940	0.00790
5.417	5.754	−1.3270	0.06910

4.6 EXAMPLE: MULTIPLE LINEAR REGRESSION

The data shown in Figure 4.14, taken from Haan (1977) by kind permission of Charles T. Haan, are from 13 small basins in Kentucky, USA, and were used to calculate a regression for use in predicting annual runoff (*RO*, measured in inches), given basin mean annual rainfall (*Prec*, in inches), basin area (*A* in square miles), and other variables, as shown, describing basin area shape and relief.

For the purposes of this example, we retain the units of inches, feet and miles used by Haan; however, it is an easy matter to convert to metric units using a program such as GENSTAT. For example, having read the variables *RO* and *Prec* in inches, these can be converted to metric units by

> CALCULATE RO = RO * 25.4: & Prec = Prec * 25.4

one inch being 25.4 mm. Similarly, an area expressed in square miles can be converted to square kilometres by multiplying it by $1.609\,344^2$, there being 1.609 344 km in a mile.

Data from 13 small basins in Kentucky, USA. Variables are as follows:
RO :   mean annual runoff (inches)
Prec : mean annual precipitation (inches)
A :   basin area (square miles)
S :   average land slope (%)
L :   axial length (miles)
P :   basin perimeter (miles)
di :   diameter of largest circle that can be drawn entirely within the basin (miles)
Rs :   Shape factor: ratio *di/do*, where *do* is the diameter of the smallest circle entirely
         enclosing the basin
F :   stream frequency : ratio of number of streams in basin to total area of basin (units:
         1/square miles)
Rr :   relief ratio: ratio of total relief to largest dimension of basin generally parallel to
         main stream (units: feet per mile)

RO	Prec	A	S	L	P	di	Rs	F	Rr
17.38	44.37	2.21	50	2.38	7.93	0.91	0.38	1.36	332
14.62	44.09	2.53	7	2.55	7.65	1.23	0.48	2.37	55
15.48	41.25	5.63	19	3.11	11.61	2.11	0.57	2.31	77
14.72	45.50	1.55	6	1.84	5.31	0.94	0.49	3.87	68
18.37	46.09	5.15	16	4.14	11.35	1.63	0.39	3.30	68
17.01	49.12	2.14	26	1.92	5.89	1.41	0.71	1.87	230
18.20	44.03	5.34	7	4.73	12.59	1.30	0.27	0.94	44
18.95	48.71	7.47	11	4.24	12.33	2.35	0.52	1.20	72
13.94	44.43	2.10	5	2.00	6.81	1.19	0.53	4.76	40
18.64	47.72	3.89	18	2.10	9.87	1.65	0.60	3.08	115
17.25	48.38	0.67	21	1.15	3.93	0.62	0.48	2.99	352
17.48	49.00	0.85	23	1.27	3.79	0.83	0.61	3.53	300
13.16	47.03	1.72	5	1.93	5.19	0.99	0.52	2.33	39

Figure 4.14   Data used in multiple regression analysis.

Some comments are required concerning the nature of the multiple regression being calculated. First, *RO* will presumably by calculated from a different number of years of record for each of the 13 basins, so that it might be more appropriate to undertake a weighted analysis of the kind described earlier; however the weights (number of years of record) are not specified, so we follow Haan in using an unweighted example. Second, *Prec* is presumably calculated as a weighted mean of annual rainfalls recorded by a network of gauges within each basin; different network densities could result in inhomogeneities which cannot be investigated in the absence of further information. Third, the variable *A* (area) is included in the multiple regression, although it is already implicit in the calculation of *RO*. After including *A* for demonstrating the calculation of a multiple regression including all nine explanatory variables, we follow Haan in noting that *A* adds no significant information to the prediction of *RO*, and drop it from subsequent analysis.

Lastly, it is clear that *RO* is related to *Prec* by the water balance equation

$$RO = Prec - \frac{\text{mean annual loss}}{\text{by evapotranspiration}} - \frac{\text{mean annual loss}}{\text{by aquifer recharge}}$$

ignoring annual changes in soil moisture storage. Evaporation and recharge 'losses' are not available in the data provided; but the variables included in the regression calculations are essentially surrogates for them. The variable *Rf*, which is related to basin relief and which, as seen below, adds significant information to the prediction of *RO*, may be regarded as a description of basin geology, and thus indirectly of aquifer recharge. Interpretation of the significance of basin perimeter *P* (see below) is more difficult; regarded as a descriptor of the tortuosity of basin boundaries, it also could be a surrogate for basin geology, although other interpretations are possible.

Returning to details of the calculation, Figure 4.15 shows the regression fitted when all nine explanatory variables are used, leaving only 3 degrees of freedom (df) for the estimation of the variance σ_ε^2 of residuals. Estimates based on such small numbers of degrees of freedom have low precision, and it is commonly advisable to base estimates of σ_ε^2 on 10 df or more. The regression coefficients $\hat{\beta}_i$ giving values of the *t*-statistic greater than 2, are those corresponding to the variates *Prec*, *P* and *Rr*; line 50 of the program listing therefore contains the directive DROP to remove from the regression the variables *A*, *S*, *L*, *di*, *Rs* and *F*, with the result shown below line 50. The estimate of σ_ε^2 has decreased slightly, from 0.4779 to 0.4734, by removal of the non-significant explanatory variables, and the values of the *t*-statistics are substantially greater (for example, the coefficient of *Prec* is now 0.4296 ± 0.0930; $t = 4.62$ with 9 df, compared with 0.450 ± 0.148; $t = 3.05$ with 3 df) and the t-tests are more powerful by virtue

```
***** Regression Analysis *****
Response variate:    RO
     Fitted terms:    Constant, Prec, A, S, L, P, di, Rs, F, Rr
*** Summary of analysis ***
                 d.f.        s.s.         m.s.
Regression        9        43.507       4.8341
Residual          3         1.434       0.4779
Total            12        44.941       3.7451
Change           -9       -43.507       4.8341
Percentage variance accounted for 87.2
*** Estimates of regression coefficients ***
                 estimate       s.e.          t
Constant        -14.74         6.78         -2.17
Prec              0.450        0.148         3.05
A                 0.19         1.42          0.13
S                -0.0182       0.0484       -0.38
L                 0.288        0.777         0.37
P                 0.985        0.366         2.69
di               -3.05         4.80         -0.64
Rs                5.67         9.03          0.63
F                 0.374        0.267         1.40
Rr                0.01288      0.00601       2.14
   46      ''
  -47      ''
   48      ''
  -49      ''
   50     DROP A,S,L,di,Rs,F
    1
   50 ......................................................................
```

Figure 4.15   GENSTAT output from fitting a multiple regression to data shown in Figure 4.14.

_____ *Continued* ___

Figure 4.15 *(continued)*

```
***** Regression Analysis *****

Response variate:   RO
        Fitted terms:   Constant, Prec, P, Rr

*** Summary of analysis ***

                    d.f.        s.s.          m.s.
Regression           3        40.680       13.5600
Residual             9         4.261        0.4734
Total               12        44.941        3.7451

Change               6         2.827        0.4712

Percentage variance accounted for 87.4

*** Estimates of regression coefficients ***

                estimate         s.e.            t
Constant         -9.64          4.44         -2.17
Prec              0.4296        0.0930         4.62
P                 0.6166        0.0748         8.24
Rr                0.01042       0.00201        5.19

    51      ''
   -52      ''
    53      RKEEP RO;RESIDUALS=Res;FITTEDVALUES=Fitvals;LEVERAGES=Levs;\
    54      VCOVARIANCE=Vcov
    55      PRINT RO,Fitvals,Res, Levs   : &   Vcov

    RO          Fitvals         Res           Levs
   17.38        17.77        -0.8096        0.5222
   14.62        14.59         0.0557        0.1917
   15.48        16.04        -1.0863        0.4458
   14.72        13.88         1.4182        0.2659
   18.37        17.86         0.8197        0.1872
   17.01        17.48        -0.7761        0.2096
   18.20        17.49         1.1931        0.2543
   18.95        19.63        -1.4034        0.4998
   13.94        14.06        -0.1957        0.2406
   18.64        18.14         0.8074        0.1866
   17.25        17.23         0.0368        0.3591
   17.48        16.87         1.0598        0.2953
   13.16        14.16        -1.8005        0.3420

            Vcov
    1       19.721
    2       -0.407        0.009
    3       -0.136        0.002        0.006
    4        0.001        0.000        0.000        0.000
                1            2            3            4

    56      ''
   -57      ''
    58      ''
   -59      ''
    60      GRAPH[NROWS=20;NCOLUMNS=70]Res;Fitvals
     1
```

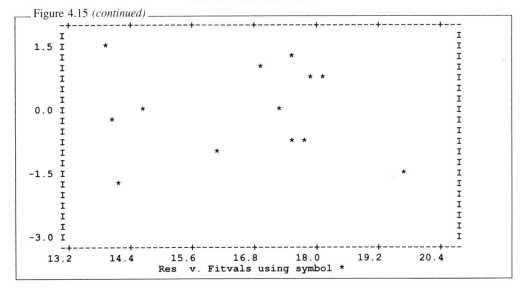

Figure 4.15 (continued)

of the larger number of degrees of freedom. The listing also shows the standardised residuals, fitted values and leverages, together with the variance-covariance matrices of the four estimates $\hat{\beta}_0$, $\hat{\beta}_1$, $\hat{\beta}_2$, $\hat{\beta}_3$. By contrast with what was observed in earlier analyses, we have no warning messages concerning large residuals or large leverages.

It is desirable, in a program for the computation of multiple regression, to be able to add new variables, and to try new variables, in the sense that, if they result in an improvement to the goodness of fit they are to be retained in the regression equation, but otherwise left out. Figure 4.16 shows how this can be done with GENSTAT. The calculation starts with the regression of *RO* on two explanatory variables, *Prec* and *P*; it is seen that this regression is

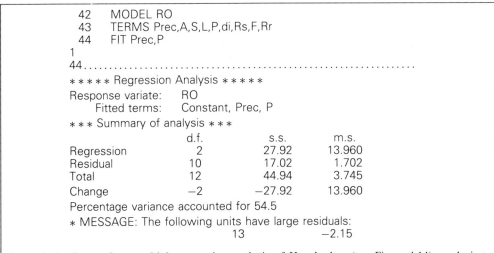

Figure 4.16   Output from multiple regression analysis of Haan's data (see Figure 4.14): analysis to demonstrate the inclusion of additional variables to see whether fit is improved.

Continued

Figure 4.16 *(continued)*

```
      * MESSAGE: The following units have high leverage:
                            8                    0.49
      * * * Estimates of regression coefficients * * *
                       estimate        s.e.           t
      Constant         -12.79          8.34         -1.53
      Prec               0.557         0.170          3.27
      P                  0.456         0.129          3.53
      45    ''
     -46    ''
      47    ''
     -48    ''
      49    "Having fitted Prec and P, now add the third variable Rr."
      50    ''
     -51    ''
      52    ADD Rr
      52.....................................................................
```

```
      * * * * * Regression Analysis * * * * *
      Response variate:    RO
          Fitted terms:    Constant, Prec, P, Rr
      * * * Summary of analysis * * *
                       d.f.           s.s.           m.s.
      Regression         3           40.680        13.5600
      Residual           9            4.261         0.4734
      Total             12           44.941         3.7451
      Change            -1          -12.761        12.7608
      Percentage variance accounted for 87.4
      * * * Estimates of regression coefficients * * *
                       estimate        s.e.           t
      Constant         -9.64          4.44          -2.17
      Prec              0.4296        0.0930         4.62
      P                 0.6166        0.0748         8.24
      Rr                0.01042       0.00201        5.19
      53    ''
     -54    ''
      55    "Now try adding S as a fourth explanatory variable."
      56    ''
     -57    ''
      58    TRY S
       1
      58.....................................................................
```

```
      * * * * * Regression Analysis * * * * *
      Response variate:    RO
          Fitted terms:    Constant, Prec, P, Rr, S
      * * * Summary of analysis * * *
                       d.f.           s.s.           m.s.
      Regression         4           41.170        10.2926
         Residual        8            3.771         0.4713
            Total       12           44.941         3.7451
         Change         -1           -0.490         0.4903
```

Figure 4.16 *(continued)*

Percentage variance accounted for 87.4

∗ MESSAGE: The following units have high leverage:
1   0.79

∗∗∗ Estimates of regression coefficients ∗∗∗

	estimate	s.e.	t
Constant	−8.12	4.67	−1.74
Prec	0.388	0.101	3.84
P	0.6683	0.0903	7.40
Rr	0.01452	0.00449	3.23
S	−0.0366	0.0358	−1.02

much poorer than the earlier regression on *Prec*, *P* and *Rr*, accounting for only 54.5% of the total variance, compared with 87.4%. Using the directive ADD *Rr*, we are able to bring *Rr* back into the regression. We can also try the effect of including basin slope, *S*, by use of the directive TRY *S*. Figure 4.16 shows the result; basin slope does not improve the regression significantly, the corresponding regression coefficient β being, rather oddly, negative, but not significantly so.

4.7 MULTIPLE REGRESSION SUBJECT TO LINEAR CONSTRAINTS: THE MUSKINGUM METHOD OF FLOOD ROUTING

The Muskingum flood-routing method uses the continuity equation

$$dS/dt = i(t) - q(t)$$

where $S(t)$ is storage within a reach, and $i(t)$, $q(t)$ are the input and output to it. The method expresses S as a weighted combination of input and output of the form

$$S(t) = S_0 + K[xi(t) + (1 - x)q(t)]$$

where K and x are constants, K with dimensions of time, and x dimensionless. We therefore obtain

$$q(t) + K(1 - x)dq/dt = i(t) - Kx \, di/dt$$

For discrete intervals of time, this equation can be written

$$q_{t+1} = \beta_1 q_t + \beta_2 i_{t-1} + \beta_3 i_t$$

where

$$\beta_1 = [K(1 - x) - \Delta t/2]/[K(1 - x) + \Delta t/2]$$

$$\beta_2 = [Kx + \Delta t/2]/[K(1 - x) + \Delta t/2]$$

$$\beta_3 = -[Kx - \Delta t/2]/[K(1 - x) + \Delta t/2]$$

so that $\beta_1 + \beta_2 + \beta_3 = 1$. We assume that the time increment Δt is small relative to K.

Where there are appreciable errors in flow measurement and lateral inflows of uncertain magnitude, the routing equation may be written

$$q_{t+1} = \beta_1 q_t + \beta_2 i_{t-1} + \beta_3 i_t + \varepsilon_t$$

where ε_t is a random variable, which, in the absence of more realistic assumptions, we assume to be normally and independently distributed with zero mean and constant variance σ_ε^2. The routing equation then becomes a multiple regression without an intercept term, and subject to the constraint that the regression coefficients satisfy $\Sigma\beta_i = 1$.

By minimising the sum of squares function subject to this constraint, the equations giving the coefficients β_1, β_2, β_3 are found to be

$$
\begin{bmatrix}
\sum q_{t-1}^2 & \sum q_{t-1}i_{t-1} & \sum q_{t-1}i_t & 1 \\
\sum q_{t-1}i_{t-1} & \sum i_{t-1}^2 & \sum i_{t-1}i_t & 1 \\
\sum q_{t-1}i_t & \sum i_t i_{t-1} & \sum i_t^2 & 1 \\
1 & 1 & 1 & 0
\end{bmatrix}
\begin{bmatrix}
\beta_1 \\
\beta_2 \\
\beta_3 \\
\lambda
\end{bmatrix}
=
\begin{bmatrix}
\sum q_t q_{t-1} \\
\sum q_t i_{t-1} \\
\sum q_t i_t \\
1
\end{bmatrix}
$$

or

$$\mathbf{Q}\boldsymbol{\beta} = \mathbf{T}$$

giving $\hat{\boldsymbol{\beta}} = \mathbf{Q}^{-1}\mathbf{T}$ and $\hat{\sigma}_\varepsilon^2 = (\sum q_t^2 - \hat{\boldsymbol{\beta}}\mathbf{T})/(N - 2)$, where N is the number of quadruplets $\{q_t, q_{t-1}, i_{t-1}, i_t\}$ used for estimation.

Having obtained the estimates $\hat{\beta}_1$, $\hat{\beta}_2$, $\hat{\beta}_3$, estimates of the routing parameters K and x may be recovered, and the standard errors of these estimates calculated. It is found that

$$x = [2\beta_2 + \beta_1 - 1]/[2(\beta_1 + \beta_2)]$$

$$K = \Delta t (\beta_1 + \beta_2)/(1 - \beta_1)$$

Soil	County	State	%RillSlo	Sand	Silt	Clay	tauC
Academy	Fresno	CA	4.5	62.7	29.1	8.2	1.82
Amarillo	Howard	TX	3.6	85.0	7.7	7.3	1.92
BarnesMN	Stevens	MN	8.3	48.6	34.4	17.0	5.70
BarnesMD	Sheridan	ND	5.8	39.5	36.0	24.5	4.53
Caribou	Aroostook	ME	8.8	47.0	40.3	12.7	7.07
Cecil	Oconee	GA	4.5	64.6	15.6	19.8	2.56
Collamer	Tompkins	NY	8.7	7.0	78.0	15.0	5.86
Frederick	Washington	MD	12.8	25.1	58.3	16.6	9.54
Gaston	Rown	NC	6.4	35.5	25.4	39.1	2.54
Grenada	Panola	MS	8.7	2.0	77.8	20.2	6.39
Heiden	Falls	TX	3.9	8.6	38.3	53.1	1.94
Hersch	Valley	NE	6.6	74.4	15.9	9.7	3.14
Hiwassee	Oconee	GA	4.0	63.7	21.6	14.7	1.94
Lewisburg	Whitley	IN	7.5	38.5	32.2	29.3	4.87
Manor	Howard	MD	8.6	43.6	30.7	25.7	6.77
Mexico	Boone	MO	3.9	5.3	68.7	26.0	1.90
Miami	Montgomery	IN	5.8	4.2	72.7	23.1	4.64
Miamian	Montgomery	OH	8.9	30.6	44.1	25.3	7.22
Nansene	Whitman	WA	6.1	20.1	68.8	11.1	3.58
Opequon	Allegany	MD	12.0	37.7	31.2	31.1	10.60
Palouse	Whitman	WA	6.5	9.8	70.1	20.1	4.35
Pierre	Jackson	SD	6.6	9.6	40.9	49.5	*
Portneuf	TwinFalls	ID	5.6	21.5	67.4	11.1	4.30
Sharpsburg	Lancaster	NE	5.7	4.8	55.4	39.8	4.46
Sverdrup	Grant	MN	4.2	75.3	16.8	7.9	1.73
Tifton	Worth	GA	4.6	86.4	10.8	2.8	2.20
Whitney	Fresno	CA	7.4	71.0	21.8	7.2	2.65
Williams	Sheridan	ND	5.1	41.6	32.4	26.0	3.81
Woodward	Harper	OK	7.1	43.7	42.4	13.9	3.61
Zahl	Roosevelt	MT	7.6	46.3	29.7	24.0	4.51

Figure 4.17 Data on critical sheer stress (*tauC*:units Pa), average rill slope (%), and particle size analysis (percentage sand, silt, and clay by weight) for 30 US soils, from Gilley *et al.* (1993).

whence, using a large-sample approximation,

$$\mathrm{var}\,\hat{K} = \Delta t^2((1+\beta_2)/(1-\beta_1)^2)^2\,\mathrm{var}\,\hat{\beta}_1 + \Delta t^2((1-\beta_1)^{-2})\,\mathrm{var}\,\hat{\beta}_2$$

$$+ 2\Delta t^2((1+\beta_2)/(1-\beta_1)^3)\,\mathrm{cov}(\hat{\beta}_1,\hat{\beta}_2)$$

$$\mathrm{var}\,\hat{x} = ((1-\beta_2)/2(\beta_1+\beta_2)^2)^2\,\mathrm{var}\,\hat{\beta}_1 + ((1+\beta_1)/2(\beta_1+\beta_2)^2)^2\,\mathrm{var}\,\hat{\beta}_2$$

$$+ 2((1-\beta_2)(1-\beta_1)/2(\beta_1+\beta_2)^4)\,\mathrm{cov}(\hat{\beta}_1,\hat{\beta}_2)$$

In these expressions, β_1 and β_2 are replaced by their estimates; approximate 95% confidence limits for K, x if required can be calculated as $\hat{K} \pm 2\sqrt{(\mathrm{var}\,\hat{K})}$, $\hat{x} \pm 2\sqrt{(\mathrm{var}\,\hat{x})}$. Alternatively, likelihood-based confidence limits can be calculated as described in Chapter 3, given an appropriate form for the probability distribution of the random components ε.

The case where the parameters β of the model are subject to a linear constraint must be distinguished from the case where the *explanatory variables* are subject to a linear constraint. Figure 4.17 shows data reproduced from Gilley *et al.* (1993) by kind permission of J. E. Gilley. The data are from 30 soils throughout the USA. The soils were studied to relate critical shear stress, at which rilling commences, to physical, chemical and biological characteristics of the soil. Three of the variables shown in Figure 4.17 are percentages of sand, silt and clay, which sum to 100%; when two of the three are included in the regression equation, the third contributes no additional information. Figure 4.18 shows the results of multiple regression analyses of critical shear stress τ_C on percentage rill slope, sand and silt percentages. The analysis proceeds in the usual way; percentage rill slope appears to be the most useful variable for explaining variation in τ_C. Soils giving large residuals in the estimation of τ_C, and points with large leverages, are identified, and standardised residuals are plotted in order, against fitted values, and against explanatory variables to seek explanations for the unusual character of the data points giving rise to warning messages. Essentially, this analysis is not different from, say, the analysis of Haan's data.

```
      * * * * * Regression Analysis * * * * *
      Response variate:     tauC
            Fitted terms:     Constant, %RillSlo, Sand, Silt
      * * * Summary of analysis * * *
                        d.f.          s.s.          m.s.
      Regression          3          131.97       43.9906
      Residual           25           17.89        0.7155
      Total              28          149.86        5.3521
      Change             -3         -131.97       43.9906
      Percentage variance accounted for 86.6
      * MESSAGE: The following units have large residuals:
                              9                    -2.31
                             27                    -2.43
      * MESSAGE: The following units have high leverage:
                              8                     0.29
                             11                     0.41
                             20                     0.30
```

Figure 4.18   Regression of rill shear stress on rill slope and particle size analysis: see data in Figure 4.17.

Continued

Figure 4.18 *(continued)*

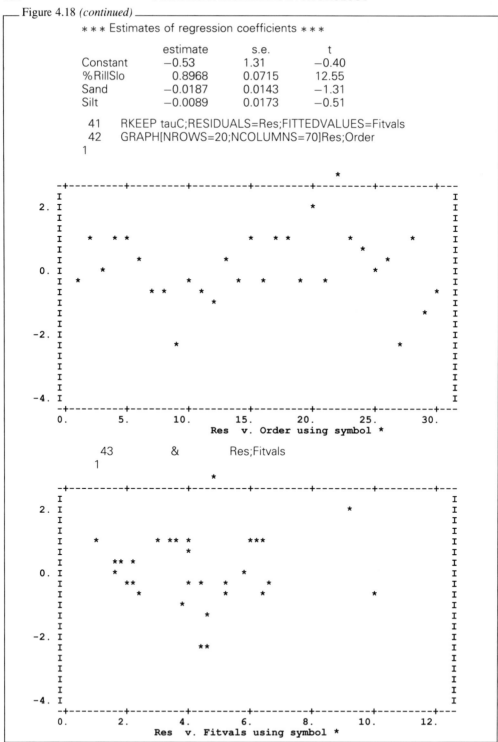

∗ ∗ ∗ Estimates of regression coefficients ∗ ∗ ∗

	estimate	s.e.	t
Constant	−0.53	1.31	−0.40
%RillSlo	0.8968	0.0715	12.55
Sand	−0.0187	0.0143	−1.31
Silt	−0.0089	0.0173	−0.51

```
41    RKEEP tauC;RESIDUALS=Res;FITTEDVALUES=Fitvals
42    GRAPH[NROWS=20;NCOLUMNS=70]Res;Order
1
```

Res   v. Order using symbol ∗

```
43          &           Res;Fitvals
1
```

Res   v. Fitvals using symbol ∗

Figure 4.18 *(continued)*

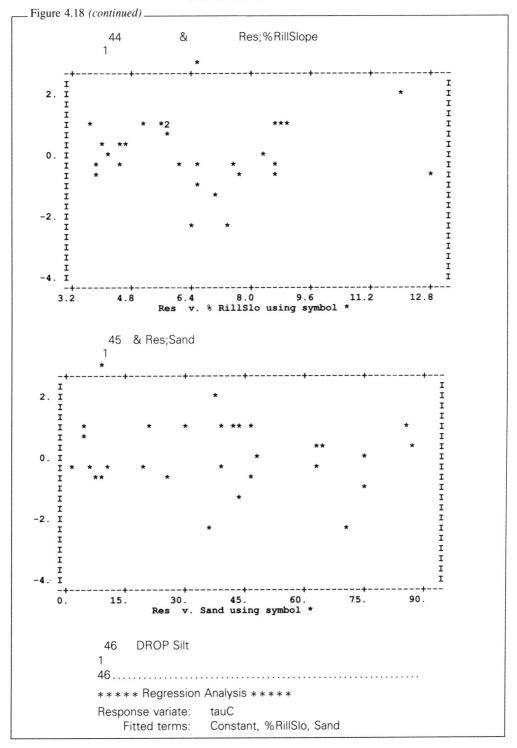

44   &   Res;%RillSlope

45   & Res;Sand

46   DROP Silt

* * * * * Regression Analysis * * * * *

Response variate: tauC
    Fitted terms: Constant, %RillSlo, Sand

Figure 4.18 *(continued)*

```
        *** Summary of analysis ***

                       d.f.        s.s.         m.s.
        Regression       2        131.79       65.8927
        Residual        26         18.07        0.6952
        Total           28        149.86        5.3521

        Change           1          0.19        0.1863

        Percentage variance accounted for 87.0

        * MESSAGE: The following units have large residuals:
                              27                -2.50

        * MESSAGE: The following units have high leverage:
                               8                 0.28
                              20                 0.23

        *** Estimates of regression coefficients ***

                       estimate      s.e.          t
        Constant        -1.118       0.600       -1.86
        %RillSlo         0.8930      0.0701       12.74
        Sand            -0.01220     0.00629      -1.94

        47    RKEEP tauC;RESIDUALS=Res;FITTEDVALUES=Fitvals
        48    GRAPH[NROWS=20;NCOLUMNS=70]Res;Order
        1
```

```
                                                    *
      -+---------+---------+---------+---------+---------+---------+---
   I                                                                  I
2. I                                              *                   I
   I                                                                  I
   I                                                                  I
   I      *    * *              *         *            *         *    I
   I           *               *        *           *               I
   I                              *                      *           I
0. I        *                                              *          I
   I   *             * *      *              *                        I
   I              *          *        *                         *    I
   I           *         *                              *           I
   I                                                         *        I
   I                                                                  I
-2. I               *                                                 I
   I                                                  *               I
   I                                                                  I
   I                                                                  I
   I                                                                  I
-4. I                                                                 I
      -+---------+---------+---------+---------+---------+---------+---
      0.        5.       10.       15.       20.       25.       30.
                   Res  v. Order using symbol *
```

Figure 4.18 *(continued)*

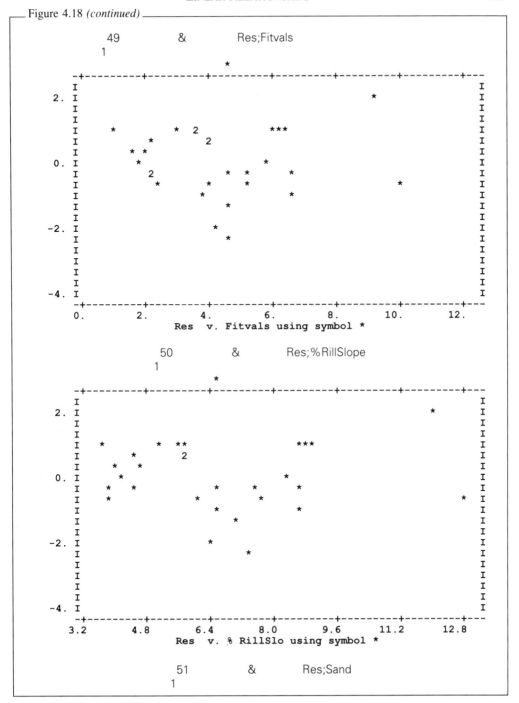

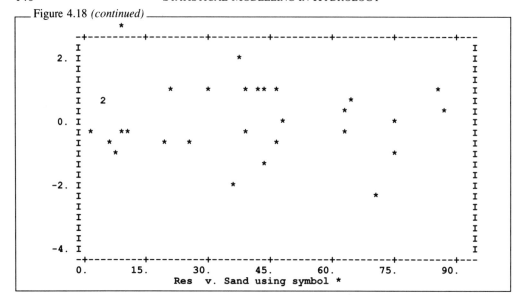
Figure 4.18 (continued)

4.8 GENERALISED LINEAR REGRESSION MODELS: RESIDUALS $\{\varepsilon_t\}$ WITH KNOWN VARIANCE-COVARIANCE MATRIX V

We discussed in Section 4.5 the case in which the random variables $\{Y_t\}$ had weights n_t, which, in the case where Y_t was the mean annual flood for a basin, represented the number of years of record from which y_t was calculated. We can express this by saying that, instead of the variance-covariance matrix of the random variables $\{Y_t\}$ being of the diagonal form diag $\{\sigma_\varepsilon^2, \sigma_\varepsilon^2, \sigma_\varepsilon^2, \ldots, \sigma_\varepsilon^2\}$, the variance-covariance matrix has the modified form $\{\sigma_\varepsilon^2/n_1, \sigma_\varepsilon^2/n_2, \sigma_\varepsilon^2/n_3, \ldots, \sigma_\varepsilon^2/n_k\}$.

The case of a weighted linear regression in which the observations $\{y_t\}$ of the response variable have variances σ_ε^2/n_t is a particular case of the more general one in which the variances of the $\{\varepsilon_t\}$ — and hence of the random variables $\{Y_t\}$ — are assumed to have a variance-covariance matrix of the form $\mathbf{V}\sigma_\varepsilon^2$, where $\mathbf{V}$ is known and with dimension $n \times n$. A straightforward modification to linear statistical theory shows that the equations for estimating the vector β, with dimension $(p + 1) \times 1$, then become

$$(\mathbf{X}^T\mathbf{V}^{-1}\mathbf{X})\beta = \mathbf{X}^T\mathbf{V}^{-1}\mathbf{y} \tag{4.20}$$

giving

$$\hat{\beta} = (\mathbf{X}^T\mathbf{V}^{-1}\mathbf{X})^{-1}(\mathbf{X}^T\mathbf{V}^{-1}\mathbf{y}) \tag{4.21}$$

The vector of residuals $\mathbf{R} = \mathbf{y} - \hat{\mathbf{y}} = \mathbf{y} - \mathbf{X}\hat{\beta}$ leads to the estimate of σ_ε^2:

$$\hat{\sigma}_\varepsilon^2 = \mathbf{R}^T\mathbf{V}^{-1}\mathbf{R}/(n - p - 1) \tag{4.22}$$

where p is the number of explanatory variables and n is the number of observations on the response variable y. Earlier algebraic expressions, and analyses of variance, are easily modified to this more general case.

What if the matrix V is unknown? An iterative calculation is often appropriate which may be illustrated by reference to the case where the matrix V is of the form

$$V = \begin{bmatrix} 1 & \rho & \rho^2 & \rho^3 & \cdots & \rho^n \\ \rho & 1 & \rho & \rho^2 & \cdots & \rho^{n-1} \\ & \cdots & & \cdots & & \cdots \\ \rho^n & \rho^{n-1} & \rho^{n-2} & \rho & \cdots & 1 \end{bmatrix} \sigma_\varepsilon^2$$

where ρ is unknown, with $-1 < \rho < 1$. The matrix V is then the variance-covariance matrix for a model in which the residuals ε_t constitute a 'lag-1 autoregression' $\varepsilon_t = \rho\varepsilon_{t-1} + a_t$, where a_t is now a sequence of independently distributed random variables with variance σ_a^2. Beginning with the model

$$y_t = \mathbf{x}_t^T \boldsymbol{\beta} + \varepsilon_t$$

we can write

$$(y_t - \rho y_{t-1}) = (\mathbf{x}_t^T\boldsymbol{\beta} - \rho\mathbf{x}_{t-1}^T\boldsymbol{\beta}) + (\varepsilon_t - \rho\varepsilon_{t-1})$$

or

$$(y_t - \rho y_{t-1}) = (\mathbf{x}_t^T - \rho\mathbf{x}_{t-1}^T)\boldsymbol{\beta} + a_t$$

Recalling that the $\{a_t\}$ are independently distributed, little information is lost if the $\boldsymbol{\beta}$ are estimated, for any fixed ρ, by minimizing

$$S = \sum_{t=2}^n \{(y_t - \rho y_{t-1}) - (\mathbf{x}_t^T - \rho\mathbf{x}_{t-1}^T)\boldsymbol{\beta}\}^2$$

(When the a_t are normally distributed $N(0, \sigma_a^2)$, this is of course equivalent to maximisation of the log-likelihood function). Hence, by taking a grid of values over the range $-1 < \rho < 1$, and minimising S with respect to $\boldsymbol{\beta}$ for each ρ-value in the grid, we find the value of ρ for which S is a minimum, thus determining also the estimates of $\boldsymbol{\beta}$ and σ_a^2. This analysis can be modified to take account of the information in a_1, which has been omitted in the above expression for S; where n is fairly large, the gain in information from including the component of the likelihood function for a_1 is likely to be small.

4.9   INFLUENCE, CONSISTENCY AND LEVERAGE

We now return to the discussion of standardized residuals and leverage, mentioned in the output from the simple linear regression of Ibirama annual floods (y) on Apiuna floods (x), in Section 4.4.1. These quantities help to identify points in the data set of pairs $\{x_t, y_t\}$ which exert a strong influence on the fitted model, or which deviate from the general distribution of points in a particular way; they therefore provide checks on the model, as well as identifying data points which may be suspect.

The concepts of influence, consistency and leverage are best illustrated by reference to the diagrams in Figure 4.19. In Figure 4.19(a), the x-value of the extreme point (denoted by the open circle) is close to the mean of all the x's; if the extreme point were omitted, the effect on the estimate of the slope β_1 would be small, although the effect on the estimate of the intercept β_0 could be substantial. Excluding the extreme point also greatly improves the model fit (as measured by the quantity, shown in the GENSTAT output, as 'Percentage variance accounted for'). In Figure 4.19(b), the extreme point is consistent with the remaining points

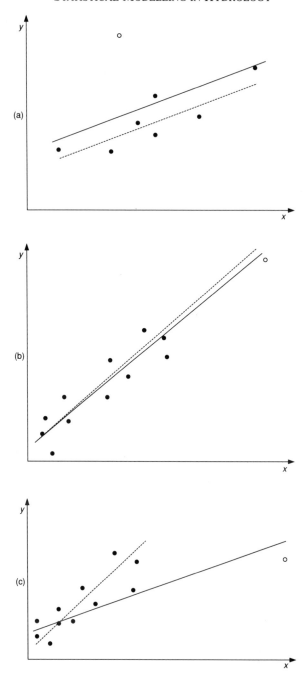

Figure 4.19   Diagram illustrating the influence of the open-circled point (○). In (a), the point has little influence on slope, but has considerable influence on the intercept and residual variance. In (b), the point is consistent with the remaining points, having little influence on slope and intercept but large influence on the residual variance. In (c), the point has large influence on slope, intercept and residual variance.

in that, if a line were fitted through them, this line would pass near to the extreme point. Inclusion of the extreme point does not greatly affect the estimate of the slope β_1, although its accuracy will be much improved. Finally, in Figure 4.19(c), omission of the extreme point would have a very great effect on the estimate of β_1; the slope of the line passing through the remaining points would be quite different from the slope obtained when the extreme point is also included.

A more formal way of expressing these ideas is to say that, in Figures 4.19(b) and 4.19(c), the extreme point has high *leverage*; in Figure 4.19(a), the point has low leverage. We can also say that, in Figure 4.19(b), the extreme point has high *consistency*, in that it lies very close to the line fitted to the remaining points although it may be far removed from it. The extreme point in Figure 4.19(c) has high influence, since the slope β_1 of the line is greatly changed by omitting it; in Figures 4.19(a) and 4.19(b) the extreme point is much less influential.

4.9.1   Leverage

To express these ideas in more concrete terms, we begin by developing the idea of leverage. The residuals, denoted by r_t, are the differences between the observed values y_t and the 'fitted values' $\hat{y}_t$ corresponding to them; thus $r_t = y_t - \hat{y}_t$ or, for the $N \times 1$ vector of residuals $\mathbf{r}$, $\mathbf{r} = \mathbf{y} - \hat{\mathbf{y}} = \mathbf{y} - \mathbf{X}\hat{\boldsymbol{\beta}}$. But $\hat{\boldsymbol{\beta}} = (\mathbf{X}^T\mathbf{X})^{-1}\mathbf{X}^T\mathbf{y}$, and substituting this in the equation $\mathbf{r} = \hat{\mathbf{y}} - \mathbf{X}\hat{\boldsymbol{\beta}}$, we find that $\mathbf{r} = (\mathbf{I} - \mathbf{H})\mathbf{y}$, where $\mathbf{I}$ is the $N \times N$ identity matrix (having ones in its leading diagonal, and zeros elsewhere) and $\mathbf{H}$ is the (so-called 'hat') matrix

$$\mathbf{H} = \mathbf{X}(\mathbf{X}^T\mathbf{X})^{-1}\mathbf{X}^T \tag{4.23}$$

The diagonal elements of the matrix $\mathbf{H}$ are of particular importance, since it is they which provide us with measures of leverage. It can be shown (Atkinson, 1987) that, for the case of simple linear regression $E[Y] = \beta_0 + \beta_1 x$, the diagonal elements of $\mathbf{H}$, denoted by h_1, $h_2, \ldots, h_N$, have minimum values $1/N$, and maximum values of 1, the latter value occurring when the extreme point exerts such strong leverage on the fitted relation that the residual r_t is zero. Hence, to identify extreme points which are exerting strong leverage on the fit, we look at the diagonal terms of the matrix $\mathbf{H}$ in equation (4.23); if a diagonal term is close to one, we expect to find that the corresponding point is exerting strong leverage on the regression fit, of the kind illustrated in Figure 4.19(c). For simple linear regression (one explanatory variable only), GENSTAT provides warning messages of high leverage where the diagonal elements in the $\mathbf{H}$ matrix exceed $2/N$. In the more general case of p explanatory variables, warning messages are printed where the diagonal elements of $\mathbf{H}$ exceed $2p/N$.

Returning to the example of the regression of Ibirama annual floods (y) on Apiuna annual floods (x), it will be recalled that the messages printed in the output gave warnings of high leverage of points $t = 50$ and $t = 51$ in the sequence. These are the extreme floods observed in the years 1983 and 1984, believed to correspond to *el Niño* events; the plots given in Figure 4.1 show that these floods lie well to the right of the other points plotted, but that they are consistent with them. The reason for their remoteness is known, and clearly we should not wish to exclude them from the analysis. However, to get an idea of how the inclusion of these points affects the estimates of β_0 and β_1, we could omit each point in turn, and recalculate the regression from the remaining points. Changes in the values of the regression coefficients will give us an idea of the influence of the omitted point on the regression, and so will the distance of the omitted point from this regression. Indeed, we can do this not only for the two *el Niño* events of 1983 and 1984, but also for each of the remaining points in the data set, thus obtaining intuitively an idea of the extent to which each point is influential. We could, in theory, take the calculation still further, by omitting every possible pair (triplet, . . .) of points in turn, and then recalculating the regression with the pair (triplet, . . .) omitted, thus obtaining an idea of the influence of that pair (triplet, . . .), on the regression as a whole; but when the number N of data points is large, the number of regressions to be calculated when the $N(N-1)/2$ pairs ($N(N-1)(N-2)/6$ triplets, ...) are omitted, becomes very large, and in practice it is sufficient to omit points singly rather than in pairs (triplets, . . .). When points are omitted one by one, the usual regression formulae can be easily modified, without the need to calculate each of the N modified regressions *ab initio*; for Atkinson (1987) shows that, if $\hat{\beta}$, s^2 are the estimates of β, σ_ε^2 using all N data points, and if $\hat{\beta}_{(i)}$, $s_{(i)}^2$ are the corresponding estimates when the ith data point is omitted,

$$\hat{\beta}_{(i)} - \hat{\beta} = -(X^TX)^{-1}x_i r_i/(1 - h_i)$$

where x_i is the $(p+1) \times 1$ vector of explanatory variables for the ith observation, r_i is the residual $y_i - \hat{y}_i$, and h_i and X are as previously defined; and

$$(n - p - 1)s_{(i)}^2 = (n - p)s^2 - r_i^2/(1 - h_i)$$

4.9.2   Standardised Residuals

We recall from the last section that the residuals, denoted by the $N \times 1$ vector $r = y - \hat{y}$, can be expressed in terms of the observations y by

$$r = (I - H)y$$

where I is the $N \times N$ identity matrix and $H = X(X^TX)^{-1}X^T$. This relation expresses each residual r_t as a linear combination of the observations y_t; and it can be shown that the variance of the residual r_t is

$$\text{var } r_t = (1 - h_t)\sigma_\varepsilon^2 \tag{4.24}$$

where h_t is the element of the matrix H on the diagonal position defined by row and column t. Equation (4.24) shows, therefore, that the residuals all have different variances, and that the residual for a point with high leverage (h_t near to one) has smaller variance than the residual for a point with low leverage. We see from equation (4.24) that it is also necessary to estimate the variance of residuals σ_ε^2, and this estimate is given by

$$S_\varepsilon^2 = R(\hat{\beta})/(N - p) \tag{4.25}$$

To examine whether the assumption of constant σ_ε^2 is valid, we need to have residuals r_t which all have constant variance when the model is 'correct'. For this purpose, we therefore calculate the 'standardised residuals', denoted by

$$r_t' = r_t/[S_\varepsilon \sqrt{(1 - h_t)}] \tag{4.26}$$

These are the quantities printed in GENSTAT output. And since

$$\text{var } r_t' = \text{var } r_t/(\sigma_\varepsilon^2[1 - h_t])$$

$$= 1$$

all standardised residuals have the same unity variance; they are therefore particularly appropriate where we need to make normal probability plots, or to check the homogeneity of variances.

4.9.3   Deletion Residuals and Cook Statistics

We have seen that the concept of leverage arises when we consider how the exclusion of one or more points, remote from the main cluster, would affect the model fit. We can extend this idea further by calculating how the residuals, and the estimates $\hat{\boldsymbol{\beta}}$, would change as a consequence of omitting the tth point of the cluster, where $t = 1, 2, \ldots, N$. In particular, we ask the following two questions:

(i)   How well does the observation y_t agree with the predicted value $\hat{y}_{(t)}$ obtained when the tth observation is omitted from the fitting procedure? (Brackets around the suffix t indicate that the tth point has been omitted, so that $\hat{y}_{(t)}$ is the estimate of y_t obtained when that observation is omitted from the model-fitting procedure.)

(ii) How do the estimates $\hat{\boldsymbol{\beta}}$ change when the tth point is excluded, thereby leading to the modified estimates $\hat{\boldsymbol{\beta}}_{(t)}$ (so that $\hat{\boldsymbol{\beta}}_{(t)}$ is the estimate of the $\boldsymbol{\beta}$ obtained when the observation y_t is omitted from the model-fitting procedure)?

In answer to question (i), we consider the difference

$$r_t^* = y_t - \hat{y}_{(t)} \tag{4.27}$$

The quantities r_t^* are known as the *deletion residuals*, since they measure the agreement between an observation and the value predicted for it when that observation has been deleted. Some algebra, which need not concern us here, shows that the deletion residuals are related to the standardised residuals r_t', defined earlier, by means of the equation

$$r_t^* = r_t'/[(N - p - r_t'^2)/(N - p - 1)]^{1/2} \tag{4.28}$$

in which p is the number of components in the vector $\boldsymbol{\beta}$ of regression coefficients (recalling that $p = 2$ for simple linear regression). The deletion residuals are important because we can use them to answer questions such as whether the r_t^* can be taken as arising from a homogeneous sample. If they can, and if the errors ε_t are normally distributed, then the normal plot discussed in the following section should be approximately a straight line; we can also use this plot to identify unduly influential or outlying observations.

In answer to question (ii), concerning how the estimates $\hat{\boldsymbol{\beta}}$ and $\hat{\boldsymbol{\beta}}_{(t)}$ differ, Cook (1977) has proposed the following statistic for detecting influential observations:

$$D_t = \{[\hat{\boldsymbol{\beta}}_{(t)} - \hat{\boldsymbol{\beta}}]^T \mathbf{X}^T \mathbf{X} [\hat{\boldsymbol{\beta}}_{(t)} - \hat{\boldsymbol{\beta}}]\}/ps_\varepsilon^2 \tag{4.29}$$

Large values of D_t indicate observations which are particularly influential on the parameter estimates. A more convenient form for calculating the D_t expresses them in terms of the standardised residuals r_t as follows:

$$D_t = (t'^2 h_t)/(1 - h_t) \qquad (4.30)$$

where h_t, as before, is the diagonal element in row t of the matrix $\mathbf{H} = \mathbf{X}(\mathbf{X}^T\mathbf{X})^{-1}\mathbf{X}^T$. However, for graphical applications, Atkinson (1987) has suggested use of the 'modified Cook statistic' given by

$$C_t = \{[(n - p)/p]h_t/(1 - h_t)\}^{1/2}|r_t^*| \qquad (4.31)$$

4.10 CHECKS ON NORMALITY OF RESIDUALS IN REGRESSION

The regression models of this chapter have assumed that the residuals ε_t have a normal distribution $N(0, \sigma_\varepsilon^2)$. This assumption is clearly one which must be checked, and the method that we use for this purpose is based upon the calculation of envelopes, of the kind used in Chapter 3. Starting with the regression model $\mathbf{y} = \mathbf{X}\boldsymbol{\beta} + \varepsilon$, we have seen that the estimate of the $(p + 1) \times 1$ vector $\boldsymbol{\beta}$ is $\hat{\boldsymbol{\beta}} = (\mathbf{X}^T\mathbf{X})^{-1}\mathbf{X}^T\mathbf{y}$, and that the residuals ε_t are then estimated by $\mathbf{e} = \mathbf{y} - \hat{\mathbf{y}} = \mathbf{y} - (\mathbf{X}^T\mathbf{X})^{-1}\mathbf{X}^T\mathbf{y} = (\mathbf{I} - \mathbf{H})\mathbf{y}$, where $\mathbf{I}$ is the $N \times N$ unit matrix, and $\mathbf{H}$ is the $N \times N$ 'hat' matrix $\mathbf{X}(\mathbf{X}^T\mathbf{X})^{-1}\mathbf{X}^T$. Putting $\mathbf{y} = \mathbf{X}\boldsymbol{\beta} + \varepsilon$ in $\mathbf{e} = (\mathbf{I} - \mathbf{H})\mathbf{y}$ gives $\mathbf{e} = (\mathbf{I} - \mathbf{H})\varepsilon$.

To check whether the ε have a normal $N(0, \sigma_\varepsilon^2)$ distribution, we generate m pseudo-random vectors (commonly with $m = 19$) with N components, using the distribution $N(0, s_\varepsilon^2)$, where s_ε^2 is the estimate of σ_ε^2. Denote these m vectors by $\varepsilon_1, \varepsilon_2, \ldots, \varepsilon_m$. Keeping the hat matrix $\mathbf{H}$ fixed, we calculate $\mathbf{e}_1 = (\mathbf{I} - \mathbf{H})\varepsilon_1, \mathbf{e}_2 = (\mathbf{I} - \mathbf{H})\varepsilon_2, \ldots, \mathbf{e}_m = (\mathbf{I} - \mathbf{H})\varepsilon_m$. The N elements of $\mathbf{e}_1, \mathbf{e}_2, \ldots, \mathbf{e}_m$ are then put into ascending order; then for the element in the ith position within each $\mathbf{e}$-vector, the maximum (over all m such elements) is found, and also the minimum. Having calculated the N maxima and N minima, these are plotted against the N quantiles of the standard normal distribution to form the upper and lower envelopes. On the same diagram, we also show the probability plot for the N values of data. If the observed residuals fall beyond or near the edge of the band defined by the two envelopes, the assumption that the ε come from a normal distribution is called into doubt. The region between the two envelopes is roughly like a confidence region for the N ordered residuals from the data sample.

We illustrate the calculation using Haan's data shown in Figure 4.14. Figure 4.20 shows the residuals, $\mathbf{e}$, calculated from the data, plotted as asterisks. Also shown are upper and lower envelopes, computed by the procedure described above, plotted as dashes. At some points, the asterisks and the dashes are too close to each other to be discriminated at the resolution afforded by the printer, so that the overlapping points are shown as colons. However, we can see that there is no suggestion that the data points wander outside the envelopes, so it is safe to conclude that the errors in the multiple regression model $RO = \beta_0 + \beta_1 Prec + \beta_2 P + \varepsilon$ can reasonably be taken as normally distributed.

A particularly useful procedure for exploring the structure of regression residuals (although not, as yet, for the construction of envelopes) is the GENSTAT library procedure RCHECK. This provides displays of various types of residual, the leverages, modified Cook statistics as simple plots against fitted values or against an index variate (that is, a plot against the order in which the data were typed for input). Normal (Q-Q) plots and half-normal plots (see Draper and Smith, 1981) can also be obtained. We illustrate the usefulness of this procedure using Haan's data (see Figure 4.14) on 13 small basins in Kentucky, from which a regression was calculated of mean annual runoff (*RO*, in inches) on a number of explanatory variables, including annual precipitation (*Prec*, in inches), basin perimeter (*P*, in miles), and relief ratio

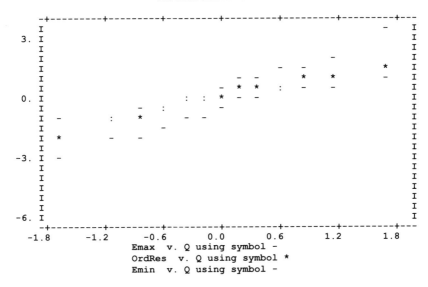

Figure 4.20   Plot to examine normality of errors in multiple regression of RO on explanatory variables *Prec* and *P* (see Haan's data in Figure 4.14). Graph shows upper and lower envelopes, denoted by horizontal dashes; and standardised residuals obtained from the data.

(*Rr*, defined as the ratio of total relief to the largest dimension of the basin generally parallel to the main stream, in units of feet per mile). Having stored the residuals about this regression, we can plot the standardised residuals against the fitted values, and against an index variate, using the directives

```
> RCHECK YSTATISTIC=residual; XMETHOD=fittedvalues
> RCHECK YSTATISTIC=residual; XMETHOD=index
```

with the results shown in Figures 4.21(a) and 4.21(b); the diagrams show no evidence of departures from randomness. The directive

```
> RCHECK YSTATISTIC=residual; XMETHOD=normal
```

gives the normal (Q-Q) plot shown in Figure 4.21(c), with no evidence of departure from linearity. We conclude that the residuals obtained when RO is regressed upon *Prec*, *P* and *Rr* can be taken as normally distributed, and that there is no evidence of variance heterogeneity.

A similar exploration of the deletion residuals, denoted above by r_i^*, can also be undertaken. We recall that the deletion residuals enable us to detect significant differences between the predicted and observed values of the response variable, for each observation. The three directives

```
> RCHECK [RMETHOD = deletion] YSTATISTIC = residual; XMETHOD\
=fittedvalues
> RCHECK [RMETHOD = deletion] YSTATISTIC = residual; XMETHOD\
=index
> RCHECK [RMETHOD = deletion] YSTATISTIC = residual; XMETHOD\
=normal
```

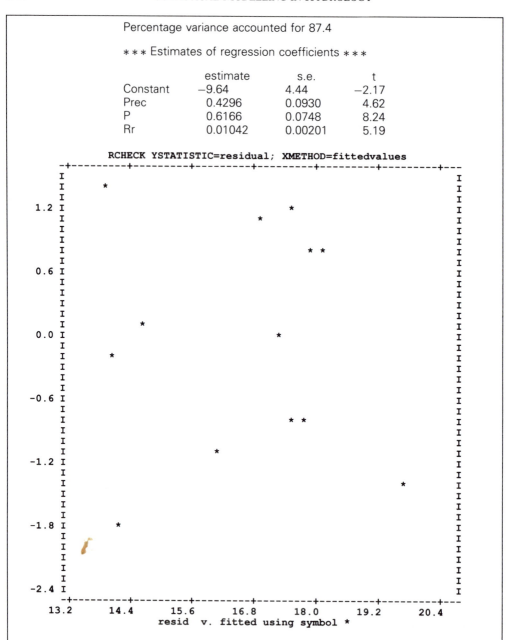

Figure 4.21   Plots of standardised residuals against (a) fitted values; (b) as index variable named 'units', giving the data order; and (c) normal distribution quantities, giving a normal or Q-Q plot; and of modified Cook statistic, for Haan's data (see Figure 4.14). Residuals after regression of mean annual runoff, RO (in inches), on the three explanatory variables $Prec$, P, Rr (see data in Figure 4.14).

Continued

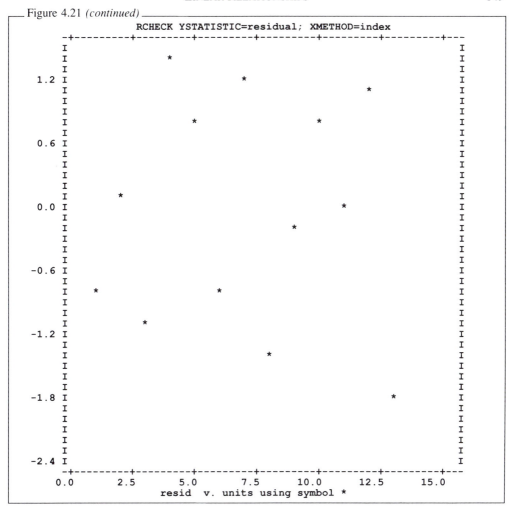

Figure 4.21 (continued)

yield three diagrams showing no evidence of departures from randomness or normality amongst the deletion residuals.

Changes in the estimates of the regression parameters β_i, due to the deletion of observations one after another, are measured by the modified Cook statistic C_t as defined in equation (4.31) above, remembering that C_t is the value of the statistic when the tth observation is deleted. As before, we can plot C_t against fitted values, and against the index variable, by means of the two directives

```
> RCHECK YSTATISTIC = Cook; XMETHOD = fittedvalues
> RCHECK YSTATISTIC = Cook; XMETHOD = index
```

These give diagrams that give no indication of any correlation between the Cook statistic and either the values predicted from the regression, or the order of the data as given by the index variable. The normal (Q-Q) plot for the C_t statistic is given by

```
> RCHECK YSTATISTIC = Cook; XMETHOD = normal
```

Figure 4.21 *(continued)*

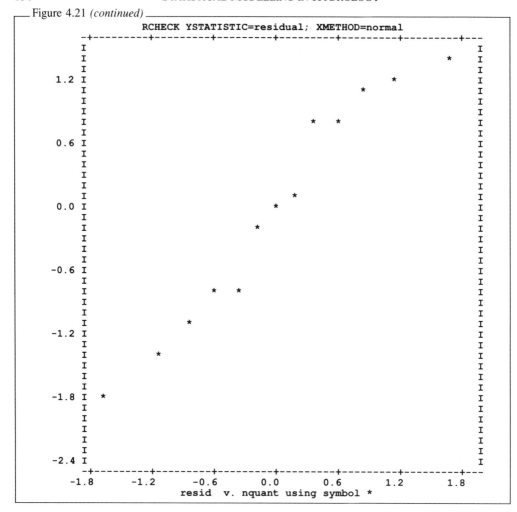

This provides giving no evidence that the Cook statistics are other than normally distributed with homogeneous variance, and we conclude that Haan's data are particularly well behaved, exhibiting none of the high leverages and extreme residuals encountered in some of the other data sets examined.

4.11   TRANSFORMATIONS IN REGRESSION ANALYSIS

In examples such as the regression of mean annual flood (as response variable y) on basin area (explanatory variable x), we transformed both variables to logarithms, without offering a justification for this procedure. We now return to discuss the matter of transformations in rather greater depth. There are really three situations in which transformation of the data may be appropriate in regression analysis (Cook and Weisberg, 1982); these are:

(i)   Where the response variable y comes from a family of distributions known to be non-normal. A transformation is selected so that the distribution of the transformed responses is sufficiently near to normal to allow us to use standard methods (t-tests, F-tests) based on normal theory.

(ii) Where the expected response $E[Y]$ is related to the explanatory variables $\mathbf{x}$ by a known non-linear function of the parameters: for example, $E[Y] = \beta_0 \exp(-\beta_1 x)$. A transformation is selected to linearise this response function; diagnostic checks are essential to ascertain how the distribution of errors is affected by the transformation.

(iii) Where neither the distribution of errors nor the functional form of the relationship between $E[Y]$ and $\mathbf{x}$ is known. This situation is the most difficult to handle since a specific rationale for choosing a transformation is lacking. Generally, we would like a transformation to result in a model with constant error variance, approximately normal errors, and a systematic component $\mathbf{X}\boldsymbol{\beta}$ which is easily interpreted in hydrological terms.

One method of proceeding in case (iii) is to specify a family of transformations $y^{(\lambda)}$, with parameter λ (possibly a vector). We then use the data to select which transformation $y^{(\lambda)}$ of the family is most appropriate, in the sense that it yields a model $y^{(\lambda)} = \mathbf{X}\boldsymbol{\beta} + \varepsilon$ having all the desirable properties.

One such family of transformations is the Box–Cox family, already mentioned in Section 3.3.1:

$$y^{(\lambda)} = \begin{cases} (y^{(\lambda)} - 1)/\lambda & \lambda \neq 0 \\ \ln y & \lambda = 0 \end{cases}$$

In our transformations of mean annual flood, MAF, and basin area A, to logarithms (see Figure 4.11), we had adopted the alternative $\lambda = 0$ on an *ad hoc* basis without really looking at whether this was justified. Furthermore, we applied the same transformation to both the response variable MAF and the explanatory variable A; there is the possibility, which we did not explore, that different transformations $MAF^{(\lambda_1)} = (MAF^{\lambda_1} - 1)/\lambda_1$ and $A^{(\lambda_2)} = (A^{\lambda_2} - 1)/\lambda_2$, might result in a model with more desirable properties (normality of error distribution, homogeneity of error variance).

The possibility of separate Box–Cox transformations to response variables and explanatory variables is by no means the only one. Added flexibility is given to the Box–Cox family by using the two-parameter *extended power family* of transformations:

$$y^{(\lambda)} = \begin{cases} [(y + \lambda_2)^{\lambda_1} - 1]/\lambda_1 & \lambda_1 \neq 0 \\ \ln(y + \lambda_2) & \lambda_1 = 0 \end{cases}$$

Where, say, the response variable is symmetric but with tails that are 'heavier' than normal, the family of *modulus transformations* is likely to be appropriate:

$$y^{(\lambda)} = \begin{cases} \text{sign}(y)[(|y| + 1)^{\lambda} - 1]/\lambda & \lambda \neq 0 \\ \text{sign}(y)[\ln(|y| + 1)] & \lambda = 0 \end{cases}$$

For variables defined over a finite range $[0, b]$, such as the degree of urbanisation in a catchment area, expressed as a proportion ($b = 1$), the family of *folded-power transformations*

$$y^{(\lambda)} = \begin{cases} [y^{\lambda} - (b - y)^{\lambda}]/\lambda & \lambda \neq 0 \\ \ln[y/(b - y)] & \lambda = 0 \end{cases}$$

is appropriate.

Faced with such numerous alternative transformations, one does well to curb one's enthusiasm by avoiding lengthy but entertaining calculations to estimate the λ-parameters of transformations, which may bring very marginal benefits at very considerable cost. It is probably better to assume a small number of λ-values (say, $\lambda = 0, \pm\frac{1}{2}$), testing the effects of each in interactive mode on the computer, instead of entering upon a sophisticated calculation to determine the optimum λ-values. Nevertheless, computer hardware and software are becoming such that even sophisticated optimisations to determine λ are within reach.

4.12 INFORMATION TRANSFER AND EXTENSION OF FLOW RECORDS

A common problem encountered in hydrological practice is that the flow records available at a site are short. If longer records of discharge exist for a gauging site nearby, then the correlation between the two records may be used to 'transfer information' from the site with a long record to the site with a short record. The transfer may take various forms. One approach, illustrated below by the GENSTAT procedure MULTMISS, is to use multiple regression to estimate the missing data at the short-record site, for which the term 'record extension' is commonly used; we discuss this topic in Chapter 7. A related approach is to utilise the correlation to estimate parameters (such as mean annual flood, when the data are annual flood series) at the site with short record. It is well known that, where the correlation between two records is low, estimates of mean annual flood may lose precision, rather than gain it, when we try to transfer information in this way. In fact, when the joint distribution of observations in the two records is bivariate normal, the estimated mean annual flood at the short-record site has greater precision only when $\rho^2 > 1/(n_y - 2)$, where n_y is the number of years of common record. Similar results can be expected for other bivariate distributions, and for the estimation of other statistics, such as the variance of annual floods at the short-record site. In practice, such results are less useful than they appear because the 'true' correlation coefficient ρ is unknown: we only have an estimate of it, say r, calculated from the available data.

A further reason why the results of much research are of limited use is that frequently we are interested not so much in the information gain (or loss) for the mean or variance of annual floods, but in the information gain in the flood with T-year return period. In principle, we can approach this problem in the following way.

Suppose that $\{y_t\}$, $\{x_t\}$ are the data at the short- and long-record sites, respectively, of lengths n_y and n_x. Given a particular bivariate distribution for the pair of random variables X_t and Y_t, we can write down the likelihood function for the $n_x + n_y$ observations. Let us suppose that the bivariate distribution contains five parameters: two means μ_x and μ_y, two variances σ_x^2 and σ_y^2, and a correlation ρ. The marginal distribution for y_t will involve μ_y and σ_y^2, and the estimated T-year flood for the short-record site will use estimates of these two parameters only, the parameters μ_x, σ_x^2 and ρ being essentially 'nuisance parameters'. If we establish a grid of values for μ_y and σ_y^2 — for example, a grid of values (μ^*, σ^{2*}) defined on the space $\mu_y > 0$ and $\sigma_y^2 > 0$ — then a T-year flood value, X, can be calculated for each grid point, using the values (μ^*, σ^{2*}) to solve for X:

$$\int_X^\infty f(y; \mu^*, \sigma^{2*}) \mathrm{d}y = 1/T$$

Further, the nuisance parameters μ_x, σ_x^2, and ρ can be estimated by maximum likelihood, and the value of the log-likelihood conditional on μ^* and σ^{2*}, namely $\ln L = l(\mu^*, \sigma^{2*})$, can be calculated. Hence, corresponding to each grid point, we have a value of the T-year flood X, and a conditional log-likelihood. By identifying the point in the grid for which the conditional log-likelihood is a maximum, the estimated T-year flood is obtained.

4.13 'UPDATING' REGRESSION ANALYSES USING ADDITIONAL DATA

The diagnostic techniques in regression analysis, discussed above, essentially examined the changes to residuals and estimated regression coefficients, brought about as a consequence of omitting each observation y_t (together with its corresponding x_t) in turn. The converse of this procedure examines how estimated regression coefficients, together with predictions made using the regression equation and their standard errors, change in consequence of the availability of *additional* observations of y_t and x_t. For example, we may have calculated a multiple regression of mean annual flood y_t — with appropriate weightings — on catchment characteristics for N drainage basins, of the form $E[Y_t] = \mathbf{x}^T\boldsymbol{\beta}$, with $\boldsymbol{\beta}$ estimated by $\hat{\boldsymbol{\beta}}_{(N)} = (\mathbf{X}_{(N)}^T\mathbf{X}_{(N)})^{-1}\mathbf{X}_{(N)}^T\mathbf{y}_{(N)}$, the suffix (N) indicating that the estimate is based upon data y and X from N basins. Subsequently, data from an additional M basins may be obtained, and it is then necessary to incorporate these additional data into the estimation procedure. Computer calculations are now so easy that it will often be straightforward to recompute the regression to obtain

$$\hat{\boldsymbol{\beta}}_{(N+M)} = (\mathbf{X}_{(N+M)}^T\mathbf{X}_{(N+M)})^{-1}\mathbf{X}_{(N+M)}^T\mathbf{y}_{(N+M)} \tag{4.32}$$

with the error variance σ_e^2 estimated by

$$\hat{\sigma}_e^2 = \sum[\mathbf{y}_{(N+M)} - \mathbf{X}_{(N+M)}\hat{\boldsymbol{\beta}}_{(N+M)}]^T[\mathbf{y}_{(N+M)} - \mathbf{X}_{(N+M)}\hat{\boldsymbol{\beta}}_{(N+M)}]/(N + M - p - 1) \tag{4.33}$$

where the matrix $\mathbf{X}$ is now of dimension $(N + M) \times (p + 1)$, with $\boldsymbol{\beta}$ and $\mathbf{y}$ having dimensions $(p + 1) \times 1$ and $(N + M) \times 1$, respectively. However, it is sometimes convenient to use 'updating' procedures based upon the following matrix relationship.

Let $\mathbf{M}$ be a $p \times p$ symmetric matrix with rank p, with $\mathbf{m}, \mathbf{n}$ $q \times p$ matrices with rank q; commonly $q \ll p$. Then where the inverses exist,

$$(\mathbf{M} + \mathbf{m}^T\mathbf{n})^{-1} = \mathbf{M}^{-1} - \mathbf{M}^{-1}\mathbf{m}^T(\mathbf{I} + \mathbf{n}\mathbf{M}^{-1}\mathbf{m}^T)^{-1}\mathbf{n}\mathbf{M}^{-1} \tag{4.34}$$

which expresses the inverse of the 'new' $p \times p$ matrix $\mathbf{M} + \mathbf{m}^T\mathbf{n}$ in terms of the inverse $\mathbf{M}^{-1}$ that is already known, of the $q \times q$ inverse of $\mathbf{I} + \mathbf{n}\mathbf{M}^{-1}\mathbf{m}^T$, and of products involving these and $\mathbf{n}, \mathbf{m}^T$. The application of equation (4.34) leads to a particularly simple result where $q = 1$; thus, suppose now that $\mathbf{M} = \mathbf{X}^T\mathbf{X}$, with $\mathbf{X}$ of dimension $N \times p$, and let $\mathbf{m} = \mathbf{x}_{(N+1)}$ be an additional $(N+1)$th row of $\mathbf{X}$, corresponding to one new observation. Put $\mathbf{n}$ also equal to $\mathbf{x}_{(N+1)}$; then on the left-hand side of equation (4.34) we have

$$[\mathbf{X}_{(N)}^T\mathbf{X}_{(N)} + \mathbf{x}_{(N+1)}^T\mathbf{x}_{(N+1)}]^{-1} \tag{4.35}$$

which is equal to $[\mathbf{X}_{(N+1)}^T\mathbf{X}_{(N+1)}]^{-1}$. But from equation (4.34) this is equal to

$$[\mathbf{X}_{(N)}^T\mathbf{X}_{(N)}]^{-1} - \lambda[\mathbf{X}_{(N)}^T\mathbf{X}_{(N)}]^{-1}[\mathbf{x}_{(N+1)}^T\mathbf{x}_{(N+1)}][\mathbf{X}_{(N)}^T\mathbf{X}_{(N)}]^{-1} \tag{4.36}$$

where λ is the scalar quantity

$$\lambda = 1 + \mathbf{x}_{(N+1)}[\mathbf{X}_{(N)}^T\mathbf{X}_{(N)}]^{-1}\mathbf{x}_{(N+1)}^T \tag{4.37}$$

Thus, given inverse $[\mathbf{X}_{(N)}^{\mathrm{T}}\mathbf{X}_{(N)}]^{-1}$ already calculated using N observations, we can easily incorporate an $(N + 1)$th observation y_{N+1} and its corresponding $(1 \times p)$ vector $\mathbf{x}_{(N+1)}$ of values for the explanatory variables.

It is easily found that the updated estimate $\hat{\boldsymbol{\beta}}_{(N+1)}$ is given by

$$\hat{\boldsymbol{\beta}}_{(N+1)} = \hat{\boldsymbol{\beta}}_{(N)} - \lambda[\mathbf{X}_{(N)}^{\mathrm{T}}\mathbf{X}_{(N)}]^{-1}[\mathbf{x}_{(N+1)}^{\mathrm{T}}\mathbf{x}_{(N+1)}]\hat{\boldsymbol{\beta}}_{(N)}$$

$$+ [\mathbf{X}_{(N)}^{\mathrm{T}}\mathbf{X}_{(N)}]^{-1}\mathbf{x}_{(N+1)}^{\mathrm{T}}y_{N+1} - \lambda(\mathbf{X}_{(N)}^{\mathrm{T}}\mathbf{X}_{(N)})^{-1}\mathbf{x}_{(N+1)}^{\mathrm{T}}y_{N+1} \qquad (4.38)$$

whilst the updated estimate of σ_e^2 is

$$\hat{\sigma}_{e(N+1)}^2 = \sum[y_{(N+1)} - \mathbf{X}_{(N+1)}\hat{\boldsymbol{\beta}}_{(N+1)}]^{\mathrm{T}}[y_{(N+1)} - \mathbf{X}_{(N+1)}\hat{\boldsymbol{\beta}}_{(N+1)}]/(N + 1 - p - 1) \qquad (4.39)$$

Updating by such methods was particularly important in the days when matrix inversions were calculated by hand. Since desk-top computers are now widespread, the methods are now only occasionally useful.

4.14   RIDGE REGRESSION

As the number of explanatory variables increases, so that the dimension of the matrix $\mathbf{X}^{\mathrm{T}}\mathbf{X}$ becomes large, numerical difficulties in the calculation of its inverse may be encountered, where two or more of the explanatory variables are highly correlated. This may easily occur in the hydrological context if, for example, two of the explanatory variables are rainfall at two rain gauge sites that are close to each other, or where the explanatory variables are powers of rainfall measured at a single site. The matrix $\mathbf{X}^{\mathrm{T}}\mathbf{X}$ then becomes very nearly singular or 'ill-conditioned', giving rise to unstable estimates of the parameters $\boldsymbol{\beta}$. Ridge regression is a controversial procedure designed to overcome these difficulties.

Suppose that each explanatory variable is 'standardised' by subtracting its mean (over all N data points) and dividing by its standard deviation; and let $\mathbf{Z}^{\mathrm{T}}\mathbf{Z}$ be the matrix corresponding to $\mathbf{X}^{\mathrm{T}}\mathbf{X}$ when the explanatory variables have been so transformed. If θ is a positive number, usually between 0 and 1, consider the estimates

$$\mathbf{b}(\theta) = (\mathbf{Z}^{\mathrm{T}}\mathbf{Z} + \theta\mathbf{I})^{-1}\mathbf{Z}^{\mathrm{T}}\mathbf{y}$$

where I is the unit matrix of dimension $(p + 1) \times (p + 1)$. Clearly when $\theta = 0$, we have effectively the familiar least-squares case; but in ridge regression, the elements of the vector $\mathbf{b}(\theta)$ are each plotted against θ for θ positive and increasing. As θ increases, the estimates $\mathbf{b}(\theta)$ become increasingly smaller, tending to zero as θ becomes infinite. The user must select an appropriate value of θ from this range, by looking at the plots to determine where the behaviour of the $\mathbf{b}(\theta)$ has 'stabilised'. This value of θ, say θ_{stab}, is used to calculate the $\mathbf{b}(\theta)$, and hence the $\hat{\boldsymbol{\beta}}(\theta)$, from the equations $\hat{\boldsymbol{\beta}}_j(\theta) = \mathbf{b}_j(\theta)/s_{jj}$, where s_{jj} is the standard deviation of the jth explanatory variable. These estimates of $\boldsymbol{\beta}$ are used in the regression equation; although they are biased, they avoid the instability inherent in least-squares estimates in such situations.

The reader is referred to the text by Draper and Smith (1981) for an excellent account of ridge regression and its limitations, including a clear definition of the situations in which the use of ridge regression can be fully justified on theoretical grounds. GENSTAT contains a library procedure of the form

```
RIDGE [PLOT=ridgetrace] y, X
```

where $\mathbf{X}$ contains the explanatory variables, which plots the $\mathbf{b}(\theta)$ as functions of θ.

REFERENCES

Cook, R. D. (1977). Detection of influential observations in linear regression. *J. Amer. Statist. Ass.*, **74** (365), 169–74.

Cook, R. D. and Weisberg, S. (1982). *Residuals and Influence in Regression*. Chapman and Hall, London.

Draper, N. and Smith, H. (1981). *Applied Regression Analysis* (2nd edn). Wiley, New York.

Gilley, J. E., Elliot, W. J., Laflen, J. M. and Simanton, J. R. (1993). Critical shear stress and critical flow rates for initiation of rilling. *Journal of Hydrology*, **142**, 251–71.

Haan, C. T. (1977). *Statistical Methods in Hydrology*. University of Iowa Press, Ames, Iowa

Kirby, C., Newson, M. D. and Gilman, K. (eds) (1991). *Plynlimon Research: The First Two Decades*. Institute of Hydrology Report No. 109. Wallingford England.

Pezzi, L. P. (1993). Aplicações de dois métodos no controle de qualidade em dados coletados por estações meteorológicas automáticas de superfície. M.Sc. thesis, Universidade Federal do Rio Grande do Sul, Pôrto Alegre RS, Brazil.

Roberts, A. M. (ed.) (1989). *The Catchment Research Data Base at the Institute of Hydrology*. Institute of Hydrology, Report No. 106. Wallingford, England.

Yevjevich, V. (1972). *Probability and Statistics in Hydrology*. Water Resources Publications, Fort Collins, CO.

EXERCISES AND EXTENSIONS

4.1. The table below shows data from NERC Institute of Hydrology Report No.33, 1976, by kind permission of the Institute of Hydrology, Wallingford OX108BB, England, from six throughfall troughs in the forested catchment area of the River Severn, part of the Institute of Hydrology's Plynlimon Study. (The data are in the file X4-1.DAT on the diskette distributed by The MathWorks.) The variates $x_2, x_3, \ldots, x_7$ are accumulated rainfall, measured in a rain gauge with orifice at forest canopy level, over periods of approximately one month; variates $y_2, y_3, \ldots, y_7$ are throughfall, measured in a trough approximately 5 m long and 0.25 m wide, during the same periods. Site details are as follows:

x_2, y_2: Upper Seven
x_3, y_3: Lower Severn
x_4, y_4: Nant Tanllwyth
x_5, y_5: Afon Hore
x_6, y_6: Afon Hore
x_7, y_7: Upper Severn.

(a) Calculate a linear regression of throughfall on rainfall for each trough, with 95% confidence limits for slopes and intercepts. Make normal plots of the residuals for each regression to test for normality.

x_2	y_2	x_3	y_3	x_4	y_4	x_5	y_5	x_6	y_6	x_7	y_7
62	20	62	28	60	37	48	10	70	50	65	27
60	48	54	42	78	42	61	15	68	32	70	47
70	53	73	47	125	70	66	21	85	30	96	60
125	40	105	38	180	82	98	35	170	78	132	70
130	75	138	22	210	108	140	12	222	62	192	87
184	95	155	51	212	110	160	10	217	68	218	82
182	98	178	76	230	112	172	40	216	75	215	95
186	102	182	62	250	126	178	37	224	92	222	107
220	120	187	60	261	110	180	32	215	121	222	116
220	115	207	56	284	150	205	24	255	82	263	87
228	110	205	62	275	148	205	18	250	100	280	122
245	95	215	77	280	136	210	17	260	99	284	140
242	125	213	82	358	190	226	37	305	128	288	154
245	168	248	92			280	48	308	158	288	158
305	162	320	130			260	80	380	187	395	205

(b) Using the dummy-variable method, test where there are significant differences between the two troughs in the Upper Severn, and between the two troughs in the Afon Hore. Where no significant differences exist, calculate pooled estimates of slope and intercept parameters together with their standard errors.

4.2. The following data, printed from file X4-2.DAT, show mean annual flood, MAF, together with number of years of record and catchment characteristics, from 18 basins in the drainage area of the River Severn, England: data from the NERC *Flood Studies Report, Volume 4*, 1975.

Column (1) Station code
Column (2) River and gauging station
Column (3) Mean annual flood, MAF ($m^3 s^{-1}$)
Column (4) Number of years of record
Column (5) Drainage area, (km^2)
Column (6) Grade of gauging station
Column (7) Main stream length, MSL (km)
Column (8) Dry valley factor (DVF)
Column (9) 10–85% stream slope, S1085 ($m km^{-1}$)
Column (10) Taylor–Schwartz slope, TAYSLO ($m km^{-1}$)
Column (11) Stream frequency, STMFRQ (junctions/km^2)
Column (12) Standard (1915–60) annual average rainfall, SAAR (mm)
Column (13) Five-year maximum two-day rainfall, M52D (mm)
Column (14) Effective mean soil moisture deficit, SMDBAR (mm)
Column (15) Rainfall minus soil moisture deficit, RSMD (mm)
Column (16) Soil index
Column (17) Fraction of catchment in urban development

(1)	(2)	(3)	(4)	(5)	(6)	(7)	(8)	(9)	(10)	(11)	(12)	(13)	(14)	(15)	(16)	(17)
54001	Severn at Bewdley	397.10	46	4330	A	206.32	0.003	0.60	2.60	0.94	945	58.6	8.6	30.3	0.347	0.003
54002	Avon at Evesham	155.95	32	2210	A2	125.37	0.020	0.80	0.90	0.66	683	46.2	12.2	20.1	0.412	0.024
54004	Severn at Stoneleigh	28.63	18	264	A	28.81	0.069	1.92	2.05	0.72	707	45.9	11.0	22.8	0.409	0.153
54005	Severn at Montford	318.98	17	2030	A	109.03	0.003	1.20	6.00	1.53	1179	70.8	4.5	43.0	0.390	0.002
54006	Stour at Kidderminster	21.76	17	324	A	33.42	0.020	2.38	2.19	0.42	721	49.2	11.0	25.1	0.172	0.138
54007	Arrow at Broom	46.94	13	319	A1	36.17	0.012	3.30	3.69	1.00	709	47.6	13.0	22.0	0.428	0.012
54008	Teme at Tenbury Wells	167.03	13	1130	B	76.68	0.007	3.04	2.61	0.90	871	54.8	8.9	29.5	0.333	0.002
54010	Stour at Alscot Park	39.01	10	319	B	38.95	0.020	2.90	2.70	0.95	711	48.8	13.0	22.8	0.379	0.000
54011	Salwarpe at Harford Hill	24.03	11	184	A	27.00	0.023	4.87	4.46	0.45	691	47.5	12.4	22.9	0.450	0.031
54012	Tern at Walcot	40.97	10	852	B	47.97	0.012	1.54	1.57	0.11	729	48.0	10.7	23.6	0.256	0.011
54013	Clywedog at Cribynan	72.84	6	57	A2	18.51	0.079	15.38	12.96	2.40	1803	97.4	1.0	67.2	0.429	0.000
54014	Severn at Abermule	341.75	8	580	A	54.58	0.010	3.60	8.90	1.83	1257	74.7	2.5	49.1	0.394	0.003
54016	Roden at Rodington	15.22	7	259	A	40.22	0.050	0.92	1.03	0.27	721	47.5	10.4	24.6	0.305	0.002

(1)	(2)	(3)	(4)	(5)	(6)	(7)	(8)	(9)	(10)	(11)	(12)	(13)	(14)	(15)	(16)	(17)
54017	Leadon, Wedderburn Bridge	20.18	8	293	C	35.18	0.013	2.11	2.09	0.95	737	57.3	9.0	33.0	0.306	0.004
54018	Rea Brook at Hookagate	24.83	6	178	A2	25.35	0.032	5.44	3.54	0.44	787	51.6	8.6	29.4	0.364	0.000
54019	Avon at Staveton	38.47	7	347	A1	56.66	0.040	1.40	1.50	0.51	676	44.9	11.0	22.0	0.359	0.000
54020	Perry at Yeaton	10.62	6	181	A1	31.85	0.063	2.45	1.65	0.70	792	48.7	9.0	26.9	0.272	0.000
54022	Severn at Plynlimon Weir	13.51	19	8.05	A	4.84	0.052	63.47	48.24	3.60	2449	119.0	1.0	81.8	0.500	0.000

(a) Calculate the weighted multiple regression of *MAF* on area, *MSL*, *DVF*, *S*1085, and *M*52*D*, using the number of years of record as weights. Which gauging stations have large residuals? Which have high leverage? Suggest explanations.

(b) Recalculate the regression after log transformation of all variables. What is the effect of the transformation on the points with high leverage?

4.3. J. M. L. Jansen and R. B. Painter published data on the relation between sediment yield, climate and topography (Predicting sediment yield from climate and topography, *J. Hydrology*, **21** (1974), 371–80). For each of 79 basins with area greater than 5000 km^2, they gave the following variables:

S: sediment (t km^{-2})
D: runoff ($\times 1000$ m^3 km^{-2})
A: basin area, (km^{-2})
H: difference between maximum and minimum altitudes (m)
R: mean channel slope (m km^{-1})
P: mean annual precipitation (mm)
T: mean annual temperature (°C)
V: index of natural vegetation (for forest, $V = 4$; for grass, $V = 3$; for steppe, $V = 2$; for desert, $V = 1$)
G: index of proneness to erosion (for Palaeozoic formations, $G = 3$; for Mesozoic, $G = 5$; for Cenozoic, $G = 6$; for Quaternary, $G = 2$)

The data from the 79 basins are given below (as well as in the file SEDIMENT.DAT on the diskette accompanying this book):

(a) Comment on the general accuracy of the data.

(b) Calculate a multiple regression of ln *S* on the remaining variables, after log-transforming them. Identify one basin with a large residual, and others having high leverage. Jansen and Painter give the regression as

$$\ln S = -2.032 + 0.100 \ln D - 0.314 \ln A + 0.750 \ln H + 1.104 \ln P + 0.368 \ln T - 2.324 \ln V + 0.786 \ln G$$

accounting for 57.9% of the variance. Do you agree with this result?

(c) Comment on the use of ln *D* as an explanatory variable.

Number	Country	Stream	Station	*S*	*D*	*A*	*H*	*R*	*P*	*T*	*V*	*G*
1	Thailand	Lam Pao	Lower Lam Pao	125.0	265	5430	200	0.7	1250	25	4	5
2	Thailand	Me Nan	Tha Pla	391.0	470	12790	1000	5.0	1250	25	4	2
3	Thailand	Me Nan	Phitsanuloke	128.0	344	25491	1000	3.3	1250	25	4	2
4	Thailand	Me Wang	Wang Krai	151.0	138	10204	1000	4.7	1750	25	4	3
5	Thailand	Me Ping	Wang Kra Chao	88.0	230	26386	1500	4.3	1750	25	4	3
6	Thailand	Me Ping	Kam Pan Petch	36.0	180	42300	1500	3.9	1750	25	4	3
7	Venezuela	Tuy	Puente San Juan	238.0	224	6610	2000	19.0	1250	25	4	3

Number	Country	Stream	Station	S	D	A	H	R	P	T	V	G
8	Brazil	Amazon	mouth, Brazil	66.0	932	6130515	6000	1.0	2000	30	4	6
9	Argentina	Parana	mouth, Argentina	39.0	204	2304121	3000	0.8	1750	20	4	5
10	Venezuela	Orinoco	mouth, Venezuela	100.0	753	949350	3000	1.5	17509	30	4	2
11	Zaire	Congo	mouth, Zaire	18.0	312	4012795	3000	0.7	1750	25	4	3
12	Nigeria	Niger	Baro	5.0	172	1113227	1000	0.3	1000	30	3	4
13	Burma	Irrawaddi	Prome	904.0	1166	366847	5000	2.5	1750	20	4	6
14	India	Mahanadi	Naraj	514.0	683	132034	2500	2.5	1250	25	4	2
15	Turkey	Yesilirmak	Ayvacik	1228.0	1	36000	2000	6.0	500	15	2	6
16	Turkey	Firat	Keban	517.0	313	63836	3000	5.0	500	15	2	6
17	Syria	Euphrate	Tabqa	36.0	253	120650	1500	2.0	500	20	3	6
18	Canada	S. Saskatchewan	Lemsford	44.0	65	116000	3000	4.0	400	5	3	5
19	Canada	S. Saskatchewan	Saskatoon	20.0	51	140000	3000	4.0	400	5	3	5
20	Canada	Saskatchewan	The Pas	12.0	65	324000	3000	3.0	500	5	4	5
21	USA	Colorado	Grand Canyon, Arizona	418.0	14	356750	4000	1.7	250	15	2	4
22	USA	Snake	Central Ferry, Washington	49.0	162	267951	2500	0.6	300	15	2	2
23	USA	Rio Grande	San Acacia, New Mexico	136.0	35	69305	3500	1.1	300	20	2	3
24	USA	Pecos	Puerto de Luna, New Mexico	265.0	52	10278	3000	4.0	350	20	2	6
25	Egypt	Nile	delta, Egypt	39.0	30	2977235	3000	0.5	200	25	2	5
26	W. Pakistan	Indus	Kotri	502.0	223	957893	5500	2.0	400	25	2	3
27	W. Pakistan	Kabul	Nowshera	288.0	237	90301	2000	3.0	300	20	1	6
28	Iraq	Tigris	Baghdad	722.0	371	79738	2000	1.3	500	20	2	4
29	Italy	Tibre (Tevere)	Corbara	242.0	276	6075	500	2.0	1000	15	4	5
30	Italy	Tibre (Tevere)	Rome	352.0	397	16545	1000	4.0	1250	20	4	6
31	Italy	Po	Boretto	275.0	973	44070	3000	8.0	1500	15	4	6
32	The Netherlands	IJssel	Westervoort	1.8	52	160000	2000	3.0	750	10	4	2
33	The Netherlands	Neder-Rijn	Arnhem	2.8	78	160000	2000	3.0	750	10	4	2
34	The Netherlands	Waal	Hulhuizen	12.7	304	160000	2000	3.0	750	10	4	2
35	The Netherlands	Rijn	Lobith	17.4	134	160000	2000	3.0	750	10	4	2
36	Japan	Tone	Matsudo et Toride	273.0	1248	12000	1000	5.0	1500	15	4	2
37	Japan	Yodo	Hirakata	20.2	897	7120	500	3.0	1500	15	4	6
38	Thailand	Me Yom	Kuang Lang	166.0	194	13214	500	2.0	1250	25	4	2
39	Germany	Main	Marktbreit	20.0	137	27225	200	9.0	750	15	4	5
40	Albania	Semani	Uraqe ucit	4150.0	674	5288	2000	12.0	1250	15	4	6
41	Albania	Drini	Can Deje	1190.0	951	12368	1500	6.0	1750	15	4	5
42	Italy	Po	Pontelagoscuro	280.0	900	54290	3000	7.0	1250	15	4	2
43	USA	Brazos	Richmond, Texas	386.0	52	90094	1500	0.8	750	20	3	5
44	USA	Alabama	Claiborne, Alabama	38.0	500	56956	500	0.5	1250	20	4	5
45	USA	San Joaquin	Vernalis, California	10.0	116	36270	3000	6.0	500	15	4	2
46	USA	Sabine	Logansport, Louisiana	58.0	625	12582	500	1.0	1000	20	4	2
47	Argentina	Uruguay	Concordia, Argentina	39.0	322	388335	3000	2.0	1500	20	4	2
48	France	Loire	Nantes	4.0	221	121005	2000	2.5	750	15	4	5
49	Bangladesh	Ganges	delta, Bangladesh	1545.0	420	1059378	5500	2.2	1250	25	4	2
50	China	Yangtze	Chikiang	541.0	671	1024298	5000	1.0	1250	15	4	3
51	Bangladesh	Brahmaputra	delta, Bangladesh	1429.0	1127	559202	6000	3.2	1750	5	1	5
52	China	Pearl-West	Wuchow	97.0	791	312739	2000	2.0	1250	20	4	3
53	North Vietnam	Red	Hanoi	1190.0	1028	119866	2500	1.5	1750	20	4	5
54	Poland	Odra (Oder)	Gozdowice	1.2	132	109400	1000	2.0	750	10	4	2

Number	Country	Stream	Station	S	D	A	H	R	P	T	V	G
55	Poland	Wisla (Vistula)	Tezew	7.4	155	193900	1000	1.0	750	10	4	2
56	Japan	Ishikari	Ebetsu	138.0	1026	12697	1500	6.0	1500	15	4	6
57	Germany	Isar	Plattling	29.0	531	8964	1000	4.4	1000	15	4	5
58	Germany	Inn	Reisach	327.0	992	9760	1000	4.4	1250	15	4	5
59	Hungary	Danube	Nagymaros	28.4	406	183262	1000	1.3	750	10	4	2
60	Hungary	Tisza	Tivadar	44.5	528	12540	1000	4.7	1000	10	4	2
61	Hungary	Raba	Arpas	24.3	178	6610	1000	5.6	750	10	4	2
62	Hungary	Tisza	Szeged	49.0	197	138408	1000	1.5	750	10	3	2
63	Canada	Pembina	Windygates	3.5	11	7800	500	3.0	400	5	4	3
64	Canada	Assiniboine	Portage de la Prairie	3.2	7	161000	500	0.4	350	5	4	3
65	Canada	Assiniboine	Headingly	2.7	8	162000	500	0.4	350	0	4	3
66	Canada	N. Saskatchewan	Prince Albert	23.0	56	119000	1000	1.3	400	0	4	5
67	Canada	Red	Ste Agathe	11.4	40	116000	500	0.5	500	0	3	3
68	Canada	Red	Lockport	6.9	21	287000	500	0.3	500	0	3	3
69	USA	Mississippi	mouth, Louisiana	107.0	175	3220592	500	0.1	750	15	4	4
70	USA	Missouri	Hermann, Missouri	175.0	45	1367457	3000	0.8	600	10	3	5
71	Canada	St. Lawrence	mouth, Canada	3.1	346	1289272	500	0.1	750	5	4	3
72	USA	Ohio	Cincinnati, Ohio	76.0	1163	198258	1000	0.7	1500	15	4	3
73	USA	Potomac	mouth, Maryland & Virginia	66.0	262	37798	1500	4.0	1000	15	4	3
74	USA	Delaware	Trenton, New Jersey	57.0	1023	17553	1000	2.5	1500	15	4	3
75	Russia	Volga	Dubovka	16.0	187	1350085	300	0.3	750	10	4	5
76	Russia	Danube	mouth, Russia	26.0	239	815504	900	0.3	750	10	4	3
77	Russia	Dnjepr	Verkhnedneprovsk	3.0	399	433693	400	0.2	600	10	4	4
78	Russia	Yenesei	Igarka	5.0	222	2470328	4000	0.7	500	0	4	3
79	Russia	Ob	Salekhard	7.0	161	2447287	4000	0.9	500	0	4	2

4.4. Integrated samples for the measurement of suspended sediment (mg l^{-1}) were collected several times a year in eight headwater basins of the upper River Uruguay. Discharge (m^3 s-1) was measured by current meter. The data given below were collected over the years 1978–87.
(a) Using dummy variables or otherwise, fit a linear model of the form

$$\ln S = \mu + \beta_i + \theta_j + \nu_k + \varepsilon_{ijk}$$

where $\ln S = \ln$ (sediment yield) $= \ln$ (Concentration) $+ \ln$ (Discharge) and μ, β_i, θ_j, ν_k, ε_{ijk} represent an overall mean, the effect of basin i, the effect of year j, the effect of month k, and a residual error. Hence derive estimates of $\mu + \beta_i$ which are free of annual and seasonal effects.
(ii) Calculate the residuals $\hat{\varepsilon}_{ijk}$ and check for their normality using a normal (Q-Q) plot.

Drainage Basin Code	Catchment 7401				
Year	1978	1978	1978	1978	1978
Month	1	4	6	8	11
Concentration	7	40	43	46	30
Discharge	444.0	251.0	198.0	445.0	812.0
Year	1979	1979	1980	1981	1981
Month	3	5	1	3	5
Concentration	15	39	47	17	20
Discharge	293.0	1680.0	679.0	752.0	332.0

Drainage Basin Code	Catchment 7401				
Year	1981	1981	1982	1982	1982
Month	8	10	1	5	8
Concentration	17	59	16	10	13
Discharge	483.0	2020.0	689.0	185.0	1860.0
Year	1982	1983	1985		
Month	11	3	3		
Concentration	76	59	14		
Discharge	5680.0	1820.0	741.0		

Drainage Basin Code	Catchment 7402				
Year	1978	1978	1978	1978	1978
Month	1	4	6	8	11
Concentration	145	45	11	23	698
Discharge	46.0	42.0	8.7	32.0	55.0
Year	1979	1979	1980	1981	1981
Month	3	5	1	3	5
Concentration	50	62	471	24	28
Discharge	25.0	81.0	114.0	36.0	19.0
Year	1981	1981	1982	1982	1982
Month	8	10	5	8	11
Concentration	23	166	9	366	34
Discharge	52.0	102.0	10.0	658.0	311.0
Year	1983	1983	1983	1983	1984
Month	3	6	10	12	7
Concentration	67	40	192	22	547
Discharge	239.0	167.0	186.0	49.0	895.0
Year	1984	1985	1985	1985	1986
Month	10	1	6	9	2
Concentration	160	46	4	38	82
Discharge	386.0	39.0	129.0	135.0	21.0
Year	1986	1986	1987		
Month	5	8	2		
Concentration	149	90	4		
Discharge	222.0	144.0	55.0		

Drainage Basin Code	Catchment 7403				
Year	1978	1978	1978	1978	1979
Month	1	4	8	11	1
Concentration	75	52	30	1270	22
Discharge	26.0	8.6	37.0	35.0	13.0
Year	1981	1981	1981	1981	1982
Month	3	5	8	11	3
Concentration	8	87	23	25	17
Discharge	12.0	10.0	13.0	15.0	12.0

Drainage Basin Code	Catchment 7403				
Year	1982	1982	1982	1983	1983
Month	5	9	12	3	5
Concentration	43	26	72	84	77
Discharge	8.2	48.0	54.0	66.0	190.0
Year	1983	1983	1984	1984	1984
Month	8	11	6	9	12
Concentration	94	31	106	148	126
Discharge	188.0	46.0	78.0	138.0	67.0
Year	1985	1985	1985	1986	1986
Month	6	10	12	2	5
Concentration	33	5	10	20	22
Discharge	71.0	36.0	15.0	17.0	85.0
Year	1987				
Month	2				
Concentration	5				
Discharge	23.0				

Drainage Basin Code	Catchment 7404				
Year	1978	1978	1978	1978	1979
Month	1	6	8	11	1
Concentration	20	39	56	562	21
Discharge	12.0	5.4	29.0	16.0	8.5
Year	1981	1981	1981	1981	1982
Month	3	5	8	11	3
Concentration	70	11	19	24	24
Discharge	7.8	6.7	6.3	13.0	8.0
Year	1982	1982	1982	1983	1983
Month	5	9	12	3	8
Concentration	94	139	74	152	175
Discharge	4.4	34.0	39.0	51.0	110.0
Year	1983	1984	1984	1984	1985
Month	11	6	9	12	6
Concentration	142	64	336	112	26
Discharge	35.0	73.0	126.0	39.0	62.0
Year	1985	1985	1986		
Month	9	12	5		
Concentration	11	6	1020		
Discharge	44.0	9.2	66.0		

Drainage Basin Code	Catchment 7405				
Year	1978	1978	1978	1978	1979
Month	1	6	8	11	1
Concentration	40	36	43	86	9
Discharge	17.0	9.0	57.0	20.0	11.0

Drainage Basin Code	Catchment 7405				
Year	1981	1981	1981	1981	1982
Month	3	5	8	11	3
Concentration	72	22	14	28	19
Discharge	14.0	11.0	9.2	19.0	7.4
Year	1982	1982	1982	1983	1983
Month	5	9	12	3	5
Concentration	37	71	93	84	19
Discharge	4.7	61.0	57.0	43.0	264.0
Year	1983	1983	1984	1984	1984
Month	8	11	6	9	12
Concentration	53	43	58	34	125
Discharge	126.0	37.0	91.0	191.0	48.0
Year	1985	1985	1985	1986	1986
Month	6	10	12	2	6
Concentration	5	13	6	22	182
Discharge	58.0	37.0	9.8	6.9	123.0
Year	1987				
Month	2				
Concentration	10				
Discharge	41.0				

Drainage Basin Code	Catchment 7406				
Year	1978	1978	1978	1978	1979
Month	1	4	6	8	1
Concentration	47	54	39	136	20
Discharge	11.0	4.7	7.1	34.0	6.9
Year	1981	1981	1981	1981	1982
Month	3	5	8	11	3
Concentration	86	43	4	25	16
Discharge	8.7	7.7	5.7	11.0	7.1
Year	1982	1982	1983	1983	1983
Month	5	9	3	5	8
Concentration	17	252	148	34	72
Discharge	3.0	51.0	47.0	281.0	47.0
Year	1983	1984	1984	1984	1985
Month	11	6	9	12	6
Concentration	24	14	388	41	25
Discharge	18.0	46.0	117.0	26.0	37.0
Year	1985	1985	1986	1986	1987
Month	9	12	6	8	12
Concentration	21	3	144	6	45
Discharge	32.0	7.9	49.0	53.0	63.0

Drainage Basin Code	Catchment 7407				
Year	1981	1981	1981	1981	1981
Month	1	3	5	8	11
Concentration	44	15	42	24	33
Discharge	1.7	2.8	1.4	1.6	3.2
Year	1982	1982	1982	1982	1983
Month	3	5	9	12	3
Concentration	93	84	63	71	49
Discharge	1.3	1.2	7.5	9.8	7.2
Year	1983	1983	1984	1984	1984
Month	5	7	6	9	12
Concentration	228	72	62	262	10
Discharge	67.0	14.0	12.0	24.0	5.5
Year	1985	1985	1985	1986	1986
Month	6	10	12	3	6
Concentration	33	11	8	36	13
Discharge	13.0	5.7	1.6	5.3	6.4
Year	1986	1986			
Month	9	12			
Concentration	8	13			
Discharge	8.7	8.0			

Drainage Basin Code	Catchment 7408				
Year	1978	1978	1978	1978	1978
Month	1	4	6	8	11
Concentration	21	82	41	110	29
Discharge	8.4	2.4	51.0	16.0	7.1
Year	1979	1981	1981	1981	1981
Month	1	3	5	8	11
Concentration	12	25	15	18	45
Discharge	3.3	6.8	3.7	4.0	7.1
Year	1982	1982	1982	1982	1983
Month	3	5	9	12	3
Concentration	21	57	402	97	44
Discharge	2.3	6.6	27.0	30.0	26.0
Year	1983	1983	1983	1984	1984
Month	5	8	11	9	12
Concentration	106	61	103	35	30
Discharge	175.0	35.0	23.0	15.0	17.0
Year	1985	1985	1985	1986	1986
Month	6	10	12	3	9
Concentration	56	20	6	65	79
Discharge	32.0	14.0	5.1	17.0	63.0

Chapter 5

Non-linear statistical models

5.1 NON-LINEARITY AS A CHARACTERISTIC OF PARAMETRIC STRUCTURE

The models of Chapter 4 expressed a response variable Y_t in terms of explanatory variables $x_1, x_2, \ldots, x_p$ in the form

$$E[Y_t] = \beta_0 + \beta_1 x_{1t} + \cdots + \beta_p x_{pt}$$

where the expression on the right-hand side is linear both in the explanatory variables x_i, and in the parameters β_i. If we write $x, x^2, \ldots, x^p$ instead of $x_1, x_2, \ldots, x_p$, the model becomes $E[Y_t] = \beta_0 + \beta_1 x_t + \beta_2 x_t^2 + \cdots + \beta_p x_t^p$; regarded as a function of x, the model is now curvilinear. From the point of view of selecting the most appropriate model and fitting it, however, the model remains linear in the parameters $\beta_0, \beta_1, \ldots, \beta_p$, and the methods of linear multiple regression remain applicable. Take, for example, the data in Figure 5.1, reproduced from Yevjevich (1972, Table 8.1) by kind permission of V. Yevjevich. The data relate discharge Q to stage H, from which a rating curve is to be calculated; a multiple regression analysis, modelling Q in terms of H and H^2, is given by the methods of Chapter 4. The results (Figure 5.2, in which H and H^2 are denoted by *Stage* and *Stage2*) show that the fitted model is $Q = 29.69 + 0.5690H + 0.007\,199H^2$, with standard errors for the three coefficients of ± 2.73, ± 0.0478, and $\pm 0.000\,130$, respectively. Two points, numbered 7 and 13 in the sequence, have large standardised residuals of 2.16 and 2.30, respectively, and the final point (number 13) also has high leverage of 0.96; for point number 13, we note its distance from the remaining points, and the large influence which this isolated point has on model fit.

Using multiple regression, it is therefore straightforward to fit curvilinear models, such as polynomials, which are linear in the model parameters. The usual close inspection of model residuals remains an essential part of good modelling practice, and we must be constantly critical of the models that we have fitted. For the fitted parabola, Figure 5.2 shows the standardised residuals plotted against the fitted values, and against data order; the large residuals for points 7 and 13 show up particularly clearly, suggesting doubts about the validity of the parabolic model. We might, instead, consider a linear regression of log discharge, say $\ln Q$, on stage; this model removes the large residual associated with points 7 in the parabolic model, but the large residual and leverage associated with point 13 remain — as, indeed, it is always likely to, when one point is so far removed from the others.

Multiple regression methods are also applicable for fitting curves by orthogonal polynomials (see Chapter 4) and for fitting surfaces where, instead of explanatory variables like $x, x^2, \ldots$ we

Stage	−23	−22	−16	−16	14
Discharge	15.55	15.46	20.07	21.99	36.11
Stage	33	46	69	88	120
Discharge	59.82	86.58	110.96	136.52	204.40
Stage	136	220	400		
Discharge	232.87	492.50	1412.48		

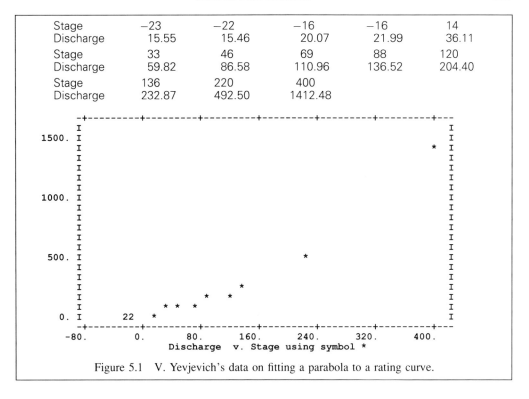

Figure 5.1   V. Yevjevich's data on fitting a parabola to a rating curve.

∗ ∗ ∗ ∗ ∗ Regression Analysis ∗ ∗ ∗ ∗ ∗

Response variate: Discharg
     Fitted terms: Constant, Stage, Stage2

∗ ∗ ∗ Summary of analysis ∗ ∗ ∗

	d.f.	s.s.	m.s.
Regression	2	1755038.5	877519.25
Residual	10	567.7	56.77
Total	12	1755606.3	146300.52

Percentage variance accounted for 100.0

∗ MESSAGE: The following units have large residuals:

	7	2.16
	13	2.30

∗ MESSAGE: The following units have high leverage:

	13	0.96

∗ ∗ ∗ Estimates of regression coefficients ∗ ∗ ∗

	estimate	s.e.	t
Constant	29.69	2.73	10.88
Stage	0.5690	0.0478	11.91
Stage2	0.007199	0.000130	55.28

Figure 5.2   Parabolic regression of *Discharge* on *Stage* for Yevjevich's data: with plots of residuals against fitted values (*Fitvals*) and against *Order*. The linear regression of log (*Discharge*), ln Q, on *Stage* is also shown.

_ Continued _

Figure 5.2 (continued)

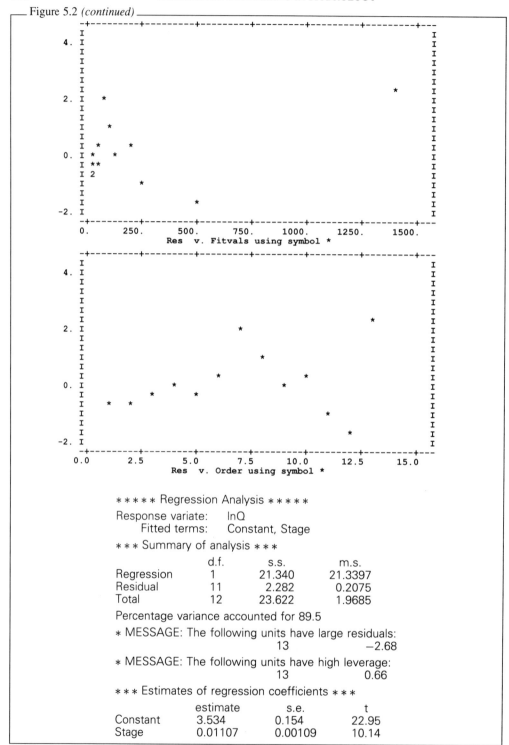

***** Regression Analysis *****

Response variate: lnQ
    Fitted terms: Constant, Stage

*** Summary of analysis ***

	d.f.	s.s.	m.s.
Regression	1	21.340	21.3397
Residual	11	2.282	0.2075
Total	12	23.622	1.9685

Percentage variance accounted for 89.5

* MESSAGE: The following units have large residuals:
                          13   −2.68

* MESSAGE: The following units have high leverage:
                          13   0.66

*** Estimates of regression coefficients ***

	estimate	s.e.	t
Constant	3.534	0.154	22.95
Stage	0.01107	0.00109	10.14

have explanatory variables representing powers of two-dimensional spatial coordinates, such as $x, y, x^2, xy, y^2, \ldots$. Where we use powers of spatial coordinates as explanatory variables we must, however, be mindful of the strong possibility of ill-conditioning in the matrix $\mathbf{X}^T\mathbf{X}$ of Chapter 4.

Having noted briefly the use of multiple regression in fitting models of this type, the remainder of the present chapter is concerned with fitting and testing models which are essentially distinct from the parabolic model considered above, in that they are non-linear in the parameters.

5.2   TRANSFORMATIONS TO LINEAR FORM

In Chapter 4 we referred to situations where transformations are appropriate, and noted that in certain cases the systematic components of models can often be transformed to yield alternative forms that are linear in new parameters. For example, a model for which the systematic component is $y = \beta_1/(\beta_2 + \beta_3 x)$ can be transformed to the form $1/y = (\beta_2 + \beta_3 x)/\beta_1$, or $Y = \beta_2^* + \beta_3^* x$, where $Y = 1/y$ and the new parameters are $\beta_2^* = \beta_2/\beta_1$ and $\beta_3^* = \beta_3/\beta_1$. A possible estimation procedure would then be to regard the new model as a simple linear regression of Y on x; this would not be very satisfactory, however, since the new model would give estimates only of the two parameters β_2^* and β_3^*, and not of the three parameters β_1, β_2 and β_3 in the original model, which may have had physical significance.

Transformations of this kind may be useful if the variable Y has very small dispersion about its expected value, when the effect of the transformation upon the random components of the model can be ignored. However, if the random components ε_t are additive and substantial in the model $y_t = \beta_1/(\beta_2 + \beta_3 x_t) + \varepsilon_t$, then the desired simplification will fail to be achieved. Similar comments are appropriate in the case of the models $y_t = \beta_1 \exp(-\beta_2 x_t) + \varepsilon_t$ (which, ignoring the errors ε_t, can be linearised by log transformation), and $y_t = x_t/(\beta_1 + \beta_2 x_t) + \varepsilon_t$ (which, ignoring the random component, can be linearised to the form $x_t/y_t = \beta_1^* + \beta_2^* x_t$), to list but two of many. Whilst such transformations may be useful in certain circumstances, their effects on the model's random component need always to be considered. Furthermore, the fact that a model may be put into a linear form, and fitted by regression, is no guarantee that the fit of the original model is satisfactory, as the example of Section 5.3 will show. Note, however, that such transformations may be useful as a means of providing rough starting estimates of model parameters, to be used in a maximum-likelihood estimation procedure taking proper account of the additive nature of the model errors ε_t.

5.3   FITTING THE GREEN–AMPT EQUATION TO DATA FROM AN INFILTROMETER: AN EXAMPLE OF A TRANSFORMATION WHICH FAILS

Figure 5.3 shows data for two variables V and t, where V is the cumulative depth of water in millimetres that has infiltrated into a soil profile at time t, measured in minutes. Inspection of the data when plotted shows little evidence of statistical fluctuations, which would probably have been more apparent if the values of V had been obtained from different infiltrometers run in parallel. The statistical fluctuations would then have indicated the extent of soil sample-to-sample variability, which cannot be deduced from the data shown in Figure 5.3.

Assessment of this variability was not, however, the primary objective of the study. The object of the analysis was to compare the fits of several equations which, from physical arguments, can be taken to represent infiltration as a function of time. Looking at the plot

t	0.00	1.55	2.55	4.55	5.55
V	0.00	0.59	0.87	1.28	1.48
t	6.55	7.55	8.55	9.55	10.55
V	1.66	1.77	1.92	2.07	2.22
t	11.55	12.55	13.55	14.55	16.55
V	2.37	2.47	2.61	2.72	2.97
t	17.55	18.55	19.55	20.55	21.55
V	3.04	3.06	3.01	3.24	3.40
t	22.55	23.55	24.55	25.55	26.55
V	3.56	3.70	3.79	3.98	4.10
t	27.55	28.55	29.55	30.55	31.55
V	4.24	4.40	4.60	4.80	4.96
t	32.55	33.55	34.55	35.55	36.55
V	5.15	5.35	5.45	5.60	5.70
t	37.55	38.55	39.55	40.55	41.55
V	5.80	5.90	6.05	6.17	6.27
t	42.55	43.55	44.55	45.55	46.55
V	6.41	6.57	6.75	6.89	6.99
t	47.55	48.55	49.55	50.55	51.55
V	7.08	7.12	7.22	7.38	7.48
t	52.55	53.55	55.55	56.55	57.55
V	7.62	7.78	8.03	8.14	8.20
t	58.55	59.55	60.55	61.55	62.55
V	8.25	8.34	8.42	8.46	8.56
t	63.55	64.55	65.55	66.55	67.55
V	8.62	8.72	8.84	8.92	9.01
t	68.55	69.55	70.55	71.55	72.55
V	9.18	9.27	9.41	9.58	9.72
t	73.55	74.55			
V	9.86	9.94			

```
        -+---------+---------+---------+---------+---------+---------+---
        I                                                               I
   12.  I                                                               I
        I                                                               I
        I                                                               I
        I                                                               I
        I                                              ***              I
        I                                           *2**                I
        I                                        2*2*2                  I
    8.  I                                     **2*                      I
        I                                  2*2*                         I
        I                               **2*                            I
        I                             **2**                             I
        I                           2**                                 I
        I                         *2*                                   I
    4.  I                      2*2                                      I
        I                    2*2*                                       I
        I                 2***                                          I
        I              *2*                                              I
        I           *2                                                 I
        I         **                                                   I
    0.  I*                                                              I
        -+---------+---------+---------+---------+---------+---------+---
         0.       15.       30.       45.       60.       75.       90.
              V   v. t using symbol *
```

Figure 5.3   Infiltrometer data: *t* is time in minutes; *V* is depth of infiltrated water in millimetres.

of V against t shown in Figure 5.3 one might note the near-linear scatter of points, and express the relation between V and t by a linear regression. However, if a *theoretical* relation exists between two hydrological variables, we normally prefer to use it. The parameters in the theoretical relation commonly have a physical interpretation, which may not be the case where a linear regression is fitted. Furthermore, whilst a linear regression may be a good description of the relationship between two variables *within the observed range of variation of the explanatory variable*, it is very unlikely that the linear relation will hold far beyond this range. A theoretical relation between two variables, on the other hand, can often be extrapolated with greater safety. Even in this case, however, the theory may be valid only within some restricted range of the explanatory variable, and it will be necessary to establish over what range the theory remains valid.

One of several theoretical equations to be fitted to the data of Figure 5.3 was the Green–Ampt equation. This has the form

$$Kt/\Delta\theta = L_f - (H_0 - H_f)\ln[1 + L_f/(H_0 - H_f)] \tag{5.1}$$

where K is the hydraulic conductivity of the soil zone through which water is transmitted (assumed spatially homogeneous), H_0 is the pressure head at the entry surface, H_f is the effective pressure head at the wetting front, L_f is the distance from the surface to the wetting front, and $\Delta\theta = \theta_t - \theta_i$, where θ_t is the wetness of the transmission zone during infiltration, and θ_i is the initial profile wetness which prevails beyond the wetting front. In the special case where θ_t is saturation, and θ_i is zero, then $V = fL_f$, where V is the cumulative infiltration and f is the soil porosity. As t increases, the second term on the right-hand side of equation (5.1) increases more and more slowly in relation to the increase in L_f, so that at very large times we can approximate the relation by $L_f = Kt/\Delta\theta + \delta$ or $V = Kt + \delta$, where δ can be effectively regarded as a constant when t is large. Hence, for large t, a linear regression of V on t is a good approximation to equation (5.1), the parameters in the regression then having a physical interpretation.

For the purposes of this analysis, we write equation (5.1) in the form

$$kt = V - s\ln(1 + V/s) \tag{5.2}$$

where k and s are parameters to be estimated from the data in Figure 5.3. From the point of view of model fitting, equation (5.2) does not have a particularly convenient form, since the infiltration V — which is the response variable — is not given as an explicit function of time t, the explanatory variable. However, if each side is differentiated with respect to t and rearranged, we have

$$dV/dt = k(1 + s/V) \tag{5.3}$$

which, in discrete form, can be written approximately

$$(V_{t+1} - V_{t-1})/2\Delta t = k(1 + s/V_t) \tag{5.4}$$

One procedure for estimating the parameters s and k might therefore be as follows:

(i) Assume a series of trial values for the parameter s.

(ii) For each value s in this series, transform to the new variables $y = (V_{t+1} - V_{t-1})/2\Delta t$, $x = (1 + s/V_t)$. Fit a linear regression $y = kx$, constrained to pass through the origin, for which the estimate of k is given by $\tilde{k} = \Sigma xy/\Sigma x^2$.

(iii)   Determine which value of s minimises the residual sum of squared deviations RSS, given by $RSS = \Sigma\, y^2 - [\Sigma\, xy]^2/\Sigma x^2$.

(iv)   Repeat for values of s at smaller intervals, until sufficient accuracy has been achieved for s, whereupon k is given by the slope of the line $y = kx$ passing through the origin.

For the purposes of this calculation, the absence of V values for $t = 3.55$, 15.55 and 54.55 minutes was ignored, and the values of V were assumed to have been measured at 72 equally spaced times.

The procedure defined by steps (i) to (iv) above is simple and easily programmed. The value of s minimising RSS can be easily found, and the corresponding estimate of k is given by $\Sigma\, xy/\Sigma\, x^2$. However, if these estimates of k and s are substituted in equation (5.2) and the fitted values of V calculated by iterative solution of that equation, the plots of V_t and of V_t (fitted), as functions of time t, are very different. The explanation is not far to seek: when the new variable, y, is plotted against the new variable $x = x(s)$, it is seen that a straight line constrained to pass through the origin is a very poor fit. A much better fit, in this instance, is given by ignoring the fact that V is the response variable depending on t, and minimising the sum of squares function

$$RSS^* = \sum [kt - V_t + s\ln(1 + V_t/s)]^2 \qquad (5.5)$$

by differentiating with respect to k and s, yielding equations $F_1(k, s) = \partial RSS^*/\partial k$ and $F_{II}(k, s) = \partial RSS^*/\partial s$, and solving the equations $F_I = 0$, $F_{II} = 0$ iteratively by a Newton–Raphson procedure. Denoting $Z_t = 1 + V_t/s$ and $W_t = \ln Z_t$, the equations $F_I = F_{II} = 0$ are

$$\sum [t - V_t/k + sW_t/k][V_t - sW_t] = 0 \qquad (5.6)$$

$$\sum [t - V_t/k + sW_t/k][W_t + Z_t^{-1} - 1] = 0$$

with solutions $\tilde{k} = 0.1072$, $\tilde{s} = 0.7796$. Figure 5.4 shows the values of V_t and of V_t (fitted) as functions of time, the fitted values having been determined by numerical solution of equation (5.2).

It should be noted that the above procedure for estimating k and s would almost certainly be unsatisfactory if there were considerable random variation in the infiltration values V_t: that is, if the values V_t showed considerable fluctuation about the trend when plotted. In this case, it would be necessary to use a numerical procedure to minimise the sum of squares function

$$RSS = \sum [V_t - V_{fit}(t; k, s)]^2 \qquad (5.7)$$

where V_t is the observed value of V at time t, and $V_{fit}(t; k, s)$ is the value of V found by numerical solution, for given k and s values, of equation (5.2). A simple approach would be to evaluate RSS at the points of a two-dimensional grid of values of k and s, locate a smaller region within which the minimum lies, and to repeat the search over a finer grid, continuing until sufficient accuracy is achieved. This method of 'direct search' is particularly convenient where one or two parameters must be estimated, but rapidly becomes laborious where models have three or more parameters; in such cases, an 'automatic search' procedure, of the kind discussed later in Section 5.5, is likely to be more convenient.

We note that the above discussion has not exhausted the possibilities for fitting equation (5.2). At least two other approaches are possible. One of these alternatives would

t	0.0	1.5	2.5	4.6	5.6
V	0.00	0.59	0.87	1.28	1.48
Vfit	0.000	0.625	0.846	1.223	1.395
t	6.6	7.6	8.6	9.6	10.6
V	1.66	1.77	1.92	2.07	2.22
Vfit	1.558	1.717	1.870	2.021	2.168
t	11.6	12.6	13.6	14.6	16.5
V	2.37	2.47	2.61	2.72	2.97
Vfit	2.312	2.454	2.595	2.733	3.006
t	17.5	18.5	19.5	20.5	21.5
V	3.04	3.06	3.01	3.24	3.40
Vfit	3.141	3.274	3.406	3.537	3.668
t	22.5	23.5	24.5	25.5	26.5
V	3.56	3.70	3.79	3.98	4.10
Vfit	3.797	3.926	4.054	4.182	4.309
t	27.5	28.5	29.5	30.5	31.5
V	4.24	4.40	4.60	4.80	4.96
Vfit	4.435	4.561	4.686	4.811	4.935
t	32.5	33.5	34.5	35.5	36.5
V	5.15	5.35	5.45	5.60	5.70
Vfit	5.059	5.183	5.306	5.429	5.551
t	37.5	38.5	39.5	40.5	41.5
V	5.80	5.90	6.05	6.17	6.27
Vfit	5.673	5.795	5.916	6.037	6.158
t	42.5	43.5	44.5	45.5	46.5
V	6.41	6.57	6.75	6.89	6.99
Vfit	6.279	6.399	6.519	6.639	6.759
t	47.5	48.5	49.5	50.5	51.5
V	7.08	7.12	7.22	7.38	7.48
Vfit	6.878	6.998	7.117	7.236	7.354
t	52.5	53.5	55.5	56.5	57.5
V	7.62	7.78	8.03	8.14	8.20
Vfit	7.473	7.591	7.827	7.945	8.063
t	58.5	59.5	60.5	61.5	62.5
V	8.25	8.34	8.42	8.46	8.56
Vfit	8.180	8.297	8.415	8.532	8.649
t	63.5	64.6	65.6	66.6	67.6
V	8.62	8.72	8.84	8.92	9.01
Vfit	8.766	8.882	8.999	9.115	9.231
t	68.6	69.6	70.6	71.6	72.6
V	9.18	9.27	9.41	9.58	9.72
Vfit	9.348	9.464	9.580	9.696	9.811
t	73.6	74.6			
V	9.86	9.94			
Vfit	9.927	10.043			

Figure 5.4   Observed and fitted values of accumulated infiltration, V and V_{fit}, against time t; with plot of V against t, for comparison with plot of V_{fit} against time t.

_ Continued _

Figure 5.4 (continued)

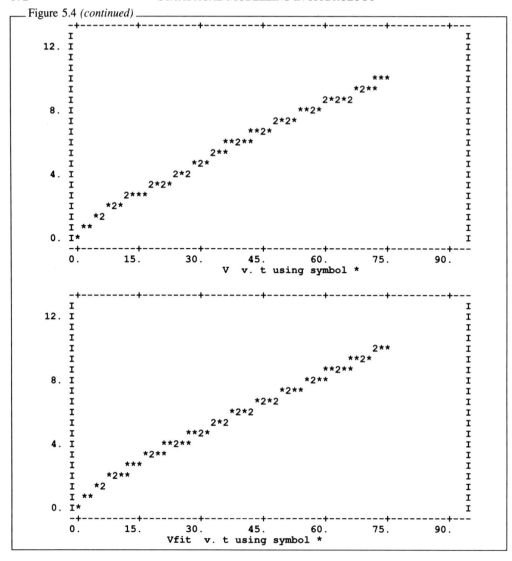

be to use a conditional optimisation method, of the kind used in Chapter 4: we could, for example, assume a series of values for the parameter s; construct, for each s-value, the new variable $y = V - s \ln(1 + V/s)$; and calculate $k = k(s)$ which minimises $RSS = \Sigma(kt - y)^2$. A second alternative would be to solve equations (5.6) for k, giving

$$\tilde{k} = \sum t_j (V_j - s W_j) / \sum t_j^2$$

as a function of s, which could be substituted to yield a single equation in the parameter s, soluble by Newton–Raphson. The important point is that alternative computational approaches exist, some of which are likely to arrive at a solution more rapidly than others, and some of which will give reliable solutions under wider conditions than others.

5.4 FITTING THE WIND VELOCITY PROFILE: A LINEARISING TRANSFORMATION WHICH SUCCEEDS (SOMETIMES)

Figure 5.5 shows wind velocities measured by anemometers at six levels, logarithmically spaced from about 0.5 m to 9 m on a sectional alloy tower of open construction, sited in open ranchland to the north of Manaus, in Amazonia. Full details of the study have been reported by Wright *et al.* (1992). Wind velocities were recorded at intervals of ten minutes, and Figure 5.5 shows the anemometer heights (left-hand column) and the velocities measured in 12 consecutive ten-minute runs on the afternoon of 9 October 1990. The data are plotted in the lower half of Figure 5.5, six to each graph, showing that wind velocity increases approximately as the logarithm of height z. The data form part of a much larger set of micro-meteorological measurements collected at six sites throughout the Amazon basin; three of these sites lie within undisturbed rainforest, and three within extensive, deforested areas. Pairs (forest, deforested) of sites are situated in the western Amazon (Ji-Paraná, Rondónia), central Amazon (Manaus) and eastern Amazon (Marabá, Pará); the data shown in Figure 5.5 are from the deforested site, Fazenda Dimona, lying about 80 km to the north of Manaus.

The 'windspeed profile equation', which assumes a horizontal and uniform fetch in the upwind direction and a neutral temperature profile, is

$$ku(z)/u_* = \ln(z - d) - \ln(z_0) \tag{5.8}$$

where the zero plane displacement d, and the roughness length z_0, are parameters describing the roughness of the surface; $u(z)$ is the wind velocity at height z; u_* is the friction velocity, and k is von Kármán's constant, with value 0.41 approximately (Shuttleworth, 1979). Wright *et al.* describe how estimates of zero plane displacement d were obtained in which triplets of height z_i, and wind velocity $u(z_i)$ $(i = 1, 2, 3)$, were substituted in equation (5.8). The parameters u_* and z_0 were then eliminated to give the equation

$$(u_1 - u_2)/(u_1 - u_3) = [\ln(z_1 - d) - \ln(z_2 - d)]/[\ln(z_1 - d) - \ln(z_3 - d)] \tag{5.9}$$

with all combinations of three measurement heights, from the six at which wind velocity was measured, taken in turn. Equations (5.9) were used to obtain $6!/(3!3!) = 20$ estimates of d, from which the arithmetic mean of d was computed. Estimates of u_* and z_0 were then obtained by substituting d in equation (5.8), and calculating the regression of $u(z)$ on $\ln(z - d)$.

An alternative estimation procedure, having the advantage of providing measures of the precision of estimates of d, u_* and z_0, would be as follows. Write equation (5.8) in the form

$$u(z) = (u_*/k) \ln(z - d) - (u_*/k) \ln z_0 + \varepsilon_z \tag{5.10}$$

or

$$u(z) = \beta_0 + \beta_1 x_z + \varepsilon_z \tag{5.11}$$

where $\beta_0 = -(u_*/k) \ln(z_0)$, $\beta_1 = (u_*/k)$, and $x = \ln(z - d)$. Let us assume as a starting point that the errors ε_z are normally distributed random variables with zero mean and variance σ_ε^2. Maximum-likelihood estimation is then equivalent to minimising the sum of squared deviations between the six measured wind speeds in a profile, and the values fitted using the model (5.11):

Height	Prof[1]	Prof[2]	Prof[3]	Prof[4]	Prof[5]	Prof[6]
0.52	1.62	1.40	1.73	1.47	1.46	1.97
0.94	1.94	1.69	2.15	1.86	1.81	2.40
1.69	2.55	2.13	2.62	2.39	2.30	2.91
2.98	3.14	2.42	3.09	2.79	2.79	3.41
5.06	3.46	2.81	3.42	3.07	3.08	3.81
9.32	3.86	2.99	3.85	3.41	3.39	4.26

Prof[7]	Prof[8]	Prof[9]	Prof[10]	Prof[11]	Prof[12]
1.46	2.06	1.78	1.50	1.55	1.32
1.74	2.52	2.18	1.78	1.96	1.60
2.22	3.25	2.76	2.32	2.49	2.01
2.54	3.79	3.40	2.73	2.96	2.41
2.80	4.10	3.83	2.99	3.19	2.66
3.11	4.45	4.38	3.29	3.42	2.91

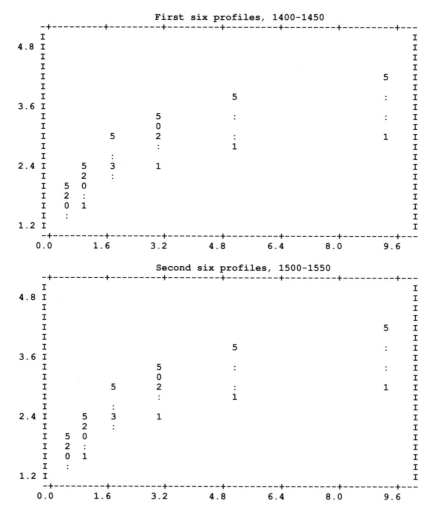

Figure 5.5 Anemometer height (first column, in metres) and measured wind velocity profiles (metres per second) at six heights, at twelve intervals of 10 minutes, on 6 October 1990, at Fazenda Dimona: ranchland, cleared from tropical forest, lying about 80 km north of Manaus, Amazonas, Brazil. Times of readings: 1400; 1410; 1420; 1430; 1440; 1450; 1500; 1510; 1520; 1530; 1540; 1550.

that is, by minimising the criterion

$$RSS = \sum [u(z) - \beta_0 - \beta_1 x_z]^2 \qquad (5.12)$$

Equation (5.11) bears a resemblance to the least-squares criterion used in fitting a simple linear regression, but the regression is not linear in the parameters u_*, z_0 and d, since x_z in equation (5.11) depends upon d. However, we can adopt a 'conditional likelihood' approach by assuming a series of trial values for d; the residual sum of squares RSS obtained by minimising equation (5.12) is then a function of d, $RSS = RSS(d)$. By evaluating $RSS(d)$ for the series of d-values, the value of d (say $\hat{d}$) can be found for which $RSS(d)$ is a minimum. The corresponding estimate of u_* can be found as $\hat{u}_* = k\hat{\beta}_1 = 0.41\hat{\beta}_1$, with variance $(0.41)^2 \text{ var } \hat{\beta}_1$. The estimate of z_0 is given by $\hat{z}_0 = \exp(-0.41\hat{\beta}_0/\hat{u}_*)$, giving $\hat{z}_0 = \exp(-\hat{\beta}_0/\hat{\beta}_1)$. An approximation to its variance can be found from the expression giving the variance of a function F of two random variables X and Y, in terms of their variances and covariance:

$$\text{var}[F(X, Y)] = (\partial F/\partial X)^2 \text{ var}[X] + (\partial F/\partial Y)^2 \text{ var}[Y]$$
$$+ 2(\partial F/\partial X)(\partial F/\partial Y) \text{ cov}[X, Y] \qquad (5.13)$$

from which we obtain

$$\text{var}[\hat{z}_0] \cong (\hat{z}_0/\hat{\beta}_1)^2 \text{ var}[\hat{\beta}_0] + (\hat{z}_0\hat{\beta}_0/\hat{\beta}_1^2)^2 \text{ var}[\hat{\beta}_1]$$
$$- 2(\hat{z}_0^2/\hat{\beta}_1^2) \text{ cov}[\hat{\beta}_0, \hat{\beta}_1] \qquad (5.14)$$

The above procedures do not yield an estimate of the variance of $\hat{d}$. If this is required, it is necessary to calculate the inverse of the matrix of second derivatives $-[\partial^2 \ln L/\partial\theta_i \partial\theta_j]$, where

$$\ln L = -6 \ln \sigma_\varepsilon - RSS \, (u_*, z_0, d, \sigma_\varepsilon)/(2\sigma_\varepsilon^2) \qquad (5.15)$$

with θ_i and θ_j the parameters u_*, z_0, d and σ_ε. The element in the third row and third column of this inverse matrix gives the variance of $\hat{d}$.

We illustrate the iterative calculation using the data from the first of the 12 ten-minute intervals shown in Figure 5.5. Having read the data for anemometer height z and for the corresponding velocity profile into variates with names z, $Profile$, we consider the trial values of d for which $RSS(d)$ is to be calculated. Clearly the value of d must be less than the first level $z(1)$; otherwise, $\ln(z - d)$ would be indeterminate. Suppose we begin by evaluating $RSS(d)$ for values of d between 0 and 0.10 by steps of 0.01 m. We enter a sequence of instructions such as

```
>FOR i=1...11
>CALCULATE d=(i−1)/100: & x= LOG(z−d)
>MODEL Profile   :   FIT x
>ENDFOR
```

and find that RSS has the following values for the values of the parameter d specified in the program:

d	RSS	d	RSS
0.00	0.010 37	0.05	0.010 19
0.01	0.010 30	0.06	0.010 20
0.02	0.010 25	0.07	0.010 23
0.03	0.010 22	0.08	0.010 28
0.04	0.010 20	0.09	0.010 34

Clearly RSS has a minimum value somewhere near $d = 0.05$. Refinement of the search grid gives $d = 0.049\dots$. At this value of d, the fitted relation is

$$u(z) = 0.7928 \ln(z - 0.049) + 2.1605$$

$$(\pm 0.0409) \qquad\qquad (\pm 0.0516)$$

with $r^2 = 98.7\%$. We therefore have $\hat{u}_* = 0.41 \times 0.7928 = 0.325 \pm 0.1678$ ms^{-1}, with $\hat{z}_0 = \exp(-2.1605/0.7928) = 0.0654$ m.

The above procedure has the disadvantage that a grid search, over the values of d, is necessary for each data set. With data from many hundreds of profiles, such a procedure would be impracticable, and a procedure is required which would effect the search automatically. An iterative procedure for this purpose is provided by the GENSTAT command NONLINEARFIT. It is implemented by the following instructions:

```
>EXPRESSION WS;   VALUE=!E(Fitted=ustar*(LOG(z−d)−LOG(z0))/0.41)
>MODEL Profile;              FITTEDVALUES=Fitted
>RCYCLE ustar,d,z0;   INITIAL=0.5,0,0.1
>FITNONLINEAR [PRINT=model,summary,estimates,correlation,\
>fittedvalues,monitoring;CALCULATION=WS]
```

The first line defines the expression to be fitted, here named WS; the second says that the profile is to be calculated for the variable *Profile*, the velocities in the first of the 12 wind profiles in Figure 5.5; the third states the parameters to be fitted are $ustar(= u_*)$, d and z_0, with initial values taken as 0.5, 0 and 0.1; and the fourth and fifth lines fit the model and control the output to be printed. Figure 5.6 shows parts of the output resulting from fitting the profile to the first six wind-speed profiles, denoted by *Prof*[1], *Prof*[2],...,*Prof*[6]; the output for *Prof*[1] shows that the estimate of d is 0.049 ± 0.227, with u_* and z_0 estimated by 0.3251 ± 0.0509 and 0.0656 ± 0.0464, respectively. Figure 5.6 gives the following values of u_*, d and z_0 for the six wind-speed profiles:

	u_*	d	z_0
Prof[1]	0.3251 ± 0.0509	0.049 ± 0.227	0.0656 ± 0.0464
Prof[2]	0.2197 ± 0.0310	0.120 ± 0.181	0.0309 ± 0.0226
Prof[3]	0.2913 ± 0.0153	0.0734 ± 0.0732	0.0401 ± 0.0105
Prof[4]	0.2459 ± 0.0234	0.196 ± 0.104	0.0293 ± 0.0144
Prof[5]	0.2627 ± 0.0330	0.130 ± 0.158	0.0423 ± 0.0258
Prof[6]	0.3264 ± 0.0167	0.0258 ± 0.0770	0.0427 ± 0.0109

***** Nonlinear regression analysis *****

Response variate: Prof[1]

*** Summary of analysis ***

	d.f.	s.s.	m.s.	v.r.
Regression	3	49.58054	16.52685	1216.23
Residual	3	0.04077	0.01359	
Total	6	49.62130	8.27022	

Percentage variance accounted for 98.2

*** Estimates of parameters ***

	estimate	s.e.
ustar	0.3251	0.0509
d	0.049	0.227
z0	0.0656	0.0464

*** Correlations ***

estimate	ref	correlations		
ustar	1	1.000		
d	2	−0.925	1.000	
z0	3	0.992	−0.952	1.000
		1	2	3

*** Fitted values and residuals ***

Unit	Response	Fitted value	Standardized residual
1	1.620	1.569	0.44
2	1.940	2.071	−1.12
3	2.550	2.554	−0.03
4	3.140	3.013	1.09
5	3.460	3.438	0.19
6	3.860	3.926	−0.57
Mean	2.762	2.762	0.00

***** Nonlinear regression analysis *****

Response variate: Prof[2]

*** Summary of analysis ***

	d.f.	s.s.	m.s.	v.r.
Regression	3	32.02749	10.675831	1768.82
Residual	3	0.01811	0.006036	
Total	6	32.04560	5.340933	

Percentage variance accounted for 98.4

*** Estimates of parameters ***

	estimate	s.e.
ustar	0.2197	0.0310
d	0.120	0.181
z0	0.0309	0.0226

Figure 5.6 Fitting of non-linear model $u = (u_*/0.41)(\ln(z - d) - \ln(z_0))$ to windspeed data by minimising the sum of squared deviations $\Sigma[u - u_*/0.41)(\ln(z-d) - \ln(z_0))]^2$. For data, see Figure 5.5. Only the first six profiles, with data $Prof[1] \ldots Prof[6]$, have been fitted.

_____ Continued _____

Figure 5.6 (continued)

*** Correlations ***

estimate	ref		correlations	
ustar	1	1.000		
d	2	−0.918	1.000	
z0	3	0.993	−0.944	1.000
		1	2	3

*** Fitted values and residuals ***

Unit	Response	Fitted value	Standardized residual
1	1.400	1.376	0.31
2	1.690	1.759	−0.88
3	2.130	2.106	0.31
4	2.420	2.427	−0.09
5	2.810	2.719	1.17
6	2.990	3.053	−0.81
Mean	2.240	2.240	0.00

***** Nonlinear regression analysis *****

Response variate: Prof[3]

*** Summary of analysis ***

	d.f.	s.s.	m.s.	v.r.
Regression	3	50.542889	16.847630	12929.79
Residual	3	0.003909	0.001303	
Total	6	50.546799	8.424466	

Percentage variance accounted for 99.8

*** Estimates of parameters ***

	estimate	s.e.
ustar	0.2913	0.0153
d	0.0734	0.0732
z0	0.0401	0.0105

*** Correlations ***

estimate	ref		correlations	
ustar	1	1.000		
d	2	−0.922	1.000	
z0	3	0.993	−0.949	1.000
		1	2	3

*** Fitted values and residuals ***

Unit	Response	Fitted value	Standardized residual
1	1.730	1.717	0.37
2	2.150	2.184	−0.95
3	2.620	2.626	−0.17
4	3.090	3.043	1.32
5	3.420	3.425	−0.15
6	3.850	3.865	−0.41
Mean	2.810	2.810	0.00

_____ Figure 5.6 *(continued)* _____

***** Nonlinear regression analysis *****

Response variate: Prof[4]

*** Summary of analysis ***

	d.f.	s.s.	m.s.	v.r.
Regression	3	40.15695	13.385650	3148.34
Residual	3	0.01275	0.004252	
Total	6	40.16970	6.694951	

Percentage variance accounted for 99.2

*** Estimates of parameters ***

	estimate	s.e.
ustar	0.2459	0.0234
d	0.196	0.104
z0	0.0293	0.0144

*** Correlations ***

estimate	ref	correlations		
ustar	1	1.000		
d	2	−0.909	1.000	
z0	3	0.992	−0.937	1.000
		1	2	3

*** Fitted values and residuals ***

Unit	Response	Fitted value	Standardized residual
1	1.470	1.447	0.36
2	1.860	1.942	−1.25
3	2.390	2.359	0.48
4	2.790	2.732	0.89
5	3.070	3.066	0.06
6	3.410	3.444	−0.53
Mean	2.498	2.498	0.00

***** Nonlinear regression analysis *****

Response variate: Prof[5]

*** Summary of analysis ***

	d.f.	s.s.	m.s.	v.r.
Regression	3	39.43932	13.146441	1880.60
Residual	3	0.02097	0.006991	
Total	6	39.46030	6.576716	

Percentage variance accounted for 98.8

*** Estimates of parameters ***

	estimate	s.e.
ustar	0.2627	0.0330
d	0.130	0.158
z0	0.0423	0.0258

Figure 5.6 *(continued)*

$*** $ Correlations $ ***$

estimate	ref	correlations		
ustar	1	1.000		
d	2	−0.917	1.000	
z0	3	0.992	−0.945	1.000
		1	2	3

$*** $ Fitted values and residuals $ ***$

Unit	Response	Fitted value	Standardized residual
1	1.460	1.429	0.37
2	1.810	1.894	−1.00
3	2.300	2.312	−0.15
4	2.790	2.698	1.10
5	3.080	3.049	0.38
6	3.390	3.448	−0.70
Mean	2.472	2.472	0.00

$***** $ Nonlinear regression analysis $ *****$

Response variate: Prof[6]

$*** $ Summary of analysis $ ***$

	d.f.	s.s.	m.s.	v.r.
Regression	3	62.396641	20.798880	14982.95
Residual	3	0.004165	0.001388	
Total	6	62.400806	10.400134	

Percentage variance accounted for 99.8

$*** $ Estimates of parameters $ ***$

	estimate	s.e.
ustar	0.3264	0.0167
d	0.0258	0.0770
z0	0.0427	0.0109

$*** $ Correlations $ ***$

estimate	ref	correlations		
ustar	1	1.000		
d	2	−0.927	1.000	
z0	3	0.994	−0.952	1.000
		1	2	3

$*** $ Fitted values and residuals $ ***$

Unit	Response	Fitted value	Standardized residual
1	1.970	1.953	0.45
2	2.400	2.440	−1.07
3	2.910	2.915	−0.15
4	3.410	3.372	1.03
5	3.810	3.795	0.39
6	4.260	4.284	−0.65
Mean		3.127	3.127	0.00

Residuals: each row corresponds to a profile:

r[1]	r[2]	r[3]	r[4]	r[5]	r[6]
0.44	−1.12	−0.03	1.09	0.19	−0.57
0.30	−0.88	0.31	−0.09	1.17	−0.81
0.37	−0.95	−0.17	1.32	−0.15	−0.41
0.36	−1.25	0.48	0.89	0.06	−0.52
0.37	−1.00	−0.15	1.10	0.38	−0.70
0.45	−1.07	−0.14	1.03	0.39	−0.65

Histogram of these residuals:

		−1.0	4	* * * *
−1.0	—	−0.5	7	* * * * * * *
−0.5	—	0.0	7	* * * * * * *
0.0	—	0.5	12	* * * * * * * * * * * *
0.5	—	1.0	1	*
1.0	—		5	* * * * *

Scale: 1 asterisk represents 1 unit.

Residuals after subtracting means at each anemometer height:

r[1]	r[2]	r[3]	r[4]	r[5]	r[6]
0.058	−0.075	−0.080	0.200	−0.150	0.040
−0.082	0.165	0.260	−0.980	0.830	−0.200
−0.012	0.095	−0.220	0.430	−0.490	0.200
−0.022	−0.205	0.430	0.000	−0.280	0.090
−0.012	0.045	−0.200	0.210	0.040	−0.090
0.068	−0.025	−0.190	0.140	0.050	−0.040

		−0.8	1	*
−0.8	—	−0.4	1	*
−0.4	—	0.0	16	* * * * * * * * * * * * * * * *
0.0	—	0.4	15	* * * * * * * * * * * * * * *
0.4	—	0.8	2	* *
0.8	—		1	*

Scale: 1 asterisk represents 1 unit.

Variances of residuals at the six anemometer heights are:

Variance[1]	Variance[2]	Variance[3]	Variance[4]	Variance[5]	Variance[6]
0.003097	0.01731	0.07668	0.2498	0.2068	0.01988

Figure 5.7   Simple examination of the residuals obtained after fitting equation (5.8) to the first six wind profiles of Figure 5.5.

We note that the values of d are not large when compared with their standard errors; taking $d \pm 2SE(d)$ as an approximate 95% confidence interval for the 'true' value, this interval includes zero in all six cases, so that the calculated values of d do not differ significantly from zero. A further point of interest arises from inspection of the (standardised) residuals from the six fitted profiles; these are shown in the 6×6 matrix, with columns denoted by $r[1], \ldots, r[6]$, in Figure 5.7. There is a clear pattern in sign which is repeated, with minor variations, in all six profiles, as the following table shows:

	Anemometer height (m)					
	0.52	0.94	1.69	2.98	5.05	9.32
Prof [1]	0.44	−1.12	−0.03	1.09	0.19	−0.57
Prof [2]	0.30	−0.88	0.31	−0.09	1.17	−0.81
Prof [3]	0.37	−0.95	−0.17	1.32	−0.15	−0.41
Prof [4]	0.36	−1.25	0.48	0.89	0.06	−0.52
Prof [5]	0.37	−1.00	−0.15	1.10	0.38	−0.70
Prof [6]	0.45	−1.07	−0.14	1.03	0.39	−0.65

The consistency of sign and magnitude in the residuals at anemometer heights 0.52 m, 0.94 m and 9.32 m is remarkable, and the explanation for this is not obvious. One possibility is that the wind velocity readings show effects particular to the anemometers used in the measurements: some anemometers tend, perhaps, to give low readings, others to give high readings, so that the corresponding residuals reflect these biases. Whatever the explanation, our analysis demonstrates once again the importance of looking very closely at the residuals obtained when a statistical model has been fitted.

Our analysis has used only the first six of the 12 profiles in Figure 5.5. If we try to repeat the analysis using the second six profiles, *Prof*[7] to *Prof*[12], difficulties are encountered in the case of *Prof*[8] and *Prof*[11] when the directive NONLINEARFIT is used. This is because the search for the optimum value of d leads into a region where $d > 0.52$ m, the height of the lowest altimeter, so that $(z - d)$ in equation (5.8) becomes negative and $\ln(z - d)$ undefined. This is telling us that, for profiles 8 and 11, the relation (5.8) does not afford a good description of how wind velocity varies with height; faced with this problem, Wright *et al.* (1992) noted that 'although many of the estimates of zero-plane displacement ... have physically anomalous values, being either below zero or above canopy height, it is not appropriate to exclude this estimation error from the evaluation of d'.

Alternative forms of the profile equation (5.8), appropriate for different meteorological conditions, are described in the hydrometeorological literature. Stewart (1989) describes an equation appropriate for non-neutral conditions of the form

$$k^2 u(z) = [\ln\{(z - d)/z_0\} - \varphi]^2 / r_a$$

where k, $u(z)$, z_0 and d are as described above; φ is a dimensionless stability function; and r_a is the aerodynamic resistance (in units s m^{-1}) to the transfer of sensible heat and water vapour from the surface to the level z. Adams *et al.* (1991) use a further modification

$$k^2 u(z) = \ln[(z - d)/z_0^m] \ln[(z - d)/z_0^h] / r_a$$

where z_0^m, z_0^h are constants ('roughness lengths' for the transfer of momentum and sensible heat, respectively). Given wind profile data $u(z)$, the computational procedures described above can be modified to fit these alternatives. Writing the Adams equation as

$$u(z) = (1/[k^2 r_a])\{\ln(z - d)\}^2 + (\ln z_0^m + \ln z_0^h)/(k^2 r_a) \ln(z - d)$$
$$+ [(\ln z_0^h) \cdot (\ln z_0^m)]/(k^2 r_a)$$

we have a quadratic equation in $\ln(z - d)$. For given d, the equation therefore becomes of the linear form

$$u(z) = A[\ln(z - d)]^2 + B \ln(z - d) + C$$

One procedure for estimating the parameters $A = 1/(k^2 r_a)$, $B = (\ln z_0^m + \ln z_0^h)/(k^2 r_a)$ and $C = [(\ln z_0^h)(\ln z_0^m)]/(k^2 r_a)$ would be to plot the residual sum of squares RSS as a function of d, and to locate the function's minimum value. Denoting the minimum by d_0, the estimate of A yields the aerodynamic resistance r_a, and the estimates of B and C yield a quadratic equation to be solved for $\ln z_0^m$ and $\ln z_0^h$. Difficulty may arise if this quadratic has imaginary roots; and it is also possible for both roots to be non-physically realistic, as occurred with the profile data fitted to equation (5.8). An alternative procedure, more suitable where the parameters r_a, z_0^h and z_0^m are to be calculated for many wind-speed profiles, would be to use the GENSTAT directive NONLINEARFIT, used as above when fitting equation (5.8).

In the discussion of the wind-speed data so far, we have made no check of the assumption that the residuals (for example, in the model (5.10)) are normally distributed, with homogeneous variances. The six residuals resulting from fitting equation (5.10) to each profile is rather a small sample for calculating normal plots and envelopes along the lines of Chapter 4, but we can use some very simple methods to get some idea of their behaviour, as Figure 5.7 shows. The first part of this figure shows that the histogram of the standardised residuals is rather 'lumpy', probably because of the closeness in magnitude, and equality of sign, of residuals particularly at the anemometer heights 0.52 m, 0.94 m and 9.32 m. If the explanation of anemometer 'bias' is a plausible explanation for the similarity of these residuals, it makes sense to subtract from each residual the mean of the residuals, for the appropriate height, over all six profiles. The point of doing this is that we can then more reasonably pool all 36 residuals together. The resulting set of modified residuals, together with their histogram, is shown in the lower part of Figure 5.7, and we see that this histogram looks more 'normal' than the upper histogram. However, problems still remain; if we calculate the variance amongst the modified residuals at each height, we find the six variances shown at the bottom of Figure 5.7. These differ by up to two orders of magnitude, suggesting that the assumption of homogeneous errors is extremely suspect, even after the removal of anemometer 'bias'. The variances follow a rather regular pattern, however, suggesting the possibility of a physical explanation.

5.5 NON-LINEAR FITTING PROCEDURES FOR SOME STANDARD MODELS

Section 5.4 demonstrated how a statistical model with one parameter occurring non-linearly (d, in the example) could be fitted by assuming particular values for it, the other parameters occurring linearly. Several other statistical models are of the same type, in that one (perhaps two) parameters are non-linear, so that these and the other parameters may be fitted by repetitive use of linear regression. Such models occur frequently, but in slightly different guises, in many areas of application. One such model is the exponential curve $y_t = \alpha + \beta \exp(-\gamma t) + \varepsilon_t$, which, after reparametrisation by writing $\rho = \exp(-\gamma)$, becomes

$$y_t = \alpha + \beta \rho^t + \varepsilon_t \tag{5.16}$$

where $0 < \rho < 1$. If β is negative (positive), this curve rises (falls) to an asymptote $y = \alpha$ from its initial value $\alpha + \beta$ at time $t = 0$. A typical hydrological example is the Horton formula for the rate of infiltration capacity f, with systematic component given by

$$f = f_c + (f_0 - f_c) \exp(-kt) \tag{5.17}$$

where f is rate of infiltration capacity; f_0 is initial rate of infiltration capacity; and f_c is the ultimate infiltration capacity (the asymptotic value). In the notation of equation (5.16), we have

$\alpha = f_c$, $\beta = f_0 - f_c$, and $\rho = \exp(-k)$; in the form (5.16) we may write $x_t = \rho^t$ and fit the linear regression $y_t = \alpha + \beta x_t + \varepsilon_t$ for a series of values of ρ between 0 and 1, choosing that value of ρ which minimises $RSS(\rho)$.

A model similar to equation (5.17) familiar to hydrobiologists is the Ricker curve (Ricker, 1958) describing the growth of fish populations. It is derived by the following argument: let N_t be the size of a fish stock at time t; let the (constant) mortality rate and the (constant) recruitment rates be i and R. Then

$$dN_t = -i N_t + R$$

Integrating, and using the initial condition that $N_t = 0$ for $t = 0$, we obtain

$$N_t = R(1 - \exp[-it])/i$$

for the size of the population at time t. Essentially the same equation is used to model the growth in fish length, by the von Bertalanffy curve $L_t = L_\infty \{1 - \exp[-k(t - t_0)]\}$ where L_∞ is the length at maturity and k is a growth parameter. Damm (1987) extends the von Bertalanffy growth curve by introducing some additional parameters, giving the curve

$$\bar{l}_t = L_\infty - (L_\infty - L_R)\exp[-Kt] \cdot \{[\exp(K + Z) - \exp(K)]/[\exp(K + Z) - 1]\}$$

for average fish length. Damm estimates the parameters K, L_∞ and Z (L_R is assumed known) by a procedure now familiar to the reader; he assumes a value for the parameter K, and for K constant the equation becomes a linear regression. The estimate of the intercept of the regression line is also an estimate of L_∞; by equating the slope of the regression line to

$$-(L_\infty - L_R) \cdot \{[\exp(K + Z) - \exp(K)]/[\exp(K + Z) - 1]\}$$

and solving for Z, this parameter is also estimated. The calculation is repeated for a series of values K until the residual sum of squares is minimised.

Statistical models which, having some parameters occurring linearly, may be fitted by similar procedures, are the following:

Exponential curves:

Ordinary exponential $y_t = \alpha + \beta\rho^t + \varepsilon_t$
Double exponential $y_t = \alpha + \beta\rho^t + \gamma x^t + \varepsilon_t$
Critical exponential $y_t = \alpha + (\beta + \gamma t)\rho^t + \varepsilon_t$
Line plus exponential $y_t = \alpha + \beta\rho^t + \gamma t + \varepsilon_t$

Logistic curves:

Ordinary logistic $y_t = \alpha + \gamma/[1 + \exp\{-\beta(t - \mu)\}] + \varepsilon_t$
Generalized logistic $y_t = \alpha + \gamma/[1 + \tau\exp\{-\beta(t - \mu)\}]^{1/\tau} + \varepsilon_t$
Gompertz $y_t = \alpha + \gamma\exp[-\exp\{-\beta(t - \mu)\}] + \varepsilon_t$

Rational functions:

Linear divided by linear $y_t = \alpha + \beta/(1 + \delta t) + \varepsilon_t$
Quadratic divided by linear $y_t = \alpha + \beta/(1 + \delta t) + \gamma t + \varepsilon_t$
Quadratic divided by quadratic $y_t = \alpha + (\beta + \gamma t)/(1 + \delta t + \phi t^2) + \varepsilon_t$

These models can all be very easily fitted by GENSTAT, and some examples are given below. The four exponential models all arise as solutions to differential equations and represent processes that increase or decrease with time, such as rate of infiltration capacity as represented by the Horton law. The three logistic curves are commonly used to describe growth of biological subjects or communities, but 'growth' curves of this type are also encountered in regional flood studies. The first of the three rational curves (linear divided by linear) is a rectangular hyperbola and occurs as the Michaelis–Menten model of chemical kinetics; the second, quadratic divided by linear, is a rectangular hyperbola with a non-horizontal asymptote; and the third, quadratic divided by quadratic, is a cubic curve having an asymmetric maximum, falling to an asymptote. The logistic curves, given above, are shaped like the soil moisture characteristic curve relating volumetric soil water content, θ, to soil moisture suction, φ; and also the curves relating percentage water loss to oven temperature for clay minerals (see, for example, Hillel, 1971).

5.5.1 Example: Fitting the Horton Infiltration Law

The first example demonstrates the fitting of the model given in equation (5.17), describing the rate of infiltration. Figure 5.8 shows measured rates at intervals of time, corresponding to three irrigation treatments, denoted in the table by $Irrig8$, $Irrig10$, and $Irrig11$. Figure 5.9 shows the output given by GENSTAT for the fitting of three curves of the form $y_t = \alpha + \beta\rho^t + \varepsilon_t$, corresponding to the three irrigation treatments. We see that the fitted curve for $Irrig8$ is

$$Infilt8_t = 0.024\,51 + 0.3070(0.8887)^t$$

$$(\pm 0.006\,72)(\pm 0.0208)(\pm 0.0169)$$

with the proportion of explained variance being 95.5%. Two data points are noted as having large residuals (points 1 and 2 in the data listing of Figure 5.8, with standardised residuals 2.77 and -3.03), and two as having high leverage (points 1 and 3, with leverages 0.78 and 0.44). All three points 1, 2 and 3 correspond to the beginning of the infiltration period, suggesting that despite the high percentage of explained variance, the curve does not describe the initial infiltration behaviour particularly well. We also have a warning message saying that it was difficult to fit the exponential curve because of the very steep slope at the beginning of the period of observation and the rapid approach to the curve's asymptotic value. This suggests

Time	Irrig8	Irrig10	Irrig11
0.50	0.340	0.520	0.240
3.00	0.190	0.098	0.550
7.50	0.176	0.050	0.152
15.00	0.073	0.033	0.076
25.00	0.055	0.024	0.043
35.00	0.043	0.025	0.027
50.00	0.021	0.018	0.026
70.00	0.030	0.015	0.028
90.00	0.020	0.015	0.026
110.00	0.022	0.012	0.021
130.00	0.018	0.012	0.017
150.00	0.022	0.018	0.018
170.00	0.026	0.011	0.013
190.00	0.013	0.010	0.020

Figure 5.8 Measured infiltration capacity rates for three irrigation treatments $Irrig8$, $Irrig10$, and $Irrig11$; data kindly provided by Professor N. L. Caicedo.

∗∗∗∗∗ Nonlinear regression analysis ∗∗∗∗∗

Response variate: Infilt[1]
    Explanatory: Time
    Fitted Curve: A + B∗R∗∗X
    Constraints: R < 1

∗∗∗ Summary of analysis ∗∗∗

	d.f.	s.s.	m.s.
Regression	2	0.113699	0.0568497
Residual	11	0.004457	0.0004052
Total	13	0.118157	0.0090890

Percentage variance accounted for 95.5

∗ MESSAGE: The following units have large residuals:

1	2.77
2	−3.03

∗ MESSAGE: The following units have high leverage:

1	0.78
3	0.44

∗∗∗ Estimates of parameters ∗∗∗

	estimate	s.e.
R	0.8887	0.0169
B	0.3070	0.0208
A	0.02451	0.00672

∗∗∗∗∗∗∗∗ Warning (Code OP 20). Statement 3 in For Loop
Command: FITCURVE Time

OP20 Fitted curve is close to, or has reached limiting form
Exponential curve approaches step function.

∗∗∗∗∗ Nonlinear regression analysis ∗∗∗∗∗

Response variate: Infilt[2]
    Explanatory: Time
    Fitted Curve: A + B∗R∗∗X
    Constraints: R < 1

∗∗∗ Summary of analysis ∗∗∗

	d.f.	s.s.	m.s.
Regression	2	0.21915	0.109575
Residual	11	0.01430	0.001300
Total	13	0.23345	0.017958

Percentage variance accounted for 92.8

∗∗∗ Estimates of parameters ∗∗∗

	estimate
R	0.705301
B	0.552200
A	0.0114443

Figure 5.9   Fit of Horton infiltration laws to infiltration measured on three irrigation treatments (see data in Figure 5.8).

Continued

Figure 5.9 *(continued)*

```
***** Nonlinear regression analysis *****
Response variate:    Infilt[3]
       Explanatory:  Time
      Fitted Curve:  A + B*R**X
       Constraints:  R < 1

*** Summary of analysis ***
                  d.f.        s.s.         m.s.
Regression         2        0.19841      0.099203
Residual          11        0.08405      0.007641
Total             13        0.28246      0.021727
Percentage variance accounted for 64.8

* MESSAGE: The following units have large residuals:
                              1          -2.97
                              2           3.31
* MESSAGE: The following units have high leverage:
                              1           0.70

*** Estimates of parameters ***
           estimate       s.e.
R          0.9169        0.0419
B          0.3831        0.0828
A          0.0159        0.0306
```

that the intervals between observations were too long at this early stage of infiltration; we have relatively many points near the asymptotic value, but few from which to estimate the rate at which the curve tends to its asymptote. The output listing also shows the infiltration curve corresponding to irrigation treatments $Irrig10$ and $Irrig11$ as:

$$Infilt10_t = 0.0114 + 0.5522(0.7053)^t$$

and

$$Infilt11_t = 0.0159 + 0.3831(0.9169)^t$$

$$(\pm0.0306)(\pm0.0828)(\pm0.0419)$$

the proportion of variance accounted for now being much lower, for $Irrig10$, at 64.8%, but high for $Irrig11$, at 95.5%. For $Irrig11$, points 1 and 2 have large residuals (-2.97 and 3.31), and point 1 has high leverage (0.70). Inspection of the data for $Irrig11$ (see Figure 5.8) shows that infiltration behaviour is not well described by a Horton curve in the first three minutes or so of observation.

It is computationally straightforward to fit Horton infiltration curves to groups of observations (such as the three irrigation treatments of this example), such that the curves have one or more parameters in common. We might wish, for example, to test whether two or more supposedly similar soil types have the same values of f_c in the Horton equation, and if so, whether the constant k of the Horton law can also be taken as equal over the soil samples. We illustrate this kind of analysis using the data from $Irrig8$, $Irrig10$ and $Irrig11$ treatments, despite the general unsatisfactoriness of the fit of Horton curves shown in Figure 5.9. Figure 5.10 shows the results of one such calculation. In lines 14–16 of the listing we have fitted Horton curves to data from $Irrig8$, $Irrig10$ and $Irrig11$, subject to the constraint that the Hortonian parameter k of equation (5.17) (or, in the notation of equation (5.16), the parameter ρ, estimated as $\hat{\rho} = r$

Time	0.50	3.00	7.50	15.00	25.00
Irrigati	8.000	8.000	8.000	8.000	8.000
Infilt	0.34000	0.19000	0.17600	0.07300	0.05500
Time	35.00	50.00	70.00	90.00	110.00
Irrigati	8.000	8.000	8.000	8.000	8.000
Infilt	0.04300	0.02100	0.03000	0.02000	0.02200
Time	130.00	150.00	170.00	190.00	0.50
Irrigati	8.000	8.000	8.000	8.000	10.000
Infilt	0.01800	0.02200	0.02600	0.01300	0.52000
Time	3.00	7.50	15.00	25.00	35.00
Irrigati	10.000	10.000	10.000	10.000	10.000
Infilt	0.09800	0.05000	0.03300	0.02400	0.02500
Time	50.00	70.00	90.00	110.00	130.00
Irrigati	10.000	10.000	10.000	10.000	10.000
Infilt	0.01800	0.01500	0.01500	0.01200	0.01200
Time	150.00	170.00	190.00	0.50	3.00
Irrigati	10.000	10.000	10.000	11.000	11.000
Infilt	0.01800	0.01100	0.01000	0.24000	0.55000
Time	7.50	15.00	25.00	35.00	50.00
Irrigati	11.000	11.000	11.000	11.000	11.000
Infilt	0.15200	0.07600	0.04300	0.02700	0.02600
Time	70.00	90.00	110.00	130.00	150.00
Irrigati	11.000	11.000	11.000	11.000	11.000
Infilt	0.02800	0.02600	0.02100	0.01700	0.01800
Time	170.00	190.00			
Irrigati	11.000	11.000			
Infilt	0.01300	0.02000			

```
   14    MODEL Infilt
   15    FITCURVE [PRINT=model,summary,estimates,correlations,fittedvalues,\
   16           monitoring]Time*Irrigation
   16 ..................................................................
```

```
******** Warning (Code. OP 20). Statement 1 on Line 16
Command: FITCURVE [PRINT=model,summary,estimates,correlations,fittedvalues,
moni
OP20 Fitted curve is close to, or has reached limiting form
Exponential curve approaches step function.
```

***** Nonlinear regression analysis *****

Response variate:	Infilt
Explanatory:	Time
Grouping factor:	Irrigati, all linear parameters separate
Fitted Curve:	A + B*R**X
Constraints:	R < 1

Figure 5.10   Output showing fit of Horton curves: with common r-value (e.g. $Infilt = a_i + b_i r^{(Time)} + e$; $i = 1, 2, 3$); with common b- and r-values ($Infilt = a_i + br^{(Time)} + e$; $i = 1, 2, 3$); with parameters a_i, b_i, r_i all distinct ($Infilt = a_i + b_i r_i^{(Time)} + e$; $i = 1, 2, 3$).

_ Continued _

Figure 5.10 *(continued)*

*** Summary of analysis ***

	d.f.	s.s.	m.s.
Regression	6	0.4823	0.080375
Residual	35	0.1574	0.004498
Total	41	0.6397	0.015602

Percentage variance accounted for 71.2

* MESSAGE: The following units have large residuals:

16	−2.42
29	−2.58
30	3.85

*** Estimates of parameters ***

		estimate	s.e.	Correlations
R		0.8581	0.0257	1.000
B	Irrigati 8	0.323026		
A	Irrigati 8	0.0287082		
B	Irrigati 10	0.406886		
A	Irrigati 10	0.00328038		
B	Irrigati 11	0.413059		
A	Irrigati 11	0.0306829		

*** Fitted values and residuals ***

Unit	Explanatory	Response	Fitted value	Standardized residual
1	0.5000	0.340	0.328	0.18
2	3.0000	0.190	0.233	−0.64
3	7.5000	0.176	0.131	0.67
4	15.0000	0.073	0.061	0.18
5	25.0000	0.055	0.036	0.29
6	35.0000	0.043	0.030	0.19
7	50.0000	0.021	0.029	−0.12
8	70.0000	0.030	0.029	0.02
9	90.0000	0.020	0.029	−0.13
10	110.0000	0.022	0.029	−0.10
11	130.0000	0.018	0.029	−0.16
12	150.0000	0.022	0.029	−0.10
13	170.0000	0.026	0.029	−0.04
14	190.0000	0.013	0.029	−0.23
15	0.5000	0.520	0.380	2.08
16	3.0000	0.098	0.260	−2.42
17	7.5000	0.050	0.132	−1.23
18	15.0000	0.033	0.044	−0.17
19	25.0000	0.024	0.012	0.18
20	35.0000	0.025	0.005	0.30
21	50.0000	0.018	0.003	0.22
22	70.0000	0.015	0.003	0.17
23	90.0000	0.015	0.003	0.17
24	110.0000	0.012	0.003	0.13
25	130.0000	0.012	0.003	0.13
26	150.0000	0.018	0.003	0.22
27	170.0000	0.011	0.003	0.12
28	190.0000	0.010	0.003	0.10
29	0.5000	0.240	0.413	−2.58

Figure 5.10 *(continued)*

30	3.0000	0.550	0.292	3.85
31	7.5000	0.152	0.162	−0.15
32	15.0000	0.076	0.072	0.06
33	25.0000	0.043	0.040	0.05
34	35.0000	0.027	0.033	−0.08
35	50.0000	0.026	0.031	−0.07
36	70.0000	0.028	0.031	−0.04
37	90.0000	0.026	0.031	−0.07
38	110.0000	0.021	0.031	−0.14
39	130.0000	0.017	0.031	−0.20
40	150.0000	0.018	0.031	−0.19
41	170.0000	0.013	0.031	−0.26
42	190.0000	0.020	0.031	−0.16
Mean	74.7143	0.075	0.075	0.00

```
 17    FITCURVE [PRINT=model,summary,estimates,correlations,fittedvalues,\
 18            monitoring]Time,Irrigation
  1
 18.........................................................................
```

```
********* Warning (Code OP 20). Statement 1 on Line 18
Command: FITCURVE [PRINT=model,summary,estimates,correlations,fittedvalues,
moni
OP20 Fitted curve is close to, or has reached limiting form
Exponential curve approaches step function.
```

***** Nonlinear regression analysis *****

Response variate:	Infilt
Explanatory:	Time
Grouping factor:	Irrigati, constant parameters separate
Fitted Curve:	A + B*R**X
Constraints:	R < 1

*** Summary of analysis ***

	d.f.	s.s.	m.s.
Regression	4	0.4768	0.119195
Residual	37	0.1629	0.004402
Total	41	0.6397	0.015602

Percentage variance accounted for 71.8

* MESSAGE: The following units have large residuals:

15	2.42
16	−2.26
29	−2.23
30	4.13

*** Estimates of parameters ***

		estimate	s.e.	Correlations
R		0.8590	0.0242	1.000
B		0.380436		
A	Irrigati 8	0.0202827		
A	Irrigati 10	0.00685409		
A	Irrigati 11	0.0351398		

Figure 5.10 *(continued)*

*** Fitted values and residuals ***

Unit	Explanatory	Response	Fitted value	Standardized residual
1	0.5000	0.340	0.373	−0.50
2	3.0000	0.190	0.261	−1.08
3	7.5000	0.176	0.142	0.51
4	15.0000	0.073	0.059	0.21
5	25.0000	0.055	0.029	0.40
6	35.0000	0.043	0.022	0.31
7	50.0000	0.021	0.020	0.01
8	70.0000	0.030	0.020	0.15
9	90.0000	0.020	0.020	0.00
10	110.0000	0.022	0.020	0.03
11	130.0000	0.018	0.020	−0.03
12	150.0000	0.022	0.020	0.03
13	170.0000	0.026	0.020	0.09
14	190.0000	0.013	0.020	−0.11
15	0.5000	0.520	0.359	2.42
16	3.0000	0.098	0.248	−2.26
17	7.5000	0.050	0.129	−1.18
18	15.0000	0.033	0.046	−0.19
19	25.0000	0.024	0.015	0.13
20	35.0000	0.025	0.009	0.25
21	50.0000	0.018	0.007	0.17
22	70.0000	0.015	0.007	0.12
23	90.0000	0.015	0.007	0.12
24	110.0000	0.012	0.007	0.08
25	130.0000	0.012	0.007	0.08
26	150.0000	0.018	0.007	0.17
27	170.0000	0.011	0.007	0.06
28	190.0000	0.010	0.007	0.05
29	0.5000	0.240	0.388	−2.23
30	3.0000	0.550	0.276	4.13
31	7.5000	0.152	0.157	−0.07
32	15.0000	0.076	0.074	0.03
33	25.0000	0.043	0.044	−0.01
34	35.0000	0.027	0.037	−0.15
35	50.0000	0.026	0.035	−0.14
36	70.0000	0.028	0.035	−0.11
37	90.0000	0.026	0.035	−0.14
38	110.0000	0.021	0.035	−0.21
39	130.0000	0.017	0.035	−0.27
40	150.0000	0.018	0.035	−0.26
41	170.0000	0.013	0.035	−0.33
42	190.0000	0.020	0.035	−0.23
Mean	74.7143	0.075	0.075	0.00

```
 19    FITCURVE [PRINT=model,summary,estimates,correlations,fittedvalues,\
 20            monitoring;NONLINEAR=separate]Time*Irrigation
20..........................................................................
```

******* Warning (Code OP 20). Statement 1 on Line 20
Command: FITCURVE [PRINT=model,summary,estimates,correlations,fittedvalues, moni
OP20 Fitted curve is close to, or has reached limiting form
Exponential curve approaches step function.

Figure 5.10 *(continued)*

***** Nonlinear regression analysis *****

Response variate: Infilt
Explanatory: Time
Grouping factor: Irrigati, all parameters separate
Fitted Curve: A + B*R**X
Constraints: R < 1

*** Summary of analysis ***

	d.f.	s.s.	m.s.
Regression	8	0.5369	0.067108
Residual	33	0.1028	0.003115
Total	41	0.6397	0.015602

Percentage variance accounted for 80.0

* MESSAGE: The following units have large residuals:

15	6.02
16	−3.89
29	−4.65
30	5.18

* MESSAGE: The following units have high leverage:

1	0.78
15	0.98
16	0.76
29	0.70

*** Estimates of parameters ***

		estimate	s.e.	Correlations		
R	Irrigati 8	0.8887	0.0469	1.000		
B	Irrigati 8	0.3070	0.0577	−0.472	1.000	
A	Irrigati 8	0.0245	0.0186	−0.397	−0.208	1.000
R	Irrigati 10	0.7053	0.0788	1.000		
B	Irrigati 10	0.5522	0.0821	−0.597	1.000	
A	Irrigati 10	0.0114	0.0168	−0.287	−0.114	1.000
R	Irrigati 11	0.9169	0.0268	1.000		
B	Irrigati 11	0.3831	0.0529	−0.420	1.000	
A	Irrigati 11	0.0159	0.0196	−0.441	−0.244	1.000

*** Fitted values and residuals ***

Unit	Explanatory	Response	Fitted value	Standardized residual	Leverage
1	0.5000	0.340	0.314	1.00	0.78
2	3.0000	0.190	0.240	−1.09	0.33
3	7.5000	0.176	0.151	0.59	0.44
4	15.0000	0.073	0.077	−0.08	0.35
5	25.0000	0.055	0.041	0.28	0.13
6	35.0000	0.043	0.029	0.25	0.09
7	50.0000	0.021	0.025	−0.08	0.10
8	70.0000	0.030	0.025	0.10	0.11
9	90.0000	0.020	0.025	−0.09	0.11
10	110.0000	0.022	0.025	−0.05	0.11
11	130.0000	0.018	0.025	−0.12	0.11
12	150.0000	0.022	0.025	−0.05	0.11
13	170.0000	0.026	0.025	0.03	0.11

Figure 5.10 *(continued)*

14	190.0000	0.013	0.025	−0.22	0.11
15	0.5000	0.520	0.475	6.02	0.98
16	3.0000	0.098	0.205	−3.89	0.76
17	7.5000	0.050	0.052	−0.04	0.28
18	15.0000	0.033	0.014	0.35	0.08
19	25.0000	0.024	0.012	0.23	0.09
20	35.0000	0.025	0.011	0.25	0.09
21	50.0000	0.018	0.011	0.12	0.09
22	70.0000	0.015	0.011	0.07	0.09
23	90.0000	0.015	0.011	0.07	0.09
24	110.0000	0.012	0.011	0.01	0.09
25	130.0000	0.012	0.011	0.01	0.09
26	150.0000	0.018	0.011	0.12	0.09
27	170.0000	0.011	0.011	−0.01	0.09
28	190.0000	0.010	0.011	−0.03	0.09
29	0.5000	0.240	0.383	−4.65	0.70
30	3.0000	0.550	0.311	5.18	0.32
31	7.5000	0.152	0.216	−1.39	0.33
32	15.0000	0.076	0.120	−1.01	0.38
33	25.0000	0.043	0.060	−0.34	0.21
34	35.0000	0.027	0.034	−0.14	0.11
35	50.0000	0.026	0.021	0.10	0.10
36	70.0000	0.028	0.017	0.21	0.11
37	90.0000	0.026	0.016	0.19	0.12
38	110.0000	0.021	0.016	0.10	0.12
39	130.0000	0.017	0.016	0.02	0.12
40	150.0000	0.018	0.016	0.04	0.12
41	170.0000	0.013	0.016	−0.06	0.12
42	190.0000	0.020	0.016	0.08	0.12
Mean	74.7143	0.075	0.075	0.05	0.21

in the output) is common to all three treatments. We see that the common r-value is estimated as $r = 0.8581 \pm 0.0257$, although with this constrained model the difficulties remained when fitting to data from the first three minutes of observation, as shown by large standardised residuals and leverages. The error variance derived from the three data sets is 0.004 498 on 35 degrees of freedom; it is reduced only slightly, to 0.004 402 on 37 degrees of freedom (line 18 of the listing), if the 'slope' parameters β of equation (5.16) — equal to the difference $f_0 - f_c$ in equation (5.17) — are constrained to be equal across the three irrigation treatments. It is further reduced, but very slightly, to 0.003 115 on 33 degrees of freedom (line 20 of the listing), when all three parameters α, β and ρ, are allowed to differ over the three data sets. We may put these results in the form of a table, analogous to the Tables 4.1, 4.2 and 4.3 of Chapter 4. Thus, with β_8, β_{10}, β_{11} denoting three parameters for *Irrig8*, *Irrig10* and *Irrig11*, and β the corresponding parameter constrained to be equal for the three treatments, we have:

	df	SS	MS
Residual after fitting ρ, α_8, α_{10}, α_{11}, β_8, β_{10}, β_{11}	37	0.1629	
Addition for fitting β_8, β_{10}, β_{11}	2	0.0055	0.002 75
Residual after fitting ρ, α_8, α_{10}, α_{11}, β	35	0.1574	
Addition for fitting ρ_8, ρ_{10}, ρ_{11}	2	0.0546	0.027 30
Residual after fitting ρ_8, ρ_{10}, ρ_{11}, α_8, α_{10}, α_{11}, β_8, β_{10}, β_{11}	33	0.1028	0.003 115

Since the fitted model is now non-linear in the fitted parameters, variance ratios such as 0.002 75/0.003 115; 0.027 30/0.003 115 are no longer distributed as F-statistics, so rigorous tests of hypotheses using F-tables are not possible. Nevertheless these ratios suggest that we have achieved very little improvement by fitting separate 'slope' parameters β for the three irrigation treatments $Irrig8$, $Irrig10$ and $Irrig11$. By contrast, there was a substantial reduction in the residual sum of squares when separate ρ parameters were fitted to data from the three treatments (measured by 0.027 30/0.003 115).

5.5.2   Example: Fitting a Line Plus Exponential to Digester Data

The data of Figure 5.11, kindly supplied by Dr Luiz Monteggia, are from a study of the performance of anaerobic reactors in the treatment of sewage sludge. The first column shows concentration of micro-organisms (gHAc) in units of grams per litre; the remaining columns show specific methogenic activity (SMA) for seven runs with different initial food allocations. The relation between SMA and gHAc is observed to rise initially, then to decline. Other runs had suggested that a curve of the form 'line plus exponential' might be an appropriate description of the relation: the line possibly accounting for the behaviour of SMA at higher concentrations, with the exponential part describing behaviour at low concentrations. This curve was therefore fitted to SMA as measured in each of the seven runs, the GENSTAT output being as shown in Figure 5.12. It is seen that the line plus exponential provides a good fit for all seven runs, the proportion of variance accounted for being 97.1%, 92.7%, 98.0%, 97.9%, 92.9%, 89.0% and 93.0% in the seven cases. This is scarcely surprising, since we are fitting a four-parameter curve to seven data points. We also see that, except for run 3, the estimate c of the parameter γ in the model $SMA = \alpha + \beta\rho^{(gHAc)} + \gamma(gHAc) + \varepsilon$ is less than twice its standard error in absolute magnitude, suggesting that inclusion of this term in the model is of doubtful value. However, the standard errors of the estimates of α and β are also very considerable in some cases. As in Section 5.5.1, where Horton infiltration curves were fitted, it would be possible to test whether, for the digester data, the parameters in the fitted curves differed significantly from run to run. For this purpose, it would be necessary (within GENSTAT) to specify 'run number' as a factor by means of

>FACTOR [LEVELS=7:VALUES=(1 ... 7)7] Run.

Use of the FITCURVE directive then allows us, for example, to fit a common ρ-parameter for all seven runs.

gHAc	SMA[1]	SMA[2]	SMA[3]	SMA[4]	SMA[5]	SMA[6]	SMA[7]
0.25	2.67	3.31	2.94	2.94	3.68	3.99	3.21
0.50	3.19	4.36	4.67	4.59	5.45	4.94	4.46
0.75	3.87	5.65	6.04	5.66	5.73	5.63	4.47
1.00	4.44	6.11	6.40	5.80	6.31	6.20	5.31
1.50	5.58	7.05	6.89	6.94	6.42	6.13	5.38
2.00	6.19	8.39	7.52	7.36	6.92	6.37	5.78
3.00	5.81	7.15	7.09	7.25	6.27	6.56	5.96
4.00	5.04	6.81	6.63	7.03	6.27	5.42	5.81

Figure 5.11   L. Monteggia's data on performance of anaerobic reactors in the treatment of sewage sludge. Data are from seven runs corresponding to different initial food allocations. SMA is specific methogenic activity; SMA is a function of concentration of micro-organisms (gHAc) in units of grams per litre.

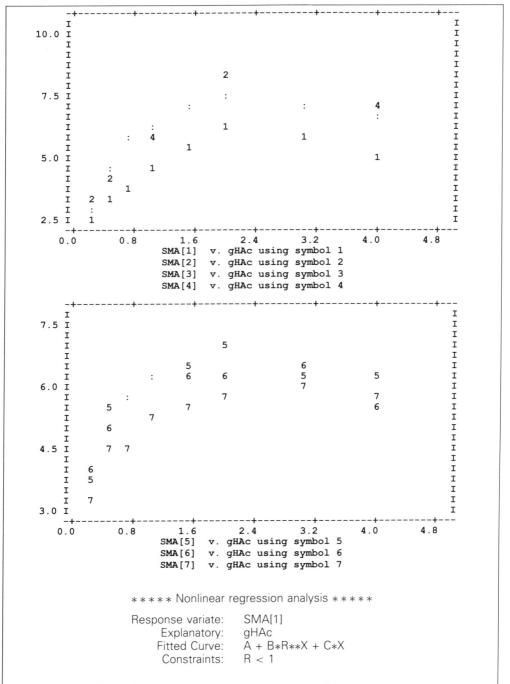

Figure 5.12 Fit of line plus exponential curves to seven runs of digester measurements of specific methogenic activity (*SMA*) as function of concentration of micro-organisms (*gHAc*, in units of grams per litre). The seven runs are denoted by *SMA*[1], ..., *SMA*[7].

___ *Continued* ___

Figure 5.12 *(continued)*

* * * Summary of analysis * * *

	d.f.	s.s.	m.s.
Regression	3	11.2274	3.74247
Residual	4	0.1903	0.04757
Total	7	11.4177	1.63110

Percentage variance accounted for 97.1

* * * Estimates of parameters * * *

	estimate	s.e.
R	0.838	0.238
B	−61.	161.
C	−6.8	12.0
A	62.	162.

* * * Fitted values and residuals * * *

Unit	Explanatory	Response	Fitted value	Standardized residual	Leverage
1	0.2500	2.670	2.504	1.44	0.72
2	0.5000	3.190	3.301	−0.60	0.29
3	0.7500	3.870	3.989	−0.64	0.28
4	1.0000	4.440	4.574	−0.75	0.34
5	1.5000	5.580	5.451	0.73	0.34
6	2.0000	6.190	5.966	1.29	0.36
7	3.0000	5.810	6.029	−1.89	0.72
8	4.0000	5.040	4.977	1.81	0.97
Mean	1.6250	4.599	4.599	0.17	0.50

18..

* * * * * Nonlinear regression analysis * * * * *

Response variate: SMA[2]
        Explanatory: gHAc
        Fitted Curve: A + B*R**X + C*X
        Constraints: R < 1

* * * Summary of analysis * * *

	d.f.	s.s.	m.s.
Regression	3	17.9879	5.9960
Residual	4	0.7795	0.1949
Total	7	18.7674	2.6811

Percentage variance accounted for 92.7

* * * Estimates of parameters * * *

	estimate	s.e.
R	0.498	0.240
B	−12.78	7.93
C	−1.73	1.69
A	14.38	8.51

Figure 5.12 *(continued)*

*** Fitted values and residuals ***

Unit	Explanatory	Response	Fitted value	Standardized residual	Leverage
1	0.2500	3.31	3.22	0.48	0.81
2	0.5000	4.36	4.50	−0.38	0.29
3	0.7500	5.65	5.51	0.39	0.32
4	1.0000	6.11	6.29	−0.50	0.35
5	1.5000	7.05	7.29	−0.66	0.31
6	2.0000	8.39	7.75	1.90	0.42
7	3.0000	7.15	7.61	−1.57	0.57
8	4.0000	6.81	6.67	1.40	0.95
Mean	1.6250	6.10	6.10	0.13	0.50

18..

***** Nonlinear regression analysis *****

Response variate: SMA[3]
        Explanatory: gHAc
        Fitted Curve: A + B*R**X + C*X
        Constraints: R < 1

*** Summary of analysis ***

	d.f.	s.s.	m.s.
Regression	3	15.7967	5.26557
Residual	4	0.1808	0.04521
Total	7	15.9775	2.28251

Percentage variance accounted for 98.0

*** Estimates of parameters ***

	estimate	s.e.
R	0.1856	0.0603
B	−8.256	0.586
C	−0.455	0.204
A	8.493	0.698

*** Fitted values and residuals ***

Unit	Explanatory	Response	Fitted value	Standardized residual	Leverage
1	0.2500	2.940	2.961	−0.33	0.91
2	0.5000	4.670	4.709	−0.23	0.36
3	0.7500	6.040	5.817	1.34	0.39
4	1.0000	6.400	6.506	−0.60	0.32
5	1.5000	6.890	7.150	−1.50	0.33
6	2.0000	7.520	7.299	1.42	0.46
7	3.0000	7.090	7.075	0.09	0.36
8	4.0000	6.630	6.663	−0.42	0.86
Mean	1.6250	6.023	6.023	−0.03	0.50

18..

***** Nonlinear regression analysis *****

Response variate: SMA[4]
        Explanatory: gHAc
        Fitted Curve: A + B*R**X + C*X
        Constraints: R < 1

┌─ Figure 5.12 *(continued)* ──

* * * Summary of analysis * * *

	d.f.	s.s.	m.s.
Regression	3	16.6395	5.54652
Residual	4	0.2012	0.05031
Total	7	16.8408	2.40583

Percentage variance accounted for 97.9

* * * Estimates of parameters * * *

	estimate	s.e.
R	0.2594	0.0910
B	−7.438	0.821
C	−0.336	0.287
A	8.41	1.07

* * * Fitted values and residuals * * *

Unit	Explanatory	Response	Fitted value	Standardized residual	Leverage
1	0.2500	2.940	3.022	−1.08	0.88
2	0.5000	4.590	4.458	0.72	0.33
3	0.7500	5.660	5.459	1.13	0.37
4	1.0000	5.800	6.149	−1.91	0.34
5	1.5000	6.940	6.928	0.07	0.31
6	2.0000	7.360	7.242	0.72	0.46
7	3.0000	7.250	7.276	−0.15	0.41
8	4.0000	7.030	7.036	−0.08	0.89
Mean	1.6250	5.946	5.946	−0.07	0.50

18..

* * * * * Nonlinear regression analysis * * * * *

Response variate: SMA[5]
        Explanatory: gHAc
      Fitted Curve: A + B*R**X + C*X
       Constraints: R < 1

* * * Summary of analysis * * *

	d.f.	s.s.	m.s.
Regression	3	6.6285	2.20950
Residual	4	0.2812	0.07030
Total	7	6.9097	0.98710

Percentage variance accounted for 92.9

* * * Estimates of parameters * * *

	estimate	s.e.
R	0.0825	0.0649
B	−5.995	0.960
C	−0.193	0.173
A	7.012	0.526

└──

Figure 5.12 *(continued)*

*** Fitted values and residuals ***

Unit	Explanatory	Response	Fitted value	Standardized residual	Leverage
1	0.2500	3.680	3.751	−1.27	0.96
2	0.5000	5.450	5.193	1.33	0.47
3	0.7500	5.730	5.944	−1.04	0.40
4	1.0000	6.310	6.324	−0.06	0.28
5	1.5000	6.420	6.581	−0.77	0.38
6	2.0000	6.920	6.586	1.64	0.41
7	3.0000	6.270	6.430	−0.72	0.30
8	4.0000	6.270	6.241	0.25	0.80
Mean	1.6250	5.881	5.881	−0.08	0.50

18...

***** Nonlinear regression analysis *****

Response variate: SMA[6]
        Explanatory: gHAc
      Fitted Curve: A + B*R**X + C*X
       Constraints: R < 1

*** Summary of analysis ***

	d.f.	s.s.	m.s.
Regression	3	4.8655	1.62182
Residual	4	0.3267	0.08169
Total	7	5.1922	0.74174

Percentage variance accounted for 89.0

*** Estimates of parameters ***

	estimate	s.e.
R	0.335	0.188
B	−5.90	1.67
C	−0.771	0.503
A	8.73	2.03

*** Fitted values and residuals ***

Unit	Explanatory	Response	Fitted value	Standardized residual	Leverage
1	0.2500	3.990	4.052	−0.57	0.86
2	0.5000	4.940	4.932	0.04	0.31
3	0.7500	5.630	5.555	0.33	0.36
4	1.0000	6.200	5.983	0.94	0.35
5	1.5000	6.130	6.429	−1.26	0.31
6	2.0000	6.370	6.525	−0.73	0.45
7	3.0000	6.560	6.194	1.75	0.46
8	4.0000	5.420	5.570	−1.78	0.91
Mean	1.6250	5.655	5.655	−0.16	0.50

18...

Figure 5.12 *(continued)*

***** Nonlinear regression analysis *****

Response variate: SMA[7]
   Explanatory: gHAc
   Fitted Curve: A + B*R**X + C*X
   Constraints: R < 1

*** Summary of analysis ***

	d.f.	s.s.	m.s.
Regression	3	5.9376	1.97920
Residual	4	0.2476	0.06189
Total	7	6.1852	0.88359

Percentage variance accounted for 93.0

*** Estimates of parameters ***

	estimate	s.e.
R	0.177	0.145
B	−3.887	0.678
C	0.018	0.230
A	5.810	0.782

*** Fitted values and residuals ***

Unit	Explanatory	Response	Fitted value	Standardized residual	Leverage
1	0.2500	3.210	3.295	−1.17	0.92
2	0.5000	4.460	4.185	1.39	0.37
3	0.7500	4.470	4.764	−1.52	0.39
4	1.0000	5.310	5.141	0.82	0.32
5	1.5000	5.380	5.548	−0.83	0.33
6	2.0000	5.780	5.725	0.30	0.46
7	3.0000	5.960	5.843	0.59	0.35
8	4.0000	5.810	5.878	−0.72	0.86
Mean	1.6250	5.048	5.047	−0.14	0.50

If it were of interest to estimate the concentration at which *SMA* is a maximum, differentiation shows that the maximum occurs where

$$gHAc = \ln[-c/(b \ln r)]/ \ln r$$

into which the estimates c, b and r may be substituted. An approximate standard error can be calculated for this estimate as follows. Setting $F(b, c, r) = \ln[-c/(b \ln r)]/ \ln r$, we have

$$\operatorname{var} F(b, c, r) \cong (\partial F/\partial b)^2 \operatorname{var} b + (\partial F/\partial c)^2 \operatorname{var} c$$

$$+ (\partial F/\partial r)^2 \operatorname{var} r + 2(\partial F/\partial b)(\partial F/\partial c) \operatorname{corr}(b, c) \sqrt{(\operatorname{var} b \operatorname{var} c)}$$

$$+ 2(\partial F/\partial b)(\partial F/\partial r) \operatorname{corr}(b, r) \sqrt{(\operatorname{var} b \operatorname{var} r)}$$

$$+ 2(\partial F/\partial c)(\partial F/\partial r) \operatorname{corr}(c, r) \sqrt{(\operatorname{var} c \operatorname{var} r)}.$$

5.5.3 Fitting Non-linear Models to Stomatal Conductances

An equation used very widely for estimating evaporation loss is the Penman–Montieth equation

$$\lambda E = [s(R_n - G) + \rho c_p D g_a]/[s + \gamma(1 + g_a/g_s)]$$

where $R_n - G$ is the available flux density (net irradiance minus soil heat flux density, W m^{-2}); s is the slope of the saturation specific humidity curve at air temperature (kg kg^{-1} °C^{-1}); ρ is the density of air at ambient air temperature (kg m^{-3}); c_p is the specific heat of air (J kg^{-1} °C^{-1}); λ is the latent heat of vaporisation at ambient air temperature (J kg^{-1}); E is the flux density of water vapour from the surface (kg m^{-2} s^{-1}); γ is the psychometric constant (c_p/λ)(°C^{-1}); D is the saturation specific humidity deficit (kg kg^{-1}); g_a is the aerodynamic conductance between the effective canopy surface and the reference height at which air temperature and D are measured (m s^{-1}); and g_s is the surface conductance (m s^{-1}). Note that g_a is the inverse of the parameter r_a used in Section 5.4 above.

Much research has recently been addressed to determining how the stomatal conductance g_s, which represents the biological control on evaporative loss, varies as a function of environmental factors, and this research has involved numerous applications of non-linear statistical modelling. It is commonly postulated, for example, that g_s is a function of quantum irradiance Q in the band (0.4–0.7 µm); of leaf temperature T_0; of leaf-surface saturation deficit D_0; of leaf water potential Ψ_1; of soil water potential Ψ_s; and of ambient CO_2 concentration C. It is customary to write g_s as a product of functions

$$g_s = g_{max} f_1(D_0) f_2(Q) f_3(T_0) f_4(\Psi_1) f_5(C)$$

where all the f_i are empirical functions commonly defined to lie between zero and one. The justification for this simplified formulation is argued on hydrometeorological grounds, although statistical considerations would suggest a more general formulation permitting interactions between the factors. For example, if one only of the factors is limiting — say, $f_i(.) = 1$ for $i = 1, 2, 3, 4$; $f_5(C) < 1$ — it would be unlikely that the change Δg_s in g_s, resulting from change ΔC in C, would be constant irrespective of the value of C, which the hydrometeorological argument implies. However, Shuttleworth (1988) reports that such interactions are of second-order magnitude. Each of the functions f_i is then given an appropriate form requiring the estimation of parameters. For example Dolman et al. (1991) set

$$f_1(D_0) = \exp(-\theta_1 D_0)$$

where θ_1 is a parameter to be estimated. Similarly

$$f_2(Q) = [Q/(\theta_2 + Q)]/[1000/(1000 + \theta_2)]$$

Shuttleworth (1988) uses a function $f_3(T)$ of temperature, containing three parameters θ_1, θ_2 and θ_3, given by

$$t_3(T) = [(T - \theta_1)(\theta_2 - T)^t]/[(\theta_3 - \theta_1)(\theta_2 - \theta_3)^t]$$

where $t = (\theta_2 - \theta_3)/(\theta_3 - \theta_1)$, and θ_1, θ_2 and θ_3 are parameters to be estimated, which are characteristic of the forest stand.

These formulations clearly involve a high degree of empiricism. The point to be made in this text, however, is that the models being fitted are essentially statistical models in which measurements of evaporative flux λE are equated to a systematic component $f(R_n -$

$G, D, T; g_s(\theta_1, \theta_2, \theta_3 \ldots))$ given by the Penman–Montieth equation, plus a random error. Given the necessary data sequences $\{\lambda E\}; \{R_n - G\}; \{D\}$ and $\{T\}$, non-linear model-fitting procedures are derived to obtain estimates of the parameters θ. Some authors (see, for example, Dolman *et al.*, 1991) quote standard errors for the parameter estimates so obtained, derived presumably from the matrix of second derivatives of the likelihood surface, evaluated at the point where its maximum occurs. The measurements of evaporation flux λE, which the approach requires, are commonly provided by the HYDRA device, which measures vertical flux of water vapour by the eddy correlation method (integration, over time, of the product of the vertical wind component and humidity).

In GENSTAT, such models can be fitted using the directive FITNONLINEAR. Taking a very simple case, in which the parameters are taken as $\theta_1 = g_a$ and $\theta_2 = g_s$, both regarded as fixed constants, let us suppose that data sequences lE ($= \lambda E$), $RnmG$ ($= R_n - G$), D and T have been read, and the variables $A = s(R_n - G)$, $B = \rho c_p D$, $C = s + \gamma$, and D calculated by GENSTAT instructions using the directive CALCULATE. Then the following GENSTAT instructions will minimise the sum of squared differences between observed and fitted λE, with respect to $theta1 = \theta_1$, $theta2 = \theta_2$:

```
>EXPRESSION PenMonSS; VALUE=!E(F=SUM(( IE−(A+B/theta1)\
>    /(C+D*theta1/theta2))**2))
>MODEL [FUNCTION=F]
>RCYCLE theta1,theta2;
>FITNONLINEAR[Print=summary, estimates,correlation,monitoring; \
>    CALCULATION=PenMonSS]
```

The quantity to be minimised, *PenMonSS*, is first defined using EXPRESSION as $\Sigma[\lambda E_{observed} - \lambda E_{modelled}(\theta_1, \theta_2)]^2$. The directive MODEL shows that this is the function requiring to be minimised; RCYCLE shows that the parameters to be varied in the minimisation are θ_1 and θ_2; and FITNONLINEAR directs the minimisation of the function *PenMonSS*.

Where the stomatal conductance g_s is not a fixed parameter, but is of the form $g_{max} f_1 f_2 \ldots f_5$, with $f_1, f_2, \ldots, f_5$ functions of parameters $\theta_1, \theta_2, \theta_3, \ldots$, the method described above is easily adapted by replacing *theta2* in the GENSTAT code by the appropriate functional form.

5.5.4 Using a Logistic Curve to Describe Soil Moisture Characteristic

The data shown in Figure 5.13, taken from Hillel (1980, p. 164) by kind permission of D. Hillel show soil suction together with soil volumetric wetness for two soils A and B. Soil A is a sandy loam, soil B is believed to be a clay loam. Hillel suggests plotting by eye of the two soil moisture characteristics, using a log transformation for soil suction. The log-transformed points are plotted in Figure 5.13; but instead of drawing a curve by eye, we fit a logistic curve to the data from each soil. This is a straightforward application of the GENSTAT directive FITCURVE, with selection of the option CURVE = logistic in the directive specification.

Figure 5.14 shows the output resulting from this directive. For both soils, a logistic curve of the form

$$Y = A + C/[1 + \exp(-B(lnSuctio - M))]$$

where A, B, C and M are parameters, gives a reasonable fit to the data, with 99% of the variation accounted for when $Y = \%WetA$, and 99.7% when $Y = \%WetB$. Estimates of all

MatSuc	SoilA	SoilB
0	44.0	52.0
10	44.0	52.0
20	43.9	52.0
50	38.0	51.0
100	22.5	48.0
300	12.5	32.0
1000	7.0	20.0
10000	5.2	13.5
20000	5.1	13.0
100000	4.9	12.8

```
      -+---------+---------+---------+---------+---------+---------+---
      I                                                                I
 60.  I                                                                I
      I                                                                I
      I B                   B  B                                       I
      I                            B                                   I
      I                               B                                I
      I A                   A  A                                       I
 40.  I                                                                I
      I                         A                                      I
      I                            B                                   I
      I                                                                I
      I                         A                                      I
 20.  I                               B                                I
      I                                                                I
      I                            A        B  B        B              I
      I                                                                I
      I                               A        A  A                    I
      I                                              A                 I
  0.  I                                                                I
      -+---------+---------+---------+---------+---------+---------+---
    -2.5       0.0       2.5       5.0       7.5      10.0      12.5
             SoilA  v.  lnSuctio using symbol A
             SoilB  v.  lnSuctio using symbol B
```

Figure 5.13   Hillel's data for suction head (in centimetres) (denoted by *MatSuc*) and soil volumetric wetness, as a percentage, for two soils A and B. The plot shows volumetric wetness on the vertical axis, log (matric suction) on the horizontal.

the parameters are large relative to their standard errors, suggesting that the fitted form is appropriate, although for both soils the fit appears to be unsatisfactory in the region where percentage of soil moisture is changing most rapidly as suction increases (soil A has large residuals where suction head is 100 and 300 cm, soil B where suction head is 300 and 1000 cm, respectively). Agreement between observed and fitted values of percentage soil wetness is good.

Why fit a logistic (or any other) curve, instead of simply fitting by eye? A fit by eye is perfectly adequate for many purposes; however, one consequence of fitting the logistic curve is that the analysis gives, as a by-product, standard errors for estimates of the curve's parameters. Corresponding to any value of volume wetness read from the curve (for example, *%WetA* for *lnSuctio* = 3), we can obtain a standard error, and an approximate confidence interval, for the estimated value. Likewise, we can obtain standard errors for estimates of mass wetness and for the water released per metre depth as suction increases. Hillel calculates curves of volumetric water capacity, $-d\theta/d\varphi$ where θ is volumetric wetness and φ is suction head, using graphical differentiation (that is, by measuring the slope of curves where wetness is plotted against

***** Nonlinear regression analysis *****

Response variate: %WetA
    Explanatory: lnSuctio
    Fitted Curve: A + C/(1 + EXP(-B*(X - M)))

*** Summary of analysis ***

	d.f.	s.s.	m.s.
Regression	3	2855.41	951.802
Residual	6	18.92	3.154
Total	9	2874.33	319.370

Percentage variance accounted for 99.0

* MESSAGE: The following units have large residuals:
5	-2.32
6	2.14

*** Estimates of parameters ***

	estimate	s.e.	Correlations			
B	-1.851	0.308	1.000			
M	4.5818	0.0947	0.039	1.000		
C	39.24	1.63	0.578	-0.135	1.000	
A	5.748	0.912	-0.362	-0.343	-0.652	1.000

*** Fitted values and residuals ***

Unit	Explanatory	Response	Fitted value	Standardized residual	Leverage
1	-2.3026	44.00	44.99	-0.78	0.49
2	2.3125	44.00	44.41	-0.28	0.34
3	3.0007	43.90	42.99	0.61	0.30
4	3.9140	38.00	36.15	1.67	0.61
5	4.6062	22.50	24.92	-2.32	0.65
6	5.7041	12.50	10.12	2.14	0.61
7	6.9079	7.00	6.27	0.46	0.21
8	9.2104	5.20	5.76	-0.36	0.26
9	9.9035	5.10	5.75	-0.43	0.26
10	11.5129	4.90	5.75	-0.56	0.26
Mean	5.4770	22.71	22.71	0.02	0.40

11..

***** Nonlinear regression analysis *****

Response variate: %WetB
    Explanatory: lnSuctio
    Fitted Curve: A + C/(1 + EXP(-B*(X - M)))

Figure 5.14   Output resulting from fitting of logistic curves to the relation between log(matric suction) and percentage wetness for two soils A and B: Hillel's data (see Figure 5.13).

_____ Continued _____

___ Figure 5.14 *(continued)* _____

$* * *$ Summary of analysis $* * *$

	d.f.	s.s.	m.s.
Regression	3	2957.532	985.844
Residual	6	6.189	1.032
Total	9	2963.721	329.302

Percentage variance accounted for 99.7

$*$ MESSAGE: The following units have large residuals:

6	−2.39
7	2.38

$* * *$ Estimates of parameters $* * *$

	estimate	s.e.	Correlations			
B	−1.577	0.145	1.000			
M	5.7300	0.0656	−0.002	1.000		
C	39.288	0.892	0.569	−0.020	1.000	
A	13.282	0.592	−0.338	−0.372	−0.739	1.000

$* * *$ Fitted values and residuals $* * *$

Unit	Explanatory	Response	Fitted value	Standardized residual	Leverage
1	−2.3026	52.00	52.57	−0.70	0.36
2	2.3125	52.00	52.39	−0.46	0.30
3	3.0007	52.00	52.05	−0.05	0.24
4	3.9140	51.00	50.45	0.62	0.25
5	4.6062	48.00	46.86	1.63	0.53
6	5.7041	32.00	33.33	−2.39	0.70
7	6.9079	20.00	18.59	2.38	0.66
8	9.2104	13.50	13.44	0.07	0.31
9	9.9035	13.00	13.34	−0.40	0.33
10	11.5129	12.80	13.29	−0.59	0.34
Mean	5.4770	34.63	34.63	0.01	0.40

suction, for different suction values; where a satisfactory soil moisture characteristic has been fitted, $-d\theta/d\varphi$ can be obtained very simply by differentiation of the fitted curve.

5.6 FITTING STANDARD NON-LINEAR CURVES TO GROUPED DATA: TESTS FOR PARALLELISM

The families of curves described in the last section are all of the general form

$$y_t = \alpha + \beta f(x_t; \gamma) + \varepsilon_t \tag{5.18}$$

with parameters α, β and γ (where γ may be a vector of parameters). Just as, in the linear regression case (discussed in Chapter 4) we can test for the parallelism of a number of linear regressions — to be followed, if appropriate, by a test of coincidence of intercepts — so, in the case of the curves (5.18), we can test hypotheses about the parameters α, β and γ. Practical examples where such analyses are likely to be necessary are the following: (i) where we need to test whether Horton infiltration curves from different irrigation treatments are coincident (see Section 5.5.1); (ii) where we have fitted ordinary logistic curves of the form $y = \alpha +$

$\beta/[1+\exp\{-\gamma_1(x-\gamma_2)\}]+\varepsilon$ to the soil moisture characteristic of different soil samples, and we wish to know if logistic curves differ significantly.

Given k 'groups' of data, for each of which a curve of the family (5.18) has been fitted, the following restricted models may need consideration:

(i)   The 'no constant' model, $y_t = \beta f(x_t; \gamma) + \varepsilon_t$. Parameters β and γ are constant over the k groups of data. The parameters to be estimated are β, γ, σ_ε^2.

(ii) The constant varies from group to group, with parameters β and γ constant over the k groups of data: $y_t = \alpha_i + \beta f(x_t; \gamma) + \varepsilon_t$ $(i = 1, 2, \ldots, k)$. The parameters to be estimated are α_i, β, γ, σ_ε^2.

(iii) The constant α is omitted, and the parameter β has some fixed and known value B: $y_t = Bf(x_t; \gamma) + \varepsilon_t$, the parameter γ being constant over the k groups. The parameters to be estimated are γ, σ_ε^2.

(iv) The 'slope' parameter β has some fixed and known value B, the parameter(s) γ are constant over the k groups, and the parameter α is to be estimated: model is $y_t = \alpha + Bf(x_t; \gamma) + \varepsilon_t$. The parameters to be estimated are α, γ, σ_ε^2.

(v)   The 'slope' parameters β_i vary over the k groups of data; the parameters α and γ are fixed over the k groups of data. The model is $y_t = \alpha + \beta_i f(x_t; \gamma) + \varepsilon_t$, and the parameters to be estimated are α, β_i, γ, σ_ε^2.

(vi) The non-linear parameters γ_i vary over the k data groups, parameters α and β being constant. The model is $y_t = \alpha + \beta f(x_t; \gamma_i) + \varepsilon_t$ and the parameters to be estimated are α, β, γ_i, and σ_ε^2.

All of the options (i) to (vi) can be explored easily by use of the GENSTAT directive NONLIN-EARFIT (Butler and Brain, 1990).

The most general model is that in which all parameters vary from group to group, namely $y_t = \alpha_i + \beta_i f(x_t; \gamma) + \varepsilon_t$. In all of the above options (i) to (vi), it has been assumed that the variance of residuals σ_ε^2 is constant over all k data groups; having fitted an appropriate model $y_t = \alpha_i + \beta_i f(x_t; \gamma_i) + \varepsilon_t$ to each group of data, the homogeneity of the residual variance estimates $s_1^2, s_2^2, \ldots, s_k^2$, based respectively on $f_1, f_2, \ldots, f_k$ degrees of freedom, can be tested by use of a Bartlett test: this requires calculation of the test statistic χ^2 with $(k-1)$ degrees of freedom, with χ^2 calculated from

$$\chi_{k-1}^2 = (f \ln \bar{s}^2 - \sum_{j=1}^{k} f_j \ln s_j^2)/C$$

where $\bar{s}^2 = \Sigma f_j s_j^2 / \Sigma f_j$, $f = \Sigma f_j$ and

$$C = 1 + [\sum(1/f_j) - 1/f]/[3(k-1)]$$

Statistical texts show that the hypothesis of homogeneity of residual variances is rejected at the 5% level of significance (for example) if the test statistic falls in the upper 5% tail of the tabulated χ^2 distribution for $(k-1)$ degrees of freedom. If the hypothesis is not rejected, the pooled estimate of the residual variance is the weighted mean $\bar{s}^2$. Care is necessary if the variances being compared come from data which are far from being normally distributed.

REFERENCES

Adams, R. S., Black, T. A. and Fleming, R. L. (1991). Evapotranspiration and surface conductance in a high elevation, grass-covered forest clearcut. *Agricultural and Forest Meteorol.*, **56**, 173–93.

Butler, R. C. and Brain, P. (1990). Parallelism in non-linear models. *GENSTAT Newsletter*, **25**, 40–46.

Damm, U. (1987). Some modifications of Ebert's method to estimate growth and mortality parameters from average lengths of a population. In Pauly, D. and Morgan, G. R. (eds) *Length-based Methods in Fisheries Research*. International Center for Living Aquatic Resources Management, Kuwait Institute for Scientific Research, 44–51.

Dolman, A. J., Gash, J. H. C., Roberts, J. and Shuttleworth, W. J. (1991). Stomatal and surface conductance of tropical rainforest. *Agricultural and Forest Meteorol.*, **54**, 303–18.

Hillel, D. (1971). *Soil and Water: Physical Principles and Processes*. Academic Press, New York.

Hillel, D. (1980). *Fundamentals of Soil Physics*. Academic Press, New York.

Ricker, W. E. (1958). *Handbook of Computation for Biological Statistics of Fish Populations*. Fisheries Research Board of Canada, Ottawa.

Shuttleworth, W. J. (1979) *Evaporation*. Institute of Hydrology Report No. 56. Institute of Hydrology, Wallingford.

Shuttleworth, W. J. (1988) Evaporation from Amazonian forest. *Proc. R. Soc. London B*, **233**, 321–46.

Stewart, J. B. (1989). On the use of the Penman–Montieth equation for determining areal evapotranspiration. In *Estimation of Areal Evapotranspiration* (Proceedings of a workshop held at Vancouver, August 1987), IAHS Publication No. 177, 3–11, Wallingford.

Wright, I., Gash, J. H. C., da Rocha, H. R., Shuttleworth, W. J., Nobre, C. A., Maitelli, G. T., Zamparoni, C. A. G. P. and Carvalho, P. R. A. (1992) Dry season micrometeorology of central Amazonian marshland. *Q. J. Roy. Meteorol. Soc.*, **118**, 1083–99.

Yevjevich, V. (1972). *Probability and Statistics in Hydrology*. Water Resources Publications, Fort Collins, CO.

EXERCISES AND EXTENSIONS

5.1.(a) Repeat the analysis of the wind profile data (Section 5.4) for profiles numbered 7 to 12.

(b) Using the GENSTAT directive CONTOUR, or otherwise, plot the contours when wind velocity is plotted as a function of time (= profile number) and anemometer height.

(c) Determine whether the pattern in residuals observed in the analysis of profiles 1 to 6 is repeated in profiles 7 to 12. Plot the residuals as in (b).

5.2. In an urban hydrology study, relations between population density, in thousands per hectare (*DensPop*) and percentage of impermeable land cover (*Imperm*) were sought for three Brazilian cities, Porto Alegre, São Paulo, and Curitiba. The data below show values of these two variables in 17, 11 and eight precincts of the three cities, respectively, and are reproduced by kind permission of Dr C. E. M. Tucciz.

(a) Using *DensPop* as explanatory variable, explore whether an exponential relation of the form

$$Imperm = \alpha + \beta \rho^{DensPop} + \varepsilon$$

describes satisfactorily how *Imperm* varies with population density. Is the relation similar for all three cities? Explore the structure of the residuals; is there evidence of inhomogeneity in their distribution?

Data for Porto Alegre					
DensPop	250.0	196.0	178.0	161.0	97.0
Imperm	69.80	63.00	59.70	62.30	46.00
DensPop	61.0	70.0	92.0	35.0	116.0
Imperm	26.00	41.00	43.60	10.80	57.00
DensPop	63.0	141.0	166.0	119.0	96.0
Imperm	25.00	58.10	58.60	59.60	51.00
DensPop	11.0	137.0			
Imperm	14.20	62.10			

Data for São Paulo

DensPop	71.40	88.10	58.30	44.90	82.30
Imperm	37.80	45.00	27.40	24.40	44.60
DensPop	110.50	82.20	141.80	62.50	124.50
Imperm	58.90	41.70	64.70	30.00	61.50
DensPop	117.30				
Imperm	59.60				

Data for Curitiba

DensPop	80.00	42.00	42.00	42.00	60.00
Imperm	36.40	22.00	24.30	21.70	32.60
DensPop	15.00	28.00	60.00		
Imperm	5.30	14.00	23.60		

Chapter 6

Generalised linear models

6.1 LINEAR MODELS AND THEIR GENERALISATION

We recall that the linear models of Chapter 4 expressed the response variable, denoted by q_t or y_t, as a *linear* function of parameters $\beta_0, \beta_1, \ldots, \beta_p$. Provided the model was linear in these parameters it did not matter, for model-fitting purposes, whether or not the explanatory variables $x_1, x_2, \ldots, x_p$ occurred linearly; thus the models $y_t = \beta_0 + \beta_1 x_1 + \beta_2 x_2 + \varepsilon_t$, $y_t = \beta_0 + \beta_1 h + \beta_2 h^2 + \varepsilon_t$ and $y_t = \beta_0 + \beta_1 \sin(2\pi t/12) + \beta_2 \cos(2\pi t/12) + \varepsilon_t$ are equivalent from the point of view of estimating the parameters β_0, β_1 and β_2, all of which occur linearly in each. The assumptions made were that the errors ε_t were normally and independently distributed, with constant variance σ_ε^2. Thus the model could be summarised in the form

$$Y_t \sim N(\beta_0 + \beta_1 x_{1t} + \beta_2 x_{2t} + \cdots + \beta_p x_{pt}; \sigma_\varepsilon^2) \tag{6.1}$$

or

$$Y_t \sim N(\mu_t, \sigma_\varepsilon^2)$$

where $\mu_t = \beta_0 + \beta_1 x_{1t} + \beta_2 x_{2t} + \cdots + \beta_p x_{pt}$.

These assumptions — particularly those of normality and homogeneity of error variances — are restrictive, particularly for hydrological application, and the present chapter is concerned with a class of models which generalise expression (6.1) in two important respects.

We introduce the first generalisation by observing that expression (6.1) can be written

$$E[Y_t] = \beta_0 + \beta_1 x_{1t} + \cdots + \beta_p x_{pt} \qquad t = 1, 2, 3, \ldots, n$$

$$= \mu_t$$

or, in the notation of vectors and matrices,

$$E[\mathbf{Y}] = \mathbf{X}\boldsymbol{\beta}$$

$$= \boldsymbol{\mu} \tag{6.2}$$

in which $\mathbf{Y}$ is an $n \times 1$ vector of response variables, $\mathbf{X}$ is a $n \times (p+1)$ matrix of explanatory variable values, $\boldsymbol{\beta}$ is a $(p+1) \times 1$ vector of unknown parameters, and $\boldsymbol{\mu}$ is the $n \times 1$ vector of means $[\mu_1, \mu_2, \ldots, \mu_n]^T$.

The first generalization consists of replacing the vector relation $\boldsymbol{\mu} = \mathbf{X}\boldsymbol{\beta}$ by

$$h(\boldsymbol{\mu}) = \mathbf{X}\boldsymbol{\beta} \tag{6.3}$$

which can be written

$$\mu = h^{-1}(\mathbf{X}\beta) = H(\mathbf{X}\beta)$$

where $h(.)$ is any function with a non-negative derivative. It is convenient to introduce the symbol η to denote $\mathbf{X}\beta$, where $\eta = \mathbf{X}\beta$ is called the *linear predictor*; the function $h(.)$ is called the *link function*, because it links the linear predictor to the vector of mean values μ. Then equation (6.3) is usually written in two-part form

$$\eta = \mathbf{X}\beta$$

$$\eta = h(\mu) \tag{6.4}$$

remembering that η and μ are $n \times 1$ vectors. In Chapter 4, the relation $\eta = h(\mu)$ was simply $\eta = \mu$, the link function in this special case being termed the *identity link*; the present chapter considers alternative forms of link function.

The second generalisation concerns the distribution of errors, which in expression (6.1) was assumed normal with constant variance. Now the normal distribution is one particular member of a family of probability distributions, all with desirable statistical properties, known as the *exponential family*; a general form, appropriate for our purposes, may be written

$$f(y; \theta, \phi) = \exp[(y\theta - b(\theta))/a(\phi) + c(y, \phi)] \tag{6.5}$$

for some specific functions $a(.)$, $b(.)$, and $c(.)$. Thus with $\theta = 1$, $a(\phi) = -\phi$, $b(\theta) = 0$, and $c(y, \phi) = \log(1/\phi)$ we have the *exponential distribution* (which we must distinguish from the exponential *family*) in the form $f(y; \phi) = \phi^{-1}\exp(-y/\phi)$. Again, if we write $\theta = \mu$, $\phi = \sigma^2$, $a(\phi) = \phi$, $b(\theta) = \theta^2/2$ and $c(y, \phi) = -\{y^2/\sigma^2 + \log(2\pi\sigma^2)\}/2$, we have the normal distribution $N(\mu, \sigma^2)$. Besides the normal and exponential distributions, the exponential family also includes other well-known distributions: the binomial, Poisson and gamma distributions, as well as less familiar ones such as the inverse normal. The second generalisation involves replacing the requirement for normality in expression (6.1) by the requirement that the distribution of y_t belong to the exponential family of distributions (6.5), of which the normal is, as we have seen, a member. The two generalisations permit us, for example, to extend ordinary multiple regression to cases in which

(i) Y_t is a Bernoulli variable (see Section A.3 of the Appendix) taking values 1 or 0, with probabilities π_t, $1 - \pi_t$;

(ii) Y_t has a binomial distribution, taking any of the integer values $0, 1, 2, \ldots, m_t$, where we allow m_t to vary with t;

(iii) Y_t has a gamma distribution with mean μ_t and constant coefficient of variation;

as well as other cases of specific hydrological interest.

6.1.1   The Variance Function, Scaled Deviance, and Deviance of Generalised Linear Models

By way of summary, the models discussed below assume the following:

(i)   We have n observations of a response variable Y_t, in the form of a vector with dimension $n \times 1$, with each Y_t being composed of a systematic component plus a random component.

The random component is such that $E[Y_t] = \mu_t$, or, in vector form, $E[\mathbf{Y}] = \boldsymbol{\mu}$. The probability distribution of Y_t is from the exponential family.

(ii)   We have a linear predictor, $\eta_t = \beta_0 + \beta_1 x_{1t} + \cdots + \beta_p x_{pt}$ or, in matrix form, $\boldsymbol{\eta} = \mathbf{X}\boldsymbol{\beta}$.

(iii) We have a link function $h(.)$ which expresses the linear predictor η_t in terms of μ_t, so that $\eta_t = h(\mu_t)$ or, in vector terms, $\boldsymbol{\eta} = h(\boldsymbol{\mu})$. The function $h(.)$ can be any monotonically increasing function of its argument.

The fact that the distribution of the response variable, y_t, is no longer restricted to being normal, but may be drawn from a much wider family of distributions, has implications for the assumption about homogeneity of variance (recall that, in the normal model (6.1), the response variable was required to have constant variance σ_ε^2). With the more general model, the variance is proportional to a function of the mean μ_t; this function itself obviously depends on the distribution which the Y_t are assumed to follow. A particular case of the general form (6.5) occurs where $a(\phi) = \phi$, where ϕ can be shown to be the constant of proportionality: then, for this more general class of models for which $\eta_t = h(\mu_t)$, it can be shown (see McCullagh and Nelder, 1991) that

$$(\text{variance of } Y_t) = \phi \cdot (\text{mean of } Y_t)$$

$$= \phi \cdot V(\mu_t), \text{ say}$$

$V(\mu_t)$ is commonly called the *variance function*, and the constant of proportionality ϕ is called the *dispersion parameter*. For the normal model (6.1), we have $\phi = \sigma_\varepsilon^2$ and $V(\mu_t) = 1$: that is, the variance function for the normal distribution is unity.

Where we assume that the response variable Y_t has a Poisson distribution, the variance function $V(\mu_t)$ can be shown to be simply $V(\mu_t) = \mu_t$, so that $\text{var}(Y_t) = \phi\mu_t$; but since, for the Poisson distribution, the mean and variance are equal, we can put the dispersion parameter $\phi = 1$. Similarly, where the response variable Y_t has a binomial distribution and takes the values $0, 1/m_t, 2/m_t, \ldots, 1$, we have $\text{var}(Y_t) = \phi\mu_t(1-\mu_t)/m_t$ and again we may take $\phi = 1$. Where y_t has a Poisson or binomial distribution, therefore, computer programs such as GENSTAT commonly set the dispersion parameter ϕ equal to one. Also, since the exponential distribution can be regarded as a special case of the gamma distribution in which the 'shape' parameter κ (see equation (3.11)) is unity, the dispersion parameter ϕ is set to one where Y_t is assumed to follow an exponential distribution. Recall that all of these distributions — Poisson, binomial, exponential, and of course the normal — belong to the exponential family of distributions.

Where Y_t has a gamma distribution, the situation is different. The variance function $V(\mu_t)$ can be shown to be $V(\mu_t) = \mu_t^2$, so that $\text{var}(Y_t) = \phi\mu_t^2$ and the dispersion parameter ϕ now needs to be estimated — unless information is available which enables us to give ϕ a specific value. The same is, of course, true for the normal distribution, where $\text{var}(Y_t) = \phi$, with ϕ usually written as σ_ε^2; in general, $\phi = \sigma_\varepsilon^2$ will be unknown, but we give an example later in which a normal model is fitted with the dispersion parameter set equal to unity (see Figure 6.6).

Having mentioned possible distributions, taken from the exponential family, which the response variable Y_t may follow, and the nature of the variance function $V(\mu_t)$, we need to consider what kinds of link function $h(\boldsymbol{\mu})$ are useful in practice. We recall from equation (6.4) that $h(\boldsymbol{\mu}) = \mathbf{X}\boldsymbol{\beta}$ (where $\boldsymbol{\mu}$ is a vector, having as many components, N, as there are data values). The simplest form of link function is the identity link, in which each of the N elements $h(\mu_t) = \eta_t$ for each of $t = 1 \ldots N$. This was the model used in Chapter 4 in our discussion of linear models. Where the response variable has a Poisson distribution, a frequently used link function is $h(\mu_t) = \ln(\mu_t)$, $t = 1 \ldots N$ (which, without risk of confusion, we shall write simply as $h(\mu) = \ln(\mu)$). This gives $\mu = \exp(X\beta)$ with elements which are always positive, a desirable characteristic since the Poisson distribution must necessarily take only positive values. Where the response variable Y_t has a binomial distribution, a frequently used link is $h(\mu) = \ln(\mu/(1 - \mu))$, the *logistic* link, as used in the section which follows. For the gamma and inverse normal distributions, the links $h(\mu) = 1/\mu$ and $h(\mu) = 1/\mu^2$ are frequently used. The link functions mentioned so far (*canonical* links) have special properties, in that a model with its canonical link always provides a unique set of parameter estimates; where link functions other than the canonical links are used, this is not necessarily the case (McCullagh and Nelder, 1991). This uniqueness of the estimates of model parameters is a good reason for using the canonical form of the link function; however, there may sometimes be sound hydrological reasons why a link function other than the canonical link is to be preferred. Other possible links include the square root $h(\mu) = \mu^{1/2}$, and more generally the power link $h(\mu) = \mu^{\alpha}$; whilst for data in the form of proportions the complementary log-log function, with $h(\mu) = \ln[-\ln(\mu)]$, is sometimes used. All of the link functions so far mentioned can be used with computer programs such as GENSTAT; if, in a GENSTAT directive, no link function is specified, it is assumed that the canonical link is to be used.

To conclude our review of the structure of these models which generalise the linear models of Chapter 4, we need to consider how the procedures, set out in that chapter, for assessing the adequacy of a fitted *linear* model, need to be modified when the model is a *generalised linear model* (GLM). We recall that the goodness of fit of a linear model was assessed by an analysis of variance, the variance ratio (mean square due to the fitted linear model divided by the residual mean square) having an F-distribution on the null hypothesis that its parameters $\beta = 0$. With generalised linear models, there is no exact distributional property of the same type. However, we can get an approximate assessment of the quality of fit from the *scaled deviance*, as used in the example of the section which follows; the scaled deviance is minus twice the ratio between the log-likelihood for the model that we have fitted, and the log-likelihood for a full model explaining all the variation in the data (the 'full' model is also sometimes called the 'saturated' model — see below). Statistical theory shows that the scaled deviance is approximately distributed as χ^2; however, the approximation is not very good when there are not many data values, and is quite poor when there are many extreme observations (zeros in a Poisson distribution, for example).

The scaled deviance is a function of the dispersion parameter ϕ which, as we have seen, must often be estimated, although for certain distributions it can be assumed equal to one. The estimate of ϕ therefore influences the distribution of the scaled deviance; to see that this is so, consider the case of the normal distribution model with known variance σ^2, for which the distribution of Y_t is

$$f(Y_t; \mu) = [1/\sqrt{(2\pi\sigma^2)}] \exp[-(Y_t - \mu)^2/(2\sigma^2)]$$

The log-likelihood for a single observation y_t is

$$l(\mu; y_t) = -\tfrac{1}{2}\ln(2\pi\sigma^2) - (y_t - \mu)^2/(2\sigma^2)$$

and the maximum achievable log-likelihood is obtained by setting $y_t = \mu$; this is

$$-\tfrac{1}{2}\ln(2\pi\sigma^2)$$

From the definition of scaled deviance given above, the scaled deviance corresponding to the single observation y_t is therefore just

$$l(\mu; y_t) - l(y_t; y_t) = (y_t - \mu)^2/\sigma^2$$

and extending the argument to a set of independently distributed data values $y_1, y_2, \ldots, y_N$, we can see that the scaled deviance becomes $\Sigma(y_t - \mu_t)^2/\sigma^2$ which clearly depends on the scale parameter $\phi = \sigma^2$. We can remove this nuisance parameter as follows. If we work with the scaled deviance multiplied by the dispersion parameter ϕ, we obtain a quantity, termed the *deviance*, which is independent of the scale parameter; and we see that, for the normal model of Chapter 4, the deviance reduces to the simple sum of squared deviations, $\Sigma(y_t - \mu_t)^2$.

When the distribution of Y_t is Poisson, binomial or gamma, the deviance takes on different forms; a simple calculation, like that used above for the normal model, shows that the deviances for the Poisson, binomial and gamma models are, respectively,

$$2\sum[y_t \ln(y_t/\mu_t) - (y_t - \mu_t)]$$
$$2\sum[y_t \ln(y_t/\mu_t) + (m_t - y_t) \ln((m_t - y_t)/(m_t - \mu_t))]$$
$$2\sum[(y_t - \mu_t)/\mu_t - \ln(y_t/\mu_t)]$$

where, for the binomial case, m_t is the number of Bernoulli 'trials' giving rise to y_t 'successes' at time t, as illustrated in the example of Section 6.2.1. For the inverse normal distribution, which has been used to model sediment routing in rainfall-runoff models of catchment behaviour, the deviance is

$$\sum[(y_t - \mu_t)^2/(y_t\mu_t^2)]$$

The deviance is also known as the *log-likelihood ratio statistic*. Using the deviance, we can assess the importance of a term, or set of terms, in the linear predictor $\eta = \mathbf{X}\boldsymbol{\beta}$, by the method illustrated in the following section; and in the particular case treated in Chapter 4 in which the response variable has a normal distribution with constant variance, the deviance is simply the sum of squared deviations $\Sigma(y_t - \mu_t)^2$. Minimising the deviance, in this case, is equivalent to minimising the sum of squared residuals.

6.2 LOGISTIC REGRESSION

Section 6.1 referred to the logistic link function and its use in generalised linear models. We now give a practical example of the use of this link function in logistic regression, in which the response variable y_t is of the Bernoulli 'all or nothing' type, $Y_t = 0$ or 1.

6.2.1 Example: Surface Runoff Studies Using Minitraps

The hydrological study of the basin of the Booro-Botou in the Ivory Coast, of which some data were presented in Chapter 1, included a study of surface runoff using an experimental device called a minitrap. This intercepted surface flow down the slope, when it occurred, and diverted it into a gutter 30 cm long; this gutter passed the surface runoff to a 60-litre storage tank at the bottom of a small ditch. Observations of runoff were restricted to three discrete values: tank empty (signifying that no runoff had occurred); tank partially full; or tank overflowing. For the purpose of this example, we consider the Bernoulli variable $Y_t = 0$, corresponding to no runoff at the minitrap, and $Y_t = 1$, corresponding to some runoff having occurred, in the tth storm. The value $Y_t = 1$ corresponds to a combination of the two cases 'tank partially full', and 'tank overflowing'. Generalisation of the analysis to include all three variables (tank empty; tank partially full; tank overflowing) will be described later in this chapter.

In the 1.36 km^2 basin, three clusters of minitraps were installed of which two, with 14 and 12 minitraps, respectively, are discussed by P. Chevallier and O. Planchon (personal communication). This example considers the analysis of just one minitrap, in which we explore whether and how the occurrence of runoff at the minitrap site depends upon storm depth, and upon time since the preceding storm. Figure 6.1 shows the data for one minitrap, reproduced by kind permission of P. Chevallier and O. Planchon; Figure 6.2 shows the non-occurrence and occurrence of runoff (given by the Bernoulli variable $Y_t = 0$, $Y_t = 1$) plotted against rainfall. We observe that many of the zeros appear to lie to the left of the figure, and most of the ones to the right. Clearly it would be inappropriate to fit a linear regression to these data as they stand.

The picture becomes clearer if we group the rainfall into classes, say from 0 to 10 mm, 10 to 20 mm, 20 to 30 mm, 30 to 40 mm, 40 to 50 mm, and 50 to 60 mm. The data are shown in Table 6.1. The proportions r/N are plotted in Figure 6.3, showing that the proportions of storms causing runoff at the minitrap increased — not surprisingly — with depth of storm

Runoff	0	0	0	0	0
Rain	3.0	4.5	7.0	7.0	5.0
TimeInt	4.0	3.0	6.0	3.0	1.0
Runoff	1	0	0	1	1
Rain	28.5	12.0	2.5	55.0	24.5
TimeInt	2.0	3.0	1.0	2.0	2.0
Runoff	1	0	0	0	1
Rain	53.5	6.5	1.5	10.5	32.0
TimeInt	4.0	3.0	0.5	1.0	1.0
Runoff	0	0	0	0	1
Rain	22.0	2.4	30.1	1.9	18.6
TimeInt	0.2	3.0	2.0	0.5	0.2
Runoff	0	0	0	0	1
Rain	13.8	2.9	1.4	11.5	9.1
TimeInt	1.0	3.0	1.0	3.0	1.0
Runoff	0	0	1	0	0
Rain	13.8	12.4	30.6	3.3	9.5
TimeInt	5.0	1.0	1.0	4.0	1.0
Runoff	0	0	0	1	1
Rain	16.7	8.6	1.9	42.0	19.6
TimeInt	2.0	0.5	1.0	1.0	2.0
Runoff	0	0	0	0	0
Rain	5.7	12.4	2.9	8.1	21.0
TimeInt	3.0	1.0	4.0	0.5	4.0
Runoff	0	0	0	0	
Rain	1.9	4.5	1.3	10.8	
TimeInt	6.0	2.0	21.0	5.0	

Figure 6.1   Minitrap data. *Runoff* represents occurrence (= 1) or absence (= 0) of runoff at the minitrap site. *Rain* is depth of rain delivered during the storm event; *TimeInt* represents the approximate time, in weeks, since the preceding storm event.

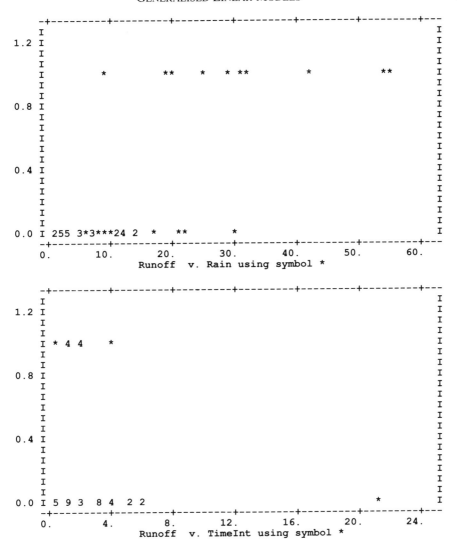

Figure 6.2   Plot of minitrap data from Figure 6.1. The first graph is of runoff occurrence (0 or 1) plotted against rainfall; the second graph is of runoff occurrence against time interval between storm events.

Table 6.1   Minitrap data of Figure 6.1, grouped by rainfall.

Rainfall class (mm)	Number of storms N	Number of runoff occurrences r	r/N
0–10	23	1	0.043
10–20	11	2	0.182
20–30	4	2	0.500
30–40	3	2	0.667
40–50	1	1	1.000
50–60	2	2	1.000

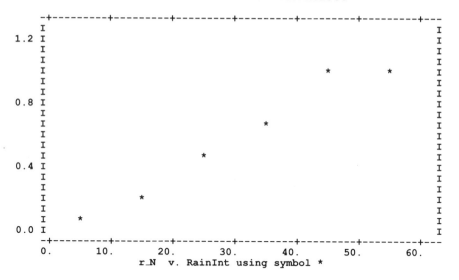

Figure 6.3   Plot of the proportions r/N (where N is the number of storms in class intervals 0–10 mm, 10–20 mm, ..., 50–60 mm, and r is the number of these storms for which runoff was observed at the minitrap site).

rainfall. But the increase is not linear, the proportions for heavier storms are based on few observations, and for the final two classes the proportions are all equal to their maximum values of one. The curve, in fact, looks rather like the cumulative distribution function of a random variable, being S-shaped, and it approximates to $E[Y|x]$, where Y is the Bernoulli variable (occurrence of surface runoff, $Y = 1$; non-occurrence of surface runoff, $Y = 0$) and x is storm rainfall depth. In simple terms, the curve says that we can expect that runoff is more likely to occur, when storm rainfall is greater. It is convenient to denote $E[Y|x]$ by $\pi(x)$, which is clearly bounded between 0 and 1. A curve having the characteristics of Figure 6.3 is the *logistic curve*

$$E[Y|x] = \exp(\beta_0 + \beta_1 x)/[1 + \exp(\beta_0 + \beta_1 x)] \tag{6.6}$$

or

$$\ln[\pi(x)/(1 - \pi(x))] = \beta_0 + \beta_1 x \tag{6.7}$$

The transformation on the left-hand side of equation (6.7) is the *logit* function, introduced earlier. In the notation of Section 6.1, we have the linear predictor, $\eta = \beta_0 + \beta_1 x$, and the link function $h(.)$ relating η to $\mu = E[Y|x] = \pi(x)$ in the form

$$\eta = h(\mu)$$

where

$$\eta = \ln(\mu/(1 - \mu)) \tag{6.8}$$

(Recall that η, μ are vectors; with the six rainfall classes in the above table, equation (6.8) is shorthand for the six equalities $\eta_i = \ln(\mu_i/(1 - \mu_i))$, where $i = 1, \ldots, 6$ represent the class intervals.

What about the distribution of the errors, ε_t? The value of the response variable Y can have just two values, 1 or 0, according as runoff occurs or does not occur, and for given rainfall x we can write

$$y = \pi(x) + \varepsilon$$

where the ε also can have two possible values. If $y = 1$, then $\varepsilon = 1 - \pi(x)$ with probability $\pi(x)$; if $y = 0$, then $\varepsilon = -\pi(x)$ with probability $1 - \pi(x)$. Thus ε has a distribution with mean zero, and variance given by

$$(-\pi(x) - 0)^2(1 - \pi(x)) + (1 - \pi(x) - 0)^2\pi(x) = \pi(x)(1 - \pi(x))$$

We can pull these various statements together, therefore, by writing, for the response variable Y,

$$y = \pi(x) + \varepsilon$$

where ε has zero mean and variance $\pi(x)(1 - \pi(x))$, where $\pi(x) = E[Y|x]$ is given by equation (6.6).

6.2.2 Estimation of Logistic Regression Parameters by Maximum Likelihood

Equation (6.6) gives the conditional probability that the response variable Y is equal to 1 (runoff occurs); that is, $\pi(x) = P[Y = 1|x]$, and we also have that $1 - \pi(x) = P[Y = 0|x]$. (The reader should not be confused by the fact that we have used the same symbol, $\pi(x)$, both for the expected value $E[Y|x]$, and for the probability $P[Y = 1|x]$. A little reflection shows that these are equal, for

$$E[Y|x] = P[Y = 1|x] \cdot 1 + P[Y = 0|x] \cdot 0$$

$$= P[Y = 1|x]$$

by definition.) The pairs of values (x_t, y_t) having $y_t = 1$ therefore each make a contribution to the likelihood function of the form $\pi(x_t)$, whilst the pairs (x_t, y_t) with $y_t = 0$ make contributions of the form $(1 - \pi(x_t))$. The likelihood function $L(\beta_0, \beta_1)$ is therefore of the form

$$L(\beta_0, \beta_1) = \prod_{t=1}^{n} \pi(x_t)^{y_t}(1 - \pi(x_t))^{1 - y_t}$$

so that the log-likelihood function is

$$l(\beta_0, \beta_1) = \sum_{t=1}^{n}[y_t \ln(\pi(x_t)) + (1 - y_t)\ln(1 - \pi(x_t))] \tag{6.9}$$

We can find the values of β_0, β_1 which maximise the log-likelihood $l(\beta_0, \beta_1)$ by differentiating equation (6.9) with respect to these parameters, giving the equations

$$\sum[y_t - \pi(x_t)] = 0$$

$$\sum x_t[y_t - \pi(x_t)] = 0$$

which can be solved iteratively for β_0 and β_1. In GENSTAT, this solution can be effected simply by using the MODEL and FIT directives — the same directives, in fact, as were used in Chapter 4 to fit linear models. We must, of course, tell the computer that we are now dealing with a generalised linear model instead of a linear one, and this is done by inserting some extra words in the MODEL directive, as follows. Having read the minitrap data from the 44 storms in Figure 6.1 into the variables *Runoff*, *Rain* and *Time*, the GENSTAT instructions

```
>MODEL [DISTRIBUTION=binomial; LINK=logit] Runoff
>FIT Rain
```

are sufficient to produce the maximum-likelihood estimates, shown in Figure 6.4. We see from this output that

$$\hat{\pi}(x) = \exp(-4.80 + 0.2117x)/[1 + \exp(-4.80 + 0.2117x)]$$

or, in terms of the logit transformation,

$$\ln[\hat{\pi}(x)/(1 - \hat{\pi}(x))] = -4.80 + 0.2117x \tag{6.10}$$

where x is storm rainfall. We also see from the output in Figure 6.4 that the standard errors of the estimates $\hat{\beta}_0$ and $\hat{\beta}_1$ are also given as ±1.32 and ±0.0664, so that both estimates are

***** Regression Analysis *****

Response variate: Runoff
Binomial totals: N
Distribution: Binomial
Link function: Logit
Fitted terms: Constant, Rain

*** Summary of analysis ***
Dispersion parameter is 1

	d.f.	deviance	mean deviance
Regression	1	*	*
Residual	42	21.64	0.5153
Total	43	*	*

* MESSAGE: The following units have large residuals:
                    25   2.46

* MESSAGE: The following units have high leverage:
                    6   0.138
                    10   0.119
                    15   0.125
                    18   0.135
                    28   0.133

*** Estimates of regression coefficients ***

	estimate	s.e.	t
Constant	−4.80	1.32	−3.64
Rain	0.2117	0.0664	3.19

* MESSAGE: s.e.s are based on dispersion parameter with value 1

Figure 6.4   Logistic regression of runoff occurrence (*Runoff* = 0 or 1) at minitrap site, on depth of storm rainfall, *Rain*. Also shown is the output for a logistic regression on two explanatory variables, *Rain* (storm rainfall depth) and *TimeInt* (approximate time, in weeks, since last storm event).

_____ *Continued* ___|

Figure 6.4 *(continued)*

*** Correlations ***

Constant	1	−1.000	
Rain	2	−0.907	1.000
		1	2

*** Fitted values and residuals ***

Unit	Binomial total	Response	Fitted value	Standardized residual	Leverage
1	1	0	0.02	−0.18	0.020
2	1	0	0.02	−0.21	0.023
3	1	0	0.04	−0.27	0.029
4	1	0	0.04	−0.27	0.029
5	1	0	0.02	−0.22	0.024
6	1	1	0.78	0.77	0.138
7	1	0	0.09	−0.46	0.040
8	1	0	0.01	−0.17	0.019
9	1	1	1.00	0.05	0.007
10	1	1	0.60	1.08	0.119
11	1	1	1.00	0.05	0.009
12	1	0	0.03	−0.26	0.027
13	1	0	0.01	−0.15	0.017
14	1	0	0.07	−0.39	0.037
15	1	1	0.88	0.54	0.125
16	1	0	0.47	−1.18	0.095
17	1	0	0.01	−0.17	0.019
18	1	0	0.83	−2.02	0.135
19	1	0	0.01	−0.16	0.018
20	1	1	0.30	1.61	0.065
21	1	0	0.13	−0.55	0.045
22	1	0	0.02	−0.18	0.020
23	1	0	0.01	−0.15	0.017
24	1	0	0.09	−0.43	0.039
25	1	1	0.05	2.46	0.034
26	1	0	0.13	−0.55	0.045
27	1	0	0.10	−0.47	0.041
28	1	1	0.84	0.63	0.133
29	1	0	0.02	−0.18	0.020
30	1	0	0.06	−0.35	0.034
31	1	0	0.22	−0.73	0.055
32	1	0	0.05	−0.32	0.032
33	1	0	0.01	−0.16	0.018
34	1	1	0.98	0.19	0.046
35	1	1	0.34	1.52	0.072
36	1	0	0.03	−0.24	0.026
37	1	0	0.10	−0.47	0.041
38	1	0	0.02	−0.18	0.020
39	1	0	0.04	−0.30	0.031
40	1	0	0.41	−1.08	0.085
41	1	0	0.01	−0.16	0.018
42	1	0	0.02	−0.21	0.023
43	1	0	0.01	−0.15	0.016
44	1	0	0.08	−0.40	0.037
Mean				−0.10	0.044

17 ....

Figure 6.4 *(continued)*

***** Regression Analysis *****

Response variate: Runoff
  Binomial totals: N
     Distribution: Binomial
    Link function: Logit
     Fitted terms: Constant, Rain, TimeInt

*** Summary of analysis ***
    Dispersion parameter is 1

	d.f.	deviance	mean deviance
Regression	2	*	*
Residual	41	19.83	0.4836
Total	43	*	*

* MESSAGE: The following units have large residuals:
                      25   2.29

* MESSAGE: The following units have high leverage:
                      6   0.179
                     16   0.215
                     18   0.178
                     20   0.221
                     40   0.281

*** Estimates of regression coefficients ***

	estimate	s.e.	t
Constant	−3.61	1.51	−2.39
Rain	0.2088	0.0699	2.99
TimeInt	−0.680	0.591	−1.15

* MESSAGE: s.e.s are based on dispersion parameter with value 1

*** Correlations ***

Constant	1	1.000		
Rain	2	−0.737	1.000	
TimeInt	3	−0.418	−0.174	1.000
		1	2	3

*** Fitted values and residuals ***

Unit	Binomial total	Response	Fitted value	Standardized residual	Leverage
1	1	0	0.00	−0.08	0.014
2	1	0	0.01	−0.14	0.021
3	1	0	0.00	−0.06	0.017
4	1	0	0.01	−0.18	0.030
5	1	0	0.04	−0.28	0.042
6	1	1	0.73	0.88	0.179
7	1	0	0.04	−0.30	0.060
8	1	0	0.02	−0.22	0.033
9	1	1	1.00	0.05	0.010
10	1	1	0.54	1.21	0.150
11	1	1	0.99	0.13	0.058

__ Figure 6.4 *(continued)* _____

12	1	0	0.01	−0.17	0.028
13	1	0	0.03	−0.23	0.044
14	1	0	0.11	−0.50	0.062
15	1	1	0.92	0.44	0.112
16	1	0	0.70	−1.75	0.215
17	1	0	0.01	−0.11	0.016
18	1	0	0.79	−1.94	0.178
19	1	0	0.03	−0.24	0.046
20	1	1	0.53	1.27	0.221
21	1	0	0.20	−0.69	0.074
22	1	0	0.01	−0.11	0.017
23	1	0	0.02	−0.19	0.029
24	1	0	0.04	−0.28	0.056
25	1	1	0.08	2.29	0.057
26	1	0	0.02	−0.19	0.079
27	1	0	0.15	−0.60	0.069
28	1	1	0.89	0.51	0.123
29	1	0	0.00	−0.08	0.015
30	1	0	0.09	−0.45	0.058
31	1	0	0.18	−0.66	0.071
32	1	0	0.10	−0.49	0.089
33	1	0	0.02	−0.20	0.031
34	1	1	0.99	0.15	0.037
35	1	1	0.29	1.65	0.093
36	1	0	0.01	−0.15	0.025
37	1	0	0.15	−0.60	0.069
38	1	0	0.00	−0.08	0.014
39	1	0	0.09	−0.47	0.085
40	1	0	0.12	−0.61	0.281
41	1	0	0.00	−0.04	0.007
42	1	0	0.02	−0.19	0.025
43	1	0	0.00	0.00	0.000
44	1	0	0.01	−0.13	0.046
Mean				−0.09	0.068

large relative to their standard errors. We discuss the use of these quantities in Section 6.2.3, commenting only at this stage that the standard error of $\hat{\beta}_1$ shows that storm depth is, unsurprisingly, a good variable for predicting the occurrence of runoff at the minitrap site.

6.2.3   Testing for the Significance of Logistic Regression

In linear regression, we have the apparatus of t- and F-tests for testing the significance of additional terms, and these tests (when the model assumptions are valid) are exact, not approximate. For logistic regression and, more generally, for all generalised linear models, the tests used are approximations; however, the principle underlying the tests is the same. To establish whether an additional term in the linear predictor is to be included, we compare the observed values of the response variable Y to the predicted values of the response variable, given by the model first without the additional term and then with it. If the improvement in prediction is 'large' (by the appropriate criterion) when the additional term is included, it is retained in the model; if not, it is discarded unless there are good, non-statistical reasons why it should be included.

To help understanding of the test procedure, we recall the idea of a *saturated* model, which is a model containing as many β parameters as data points, 44 in the case of the minitrap data. This model would 'fit' each of the points exactly, so that its log-likelihood has the same form as equation (6.9), but with y_t replacing $\pi(x_t)$. The ratio:

$$\frac{\text{likelihood for the current model (with linear predictor } \beta_0 + \beta_1 x)}{\text{likelihood for the saturated model, with each } \pi(x_t) = y_t}$$

is called the *likelihood ratio*; the statistic

$$D = -2 \ln(\text{likelihood ratio}) \tag{6.11}$$

is called the *log-likelihood ratio statistic*, and tests based upon it, which we shall use, are called *likelihood ratio tests*. The explicit form for D is

$$D = -2 \sum_{t=1}^{n} \{y_t \ln(\hat{\pi}_t/y_t) + (1 - y_t) \ln[(1 - \hat{\pi}_t)/(1 - y_t)]\} \tag{6.12}$$

The statistic D is the *deviance*, introduced earlier in Section 6.1.1, where it was written in the slightly different form

$$2 \sum [y_t \ln(y_t/\mu_t) + (m_t - y_t) \ln((m_t - y_t)/(m_t - \mu_t))]$$

The deviance in logistic regression plays the same role as the residual sum of squares plays in ordinary linear regression. To determine the significance of an explanatory variable x for predicting the response variable Y, we compare the values of D with and without the explanatory variable included in the equation. The change in D resulting from including the variable is

$$G = (D \text{ for the model without the variable}) - (D \text{ for the model including the variable})$$

or

$$G = -2 \ln [(\text{likelihood without the variable})/(\text{likelihood with the variable})] \tag{6.13}$$

since the likelihood for the saturated model is common to both and cancels. Where we are testing the linear predictor $\eta = \beta_0 + \beta_1 x$ against the linear predictor $\eta = \beta_0$, we have explicitly:

$$G = 2 \sum_{t=1}^{n} [y_t \ln \hat{\pi} + (1 - y_t) \ln(1 - \hat{\pi}_t)] - 2[n_1 \ln(n_1) + n_0 \ln(n_0) - n \ln(n)] \tag{6.14}$$

where $n_1 = \Sigma\, y_t$, $n_0 = \Sigma(1 - y_t)$, and of course $n = n_0 + n_1$.

In the case of the minitrap data, the log-likelihood is given by equation (6.9) and has value -13.07. We then find from equation (6.14) that the value of G is $2\{-13.07 - [14 \ln(14) + 30 \ln(30) - 44 \ln(44)]\} = 28.90$.

Under the null hypothesis that $\beta_1 = 0$ (that is, that storm rainfall has no value for predicting runoff occurrence at the minitrap site), the statistic G is approximately distributed as χ^2 with 1 degree of freedom. By contrast with the deviance itself, for which the χ^2 distribution is not a particularly good approximation, the χ^2 distribution affords a much better approximation to the distribution of the statistic G, the change in deviance. We find from tables (or from the GENSTAT function CHISQ(28.90;1)) that, if the null hypothesis is true, the probability of obtaining a χ^2 as large as or larger than 28.90 is much less than 0.001. Such a small probability justifies rejection of the null hypothesis, and we conclude that storm rainfall is a valuable predictor of storm runoff occurrence, confirming intuition.

Just as we have considered in the above a simple logit regression of runoff occurrence on the variable *Rain* (depth of storm rainfall), it would also be possible to look at the simple logit regression of runoff occurrence on *TimeInt*, the interval since the last storm. In view of the significance of *Rain*, it is probably of greater interest to examine whether the variable *TimeInt* further improves the prediction of runoff occurrence significantly, when the variable *Rain* is already included. For this, we require the general test statistic given in equation (6.13) to test the null hypothesis that $\beta_2 = 0$. When the variable *TimeInt* is excluded, we find by maximum likelihood, as before,

$$\hat{\beta}_0 = -4.80; \ \hat{\beta}_1 = 0.2117; \ \ln L = -13.071\,128$$

When both variables, *Rain* and *TimeInt* are included, we have, from GENSTAT or otherwise,

$$\hat{\beta}_0 = -3.61; \ \hat{\beta}_1 = 0.2088; \ \hat{\beta}_2 = -0.680; \ \ln L = -12.206\,295$$

The test statistic G is therefore

$$G = -2[-13.071\,128 - (-12.206\,295)] = 1.7297.$$

We find from χ^2 tables, or by the use of the GENSTAT function CHISQ(1.7297;1) or its equivalent, that the probability of obtaining a χ^2 statistic as large or larger than 1.7297 is about 0.1884, not sufficiently small to warrant rejection of the null hypothesis that $\beta_2 = 0$. We conclude that, for this particular set of minitrap data, the variable *TimeInt* gives no significant improvement in the prediction of runoff occurrence, when *Rain* is already being used.

The argument developed in this section applies also to the case where p variables $x_1, x_2, \ldots, x_p$ have been included in the logistic regression, and we wish to ascertain whether q additional variables $x_{p+1}, x_{p+2}, \cdots, x_{p+q}$ should also be included. We then require to test the null hypothesis $\beta_{p+1} = \beta_{p+2} = \cdots = \beta_{p+q} = 0$. We calculate the log-likelihood, using equation (6.9), first with just the variables $x_1, x_2, \ldots, x_p$ and second with all the variables $x_1, x_2, \ldots, x_p, x_{p+1}, \cdots, x_{p+q}$. Substitution in the statistic G is followed by a χ^2-test with q degrees of freedom.

6.2.4   The Information Matrix, and Wald Tests

At the end of Section 6.2.2 we made brief use of the standard errors of the estimates of β_0 and β_1, taken from the GENSTAT output, without addressing the method used to calculate them. Likelihood theory shows that, provided the number of observations is not too small, the variances and covariances of the estimates $\hat{\beta}_j$ are given by the inverse of the information matrix $\mathbf{I}$, where

$$\mathbf{I} = [-\partial^2 \ln L / \partial \beta_i \, \partial \beta_j] \tag{6.15}$$

In the case of logistic regression, the element in the ith row and ith column of the matrix $\mathbf{I}$ is

$$\partial^2 \ln L / \partial \beta_i^2 = -\sum_{k=1}^{n} x_{ki}^2 \pi_k (1 - \pi_k)$$

and in the ith row and jth column

$$\partial^2 \ln L / \partial\beta_i \partial\beta_j = -\sum_{k=1}^{n} x_{ki} x_{kj} \pi_k (1 - \pi_k)$$

The variance of $\hat{\beta}_i$, for example, is given by the element in the ith row and column in $\mathbf{I}^{-1}$; the covariance of $\hat{\beta}_i$ and $\hat{\beta}_j$ is given by the element in the ith row and jth column of $\mathbf{I}^{-1}$. A useful way of writing the information matrix $\mathbf{I}$ is in the form of the product $\mathbf{I} = \mathbf{X}^T \mathbf{V} \mathbf{X}$, where $\mathbf{X}$ is the matrix of values of the explanatory variable, given in general by

$$\mathbf{X} = \begin{bmatrix} 1 & x_{11} & x_{12} & \cdots & x_{1p} \\ 1 & x_{21} & x_{22} & \cdots & x_{2p} \\ & \cdots & & & \cdots \\ & \cdots & & & \cdots \\ 1 & x_{n1} & x_{n2} & \cdots & x_{np} \end{bmatrix}$$

and the matrix $\mathbf{V}$ is

$$\mathbf{V} = \begin{bmatrix} \pi_1(1-\pi_1) & 0 & 0 & \cdots & 0 \\ 0 & \pi_2(1-\pi_2) & 0 & \cdots & 0 \\ 0 & 0 & \pi_3(1-\pi_3) & \cdots & 0 \\ & & \cdots & & \\ 0 & 0 & 0 & & \pi_n(1-\pi_n) \end{bmatrix}$$

The square roots of the variances of $\hat{\beta}_1$ (given by the terms in the leading diagonal of $\mathbf{I}^{-1}$) are their standard errors, and it is these that are supplied by the GENSTAT output. Thus when both variables *Rain* and *TimeInt* are fitted in the example of Section 6.2.3, we find that

$$\hat{\beta}_0 = -3.61 \pm 1.51; \quad \hat{\beta}_1 = 0.2088 \pm 0.0699; \quad \hat{\beta}_2 = -0.680 \pm 0.591$$

An approximate test, due to Wald, consists of comparing the ratio

$$\hat{\beta}_i / (\text{standard error of } \hat{\beta}_i)$$

as a standard normal deviate, having distribution $N(0, 1)$. Thus a Wald test of the null hypothesis that $\hat{\beta}_2 = 0$ is obtained by calculating $u = -0.680/0.591$ which, being less in absolute value than 1.96, lies in the acceptance region for the null hypothesis (we recall that the region $u = \pm 1.96$ includes 95% of the probability for the standard normal distribution). Hence we conclude, as previously, that inclusion of the explanatory variate *TimeInt* does not improve significantly the prediction of runoff occurrence, when *Rain* is already being used in the logistic regression.

Standard errors of logit regression coefficients in a computer listing give a rapid means of identifying coefficients β_i that are likely to be of particular importance for prediction purposes. However, there is some evidence that the Wald test may at times behave in an aberrant manner, and the likelihood ratio test, although itself an approximation, is recommended for general use.

6.2.5 Interpretation of the Coefficients β_i in Logistic Regression

Our discussion so far has been concerned with aspects of the calculation of logistic regression equations. Once the parameters β in the regression have been estimated, it becomes important to interpret the estimates to find out what they are telling us about relations between the variables being analysed. To illustrate the argument, we begin with the simple case given by equation (6.7) in which the right-hand side contains just two parameters β_0 and β_1, as in the case where the one explanatory variable, *Rain*, is used in the minitrap data. Sometimes the intercept β_0 will be of interest but this is the exception rather than the rule; more commonly we are interested in interpreting the slope parameter β_1.

For an ordinary linear regression, the slope parameter β_1 represents the change in the response variable associated with unit change in the explanatory variable x. That is, β_1 is equal to the difference between the value of the response variable at $x + 1$ and the value of the response variable at x; thus if $y(x) = \beta_0 + \beta_1 x$, we have $\beta_1 = y(x + 1) - y(x)$.

For a logistic regression, interpretation of β_1 is a little more general; we have, from equation (6.7),

$$\beta_1 = \ln[\pi(x + 1)/(1 - \pi(x + 1))] - \ln[\pi(x)/(1 - \pi(x))]$$

which is the difference between the logits corresponding to unit change in the explanatory variable x. The interpretation of β_1 is now that it represents the change in the logarithm of the odds in favour of a 'success' for the Bernoulli variable, corresponding to a change of one unit in the explanatory variable. In the case of the minitrap data, $\exp(\beta_1)$ is therefore the factor multiplying the odds in favour of the occurrence of storm runoff, per millimetre increase in storm rainfall. Put another way, storm runoff is $\exp(10\beta_1)$ times more likely to be observed at the particular minitrap being studied, for each centimetre increase in storm rainfall. One means of calculating an approximate confidence interval for these odds, is to calculate

$$\exp[10\hat{\beta}_1 \pm u_{1-\alpha/2} \times 10 SE(\hat{\beta}_1)]$$

where u is the normal deviate defining a probability of $\alpha/2$ in the upper tail of the distribution.

Hitherto, in the discussion concerning interpretation of coefficients β in a logistic regression, the explanatory variable x has been *continuous*, like *Rain* in the minitrap example. However, in many applications of logistic regression, the explanatory variable is *discrete*; suppose, for example, that x were the presence ($x = 1$) or absence ($x = 0$) of soil cultivation at the minitrap site in the ten-day period preceding the storm event. We should then be interested in establishing whether the odds in favour of surface runoff occurring were different where soil cultivation had occurred. We can then think of the probability of surface runoff occurring as given in the following two-way table:

	No soil cultivation $(x = 0)$	Soil cultivation $(x = 1)$
Runoff absent $(y = 0)$:	$1 - \pi(0) =$ $1/[1 + \exp(\beta_0)]$	$1 - \pi(1) =$ $1/[1 + \exp(\beta_0 + \beta_1)]$
Runoff present: $(y = 1)$:	$\pi(0) =$ $\exp(\beta_0)/[1 + \exp(\beta_0)]$	$\pi(1) =$ $\exp(\beta_0 + \beta_1)/[1 + \exp(\beta_0 + \beta_1)]$

Thus the odds of runoff occurring, where the soil has not been cultivated, are $\pi(0)/[1 - \pi(0)]$; the odds of runoff occurring, where soil has been cultivated, are $\pi(1)/[1 - \pi(1)]$. The odds ratio — that is, the ratio

$$\frac{\text{odds that runoff occurs when soil has been cultivated } (x = 1)}{\text{odds that runoff occurs when soil has not been cultivated } (x = 0)}$$

— is therefore

$$\frac{\pi(1)/[1 - \pi(1)]}{\pi(0)/[1 - \pi(0)]} = \frac{\{\exp(\beta_0 + \beta_1)/[1 + \exp(\beta_0 + \beta_1)]\} \cdot \{1/[1 + \exp(\beta_0 + \beta_1)]\}}{\{\exp(\beta_0)/[1 + \exp(\beta_0)]\} \cdot \{1/[1 + \exp(\beta_0)]\}}$$

$$= \exp(\beta_1)$$

Hence β_1 is the log of the odds ratio. It therefore shows (in terms of the hypothetical example) how much more likely (or unlikely) it is for surface runoff to occur in those storms for which $x = 1$ (soil cultivation having occurred within the last ten days) than in those storms for which $x = 0$ (no cultivation having occurred within the last ten days).

The last paragraph has discussed in some detail the case where the explanatory variable x has just two values: presence $(x = 1)$ and absence $(x = 0)$. The argument is easily extended to the case where x is a qualitative variable having several 'dummy' values; pursuing further our hypothetical example, we might wish to use, as an explanatory variable, the variable x with values $x = 2$ (soil cultivated recently); $x = 1$ (soil cultivated, but not recently); $x = 0$ (no soil cultivation). Consideration of a 2×3 table, similar to that given above, would then yield estimates of the likelihood, or otherwise, of surface runoff occurring after recent cultivation, after less recent cultivation, or in the absence of soil cultivation.

6.2.6   Other Applications of Logistic Regression

The example of the minitrap data used observations from one minitrap only, although it was mentioned that the study involved the use of two clusters of 12 and 14 minitraps, respectively. It would be possible to repeat the analysis on each minitrap within a cluster separately; it would also be possible to analyse data from all minitraps within a cluster together, perhaps including, as a further variable, distance from the nearest downslope watercourse, to examine whether minitraps further downslope contributed more frequently to runoff than minitraps that were further upslope.

It was also mentioned in Section 6.2.1 that the state of each minitrap, after each storm, fell into one of three categories: empty, partially full, or overflowing. Denoting these states by 0, 1 and 2, the analysis given in the above example was made after combining the second and third categories to give two new ones: state 0, when no runoff had occurred, and state 1, when runoff had occurred. The extension necessary to deal with the three-state case is straightforward; denoting by Y the variable for runoff occurrence, with values 0, 1 and 2, we define

$$g_1(x) = \ln\{P[Y = 1|x]/P[Y = 0|x]\}$$

$$= \beta_{10} + \beta_{11}x_1 + \cdots + \beta_{1p}x_p$$

and

$$g_2(x) = \ln\{P[Y = 2|x]/P[Y = 0|x]\}$$

$$= \beta_{20} + \beta_{21}x_1 + \cdots + \beta_{2p}x_p$$

It can be shown that the log-likelihood now becomes

$$\ln L = \sum_{t=1}^{n}[y_{1t}g_1(x_t) + y_{2t}g_2(x_t) - \ln\{1 + \exp(g_1(x_t)) + \exp(g_2(x_t))\}]$$

and the maximum-likelihood procedure for fitting the parameters by calculating the two sets of estimates $\hat{\beta}_{10}, \hat{\beta}_{11}, \ldots, \hat{\beta}_{1p}$ and $\hat{\beta}_{20}, \hat{\beta}_{21}, \ldots, \hat{\beta}_{2p}$ continues.

The two preceding paragraphs in this section have described some further possible developments using the minitrap data from the basin of the Booro-Borotou; the applications of logistic regression are, of course, much wider. One such application concerns the study of time trends in the occurrence of rainfall; given, say, 50 years of record, we could count the number of days in each month on which significant rain fell, and explore a logistic regression of the form

$$\ln[\pi(x_{ij})/(1 - \pi(x_{ij}))] = \sum \beta_i x_i + \sum \beta_j x_j$$

where $i = 1, 2, \ldots, 50$; $j = 1, 2, \ldots, 12$. The calculation involves no new principle, although we are now fitting 62 parameters. Where rainfall varies seasonally throughout the year we should expect the month parameters β_j to be significantly different from zero; if the year parameters are also significantly different from zero, we may have evidence of a time trend.

The use of minitraps for the detection of surface runoff has been described; a similar technique, with similar analysis, may be appropriate for identifying sites where soil erosion is active.

Coe and Stern (1982) and Stern and Coe (1984) used two generalised linear models, one of which was a logistic regression, to model daily rainfall data. Histograms of daily rainfall data are usually skewed to the right, with a 'spike' at the origin. This form of distribution suggests (McCullagh and Nelder, 1991) that daily rainfall data might be modelled in two stages, one stage being concerned with the pattern of occurrence of wet and dry days, and the second stage being concerned with the depth of rain falling on wet days. Coe and Stern (1982) and Stern and Coe (1984) modelled the first stage, which involves a random variable Y taking values 1 (when rain occurs) and 0 (no rain), as a stochastic process in which the probability that $Y = 1$ on day t depends on the past history of rainfall up to day $t - 1$: note that in this case $P[Y_t = 1]$ is not independent of $P[Y_{t-1} = 1]$. Stern and Coe found that, in many cases, first-order dependence corresponding to a *Markov chain* gave a satisfactory model for the occurrence and non-occurrence of rain. In the second stage, a model of rainfall depth was required which was defined for positive values only, positively skewed, and had variance increasing with the mean. The gamma distribution was used for this purpose.

With this model structure, the data for n years form an $n \times 365$ table of rainfall amounts, if leap years are ignored. Numbering the days within each year from 1 to 365, then each pair of consecutive days $t - 1$ and t gives rise, over the n years, to a two-way table of the form

| | | Day t | | |
		Dry	Wet	Total
Day $t - 1$	Dry	n_{00}	n_{01}	$n_{0.}$
	Wet	n_{10}	n_{11}	$n_{1.}$
Total		$n_{.0}$	$n_{.1}$	$n_{..}$

where the overall total $n_{..}$ will equal n, the number of years of record, if all n years have complete data for days $t - 1$ and t; but an advantage of the Stern–Coe model is that tables similar to the above can be drawn up, whether or not the n years of record are complete, so that 'gappy' data and parts of years present no difficulty and it is unnecessary to 'infill' the record to estimate any missing values.

Now let $\pi_1(t)$ be the probability that day t is wet, given that day $t - 1$ was also wet; let $\pi_0(t)$ be the probability that day t is wet, given that day $t - 1$ is dry. Then the likelihood is (apart from a small correction due to the ends of the record) the product over all the t of terms such as

$$\pi_1(t)^{n_{11}}[1 - \pi_1(t)]^{n_{10}}\pi_0(t)^{n_{01}}[1 - \pi_0(t)]^{n_{00}}$$

Using the GLIM package mentioned in Chapter 1, Stern and Coe maximised this likelihood, expressing $\pi_1(t)$ and $\pi_0(t)$ in GLM form as

$$\text{logit}[\pi_0(t)] = \alpha_0 + \alpha_{01} \sin(2\pi t/365) + \beta_{01} \cos(2\pi t/365)$$

$$\text{logit}[\pi_1(t)] = \alpha_1 + \alpha_{11} \sin(2\pi t/365) + \beta_{11} \cos(2\pi t/365)$$

giving a model with six parameters.

This model for the occurrence and non-occurrence of rain can be extended in various ways. A second-order model, for example, takes account of the presence and absence of rain on two preceding days; instead of the 2×2 table given above, a 4×4 table is then required, and four time-dependent probabilities must be modelled. Stern and Coe (1984) illustrate this, and show how GLIM may be used to test, very simply, whether a second-order model is necessary, or whether a first-order model is sufficient. McCullagh and Nelder (1991) point out that if it is suspected that a time trend exists in the rainfall record, it is preferable to consider the data as $365n$ Bernoulli observations indexed by day, year, and previous day wet/dry; this increases the computational burden (which would probably be inappropriate for a desk-top computer) but introduces no new problems.

For modelling rainfall depths on wet days, Stern and Coe used gamma distributions with time-dependent means $\mu(t)$, and $\ln[\mu(t)]$ was expressed as a linear function involving harmonic components. Here, $\mu(t)$ is the mean rainfall on day t, conditional upon day t being wet. Assuming that no time trends exist in rainfall amounts over the period of record, fitting a generalised linear model uses the mean values for each day, with the number of days (out of the n years) for which day t was wet being used as weights in the model. If a time trend over the n years is suspected, the individual daily rainfall amounts are modelled along the lines mentioned in the preceding paragraph.

In their remarkable paper, Stern and Coe emphasise the uses of models of this type. They show how the fitted logistic model can be used to calculate the probabilities of runs of dry and wet days, of any length, beginning on any day of the year. Given some very simple assumptions about field capacity, they show how to calculate the probability distribution of soil water within the profile, and hence how an optimum planting date may be determined for crops in semi-arid regions. The Stern–Coe model has been widely applied to answer questions about planting dates, and about the probabilities of dry spells at critical phases in crop development, throughout Africa and the Indian subcontinent.

6.3 REGRESSION WITH CONSTANT COEFFICIENT OF VARIATION: AN EXTENSION OF THE THOMAS–FIERING MODEL

The Thomas–Fiering model for the simulation of sequences of monthly river flow has been mentioned several times in earlier chapters. In its simplest form, a simple linear regression is calculated of each month's flow on the flow of the month preceding it; errors are assumed normally distributed, and artificial sequences of monthly flow are generated by taking the predicted value for each month's flow from the appropriate regression; adding to it a normally distributed random component with the appropriate variance, to give the generated flow for the month; and using this generated flow to generate flow in the following month, employing the same procedure. The simplest form of the Thomas–Fiering model can be extended in various ways, perhaps most easily by including more explanatory variables: perhaps flows in two preceding months instead of one, perhaps rainfall in the preceding month, and so on.

Not infrequently, uncritical use of the Thomas–Fiering model leads to problems with the artificial sequences generated by the model: in particular, negative monthly flows can be generated. Where these are not too frequent, a common solution is to set the negative flows equal to zero or some small positive value, but the fact that they occur is really an indication that the model is inappropriate: that the data are being forced into a model of unsuitable form. An alternative is to transform flows by taking logarithms, but complications may then arise after detransformation of generated flows to the original scale. What is really required is a model which works in the units of flow, which avoids negative values in generated series, and which avoids problems when we come to consider the distribution of seasonal or annual flows.

Suppose, for simplicity, that we restrict attention, for the time being, to regressions with just one explanatory variable, flow in the preceding month. Thus for month t, the simple Thomas–Fiering model is

$$E[Q_t] = \beta_0 + \beta_1 q_{t-1} \tag{6.16}$$

and the log-transformed model is

$$E[\ln Q_t] = \beta_0 + \beta_1 q_{t-1} \tag{6.17}$$

A model having one of the necessary characteristics (mean value always positive) is

$$\ln\{E[Q_t]\} = \beta_0 + \beta_1 q_{t-1} \tag{6.18}$$

which, of course, is not the same as model (6.17). In the notation of Section 6.1, we have a linear predictor $\eta = \beta_0 + \beta_1 q_{t-1}$, and a link function relating η to $E[Q_t]$ which is $\eta = \ln\{E[Q_t]\}$, a logarithmic link function.

A further shortcoming of the Thomas–Fiering is its assumption of normally distributed errors with constant variance. The consequence is that the variance of the generated flows using equation (6.16) is the same regardless of the value of q_{t-1}, and this is often an assumption not supported by the data. Indeed, negative monthly flows are generated by the model (6.16) precisely because the dispersion of generated values about the fitted regression is as large for small values of the explanatory variable q_{t-1} as for large values. For many records of monthly flow, simple graphical displays like those of Chapter 2 show clearly that dispersion commonly increases as q_{t-1} increases.

To illustrate the use of generalised linear models in such situations, we consider 47 years of monthly flow record from a gauging site on the Rio Lava Tudo at Fazenda Mineira, a

```
***** Regression Analysis *****

Response variate:    mar
      Fitted terms:   Constant, feb

*** Summary of analysis ***

                    d.f.          s.s.          m.s.
Regression           1           2231.        2231.0
Residual            45           8015.         178.1
Total               46          10246.         222.7

Percentage variance accounted for 20.0

* MESSAGE: The following units have large residuals:
                     29                    3.50

* MESSAGE: The following units have high leverage:
                     24                    0.199
                     39                    0.100

*** Estimates of regression coefficients ***

                  estimate         s.e.          t
Constant            8.42          3.61         2.33
feb                 0.424         0.120        3.54
```

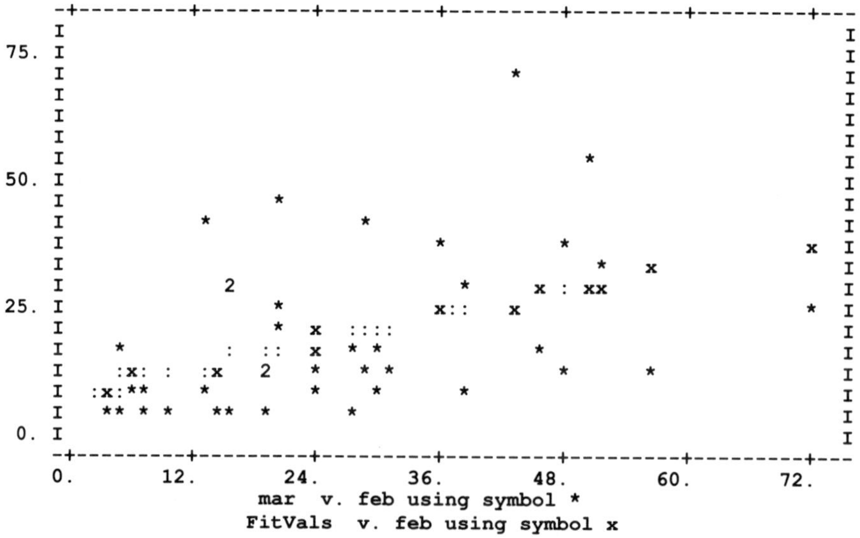

```
     -+---------+---------+---------+---------+---------+---------+---
      I                                                              I
75.   I                                                              I
      I                                       *                      I
      I                                                              I
      I                                                              I
      I                                                              I
      I                                            *                 I
50.   I                                                              I
      I               *                                              I
      I          *         *                                         I
      I                         *           *                   x  I
      I                                         *    x              I
      I          2         *                  x : xx               I
25.   I                *              *                          *  I
      I                   x    ::::   x::     x                      I
      I     *        :   ::  x  * *               *                  I
      I   :x:  :  :x   2   *   * *          *        *               I
      I  :x:**      *         *     *    *                           I
      I   ** * *   **    *         *                                 I
 0.   I                                                              I
     -+---------+---------+---------+---------+---------+---------+---
      0.       12.       24.       36.       48.       60.       72.
                      mar  v.  feb using symbol *
                      FitVals  v.  feb using symbol x
```

Figure 6.5   Thomas–Fiering model: regression using two months of data (March and February, denoted by *mar, feb*) from 47 years of record from Rio Lava Tudo at Fazenda Mineira.

tributary of the Rio Uruguay, southern Brazil. We consider only one pair of months, February and March, from these data; Figure 6.5 shows a plot of these monthly flows, together with the linear regression that would be used by the simple Thomas–Fiering model. The linear regression is shown by the symbol $\times$, the observed monthly flows by the symbol $*$; the colon, :, occurs where points are near to coincidence. The linear regression $q_{mar} = 8.42 + 0.424 q_{feb}$ accounts for only 20% of the total variance in q_{mar}, although the coefficient $\hat{\beta}_1 = 0.424 \pm 0.121$ differs significantly from zero ($t = 3.54$ with 45 degrees of freedom). Despite the coarseness of the plot in Figure 6.5, it is clear that the dispersion is less where February flows are small, so we might expect that assuming constant variance, independent of the magnitude of q_{feb}, would lead to difficulty, and this is what happened in practice: an appreciable number of negative flows were generated when the simple Thomas–Fiering model was used to generate synthetic sequences of monthly flows. Section 6.3.1 explores the use of a generalised linear model, with monthly flows in March having a gamma distribution with mean value a function of February flows — the functional relation being expressed by means of the link function.

6.3.1 Generalised Linear Model with Gamma-distributed Errors and Log Link Function

The rationale of fitting the model

$$\ln E[Y|x] = \beta_0 + \beta_1 x \qquad (6.19)$$

with Y having a gamma distribution, is identical to that for the logistic regression considered in Section 6.2. We calculate the deviance, which is minus twice the logarithm of the ratio (likelihood for model given by equation (6.19))/(likelihood of saturated model). Some straightforward algebra shows that this is

$$D = 2\sum\{-\ln(y_t/\mu_t) + (y_t - \mu_t)/\mu_t\}$$

where

$$\mu_t = \exp(\beta_0 + \beta_1 x_t)$$

To test the significance of the generalised regression on x, we calculate the statistic

$$G = D(\text{for the model with } \beta_1 = 0) - D(\text{for the model with } \beta_1 \text{ included})$$

Under the hypothesis that $\beta_1 = 0$, G is approximately distributed as χ^2 with one degree of freedom.

The above calculation is much simplified by the use of GENSTAT or similar packages. In GENSTAT, we use forms of the MODEL and FIT directives used earlier for linear regressions. To inform GENSTAT of the nature of the distribution and of the link function, the MODEL directive becomes (using 'mar' and 'feb' for the monthly flows being analysed)

```
>MODEL [DISTRIBUTION=gamma; LINK=log] mar
```

followed by

```
>FIT [PRINT=model, summary, estimates,correlations,\
fittedvalues, monitoring] feb
```

Figure 6.6 shows the output resulting from these directives. We see that the model, in which q_{mar} is gamma-distributed with

$$\ln\{E[Q_{mar}]\} = 2.268 + 0.024\,30 q_{feb}$$

***** Regression Analysis *****

Response variate: mar
        Distribution: Gamma
      Link function: Log
       Fitted terms: Constant, feb

*** Summary of analysis ***

	d.f.	deviance	mean deviance
Regression	1	*	*
Residual	45	19.84	0.4408
Total	46	*	*

* MESSAGE: The following units have high leverage:
                          24   0.199
                          39   0.100

*** Estimates of regression coefficients ***

	estimate	s.e.	t
Constant	2.268	0.180	12.63
feb	0.02430	0.00596	4.08

*** Correlations ***

Constant	1	1.000	
feb	2	−0.842	1.000
		1	2

*** Fitted values and residuals ***

Unit	Response	Fitted value	Standardized residual	Leverage
1	3.15	10.40	−1.55	0.062
2	8.21	10.90	−0.42	0.055
3	4.57	18.66	−1.74	0.022
4	16.06	19.90	−0.32	0.023
5	10.16	17.40	−0.75	0.021
6	15.59	14.29	0.13	0.028
7	17.49	10.78	0.82	0.056
8	28.04	14.28	1.16	0.028
9	16.90	29.11	−0.77	0.053
10	5.65	13.57	−1.17	0.032
11	5.84	14.14	−1.18	0.029
12	14.16	15.20	−0.11	0.025
13	8.46	20.01	−1.15	0.023
14	12.86	31.19	−1.20	0.063
15	37.93	23.30	0.81	0.031
16	41.28	19.70	1.29	0.023
17	13.42	17.50	−0.39	0.021

Figure 6.6 Thomas–Fiering model with March flows (denoted by *mar*) having a gamma distribution: that is, the model is $\log E[mar] = b_0 + b_1 feb$, where February flows are denoted by the variable named *feb*. The graph shows observed monthly flows (denoted by $\bigcirc$) for March plotted against observed monthly flows for February; 'f' denotes fitted values given by the model $\log E[mar] = b_0 + b_1 feb$.

_____ Continued ___

Figure 6.6 *(continued)*

18	19.39	15.80	0.32	0.023
19	44.33	15.90	1.88	0.023
20	4.25	10.86	−1.26	0.055
21	54.55	32.95	0.86	0.072
22	9.83	13.30	−0.44	0.033
23	7.58	11.51	−0.60	0.048
24	25.03	56.01	−1.20	0.199
25	37.27	31.02	0.29	0.062
26	7.46	10.36	−0.48	0.062
27	31.03	30.84	0.01	0.062
28	10.60	12.11	−0.20	0.042
29	72.21	27.44	1.78	0.046
30	13.25	19.48	−0.55	0.022
31	6.06	15.24	−1.22	0.025
32	30.82	24.26	0.38	0.034
33	4.33	11.43	−1.29	0.049
34	42.61	13.14	2.24	0.034
35	23.19	23.96	−0.05	0.033
36	15.91	18.89	−0.25	0.022
37	9.24	11.14	−0.28	0.052
38	28.62	14.26	1.20	0.028
39	12.24	38.30	−1.52	0.100
40	11.96	15.32	−0.36	0.025
41	35.36	34.09	0.06	0.078
42	20.66	20.75	−0.01	0.024
43	26.50	15.94	0.85	0.023
44	3.38	12.19	−1.63	0.041
45	8.59	24.23	−1.36	0.034
46	15.12	14.14	0.10	0.029
47	10.59	20.80	−0.93	0.024
Mean	19.19	19.49	−0.22	0.043

```
     -+---------+---------+---------+---------+---------+---------+---
      I                                                              I
75.   I                                                              I
      I                                                              I
      I                               o                              I
      I                                                              I
      I                                                              I
      I                                            o              f  I
50.   I                                                              I
      I                o                                             I
      I            o                o                                I
      I                          o           o      f               I
      I                                         f:                   I
      I              :                 o    f f  :               o   I
25.   I                  o         f::                                I
      I                  o      f:::                                  I
      I       o       :   ::   :   : o            o                   I
      I     :f:  :  :f:  :    o    o o            o      o            I
      I    :fooo    o         o    o   o                              I
      I     oo o o     oo   o      o                                  I
 0.   I                                                              I
     -+---------+---------+---------+---------+---------+---------+---
      0.        12.       24.       36.       48.       60.       72.
                  mar   v.  feb using symbol o
                  FitVals   v. feb using symbol f
```

the standard errors of $\hat{\beta}_0$, $\hat{\beta}_1$ being ± 0.180 and $\pm 0.005\,96$, respectively. The output in Figure 6.6 also shows the fitted values given by the generalised linear model, denoted by the symbol 'f'; the observed monthly flows for March are also plotted against the observed monthly flows for February, using the symbol $\bigcirc$. Clearly the fit given by the generalised linear model is reasonable, and the fact that dispersion about the fitted model is not constant, but increases with February flow (recall that the variance function is $V(\mu_t) = \phi \mu_t^2$ for the gamma model, with $\mu_t = \exp(\hat{\beta}_0 + \hat{\beta}_1 q_{feb}))$, eliminates the incidence of negative flows in the generated flow sequences.

6.3.2   Estimation of the Dispersion Parameter κ

The model for q_{mar} in terms of q_{feb} can be written

$$f(q_{mar}) = \Gamma(\kappa)^{-1}(\kappa/\hat{\mu})^{\kappa} q_{mar}^{\kappa-1} \exp(-\kappa q_{mar}/\hat{\mu}) \tag{6.20}$$

where $\hat{\mu} = \exp(2.268 + 0.024\,30 q_{feb})$. To generate the value of q_{mar} from this distribution, given q_{feb}, we require an estimate of the dispersion parameter κ to specify the distribution completely, and a method for generating a pseudo-random variable with the distribution given by equation (6.20). We now deal with the first of these problems.

There are various procedures leading to estimates of κ. The simplest is given by McCullagh and Nelder (1991) as

$$1/\hat{\kappa} = \sum[(q_{mar} - \hat{\mu})/\hat{\mu}]^2/(n - p) \tag{6.21}$$

where p is the number of parameters fitted in the linear predictor, in the present case 2. Substituting the values $\hat{\mu} = \exp(2.268 + 0.024\,30 q_{feb})$ in this relation, we obtain the estimate of κ as $\hat{\kappa} = 2.014\,346\ldots$.

This estimate is an approximation to the maximum-likelihood estimate. The maximum-likelihood estimate itself can be found, straightforwardly, by the following procedure, which is essentially a conditional-likelihood method, already used in earlier chapters:

(i)   Assume a series of trial values for κ: say, 0.5, 1.0, 1.5, 2.0, ...

(ii) Using the values of $\hat{\beta}_0, \hat{\beta}_1, \ldots, \hat{\beta}_p$ already calculated by maximum likelihood, calculate the log-likelihood as:

$$\ln L_{Gamma} = n\kappa \ln(\kappa) - n \ln(\Gamma(\kappa)) - \kappa \sum \ln(\hat{\mu}_t) - \kappa \sum(y_t/\hat{\mu}_t) + (\kappa - 1)\sum \ln(y_t) \tag{6.22}$$

where $\hat{\mu}_t = \exp(\hat{\beta}_0 + \hat{\beta}_1 x_{1t} + \cdots + \hat{\beta}_p x_{pt})$. In the case of the example, $p = 1$, y_t is the monthly runoff for March in the tth year of the sequence, and x_{1t} is the monthly runoff for February in the tth year of record.

(iii) Identify the value of κ for which the log-likelihood is a maximum.

Applying this procedure to values of κ from 0.5 to 5.0 in steps of 0.5, we easily find the following:

κ	$\ln L_{\text{Gamma}}$	κ	$\ln L_{\text{Gamma}}$
0.5	-197.3553	2.42	-173.6894
1.0	-182.6113	2.44	-173.6807
1.5	-176.7953	2.46	-173.6739
2.0	-174.3476	2.48	-173.6688
2.5	-173.6655	2.50	-173.6655
3.0	-174.0692	2.52	-173.6638
3.5	-175.2079	2.54	-173.6638
4.0	-176.8771	2.56	-173.6655
4.5	-178.9471	2.58	-173.6687
5.0	-181.3305	2.60	-173.6736

so that the maximum-likelihood estimate is in the region of 2.53. We can continue to explore over a finer grid, or we can use the Nelder–Mead algorithm to maximise the log-likelihood surface (remembering that this algorithm gives the minimum of a function, so to maximise $\ln L_{\text{Gamma}}$ we need to minimise $-\ln L_{\text{Gamma}}$).

We have thus obtained a complete specification of the generalised linear model fitting the data for the month of March. It is, of course, necessary to model all 12 months. There is no reason why the model should necessarily be a generalised linear model for every month, although the dispersion characteristics — increased dispersion for larger monthly flows — suggest that generalised linear models are more appropriate than a simple Thomas–Fiering model for these particular data. It is also of interest, for the March monthly flows, to calculate the log-likelihood under the two alternative models. For the Thomas–Fiering model, the log likelihood is

$$\ln L_{\text{TF}} = -47 \ln \hat{\sigma} - (47/2) \ln(2\pi) - RSS/(2\hat{\sigma}^2)$$

with value $-187.476\,624$. For the gamma model, the log-likelihood is given by expression (6.22) with $\hat{\mu}_t = \exp(2.268 + 0.024\,30 q_{\text{feb},t})$ and summation is over the 47 data values. The value of $\ln L_{\text{Gamma}}$ is found to be in the region of -173, greater than the value of $\ln L_{\text{TF}}$ but not considerably so. The gamma model is therefore to be preferred, for these particular data, over the Thomas–Fiering model.

6.3.3 Generation of Gamma-distributed Pseudo-random Variates

Having fitted a modified Thomas–Fiering model with gamma-distributed response variate, in most water resource simulation studies it will be necessary to generate artificial sequences of monthly runoff, and for this purpose it will be necessary to use a algorithm for generating gamma-distributed pseudo-random variates.

A universally applicable method for generating pseudo-random variables, from any probability distribution $f(x; \theta)$, is the following:

(i) Tabulate the cumulative distribution function

$$F(x; \theta) = \int_{-\infty}^{x} f(u; \theta)\,du$$

for a series of x-values, $x = -N\Delta x, -(N-1)\Delta x, \ldots, (M-1)\Delta x, M\Delta x$, with M and N large and positive, Δx small.

(ii)   Generate a pseudo-random variable y, uniformly distributed on the interval $[0, 1]$.

(iii) Calculate the x which satisfies $y = F(x; \theta)$: that is, calculate $x = F^{-1}(y; \theta)$.

This procedure may be used to generate pseudo-random variables from the gamma distribution (6.20). However, more efficient methods are available, particularly if the shape variable κ is integral or nearly so; for with κ integral, the gamma distribution $f(x) = x^{\kappa-1}\exp(-x)/\Gamma(\kappa)$ is simply the sum of κ exponentially distributed random variables. If U is a pseudo-random variable with a uniform distribution on $[0, 1]$, then $V = -\ln U$ has the exponential distribution $\exp(-x)$; and if $W = U_1 + U_2 + \cdots + U_\kappa$ with $U_1, U_2, \ldots, U_\kappa$ uniformly distributed, then W has the distribution $f(x) = x^{\kappa-1}\exp(-x)/\Gamma(\kappa)$. Change of variable to $T = \kappa W/\mu$ then gives a variable T having the distribution (6.20). This is the method used by GENSTAT, in the routine GRGAMMA.

Where the shape parameter κ in expression (6.20) is not integral or nearly so, a more complicated procedure is required (although it is still more efficient than the general procedure described above) as follows:

(i)   Compute k, the largest integer in κ.

(ii) Compute $\gamma = \kappa - k$.

(iii) Generate k variates from the exponential distribution and compute their sum V (that is, generate $U_1, U_2, \ldots, U_k$ with a uniform distribution on $[0, 1]$, transform to $V_i = -\ln U_i$, and compute $V = V_1 + V_2 + \cdots + V_k$).

(iv) Generate Y from a beta distribution $B(\gamma, 1 - \gamma)$, where

$$B(a, b) = \Gamma(a + b)x^{a-1}(1 - x)^{b-1}/[\Gamma(a)\Gamma(b)] \qquad (0 \le x \le 1)$$

and Z from the exponential distribution $\exp(-x)$ (that is, $Z = -\ln U$, U being uniform on $[0, 1]$).

(v)   Compute $X = \beta(V + YZ)$. Then X has the gamma distribution in the form $x^{\kappa-1}\exp(-x/\beta)/\{\Gamma(\kappa)\beta^\kappa\}$.

The problem then becomes one of generating pseudo-random variables from the beta distribution $B(a, b)$. This can be done in a number of ways, their computational efficiency varying in different regions of the parameter space, and the reader is advised to take specialist advice on this matter. We describe one method only, of the several that exist.

First, we consider the case where we need to generate pseudo-random variables from $B(a, b)$, where a and b are integers (GENSTAT routine, GRBETA, is appropriate for this case). The procedure is as follows:

(i)   Generate $U_1, U_2, \ldots, U_{a+b}$ uniformly distributed on $[0, 1]$.

(ii) Form $X_1 = -\ln(\Pi_{j=1}^a U_j)$ and $X_2 = -\ln(\Pi_{j=a+1}^b U_j)$.

(iii) Form $X = X_1/(X_1 + X_2)$; then X is distributed as $B(a, b)$.

We can now use this procedure to generate pseudo-random variables from $B(a, b)$, where a and b are non-integral, as follows.

(i)   Compute k_1 and k_2, the largest integers in a and b, respectively;

(ii)   Compute $\gamma_1 = a - k_1$ and $\gamma_2 = b - k_2$.

(iii) Generate k_1 variables having distribution $\exp(-x)$, and calculate their sum X_1'.

(iv)   Generate k_2 variables having distribution $\exp(-x)$ and calculate their sum X_2'.

(v)   Generate four random variables Y_1, Y_2, Z_1 and Z_2 from the distributions $B(\gamma_1, 1 - \gamma_1)$, $B(\gamma_2, 1 - \gamma_2)$, $\exp(-x)$, and $\exp(-x)$, respectively.

(vi)   Compute $X = (X_1' + Y_1 Z_1)/(X_1' + X_2' + Y_1 Z_1 + Y_2 Z_2)$; then X has the distribution $B(a, b)$, with a, b non-integral.

Thus, generation of pseudo-random variables having the gamma probability distribution (6.20) is achieved (in the case of κ non-integral) through the generation of pseudo-random variables from the uniform distribution, which are transformed to pseudo-random variables having exponential and beta distributions.

6.4   OTHER FORMS OF GENERALISED LINEAR MODEL

So far we have considered only those cases where the response variable Y, given the explanatory variable x, has a binomial distribution (Section 6.2) or a gamma distribution (Section 6.3). The normal distribution is also a particular case, in which case the likelihood structure used in the fitting of generalised linear models reduces to familiar least squares, the deviance reducing to the familiar residual sum of squares. Two other distributions within the exponential family having particular hydrological application are the Poisson and the inverse normal.

Within GENSTAT, and some other software packages, the permissible distributions are the normal, binomial, Poisson, gamma and inverse normal; GLIM also includes them except for the inverse normal. In GENSTAT, the permissible link functions — apart from the logit and log functions used in Sections 6.2 and 6.3 — are the reciprocal, $\eta = 1/\mu$; the power function, $\eta = \mu^\alpha$, where the index α can be calculated by the software; the square root, $\eta = \mu^{1/2}$; and two link functions widely used in biological assay, the probit and complementary log-log. These are all standard models; a GENSTAT library procedure, GLM, also permits the user to specify his own distribution of the response variable, and his own link function. A wide range of generalised linear models can therefore be accommodated.

As further examples of the application of generalised linear models, we now present two cases where the response variable is assumed to have a Poisson distribution (Section 6.4.1); and where a normal model with fixed dispersion parameter ϕ is fitted together with a gamma model (Section 6.4.2). The latter example develops an alternative means of modelling the heterogeneous variance encountered in the application of the Thomas–Fiering model to monthly flow sequences from the Rio Lava Tudo at Fazenda Mineira.

6.4.1   A Generalised Linear Model with the Response Variable Poisson-distributed: Occurrence of Peak Flows on Rio Itajaí-Açú at Apiuna

Figure 6.7 begins with the numbers of 'peak flows' (more precisely, the numbers of mean daily discharges q_t satisfying the conditions $q_{t-1} < q_t < q_{t+1}$, with $q_t \geq 100 \ \mathrm{m^3 \ s^{-1}}$) in each of 36 consecutive months on the Rio Itajaí-Açú at Apiuna. The observed number of peak flows in each month varies between zero and six in this period of record; for the purposes of this example, we ignore the fact that the months have different numbers of days. If peak flows occur in time according to a Poisson process (such that the probability

Peak_No	3	6	2	5	2	0	2	2	2
Month	0	1	2	3	4	5	6	7	8
Peak_No	3	0	2	0	1	4	0	0	1
Month	9	10	11	0	1	2	3	4	5
Peak_No	1	2	3	3	3	3	4	0	0
Month	6	7	8	9	10	11	0	1	2
Peak_No	0	1	4	4	3	3	2	3	2
Month	3	4	5	6	7	8	9	10	11

```
16    MODEL [DISTRIBUTION=poisson;DISPERSION=*]Peak_No
17    TERMS cs,sn
18    FIT [PRINT=model,summary,estimates,correlations,fittedvalues,\
19              monitoring] cs
19 ...........................................................................
```

***** Regression Analysis *****

Response variate: Peak_No
    Distribution: Poisson
  Link function: Log
   Fitted terms: Constant, cs

*** Summary of analysis ***

	d.f.	deviance	mean deviance
Regression	1	0.13	0.131
Residual	34	52.53	1.545
Total	35	52.66	1.505
Change	−1	−0.13	0.131

*** Estimates of regression coefficients ***

	estimate	s.e.	t
Constant	0.746	0.143	5.23
cs	0.059	0.202	0.29

*** Correlations ***

Constant	1	1.000	
cs	2	−0.042	1.000
		1	2

*** Fitted values and residuals ***

Unit	Response	Fitted value	Standardized residual	Leverage
1	3.00	2.24	0.41	0.085
2	6.00	2.22	1.74	0.070
3	2.00	2.17	−0.10	0.041
4	5.00	2.11	1.38	0.028
5	2.00	2.05	−0.03	0.042
6	0.00	2.00	−1.67	0.069
7	2.00	1.99	0.01	0.082

Figure 6.7 Fitting a Poisson model to numbers of peak flows, per month, greater than 100 m^3 s^{-1}, Rio Itajaí-Açú at Apiuna, 1934–36. Explanatory variables are harmonic functions $\cos(2\pi t/12)$ and sin $(2\pi t/12)$, where t is the number of the month ($t = 0, 1, \ldots, 11, 0, \ldots, 11, 0, \ldots, 11$).

Continued

Figure 6.7 *(continued)*

8	2.00	2.00	0.00	0.069
9	2.00	2.05	−0.03	0.042
10	3.00	2.11	0.47	0.028
11	0.00	2.17	−1.71	0.041
12	2.00	2.22	−0.12	0.070
13	0.00	2.24	−1.78	0.085
14	1.00	2.22	−0.77	0.070
15	4.00	2.17	0.91	0.041
16	0.00	2.11	−1.68	0.028
17	0.00	2.05	−1.66	0.042
18	1.00	2.00	−0.66	0.069
19	1.00	1.99	−0.65	0.082
20	2.00	2.00	0.00	0.069
21	3.00	2.05	0.51	0.042
22	3.00	2.11	0.47	0.028
23	3.00	2.17	0.44	0.041
24	3.00	2.22	0.41	0.070
25	4.00	2.24	0.89	0.085
26	0.00	2.22	−1.76	0.070
27	0.00	2.17	−1.71	0.041
28	0.00	2.11	−1.68	0.028
29	1.00	2.05	−0.67	0.042
30	4.00	2.00	1.03	0.069
31	4.00	1.99	1.05	0.082
32	3.00	2.00	0.55	0.069
33	3.00	2.05	0.51	0.042
34	2.00	2.11	−0.06	0.028
35	3.00	2.17	0.44	0.041
36	2.00	2.22	−0.12	0.070
Mean	2.11	2.11	−0.16	0.056

```
  20 ADD sn
20................................................................
```

***** Regression Analysis *****

Response variate:	Peak_No
Distribution:	Poisson
Link function:	Log
Fitted terms:	Constant, cs, sn

*** Summary of analysis ***

	d.f.	deviance	mean deviance
Regression	2	1.96	0.982
Residual	33	50.70	1.536
Total	35	52.66	1.505
Change	−1	−1.83	1.832

*** Estimates of regression coefficients ***

	estimate	s.e.	t
Constant	0.734	0.144	5.10
cs	0.059	0.202	0.29
sn	−0.221	0.203	−1.09

of a peak occurring in any short time interval δt is $\lambda \, \delta t$) then this affords one means by which we can derive the probabilities of extreme floods, using the 'peaks over a threshold' approach — see, for example NERC (1975) for details and references. In this example, the threshold has been set for illustrative purposes at the rather low value of 100 m^3 s^{-1}; in practice, a higher threshold would be used, such that a small number n_t of peaks exceed the threshold in year t. Given the parameter λ, the number N of peak flows occurring in any period of length T (say, 30 days) is a Poisson variable with $P[N = n] = (\lambda T)^n \exp(-\lambda T)/n!$ and mean value λT.

This simplest form of the Poisson model for the occurrence of peak flows assumes that the rate parameter, λ, is constant in time. In practice, peak flows may occur more frequently at certain seasons, and in this section we explore the possibility that the mean value λT varies seasonally. A simple way to explore whether seasonality exists is to explore whether the coefficients β_1, β_2 of harmonic terms $\cos(2\pi t/12)$, $\sin(2\pi t/12)$ in the relation $\lambda T = \beta_0 + \beta_1 \cos(2\pi t/12) + \beta_2 \sin(2\pi t/12)$
are statistically significant; more complex kinds of seasonal variability can obviously be introduced by inclusion of terms such as $\cos(4\pi t/12)$, $\sin(4\pi t/12), \ldots$ in the expression for λT. However, this simple model has the disadvantage that it takes no account of the fact that the number of events occurring in the interval $[0, T)$ must be non-negative; we can satisfy this requirement by casting the problem in the following, generalised linear model, form:

(i)   The number of peaks n_t occurring in month t is a Poisson variable with mean λT.

(ii)   The linear predictor is $\eta = \beta_0 + \beta_1 \cos(2\pi t/12) + \beta_2 \sin(2\pi t/12)$.

(iii) The link function is the logarithm defined by $\ln(\lambda T) = \eta$.

Fitting the model thus formulated is straightforward using GENSTAT, with the output as shown in Figure 6.7. The variate n_t is denoted by *Peak_No*, and the cosine and sine terms by *cs, sn*, respectively. The output shows first the result of fitting the cosine term alone, and then the result of adding in the sine term also. Fitting the cosine term alone gives a very small reduction in deviance of 0.13; when both terms are included, the deviance is reduced, again slightly, by 1.83. Whether we use an approximate F-test based upon the mean deviances shown in the output, or whether we compare the coefficients β_1, β_2 with their standard errors ($\hat{\beta}_1 = 0.059 \pm 0.202$; $\hat{\beta}_2 = -0.221 \pm 0.203$), we see no evidence of seasonality in the occurrence of peak flows in excess of 100 m^3 s^{-1}. Thus, the data show that we have no evidence on which to reject the simple model $\lambda = $ constant for the occurrence of peak flows in excess of 100 m^3 s^{-1} on the Rio Itajaí-Açú at Apiuna.

6.4.2 An Alternative Method for Modelling Heterogeneous Variance in Linear Regression: Monthly Flows from Rio Lava Tudo at Fazenda Mineira

In Section 6.3.1 we discussed a modification to the standard Thomas–Fiering model in which monthly flow was assumed to be gamma-distributed. Because the variance function $V(\mu_t)$ in the gamma model is of the form $V(\mu_t) = \phi \mu_t^2$, the model allows for the increased dispersion about expected values as the explanatory variable increases. The gamma model is not, however, the only way in which heterogeneity of dispersion may be dealt with; one alternative would be to retain a linear model relating (say) March flow to February flow, of the form $q_{\text{mar}} = \beta_0 + \beta_1 q_{\text{feb}} + \varepsilon$, and to include a description of the increased dispersion for large values of the explanatory variable, in this case q_{feb}. Thus one possibility, with obvious notation, is

$$E[Q_{mar}] = \beta_0 + \beta_1 q_{feb}$$

for which

$$q_{mar} = \beta_0 + \beta_1 q_{feb} + \varepsilon$$

with

$$\sigma_\varepsilon^2 = \exp[\lambda_0 + \lambda_1 q_{feb}]$$

where the four parameters, $\beta_0, \beta_1, \lambda_0, \lambda_1$ are now to be fitted. The exponential form is adopted for σ_ε^2 because it ensures that the variance is always positive, but other options are also possible. As with the classical form of the Thomas–Fiering model, further explanatory variables (possibly q_{jan}, for example) could be included either in the linear model for q_{mar}, or in the model for the error variances.

It would be possible to write a computer code for the log-likelihood function, and to determine values of $\beta_0, \beta_1, \lambda_0, \lambda_1$ at which this function is maximised: either by Newton–Raphson solution of the four equations $\partial \ln L / \partial \beta_0 = \partial \ln L / \partial \beta_1 = \partial \ln L / \partial \lambda_0 = \partial \ln L / \partial \lambda_1 = 0$; or by iterative search over the log-likelihood surface for the maximum point; or by some other means. However, following the method given by Aitkin (1987), the joint estimation of the four parameters is much more straightforward. Aitkin proposes the following, iterative procedure, which he shows to be equal to the maximisation of the log-likelihood surface:

(i)   Calculate initially an unweighted, normal regression of q_{mar} on q_{feb}.

(ii) Obtain the simple residuals, q_{mar} (observed)$-q_{mar}$ (fitted), from this regression, and square them to give a new variable, denoted by, say, *ResSqd*. This new variable has a gamma distribution with scale parameter $\phi = 2$.

(iii) Fit the linear predictor $\eta = \lambda_0 + \lambda_1 q_{feb}$ using a log link function: obtain the corresponding deviance.

(iv) Now fit a normal, but weighted, regression of q_{mar} on q_{feb}, with scale parameter $\phi = 1$, the weights being given by the reciprocals of the fitted values from the gamma model.

(v)   The squared residuals are again calculated, and are fitted using the gamma model, and the normal and gamma models are alternated until the deviance converges.

Figure 6.8 shows the output from a small GENSTAT program which performs this iterative calculation. We see that the residual deviance in the gamma model has the following values at the iteration stated:

Iteration	Residual deviance
1	65.07
2	69.15
3	69.75
4	69.77
5	69.77

"First initialise the calculation, using a simple, unweighted linear regression of qmar on qfeb. We specify that the distribution of qmar is Normal, and that the link function is the identity function. However these options could have been omitted, as they are the default values when the directive MODEL is called."

```
    27     ''
   -28     ''
    29     MODEL [DISTRIBUTION=normal;LINK=identity] mar;FITTEDVALUES=FvalsN
    30     FIT[PRINT=model,summary,estimates,correlations,\
    31     monitoring]feb
31 .........................................................................
```

***** Regression Analysis *****

Response variate: mar
    Fitted terms: Constant, feb

*** Summary of analysis ***

	d.f.	s.s.	m.s.
Regression	1	2231.	2231.0
Residual	45	8015.	178.1
Total	46	10246.	222.7

Percentage variance accounted for 20.0

* MESSAGE: The following units have large residuals:
                         29   3.50

* MESSAGE: The following units have high leverage:
                         24   0.199
                         39   0.100

*** Estimates of regression coefficients ***

	estimate	s.e.	t
Constant	8.42	3.61	2.33
feb	0.424	0.120	3.54

*** Correlations ***

Constant	1	1.000	
feb	2	−0.842	1.000
		1	2

```
    32     ''
   -33     ''
    34     "Now we calculate the ordinary residuals, qmar(obs)-qmar(fitted),
              and their
   -35     squares which are put into the array ResSqd. The squared residuals
   -36     are later modelled using a log link function."
    37     ''
   -38     ''
    39     CALCULATE Res=mar-FvalsN        :    &      ResSqd=Res*Res
    40              &    Dev=SUM(ResSqd)
    41     ''
   -42     ''
```

Figure 6.8 Modified Thomas–Fiering model with heterogeneous variance: fitting for one pair of months only: $q_{mar} = b_0 + b_1 q_{feb} + e$, where the variance of the residual e is taken to be of the form $\text{var}(e) = \exp(l_0 + l_1 q_{feb})$, l_0 and l_1 being constants to be estimated (in addition to b_0 and b_1).

_____ Continued ___

_____ Figure 6.8 *(continued)* _____

```
 43  ''
-44  ''
 45  ''Now start the iterative calculation between
-46       (a) modelling of the squared residuals, ultimately to obtain
-47            l0 and l1 in the variance model for for (e);
-48       (b) modelling of the regression of qmar on qfeb, using
-49            a scale factor of one, and weights equal to the
-50            inverses of the fitted values obtained instage (a).''
 51  ''
-52  ''
 53  ''
-54  ''
 55  FOR i=1...5
 56  ''
-57  ''
 58  MODEL [DISTRIBUTION=gamma;LINK=log;DISPERSION=2]ResSqd;FITTEDVALUES=\
 59        FValsG
 60  FIT [PRINT=model,summary,estimates,correlations,monitoring] feb
 61  CALCULATE   wts=1/FValsG
 62  ''
-63  ''
 64  MODEL [DISTRIBUTION=normal;LINK=identity;DISPERSION=1;\
 65        WEIGHTS=wts] mar; FITTEDVALUES=FvalsN
 66  FIT [PRINT=model,summary,estimates,correlations,monitoring] feb
 67  ''
-68  ''
 69  CALCULATE Res=mar-FvalsN          :        & ResSqd=Res*Res
 70  ''
-71  ''
 72  ENDFOR
72...........................................................................
```

* * * * * Regression Analysis * * * * *

Response variate:	ResSqd
Distribution:	Gamma
Link function:	Log
Fitted terms:	Constant, feb

* * * Summary of analysis * * *
     Dispersion parameter is 2.00

	d.f.	deviance	mean deviance
Regression	1	*	*
Residual	45	65.07	1.446
Total	46	*	*

* MESSAGE: The following units have large residuals:

6	−2.45
34	2.47
46	−3.13

* MESSAGE: The following units have high leverage:

24	0.199
39	0.100

Figure 6.8 *(continued)*

*** Estimates of regression coefficients ***

	estimate	s.e.	t
Constant	4.043	0.383	10.57
feb	0.0371	0.0127	2.93

* MESSAGE: s.e.s are based on dispersion parameter with value 2.00

*** Correlations ***

Constant	1	1.000	
feb	2	−0.842	1.000
		1	2

72 ....

***** Regression Analysis *****

Response variate: mar
   Weight variate: wts
     Fitted terms: Constant, feb

*** Summary of analysis ***
   Dispersion parameter is 1

	d.f.	s.s.	m.s.
Regression	1	15.14	15.138
Residual	45	46.64	1.036
Total	46	61.78	1.343

Percentage variance accounted for 22.8

* MESSAGE: The following units have large residuals:

19	2.50
29	2.69
34	3.14

*** Estimates of regression coefficients ***

	estimate	s.e.	t
Constant	6.86	2.77	2.48
feb	0.496	0.128	3.89

* MESSAGE: s.e.s are based on dispersion parameter with value 1

*** Correlations ***

Constant	1	1.000	
feb	2	−0.808	1.000
		1	2

72 ....

***** Regression Analysis *****

Response variate: ResSqd
    Distribution: Gamma
   Link function: Log
     Fitted terms: Constant, feb

*** Summary of analysis ***
   Dispersion parameter is 2.00

___ Figure 6.8 *(continued)* ___

	d.f.	deviance	mean deviance
Regression	1	*	*
Residual	45	69.15	1.537
Total	46	*	*

* MESSAGE: The following units have large residuals:

27	−2.61
34	2.60
46	−2.29

* MESSAGE: The following units have high leverage:

24	0.199
39	0.100

* * * Estimates of regression coefficients * * *

	estimate	s.e.	t
Constant	3.995	0.383	10.44
feb	0.0387	0.0127	3.05

* MESSAGE: s.e.s are based on dispersion parameter with value 2.00

* * * Correlations * * *

Constant	1	1.000	
feb	2	−0.842	1.000
		1	2

72 ....

* * * * * Regression Analysis * * * * *

Response variate:	mar
Weight variate:	wts
Fitted terms:	Constant, feb

* * * Summary of analysis * * *
    Dispersion parameter is 1

	d.f.	s.s.	m.s.
Regression	1	15.25	15.253
Residual	45	47.00	1.044
Total	46	62.25	1.353

Percentage variance accounted for 22.8

* MESSAGE: The following units have large residuals:

19	2.52
29	2.66
34	3.18

* MESSAGE: The following units have high leverage:

26	0.099

* * * Estimates of regression coefficients * * *

	estimate	s.e.	t
Constant	6.82	2.73	2.50
feb	0.499	0.128	3.91

* MESSAGE: s.e.s are based on dispersion parameter with value 1

—— Figure 6.8 *(continued)* ——

∗ ∗ ∗ Correlations ∗ ∗ ∗

Constant   1   1.000
feb   2   −0.807   1.000
                        1   2
72 ....

∗ ∗ ∗ ∗ ∗ Regression Analysis ∗ ∗ ∗ ∗ ∗

Response variate: ResSqd
     Distribution: Gamma
    Link function: Log
     Fitted terms: Constant, feb

∗ ∗ ∗ Summary of analysis ∗ ∗ ∗
     Dispersion parameter is 2.00

	d.f.	deviance	mean deviance
Regression	1	∗	∗
Residual	45	69.75	1.550
Total	46	∗	∗

∗ MESSAGE: The following units have large residuals:
                        27   −2.68
                        34   2.60

∗ MESSAGE: The following units have high leverage:
                        24   0.199
                        39   0.100

∗ ∗ ∗ Estimates of regression coefficients ∗ ∗ ∗

	estimate	s.e.	t
Constant	3.994	0.383	10.44
feb	0.0387	0.0127	3.05

∗ MESSAGE: s.e.s are based on dispersion parameter with value 2.00

∗ ∗ ∗ Correlations ∗ ∗ ∗

Constant   1   1.000
feb   2   −0.842   1.000
                        1   2
72 ....

∗ ∗ ∗ ∗ ∗ Regression Analysis ∗ ∗ ∗ ∗ ∗

Response variate: mar
   Weight variate: wts
     Fitted terms: Constant, feb

∗ ∗ ∗ Summary of analysis ∗ ∗ ∗
     Dispersion parameter is 1

	d.f.	s.s.	m.s.
Regression	1	15.25	15.253
Residual	45	47.00	1.044
Total	46	62.25	1.353

Percentage variance accounted for 22.8

Figure 6.8 (continued)

```
* MESSAGE: The following units have large residuals:
                              19              2.52
                              29              2.66
                              34              3.18

* MESSAGE: The following units have high leverage:
                              26              0.099
```

*** Estimates of regression coefficients ***

	estimate	s.e.	t
Constant	6.82	2.73	2.50
feb	0.499	0.128	3.91

* MESSAGE: s.e.s are based on dispersion parameter with value 1

*** Correlations ***

Constant	1	1.000	
feb	2	−0.807	1.000
		1	2

72 ....

***** Regression Analysis *****

Response variate: ResSqd
      Distribution: Gamma
      Link function: Log
      Fitted terms: Constant, feb

*** Summary of analysis ***
      Dispersion parameter is 2.00

	d.f.	deviance	mean deviance
Regression	1	*	*
Residual	45	69.77	1.550
Total	46	*	*

```
* MESSAGE: The following units have large residuals:
                              27             −2.68
                              34              2.60

* MESSAGE: The following units have high leverage:
                              24              0.199
                              39              0.100
```

*** Estimates of regression coefficients ***

	estimate	s.e.	t
Constant	3.994	0.383	10.44
feb	0.0387	0.0127	3.05

* MESSAGE: s.e.s are based on dispersion parameter with value 2.00

*** Correlations ***

Constant	1	1.000	
feb	2	−0.842	1.000
		1	2

72 ....

___ Figure 6.8 *(continued)* _____

```
***** Regression Analysis *****

Response variate:    mar
    Weight variate:  wts
      Fitted terms:  Constant, feb

*** Summary of analysis ***
    Dispersion parameter is 1
```

	d.f.	s.s.	m.s.
Regression	1	15.25	15.253
Residual	45	47.00	1.044
Total	46	62.25	1.353

Percentage variance accounted for 22.8

```
* MESSAGE: The following units have large residuals:
                        19              2.52
                        29              2.66
                        34              3.18

* MESSAGE: The following units have high leverage:
                        26              0.099

*** Estimates of regression coefficients ***
```

	estimate	s.e.	t
Constant	6.82	2.73	2.50
feb	0.499	0.128	3.91

```
* MESSAGE: s.e.s are based on dispersion parameter with value 1

*** Correlations ***

Constant     1        1.000
feb          2       −0.807          1.000
                        1               2
72 ..........................................................................
***** Regression Analysis *****

Response variate:    ResSqd
      Distribution:  Gamma
     Link function:  Log
      Fitted terms:  Constant, feb

*** Summary of analysis ***
    Dispersion parameter is 2.00
```

	d.f.	deviance	mean deviance
Regression	1	*	*
Residual	45	69.77	1.550
Total	46	*	*

```
* MESSAGE: The following units have large residuals:
                        27              −2.68
                        34              2.60

* MESSAGE: The following units have high leverage:
                        24              0.199
                        39              0.100
```

___ Figure 6.8 *(continued)* _____

∗∗∗ Estimates of regression coefficients ∗∗∗

	estimate	s.e.	t
ConstanE	3.994	0.383	10.44
feb	0.0387	0.0127	3.05

∗ MESSAGE: s.e.s are based on dispersion parameter with value 2.00

∗∗∗ Correlations ∗∗∗

Constant	1	1.000	
feb	2	−0.842	1.000
		1	2

72 ....

∗∗∗∗∗ Regression Analysis ∗∗∗∗∗

Response variate: mar
  Weight variate: wts
   Fitted terms: Constant, feb
∗∗∗ Summary of analysis ∗∗∗
   Dispersion parameter is 1

	d.f.	s.s.	m.s.
Regression	1	15.25	15.253
Residual	45	47.00	1.044
Total	46	62.25	1.353

Percentage variance accounted for 22.8

∗ MESSAGE: The following units have large residuals:

19	2.52
29	2.66
34	3.18

∗ MESSAGE: The following units have high leverage:

26	0.099

∗∗∗ Estimates of regression coefficients ∗∗∗

	estimate	s.e.	t
Constant	6.82	2.73	2.50
feb	0.499	0.128	3.91

∗ MESSAGE: s.e.s are based on dispersion parameter with value 1

∗∗∗ Correlations ∗∗∗

Constant	1	1.000	
feb	2	−0.807	1.000
		1	2

and after the fifth iteration, the model for q_{mar} is

$$q_{mar} = 6.82 + 0.499 q_{feb}$$

$$(\pm 2.73) \quad (\pm 0.128)$$

and the variance model is

$$\sigma_\varepsilon^2 = \exp(3.994 + 0.0387 q_{\text{feb}})$$

$$(\pm 0.383) \quad (\pm 0.0127)$$

We see that all four coefficients $\hat{\beta}_0, \hat{\beta}_1, \hat{\lambda}_0, \hat{\lambda}_1$ are more than twice their standard errors; in particular, the value of $\hat{\lambda}_1 = 0.0387 \pm 0.0127$ demonstrates the necessity for modelling the heterogeneous variance.

6.5   DIAGNOSTIC METHODS FOR GENERALISED LINEAR MODELS

Chapter 4 showed how standardised residuals, deletion residuals, and leverages were used to check or 'diagnose' the suitability of the model as a descriptor of the data, when the model was linear. Similar methods are available for checking generalised linear models. We recall that the simple residual, $r_t = y_t - \hat{\mu}_t$, was standardised by dividing it by a factor that makes its variance for all t; these standardised residuals are of the form

$$(y_t - \hat{\mu}_t)/[\sqrt{(1 - h_t)}]$$

where the h_t are the diagonal elements of the 'hat' matrix $\mathbf{H} = \mathbf{X}(\mathbf{X}^{\mathrm{T}}\mathbf{X})^{-1}\mathbf{X}^{\mathrm{T}}$; these diagonal elements are measures of the remoteness of the tth point from the remaining points, in the space within which the explanatory variables vary. In Chapter 4, in fact, we also divided $(y_t - \hat{\mu}_t)/[\sqrt{(1 - h_t)}]$ by $\hat{\sigma}_\varepsilon$, to give standardised residuals not only with constant variance, but with unit variance too:

$$r_t' = (y_t - \hat{\mu}_t)/[\hat{\sigma}_\varepsilon \sqrt{(1 - h_t)}]$$

We also defined the deletion residuals, r_t^*, by

$$r_t^* = (y_t - \hat{\mu}_t)/[\hat{\sigma}_{\varepsilon(t)} \sqrt{(1 - h_t)}]$$

where $\hat{\sigma}_{\varepsilon(t)}$ is the estimate of σ_ε, the standard deviation of the errors ε in the linear model, when the tth data point is omitted from its calculation. We saw that the deletion residual r_t^* measures the deviation of y_t from the value predicted by the model fitted to the remaining points, standardised like r_t'; the two residuals r_t' and r_t^* being related by

$$r_t^* = r_t' \hat{\sigma}_\varepsilon / \hat{\sigma}_{\varepsilon(t)}$$

All of the above refers to the residuals from linear models. When we move from linear models to generalised linear models, each value y_t of the response variable contributes a quantity d_t to the deviance introduced in Section 6.1.1, so that $\Sigma d_t = D$. The *deviance residual*, r_D, is therefore defined by

$$r_{\mathrm{D},t} = \text{sign}(y_t - \hat{\mu}_t)\sqrt{(d_t)}$$

where we have added a suffix t to indicate that the left-hand side is the deviance residual for the tth observation. The contributions d_t will be different for each distribution in the exponential family; we recall that for the Poisson, binomial and gamma distributions the deviances are (see Section 6.1.1):

$$2\sum[y_t \ln(y_t/\mu_t) - (y_t - \mu_t)]$$

$$2\sum[y_t \ln(y_t/\mu_t) + (m_t - y_t)\ln((m_t - y_t)/(m_t - \mu_t))]$$

$$2\sum[(y_t - \mu_t)/\mu_t - \ln(y_t/\mu_t)]$$

respectively. Hence the deviance residuals are, respectively,

$$r_{D,t} = \text{sign}(y_t - \hat{\mu}_t)\sqrt{\{y_t \ln(y_t/\hat{\mu}_t) - (y_t - \hat{\mu}_t)\}}$$

$$r_{D,t} = \text{sign}(y_t - \hat{\mu}_t)\sqrt{\{y_t \ln(y_t/\hat{\mu}_t) + (m_t - y_t)\ln((m_t - y_t)/(m_t - \hat{\mu}_t))\}}$$

$$r_{D,t} = \text{sign}(y_t - \hat{\mu}_t)\sqrt{\{(y_t - \hat{\mu}_t)/\hat{\mu}_t - \ln(y_t/\hat{\mu}_t)\}}$$

for these three distributions.

From the deviance residuals, *standardised deviance residuals* can be calculated from

$$r'_{D,t} = r_{D,t}/\sqrt{[\hat{\phi}(1 - h_t)]}$$

where $\hat{\phi}$ is the estimate of the dispersion parameter, and h_t is the diagonal term in a modified 'hat' matrix $\mathbf{H}$, now generalised to

$$\mathbf{H} = \mathbf{W}^{1/2}\mathbf{X}(\mathbf{X}^\mathrm{T}\mathbf{W}\mathbf{X})^{-1}\mathbf{X}^\mathrm{T}\mathbf{W}^{1/2}$$

where $\mathbf{W}$ is the $N \times N$ diagonal matrix whose tth element is defined by

$$W_t = 1/[(\mathrm{d}\eta_t/\mathrm{d}\mu_t)^2 V(\mu_t)]$$

where $\eta(\mu)$ is the linear predictor defined in equation (6.4), and $V(\mu_t)$ is the variance function for the relevant exponential-family distribution ($V(\mu) = 1$ for the normal distribution; $V(\mu) = \mu$ for the Poisson; $V(\mu) = \mu(1 - \mu)$ for the binomial; $V(\mu) = \mu^2$ for the gamma).

Fortunately, calculation of the leverages and standardised deviance residuals is a standard feature of GENSTAT and some other programs. However, the exact calculation of these quantities is time-consuming, and it has become common practice to approximate the quantities used in the standardised deviance residuals using one-step approximations instead of iterating through to convergence (McCullagh and Nelder, 1991; GENSTAT, 1992; Williams, 1987). Chapter 4 showed how the GENSTAT library procedure RCHECK could be used to plot residuals, leverages and Cook statistics, given by linear models, against index variables, or against fitted values, or as a normal or half-normal plot; RCHECK can also be used to plot generalised linear model residuals.

6.6 CONCLUDING REMARKS

The examples discussed in this chapter have barely demonstrated the wealth of statistical models included within the generalised linear model formulation. The link functions used have been those already available within the GENSTAT program, but it is possible for the user to extend this selection by incorporating his own link function. It is also possible, in theory, to define a link function relating the mean value μ_t of, say, a gamma-distributed daily flow at a gauging site, in terms of basin precipitation and climatological variables as explanatory variables, and to use the GENSTAT MODEL and FIT directives to explore model structure. We return to the subject of rainfall-runoff models in Chapter 8.

REFERENCES

Aitkin, M. (1987). Modelling variance heterogeneity in normal regression using GLIM. *Appl. Statist.*, **36**, 332–9.

Coe, R. and Stern, R. D. (1982). Fitting models to daily rainfall. *J. Appl. Meteorol.*, **21**, 1024–31.

GENSTAT 5 Committee, Rothamsted Experimental Station (1992) *GENSTAT 5 Reference Manual.* Clarendon Press, Oxford.

McCullagh, P. and Nelder, J. A. (1991). *Generalized Linear Models* (2nd edition). Chapman and Hall, London.

NERC (1975). *Flood Studies Report Vol. 1: Hydrological Studies.* Natural Environment Research Council, London.

Stern, R. D. and Coe, R. (1984). A model fitting analysis of daily rainfall data. *J. R. Statist. Soc. A*, **147**, 1–34.

Williams, D. A. (1987). Generalized linear model diagnostics using the deviance and single-case deletions. *Appl. Statist.*, **36**, 181–91.

EXERCISES AND EXTENSIONS

6.1. The data below are mean monthly flows (m^3 s^{-1}) on the River Komadougou at Gueskerou, Nigeria, for the period 1957–58 to 1976–77 (data taken from G. Vuillaume, 'Bilan hydrologique mensuel et modelisation sommaire du régime hydrologique du lac Tchad', *Cah. ORSTOM* ser. Hydrol., **XVIII**, 1 (1981), and used by kind permission of ORSTOM).

(a) Explore the use of a Markov chain model for the occurrence/non-occurrence of monthly flow, ignoring any time trend in the data, and fit logistic regressions with harmonic terms $\cos(2\pi t/12)$, $\sin(2\pi t/12)$.

(b) Explore the use of a gamma model to represent mean monthly discharge in months when flow occurred.

(c) The River Komadougou flows into Lake Chad, and it is believed that flows have diminished because of increased abstractions for irrigation, and because of impoundments in its headwaters. Using generalised linear models for the occurrence of monthly discharge, and for the quantity of mean monthly discharge, explore whether the data support this hypothesis.

Year	May	June	July	Aug	Sept	Oct	Nov	Dec	Jan	Feb	Mar	Apr
1957–58	0	3.52	17.4	22.4	26.0	28.9	30.8	32.9	35.7	33.7	9.95	2.14
1958–59	0.52	0.03	12.2	26.5	31.1	31.9	32.7	34.4	0	0	0	0
1960–61	0	0	5.89	19.0	27.0	0	30.3	31.1	0	0	0	0
1961–62	0	0	0	22.3	30.0	30.7	31.8	32.9	0	4.4	1.33	0.22
1962–63	0	0	7.92	23.1	28.8	30.4	32.0	34.9	37.3	18.6	3.34	1.25
1963–64	0.30	0	3.32	17.6	24.8	28.6	30.4	31.8	14.4	2.48	0.8	0.10
1964–65	0	0	6.34	21.4	29.2	31.0	32.6	37.1	39.7	21.7	4.08	1.41
1965–66	0.41	0.01	6.13	22.4	20.0	30.7	31.9	33.0	22.4	4.02	1.28	0.33
1966–67	0.01	1.23	17.3	24.1	25.9	27.7	29.5	30.8	27.3	6.09	1.69	0.42
1967–68	0.01	0	0	14.9	24.7	27.5	29.6	30.3	9.49	2.35	0.72	0.04
1968–69	0	2.54	17.5	24.7	27.2	28.8	29.9	12.4	1.95	0.64	0.24	0
1969–70	0	0	0	0	0	0	0	0	18.6	3.15	1.02	0.13
1970–71	0	0	0	11.1	25.7	30.4	33.7	35.6	27.8	4.54	1.46	0
1971–72	1.09	0	0	0	0	0	0	0	9.34	2.33	0.66	0.04
1972–73	0	0	2.89	15.6	23.9	27.0	26.3	4.13	0.74	0.03	0	0
1973–74	0	0	0	9.14	21.9	26.5	12.4	0	0	0	0	0
1974–75	0	0	0	0	0	0	0	0	2.19	0.58	0	0
1975–76	0	0	0	8.22	22.6	25.5	28.6	24.2	3.14	0.66	0.01	0
1976–77	0	0	0.83	15.8	23.3	27.4	29.3	8.85	1.01	0.03	0	0

6.2. The file X6.2DAT on the diskette distributed by The MathWorks gives daily rainfall data from a rain gauge with site code 02950020, in the basin of the upper Uruguay River. The period of record is 1941–79, and missing values are denoted by 777. Explore the use of a Stern–Coe model (logistic regression for rainfall occurrence; gamma distribution with time varying mean for rainfall depths) to represent these data. Consult the 1984 paper by Stern and Coe, to evaluate probabilities of dry spells of length 10, 20 days, beginning on days 10, 20, . . . , 360 throughout the year.

6.3. The data below are estimates of mean monthly rainfall (mm) falling on Lake Chad for the period May 1968 to April 1977 (taken from G. Vuillaume, 'Bilan hydrologique mensuel et modelisation sommaire du régime hydrologique du lac Tchad', *Cah. ORSTOM* ser. Hydrol., **XVIII**, 1 (1981), and used by kind permission of ORSTOM).

Model the rainfall for the period May to November by generalised linear model using a gamma distribution with time-dependent mean.

Year	May	June	July	Aug	Sept	Oct	Nov	Dec	Jan	Feb	Mar	Apr
1968	8	48	80	96	23	1	0	0	0	0	0	5
1969	12	31	35	76	31	8	0	0	0	0	0	3
1970	22	21	95	110	65	0	0	0	0	0	0	0
1971	3	0	58	109	68	0	0	0	0	0	0	0
1972	3	8	26	52	86	20	8	0	0	0	0	0
1973	0	2	17	25	89	24	0	0	0	0	0	0
1974	1	12	10	93	153	26	7	0	0	0	0	0
1975	0	1	0	91	166	31	0	0	0	0	0	0
1976	2	4	14	63	87	27	34	0	0	0	0	0

Chapter 7

Multivariate models

7.1 MODELS WITH MULTIPLE RESPONSE VARIABLES

Earlier chapters have discussed models with several explanatory variables (rainfall, evaporation) and *one* response variable. In some cases, the explanatory variables were *causative*, as in the case of rainfall causing streamflow; in other cases the explanatory variables were not directly causative, as where monthly streamflow q_t is represented in terms of harmonic functions of time, $q_t = \beta_0 + \beta_1 \cos(2\pi t/12) + \beta_2 \sin(2\pi t/12) + \varepsilon_t$; or where a relation was sought expressing, say, March streamflow at a gauging station in terms of February streamflow.

The present chapter discusses models in which there are *several* response variables which can be displayed as a vector. The simplest form for such a model is

$$\mathbf{y}_t = \boldsymbol{\mu} + \boldsymbol{\varepsilon}_t$$

where $\mathbf{y}_t$ is a vector of size $G \times 1$; $\boldsymbol{\mu}$ is a $G \times 1$ vector of parameters; and $\boldsymbol{\varepsilon}_t$ is a vector of random variables, also $G \times 1$, with $E[\boldsymbol{\varepsilon}_t] = \mathbf{0}$ and variance-covariance matrix $\boldsymbol{\Sigma}$. We shall have a number of observations of the vector $\mathbf{y}_t$, say $\{\mathbf{y}_t\}$ for $t = 1, \ldots, P$; but almost always, some of the observations of some of the variables in the vector $\mathbf{y}_t$ will be missing, because hydrological records begin at different times, because readings are made using fallible instruments, and because field operators are not always available to take readings. Typically, therefore, the data available for making inferences about the multivariate model $\mathbf{y}_t = \boldsymbol{\mu} + \boldsymbol{\varepsilon}_t$ (and about more sophisticated models in which the constant $\boldsymbol{\mu}$ is replaced by something more complicated) will look like the following array:

	Column							
	1	2	3	...	...	...	$P - 1$	P
Row 1	*	*	y_{13}	...	...	...	$y_{1, P-1}$	y_{1P}
Row 2	*	y_{22}	y_{23}	...	...	...	*	y_{2P}
...								
Row G	*	y_{G2}	y_{G3}	...	...	...	$y_{G,P-1}$	*

An example is the sequences of annual maximum floods at several gauging sites in adjacent drainage basins; commonly our interest will be concentrated on just one of the G sequences in the array, and we seek good estimates of the parameters describing this sequence. However, if the P columns are months, the G rows are gauging sites, and the elements of $\mathbf{y}_t$ are measures of monthly runoff, we might wish to use the data to provide a model capable of generating artificial sequences of monthly flow, for the purpose of simulating a water resource system. In such an application, we should be interested in estimating all the elements of the vector of systematic components $\boldsymbol{\mu}$, and of the variance-covariance matrix $\boldsymbol{\Sigma}$.

Usually, there will be no causative relation between the variables in the vector $\mathbf{y}_t$, so that any statistical model relating them will be solely descriptive and generally incapable of interpretation in physical terms. We have used the qualifier 'usually' because the data in two (or more) rows of the above array could, in some applications, come from gauging sites, one of which is upstream of the other(s); it is then conceivable that the event which resulted in, say, the annual flood at the upstream site produced a hydrograph which, when routed along the river channel, caused the annual flood at the downstream site(s). In this instance, the upstream annual flood may be causally related to the downstream annual flood; the discussion of the present chapter includes such cases, but in general we shall not consider cause and effect relationships amongst the variables in the vector $\mathbf{y}_t$.

The material which follows in this chapter consists of two parts. In the first, Section 7.2, we consider the case where we have several sequences of response variables, some with missing values: for example, we may have annual streamflow data from a number of gauging stations, with missing values in some of the records, so that some years are incomplete — a not uncommon feature of runoff records in areas difficult of access, with low population densities and generally unhealthy environments, such as the Amazon basin. A second example is the case where we have rainfall records at several sites, but with incomplete records at one or more sites. The objectives of the analysis in both these examples may be twofold. First, we may wish to complete the series by supplying estimates of the missing values, caused perhaps by gauge failure or observer absence. And second, we may wish to 'improve' estimates of parameters that summarise some of the more important characteristics of the sequences: perhaps the mean and variance of annual streamflow, in the first example. However, improved estimates of parameters are usually but a first step to obtaining an improved estimate of some function of them, such as the annual flood with T-year return period. This problem we also consider: in particular, we discuss the derivation of confidence limits for the flood with T-year return period at one site, utilising flood records at other sites.

The second part of this chapter, Section 7.3, deals with some aspects of the problem in which we need to simulate streamflow sequences at several sites. We restrict consideration to models at P sites, such that the random components $\varepsilon_t^{(1)}, \varepsilon_t^{(2)}, \ldots, \varepsilon_t^{(P)}$ are serially independent — that is, we assume that $\varepsilon_t^{(m)}, \varepsilon_{t+k}^{(n)}$ are statistically independent for all $k \neq 0$, and for all m, n. However, we allow the possibility of correlation between $\varepsilon_t^{(m)}, \varepsilon_t^{(n)}$, this correlation being of course unity when $m = n$. Fitting models for the simulation of multivariate streamflow sequences is a subject with an immense literature, and we shall not attempt to deal exhaustively with all that has appeared in it. Indeed, it is possible to argue that much effort in this area has been misplaced; a large part of it has been addressed to problems of fitting multivariate models of various kinds, and not nearly enough attention has been paid to the problems of ascertaining whether the model is appropriate. That is to say, the literature contains a very large number of models, often with accounts of alternative procedures for estimating the parameters of each, and often with Monte Carlo studies of the biases and variances of alternative estimates and the extent to which the model 'preserves' certain statistical characteristics, held to be desirable, in artificial sequences generated by its use. There is, however, a dearth of literature on how to establish whether or not any given model is consistent with the hydrological data. It is quite possible that the reported failure of certain models to preserve characteristics observed in the historic record has been compounded by failure to apply standards of statistical criticism of the selected model, in terms of parsimony of parameters, and the checking of assumptions about its residuals. This chapter therefore repeats the theme of earlier chapters, namely that study of residuals is of the highest importance in establishing whether a candidate model is satisfactory or otherwise.

7.2   MISSING DATA

The two objectives of missing-value analysis defined in Section 7.1 — filling in missing values in sequences and improving estimates of quantities that describe the characteristics of sequences — apply in different hydrological contexts, and the basic time unit of the data sequence will often determine which objective is the more appropriate.

For example, given annual flood sequences from several gauging stations with missing values, we may regard the data as being displayed as an incomplete rectangular array with years as columns and gauge sites as rows. Interest centres less on the need to 'complete the table' of data by estimating values that are missing from it, than on the need to improve estimates of mean and variance of mean annual flood at each site. This aim is to be achieved by the transfer, where appropriate, of information from sites with complete records to those where records are incomplete.

Again, given annual runoff at several sites with some values missing, we are generally more interested in estimating mean annual runoff and its variance (possibly together with other characteristics such as skewness, l-moments, . . .) than in completing the two-way table of data in which individual data values are classified by gauge sites (rows) and years (columns).

To take another example, suppose that we wish to simulate the operation of an irrigation development which is supplied by abstracting flow in excess of some specified threshold, from one or more rivers. Suppose also that monthly flow records from the gauge sites on these rivers have values missing: that is to say, the two-way table with months as columns (numbering 12 times the number of years of record) and gauge sites as rows is incomplete. Simulation of the availability of water for irrigation will require knowledge of the sequences of monthly streamflow; in this context, the means and variances of flows in each month of the hydrological year will be of less interest than the sequences themselves. An important decision must be made concerning whether the missing values are to be substituted by their 'best' estimates, or whether these estimates are to be perturbed by adding to them disturbances drawn at random from the appropriate probability distribution. If we simply substitute the 'best' estimates, the variability of monthly flow, in those months for which missing values were estimated, will be underestimated.

Where we have sequences of mean *daily* streamflow with missing values, and the sequences are to be used to simulate, say, control rules for the operation of a flood control reservoir, we shall be interested in completing the sequences by filling in the values missing from them, and not in the better estimation of parameters by information transfer. For this purpose, we should resort to a rainfall-runoff model of the kind described in Chapter 8, seeking to use causative explanatory variables to obtain estimates of missing values. There would again be the need to decide whether the 'best' estimates of the missing mean daily flows are required, or whether these best estimates are to be disturbed by adding appropriate random errors to them. If the latter, decisions must be made concerning the probability distributions from which these random errors are to be drawn. It is extremely likely, also, that the distributions of the random components will be different for the rising limbs of hydrographs, and for recessions.

As mentioned above, this chapter assumes that the random components ε_t in the data are serially uncorrelated. Where months of high flow are preceded by months when flow was also high (and conversely), we assume that such dependence can be satisfactorily modelled by some linear predictor having flow in one or more preceding months $q_{t-1}, \ldots$ as explanatory variables. Where this approach is inadequate — that is, where serial correlations in the ε_t continue to be evident — more complicated time-series models, as described by Box and Jenkins (1970), are appropriate. These are beyond the scope of the present text, although software exists for fitting them in GENSTAT and some other statistical packages. The assumption of serial independence of the random components of each series is almost certainly a reasonable assumption in the following cases:

(i)   Where the sequences concerned are of annual maximum discharge or annual maximum mean daily discharge. The serial independence of such sequences was essential to the methods of Chapter 3, and the assumption is probably valid for basins of any size.
(ii) Where the sequences are of annual runoff from basins without large year-to-year storage: for example, small upland basins with shallow soils; basins with large impervious areas, from which runoff is rapid.

Conversely, the assumption of serial independence of random components is to be questioned in the following cases:

(i)   Where the sequences are of annual runoff from very large basins, or from smaller basins which are highly permeable and with large storage capacity.
(ii) Where the sequences describe certain drought characteristics, such as the length of time for which discharge is less then some specified critical level, for basins with large storage capacity. Severe depletion of storage in one year, followed by insufficient recharge to restore to field capacity, may cause discharges in the subsequent year to be less than normal.

7.2.1   Sequences with Parts of Years Missing: Use of Censored Data in Flood Estimation

We start with a particular case of missing data, common in remote and unhealthy areas of developing countries. Where, in certain years, some daily observations in those years are missing for part of the year, we can determine which was the largest mean daily flow, say Y_0, in that proportion of each year for which daily observations exist. This largest value may or may not have been the largest for the whole year, since we do not know what happened in the period for which data are missing. What we can say is that, for such a year, the annual maximum mean daily discharge must have been greater then or equal to Y_0 : it certainly cannot have been less than that value.

We illustrate one approach to the calculation for the univariate case only; the multivariate case introduces no new matters of principle. Consider the data shown at the top of Figure 7.1 showing annual floods (annual maximum mean daily discharges) for 39 years at Rio do Sul, on the Rio Itajaí-Açú in Santa Catarina, southern Brazil. In four of these years, breaks occurred which resulted in the loss of part, but not all, of the year's record; in the first of the incomplete years, it is known that the annual flood was greater than or equal to 254 m³ s⁻¹, and similarly for the other three values listed with negative signs. In these three years, it is known that the annual floods were greater than or equal to 1150, 127 and 720 m³ s⁻¹, respectively.

R_doSul	−254	465	1090	324	270
R_doSul	801	645	1080	338	922
R_doSul	675	518	780	1470	846
R_doSul	730	1190	666	535	682
R_doSul	1020	801	−1150	−127	−720
R_doSul	1180	441	532	823	637
R_doSul	1000	1210	1120	458	1050
R_doSul	735	969	750	668	

```
 31     CALCULATE ind=R_doSul.LT.0  :  & R_doSul=ABS(R_doSul)

 34     "The variate ind now has ones in those years where the flood
−35     record is incomplete, zeros elsewhere; and R_doSul contains
−36     only positive annual maximum flows."

 39     "Now calculate the starting values for the maximum likelihood
−40     calculation. Initial estimates are the mean and variance of log
−41     annual floods with no data missing."

 44     CALCULATE logRSul=LOG(R_doSul)  :  & mind=1-ind  :  & n1=SUM(mind)
 45     &          muinit=SUM(logRSul*mind)/n1
 46       & siginit=SQRT(SUM((logRSul-muinit)**2)/(n1−1))
 47     PRINT muinit,siginit

        muinit   siginit
        6.591    0.5399

 50     "Now construct the likelihood function to be maximised."

 53     CALCULATE n2=SUM(ind)
 54       & const=−n1*0.5*LOG(2*3.14159265)
 55     MODEL [FUNCTION=logL]
 56     EXPRESSION [VALUE=(S1=−n1*LOG(sig)+const−SUM(mind*logRSul)−\
 57                 SUM(mind*(logRSul-mu)**2)/(2*sig**2))]            e[1]
 58       &        [VALUE=(S2=SUM(ind*(1-NORMAL(logRSul-mu)/sig)))]   e[2]
 59       &        [VALUE=(logL=−(S1+S2))]                            e[3]
 60     RCYCLE mu,sig; INITIAL=muinit,siginit
 61     FITNONLINEAR [CALCULATION=e]
 1
 61.............................................................................
```

***** Results of optimization *****

*** Minimum function value: ***
        247.398

*** Estimates of parameters ***

	estimate	sq. root of 2nd derivs
mu	6.6035	0.0705
sig	0.4179	0.0511

Figure 7.1 Fitting the two-parameter log-normal distribution to annual flood sequence, with some years' data incomplete: data for Rio Itajaí-Açí at Rio do Sul. Negative values show where the record for a year is incomplete; the value −254, for example, shows that the year was incomplete, and that the largest mean daily flow observed in that year was 254 m^3 s^{-1}.

Data of this kind are said to be *censored*; we do not have a specific value for the annual flood in these years, but know that it exceeded a certain value in each of the incomplete years. The data in Figure 7.1 are said to be *censored on the right*; other kinds of censoring occur in which data are *censored on the left*, for example where annual minimum flows are known to be less than some given value. One way in which censored data can occur in practice, is where the pen of a chart recorder hits the edge of the chart during the rising stage of a flood hydrograph, and remains there until water levels fall again. Censoring on the left can occur when water level falls at the beginning of a dry period, causing the pen to run against the chart edge, where it remains until levels rise again at the end of the dry period. Local knowledge of floods can also provide a kind of censored data — where, for example, it is known that K of the annual floods witnessed by the oldest inhabitant exceeded a fixed mark. The kind of censoring illustrated by these examples is easier to deal with because the censoring point is fixed; in the case of incomplete years, the censoring point is itself a random variable (if we can assume that the break in record occurred randomly in time).

To utilise the information contained in the censored data, we make use of the likelihood function, for which we first require the *survivor function* defined by $S(y) = P[Y > y] = 1 - F(y)$, where $F(y)$ is the cumulative probability distribution of the random variable Y. Suppose now that we try to fit a two-parameter log-normal distribution to the data in Figure 7.1; if we ignore the fact that the censoring points are themselves random variables, the likelihood function is

$$L = \left(1/\sum_{t=1}^{35} y_t\right)(1/(2\pi\sigma^2))^{35/2} \exp\left[-\sum_{t=1}^{35}(\ln y_t - \mu)^2/(2\sigma^2)\right]$$
$$S(254)S(1150)S(127)S(720) \tag{7.1}$$

where $S(y_t) = 1 - F(y_t)$ and $F(y_t)$ is given by

$$F(y_t) = \Phi[(\ln y_t - \mu)/\sigma]$$

with $\Phi(.)$ the standard $N(0, 1)$ cumulative probability distribution, $\Phi(y) = \int_{-\infty}^{y} \exp(-t^2/2)dt$. Thus the log-likelihood function, $\ln L = l(\mu, \sigma)$, is given by

$$l(\mu, \sigma) = -(35/2)\ln(2\pi\sigma^2) - \sum_{t=1}^{35}[(\ln y_t - \mu)^2/(2\sigma^2)] - \sum_{t=1}^{35} y_t$$
$$+ \ln[1 - \Phi((\ln 254 - \mu)/\sigma)] + \ln[1 - \Phi((\ln 1150 - \mu)/\sigma)]$$
$$+ \ln[1 - \Phi((\ln 127 - \mu)/\sigma)] + \ln[1 - \Phi((\ln 720 - \mu)/\sigma)] \tag{7.2}$$

To find estimates of μ, σ which maximise $l(\mu, \sigma)$, we need some initial estimates of μ and σ. Suitable values are the mean and standard deviation of those $\ln y_t$ for which y_t is not censored; these give $\tilde{\mu} = 6.591$, $\tilde{\sigma} = 0.5399$. Using the Gauss–Newton algorithm to minimise $-l(\mu, \sigma)$, we find the maximum-likelihood estimates $\hat{\mu} = 6.6035 \pm 0.0705$, $\hat{\sigma} = 0.4179 \pm 0.0511$ as shown in the lower part of Figure 7.1, with $l(\mu, \sigma) = -247.398$. When a Gumbel distribution is fitted to the same data, the maximum-likelihood estimates of the Gumbel parameters u and a are 650.2 ± 45.7 and $0.003\,919 \pm 0.000\,506$, respectively, with a maximised likelihood of -245.108. If the censored data are omitted from the two calculations, the log-normal parameters μ and σ are estimated as 6.5908 ± 0.0673 and 0.3979 ± 0.0476, with a log-likelihood that is slightly smaller, -248.084; for the Gumbel, the parameters u

and a are estimated as 645.5 ± 45.3 and $0.003\,952 \pm 0.000\,509$, with a slightly lower log-likelihood, -247.757. The smaller log-likelihoods indicate the small gain from the inclusion of the censored data.

The above analysis becomes considerably more complicated if the censoring points (denoted by Y_0) are taken to be random variables with their own probability distributions. It may also be necessary to take account of the facts that, first the probability of the annual flood exceeding the censoring point in any incomplete year is dependent upon the length of period for which the record is missing; and second, that the periods for which records are missing are also random variables, with their own probability distributions. The likelihood method used above, which ignores these complications, may therefore be greatly oversimplified. Where many years are incomplete, we should almost certainly need to adopt a rather different approach, considering the periods with and without records as an alternating renewal process, with peaks over a threshold observed during periods when records were available. Some simple alternatives to the maximisation of equation (7.2) also suggest themselves; assuming a probability distribution for annual floods $f(y; \theta)$, we might pursue the following estimation procedure:

(i)   Estimate the parameters θ initially by ignoring the censored data, to give estimates θ_0.

(ii) Where there are K incomplete years, with annual floods greater than or equal to $Y_1, Y_2, \ldots, Y_K$, take

$$\int_{Y_1}^{\infty} yf(y; \theta_0)\mathrm{d}y/D_{11}, \quad \int_{Y_2}^{\infty} yf(y; \theta_0)\mathrm{d}y/D_{21}, \ldots, \quad \int_{Y_K}^{\infty} yf(y; \theta_0)\mathrm{d}y/D_{K2}$$

where

$$D_{11} = \int_{Y_1}^{\infty} f(y; \theta_0)\mathrm{d}y; \quad D_{21} = \int_{Y_2}^{\infty} f(y; \theta_0)\mathrm{d}y; \ldots ; \quad D_{K2} = \int_{Y_K}^{\infty} f(y; \theta_0)\mathrm{d}y$$

(that is, the mean values of the random variable over the censored regions) as additional observations, and recompute estimates of θ, using these additional observations together with the uncensored data, to give θ_1.

(iii) Now take

$$\int_{Y_1}^{\infty} yf(y; \theta_1)\mathrm{d}y/D_{12}, \quad \int_{Y_2}^{\infty} yf(y; \theta_1)\mathrm{d}y/D_{22}, \ldots, \quad \int_{Y_K}^{\infty} yf(y; \theta_1)\mathrm{d}y/D_{K2}$$

where

$$D_{12} = \int_{Y_1}^{\infty} f(y; \theta_0)\mathrm{d}y; \quad D_{22} = \int_{Y_2}^{\infty} f(y; \theta_1)\mathrm{d}y; \ldots ; \quad D_{K2} = \int_{Y_K}^{\infty} f(y; \theta_1)\mathrm{d}y$$

as the recomputed estimates of the censored data, leading to estimates θ_2.

(iv) Repeat until the estimates $\theta_0, \theta_1, \theta_2, \ldots$ converge.

Such alternative procedures for the analysis of censored data do not appear to have been explored in the hydrological literature.

The general form of the likelihood function, of which equations (7.1) and (7.2) above are a particular case, is obtained as follows. The annual flood in each of the n years of record is described by a pair of random variables $\{q_t, w_t\}$, such that $w_t = 0$ if the year is incomplete, and $w_t = 1$ when it is complete so that the annual flood is known. With $f(y; \theta)$ the probability distribution of annual floods y_t, the likelihood function is proportional to

$$L \sim \prod_{t=1}^{n} f(y_t; \theta)^{w_t} [S(Y_t)]^{1-w_t}$$

where Y_t are the censoring points and $S(.)$ is the survival function. Where we have two cross-correlated records of annual floods, say $\{x_t\}$ and $\{y_t\}$ with incomplete years in both records, generalisation is straightforward but computationally much heavier, requiring the evaluation of double integrals.

7.2.2 Data Sequences with Entire Years Missing: the Beale–Little Algorithm

A more common type of missing-data problem is that in which entire years of record are missing. In the case of flood records, this may be because of difficulties in finding reliable observers; or because gauging stations were abandoned; or simply because some records from some gauging stations commenced well before others.

Provided that the context is appropriate for its use, a statistical method exists for this problem. It is an iterative calculation due to Beale and Little (1975), but before describing its use we make some general comments about missing-data problems and return to the phrase which begins this paragraph, concerning the context in which the Beale–Little algorithm is appropriate.

Strictly, the Beale–Little algorithm is appropriate *only where the missing data are missing at random*; in many hydrological applications, this cannot be assumed. It may be a valid assumption where entire years are missing within the course of a record because of observer absence; it will not be true where years are absent because some gauging sites were installed earlier, or came out of service later, than others. Where, for example, the lower regions of a river basin are colonised first because of easier access, gauging stations in those regions are likely to have been established before settlement begins in steeper, less accessible headwater basins. But a concomitant of lowland settlement may be deforestation, agricultural development and urbanisation, all of which may modify the annual maximum discharge and other quantities derived from the hydrograph. If, subsequently, a gauging station is established in a headwater basin and we try to extend its short flood record by correlation with longer records from lowland stations, our estimates of mean and variance of annual floods in the headwater basin may well be biased. This is because we may be extrapolating a statistical model (say, a linear regression) to values of the explanatory variables where linearity does not hold. Denote the long, lowland flood record by x, and the short, upland flood record by y. If the period of common record is for a period when urbanisation had increased annual peak discharges in x, then a regression of y on x, extrapolated using the earlier x-floods where urbanisation effects were smaller, will underestimate annual peak discharges from the upland basin.

Whilst, strictly, use of the Beale–Little algorithm may be inappropriate for making inferences about the parameters of flow records with missing values, the same is true of *any* method used to estimate missing values in hydrological series, or to make inferences about their parameters. So provided we bear clearly in mind that any missing data procedure is strictly valid only where observations are missing at random, and may lead to unsuspected biases otherwise, we may proceed to use it with caution.

As regards calculation, suppose that we have a $G \times P$ matrix representing annual floods from G gauge sites in P years, G and P referring to the rows and columns of the table. Let x_{ij} be the observation of the annual flood in year j, at gauge i; we shall use x_i to denote the available flood record for gauge i, and shall refer to the regression of x_i on $x_1, x_2, \ldots, x_{i-1},$ $x_{i+1}, \ldots, x_G$. The calculation then proceeds as follows:

(i) For the subset of years for which all G sites have data, calculate the $(G \times 1)$ vector of mean values, denoted by $\hat{\mu}$, and the $(G \times G)$ matrix of sums of squares and products, denoted by $\hat{\Sigma}$. Using $\hat{\mu}$ and $\hat{\Sigma}$, calculate the multiple regression of x_1 on $x_2, x_3, \ldots, x_G$; the regression of x_2 on $x_1, x_3, \ldots, x_G, \ldots$; the regression of x_G on $x_1, x_2, \ldots, x_{G-1}$.

(ii) For each of rows $1, 2, 3, \ldots, G$, substitute the appropriate element of the vector $\hat{\mu}$ wherever a missing value occurs.

(iii) Using the multiple regression of x_1 on $x_2, x_3, \ldots, x_G$ calculated in step (i) above, estimate the missing values for row 1. Do not replace the means (entered at step (ii)) yet, but store them for future use. Record, for each missing value in row 1, the deviation between the 'old' estimate for the missing value (the element of $\hat{\mu}$ entered at step (ii)) and the 'new' estimate for the missing value, given by the multiple regression. If, in row 1, a missing value occurred in year k, denote the deviation (old value minus new value) by d_{1K}.

(iv) Repeat step (iii) for row 2, using the regression of x_2 on $x_1, x_3, \ldots, x_G$, calculated at step (i); record the deviations (old value minus new value), denoting these by d_{2k}. Repeat the calculation for row 3, row 4, $\ldots$, row G.

(v) At this stage, we have computed, for each gauge site (row): (a) new estimates for the missing values, to replace the old ones (which, at this stage, were simply the row means for years when all records were complete); (b) the deviations between the old estimates and the new ones. We now use these quantities to recalculate $\hat{\Sigma}$ as follows. If, in year k, observations are missing in both rows i and j (that is, observations are lacking in year k for both gauge sites i and j), add the product of deviations $d_{ik}d_{jk}$ to the (j, k)th element of the matrix of sums of squares and products $\hat{\Sigma}$. Nothing is added to the (i, j)th element of $\hat{\Sigma}$ if observations are available for one, or for both, gauge sites in year k.

(vi) Using $\hat{\mu}$ and the new $\hat{\Sigma}$, recalculate the multiple regressions of x_1 on $x_2, x_3, \ldots, x_G$; of x_2 on $x_1, x_3, \ldots$; of x_G on $x_1, x_2, \ldots, x_{G-1}$.

(vii) Now replace the 'old' estimates of the missing values by the 'new' ones. Return to step (iii), using the new multiple regression equations to recompute missing values. Continue iterations between steps (iii) and (vii) until all changes to the missing values become smaller than some preset criterion for convergence, and the changes to the matrix $\hat{\Sigma}$ become negligible.

Before moving to an example, five points must be made about the above calculation. First, Beale and Little (1975), who developed the algorithm, state that it can be improved if the vector $\hat{\mu}$ is also recalculated at each cycle, by summing over all P years, and not just over the subset of years for which all gauge sites were complete. This modification requires more computation than the algorithm as described above, but it can be used when there are no years in which all gauge sites are complete, and is correspondingly more general.

Second, the procedure makes no assumptions about the probability distributions of the data: it can be used irrespective of the joint distribution of the variables $x_1, x_2, \ldots, x_G$. However, if the x_i, or their logarithms, have a multivariate normal distribution, the algorithm gives maximum-likelihood estimates of the parameters μ, Σ of this distribution. When the distribution is not multivariate normal, the best estimate of a missing value is not necessarily a *linear* combination of the existing observations, which is what we effectively are using in the above algorithm; a non-linear function of the observations may be more appropriate. On the other hand, we cannot say what alternative would be more appropriate, without specifying the form of the multivariate distribution from which the data are a sample.

Third, it might be thought that an appropriate strategy for dealing with missing flow data would be to take as many gauge sites as possible, taking a very large number of rows, obtaining estimates for missing values at a very large number of sites. However, where correlations between records at different sites (that is, between the rows of the $G \times P$ data matrix) are small, a loss of information will result. The explanation for this can be shown as follows. Suppose we calculate the multiple regression of x_1 on $x_2, x_3, \ldots, x_{G1}$, where all these explanatory variables are highly correlated with x_1. If the period for which all records are complete is $P1$ years, the residual variance $(= RSS/(P_1 - G_1 - 1))$ will have $P1 - G1 - 1$ degrees of freedom. Suppose now that we include in the multiple regression a further G_2 explanatory variables, all of which have very low correlation with x_1. The multiple regression will now have $G_1 + G_2$ explanatory variables; since the additional G_1 variables all have low correlation with x_1, the additional sum of squares accounted for by including these variables will be small, and the residual sum of squares will be reduced by very little. However, the degrees of freedom remaining for estimating the error variance will be reduced from $P_1 - G_1 - 1$ to $P_1 - G_1 - G_2 - 1$. The consequence is that including the additional G_2 records, poorly correlated with x_1, may actually *increase* the error variance, since its numerator, RSS, has been reduced only slightly by including the G_2 variables, whilst its denominator, the residual degrees of freedom, has been reduced substantially. Before undertaking any application of the Beale–Little algorithm, it is therefore advisable by means of a preliminary analysis to ensure that all correlations amongst the G records are substantial.

Fourth, recalling that the data matrix has G rows and P columns, corresponding to up to P years of record from G sites, the algorithm assumes the observations in each *column* are correlated, but that no correlations exist between observations in the same *row*: that is, between observations at the same site in different years. This assumption is not unreasonable where, say, the data are annual floods, or where the data are total annual flows for small basins which respond quickly to rainfall; it may well not be appropriate where the data are annual low flows defined by taking the smallest value in each year of a 30-day running mean. Where missing values are to be calculated for annual low flows, it will be necessary to check for the existence of serial correlation in the records; if significant serial correlation exists, a more complicated, time-series model must be used, and this is beyond the scope of the present work.

A particular case of interest occurs where we need to estimate missing data for sequences of monthly flows. Except for some special cases, it will not be appropriate to apply the Beale–Little algorithm to all February flows taken separately, then to all March flows, ..., since this will not allow for the fact that high flows in one month are more likely to occur where flows in the preceding month were also high. This dependence could be allowed for in some cases by including, together with the March flow records for the G sites, the February flow records for the G sites, as additional explanatory variables. As with the Thomas–Fiering model, complications can be expected if flow in some months is small, when estimated missing values may be negative; and if variances are heterogeneous.

Finally, a considerable amount of computer time will be required if the number of missing values is substantial. Beale and Little recommend setting the maximum number of iterations at 100, although they report a case in which more than 170 iterations were required. They suggest that the tolerance, determining when iteration stops, should be 0.01: that is, iteration ceases when all changes to the means in the vector $\hat{\mu}$ are less than 0.01 of the current absolute value.

7.2.3 A Numerical Example of the Beale–Little Algorithm: Annual Flood Data from Four Sites (Rio do Sul, Ibirama, Apiuna, Indaial) on the Rio Itajaí-Açú

The calculation outlined in Section 7.2.2, although heavy, can be readily undertaken by modern statistical packages; for example, GENSTAT has a library routine, MULTMISS, for this purpose. We take as an example the data shown in Figure 7.2; these data are annual flood data (more exactly, annual maximum mean daily discharges) for the period 1934–84 from four sites in the basin of the Rio Itajaí-Açú in Santa Catarina, southern Brazil. The four columns in the table show data from the Rio Itajaí-Açú at Rio do Sul (denoted by *R_doSul*); the Rio Hercílio at Ibirama (*Ibirama*); the Rio Itajaí-Açú at Apiuna (*Apiuna*); and the Rio Itajaí-Açú at Indaial

Data for the four sites (asterisks represent missing values):

R_doSul	*	*	*	*	*
Ibirama	*	1342	625	619	797
Apiuna	1111	1914	1507	950	1111
Indaial	*	2684	1913	1279	1995
R_doSul	*	*	*	465	1090
Ibirama	1250	271	263	566	649
Apiuna	1742	1033	918	881	1960
Indaial	2596	1256	996	1410	2420
R_doSul	324	270	801	645	1080
Ibirama	236	474	763	592	981
Apiuna	495	566	1280	1100	2250
Indaial	645	748	1755	1256	2764
R_doSul	338	922	675	518	780
Ibirama	438	281	556	393	726
Apiuna	702	1680	1260	909	1620
Indaial	819	2308	1545	1332	2724
R_doSul	1470	846	730	1190	666
Ibirama	897	969	566	1300	526
Apiuna	2630	1890	*	3090	1220
Indaial	3922	3060	1079	5468	1079
R_doSul	535	682	1020	801	*
Ibirama	520	487	897	582	510
Apiuna	936	1240	2160	1550	1750
Indaial	1126	1425	2468	1740	2010
R_doSul	*	*	1180	441	532
Ibirama	*	708	998	477	298
Apiuna	648	1460	1930	727	562
Indaial	*	*	*	1256	760

Figure 7.2   Calculation of estimates of missing values in four annual flood sequences in the basin of the Rio Itajaí-Açú: 51 years of parallel record ($m^3\ s^{-1}$) at the gauging sites Rio do Sul, Ibirama, Apiuna and Indaial. Missing values are obtained by the Beale–Little algorithm, using the GENSTAT library procedure MULTMISS.

_____ Continued _____

Figure 7.2 *(continued)*

R_doSul	823	637	1000	1210	1120
Ibirama	872	483	1040	1010	1240
Apiuna	1730	1020	2030	2210	2310
Indaial	2364	1338	2356	2340	2900
R_doSul	458	1050	735	969	750
Ibirama	697	1406	801	741	1002
Apiuna	951	2760	1575	1640	2156
Indaial	1256	2980	*	1995	2836
R_doSul	668	*	*	*	*
Ibirama	1090	*	589	490	2475
Apiuna	1847	3086	927	1539	4327
Indaial	2308	3700	1197	1920	4791
R_doSul	*				
Ibirama	2125				
Apiuna	4314				
Indaial	5026				

Means, over existing years of record, for the four sites are:

mR_doSul	mIbirama	mApiuna	mIndaial
783	784	1624	2112

* * * Correlation matrix * * *

R_doSul	1.000			
Ibirama	0.686	1.000		
Apiuna	0.911	0.867	1.000	
Indaial	0.821	0.786	0.928	1.000
	R_doSul	Ibirama	Apiuna	Indaial

Records for four sites after estimation of missing values by use of MULTMISS:

R_doSul	614	732	807	498	470
Ibirama	546	1342	625	619	797
Apiuna	1111	1914	1507	950	1111
Indaial	1462	2684	1913	1279	1995
R_doSul	661	688	639	465	1090
Ibirama	1250	271	263	566	649
Apiuna	1742	1033	918	881	1960
Indaial	2596	1256	996	1410	2420
R_doSul	324	270	801	645	1080
Ibirama	236	474	763	592	981
Apiuna	495	566	1280	1100	2250
Indaial	645	748	1755	1256	2764
R_doSul	338	922	675	518	780
Ibirama	438	281	556	393	726
Apiuna	702	1680	1260	909	1620
Indaial	819	2308	1545	1332	2724
R_doSul	1470	846	730	1190	666
Ibirama	897	969	566	1300	526
Apiuna	2630	1890	1172	3090	1220
Indaial	3922	3060	1079	5468	1079
R_doSul	535	682	1020	801	1003
Ibirama	520	487	897	582	510
Apiuna	936	1240	2160	1550	1750
Indaial	1126	1425	2468	1740	2010

Figure 7.2 *(continued)*

R_doSul	438	747	1180	441	532
Ibirama	328	708	998	477	298
Apiuna	648	1460	1930	727	562
Indaial	905	1881	2056	1256	760
R_doSul	823	637	1000	1210	1120
Ibirama	872	483	1040	1010	1240
Apiuna	1730	1020	2030	2210	2310
Indaial	2364	1338	2356	2340	2900
R_doSul	458	1050	735	969	750
Ibirama	697	1406	801	741	1002
Apiuna	951	2760	1575	1640	2156
Indaial	1256	2980	2072	1995	2836
R_doSul	668	1374	502	879	1707
Ibirama	1090	1475	589	490	2475
Apiuna	1847	3086	927	1539	4327
Indaial	2308	3700	1197	1920	4791
R_doSul	1810				
Ibirama	2125				
Apiuna	4314				
Indaial	5026				

*** Correlation matrix ***

R_doSul	1.000			
Ibirama	0.759	1.000		
Apiuna	0.936	0.906	1.000	
Indaial	0.851	0.848	0.943	1.000
	R_doSul	Ibirama	Apiuna	Indaial

(*Indaial*). As usual, asterisks indicate missing data. Beneath the listed data, the correlation matrix of the four flood sequences is shown; all correlations are quite high, none being less than the 0.686 between Ibirama and Rio do Sul. In the program, the names of the four variables (*R_doSul*, *Ibirama*, *Apiuna* and *Indaial*) are read into a data structure called a pointer, with name P_p. The directive MULTMISS P_p goes through essentially the same calculation as described in the last section, and a PRINT directive gives the four columns of data with the missing annual floods inserted, in place of asterisks. The directive CORRELATE gives the revised correlation matrix. The smallest correlation, between *R_doSul* and *Ibirama*, has increased to 0.759.

A particularly important point to be borne in mind where missing values are estimated by regression techniques is the following. After convergence is achieved, the missing values have been substituted by expected values. If the variation, over years, is important for the problem at hand, the variance of the completed sequence will be underestimated. This can easily be seen by considering the use of a regression between a short series $\{y_t\}$ and a long series $\{x_t\}$ to fill in the missing values in the $\{y_t\}$ sequence. Using the regression, all the estimates $\hat{y}_t$ of the missing values will lie on the regression line itself; these estimates therefore contain no random component due to variation about the regression line. Hence the total variance, calculated by putting the data $\{y_t\}$ and the estimated $\{\hat{y}_t\}$ together, will underestimate the true variance. We can restore matters by adding to each estimated $\hat{y}_t$ a random error drawn from the distribution of residuals about the regression; however, the quantities substituted for the missing values are only one of an infinite number of possible substitutions. Great care is therefore needed in the publication of hydrological records for which missing values have been filled in; the published records must make it quite clear how the missing values were estimated, and where they occurred in the record. Indeed, it is often preferable to publish data with their missing values, together with instructions on how they should be completed for different hydrological usages.

7.2.4 Data Sequences with Entire Years Missing: the EM Algorithm

A method related to the Beale–Little algorithm is the EM algorithm, which involves a two-step iterative calculation, the two steps being an *expectation* step followed by a *maximisation* step (hence the letters EM). The algorithm is remarkable for the generality of the underlying theory, and because of the wide range of missing-data problems with which it can deal.

In its most general form, the EM algorithm works as follows. Consider a simple example with two gauge sites, with data from three years:

	Year 1	Year 2	Year 3
Gauge site 1	x_1	X_2	x_3
Gauge site 2	y_1	y_2	Y_3

in which upper-case letters $(X_2; Y_3)$ indicate values which are missing. Let θ be the vector of parameters defining the probability distribution of the vector $[x, y]^\mathrm{T}$, where we assume — as in the Section 7.2.3 — that the columns of data for each year are statistically independent of data in other years (that is, are serially independent, as is commonly the case for annual floods), although this assumption is not essential to the argument. Let $p(x, y; \theta)$ be the joint probability distribution of (x, y), so that the log-likelihood function is

$$\ln L = \ln p(x_1, y_1; \theta) + \ln p(X_2, y_2; \theta) + \ln p(x_3, Y_3; \theta) \qquad (7.3)$$

We should like to find the θ which maximises this quantity, but we do not know X_2 and Y_3. However, given an estimate of θ, say $\theta^{(k)}$, we can maximise the expected value of $\ln L$, given the known data x_1, y_1, y_2, x_3, and the estimate $\theta^{(k)}$. Then we choose a new estimate $\theta^{(k+1)}$ which maximises this expected value Q, given by

$$Q(\theta|\theta^{(k)}) = \int_{X_2} \int_{Y_3} \ln L \cdot f \, \mathrm{d}X_2 \, \mathrm{d}Y_3 \qquad (7.4)$$

with respect to θ, where f is the distribution of X_2 and Y_3 conditional on the data. Thus f is given by the expression

$$p(x_1, y_1; \theta^{(k)}) p(X_2|y_2, \theta^{(k)}) p_\mathrm{M}(y_2; \theta^{(k)}) p(Y_3|x_3, \theta^{(k)}) p_\mathrm{M}(x_3; \theta^{(k)}) \qquad (7.5)$$

where $p_\mathrm{M}(.)$ refers to the relevant marginal distribution.

The EM algorithm therefore consists of the following:

(i) E-step: Given an estimate $\theta^{(k)}$ of the parameters θ, calculate $Q(\theta|\theta^{(k)})$.

(ii) M-step: Choose as the new estimate $\theta^{(k+1)}$ of θ, that value of θ which maximises $Q(\theta|\theta^{(k)})$. Go back to the E-step, and continue until the convergence criteria are satisfied.

The above procedure is considerably simplified where the joint distribution $p(x, y; \theta)$ (or, in the more general case, $p(\mathbf{x}; \theta)$ where $\mathbf{x}$ is a $G \times 1$ vector) belongs to the exponential family of distributions, introduced in the last section, with general form

$$p(\mathbf{x}; \theta) = \exp\left[\sum_{j=1}^{k} A_j(\theta) B_j(\mathbf{x}) + C(\mathbf{x}) + D(\theta)\right] \tag{7.6}$$

In this case, the k statistics $t_j = B_j(\mathbf{x})$, $j = 1, 2, \ldots, k$ are *sufficient statistics* for the $A_j(\theta)$, the simplest case being when $A_j(\theta) = \theta_j$, the jth element of the vector θ of parameters. The simplification, together with proofs for the general case that the calculation yields maximum-likelihood estimates, is described by Dempster *et al.* (1977).

7.2.5 Box–Cox Power Transformations in the Multivariate Case

The general case of the EM algorithm described in Section 7.2.4 requires knowledge of the joint distribution $p(\mathbf{x}; \theta)$ of the $G \times 1$ vector x. Apart from the multivariate normal and log-normal, multivariate distributions tend to be unfamiliar and difficult to handle. A reasonable question to ask is therefore the following: is it possible, by means of changes of scale in the data, to recast them in multivariate normal form? The advantages of doing this are considerable, not the least being that sufficient statistics exist for the multivariate normal distribution, so that on the transformed scale the desirable properties of maximum-likelihood estimates persist for all sample sizes, and not just in large samples. Furthermore, where interest centres on the parameters of the marginal distribution at one particular site only (say, for the purpose of calculating a flood with given return period), this marginal distribution is easily obtained; also the T-year flood can easily be obtained on the original scale, by detransforming the T-year flood calculated on the transformed, normal scale; and for any size of sample, the detransformed T-year flood will estimate its expected value with minimum attainable variance. Transformation of multivariate data to a normal scale therefore has important advantages.

In the case of univariate data (for example, flood sequences at a single site), we recall that a transformation that often secures normality is the Box–Cox transformation, from x to y, where

$$y = \begin{cases} (x^\lambda - 1)/\lambda & \lambda \neq 0 \\ \ln x & \lambda = 0 \end{cases}$$

Atkinson (1987) suggests a modification to this transformation, putting it in the following form:

$$y = \begin{cases} (x^\lambda - 1)/(\lambda \dot{x}^{\lambda-1}) & \lambda \neq 0 \\ \dot{x} \ln x & \lambda = 0, \end{cases}$$

where $\dot{x}$ is the geometric mean of the observations. Using the Atkinson modification, the partially maximised log-likelihood becomes

$$\ln L_{\max} = -(n/2) \ln\{RSS(\lambda)/n\} \tag{7.7}$$

where $RSS(\lambda)$ is the sum of squared deviations amongst the ys, given by

$$RSS(\lambda) = (\mathbf{y} - \mu)^{\mathrm{T}}(\mathbf{y} - \mu) \tag{7.8}$$

To estimate the parameter λ of the transformation which most nearly achieves normality, we plot $\ln L_{\max}$ as a function of λ, selecting that which maximises this function. Taking, for

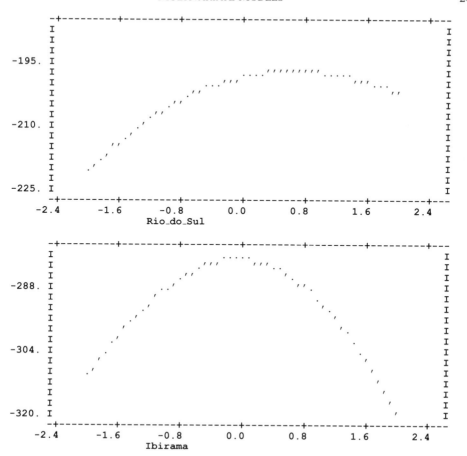

Figure 7.3   Plot of log-likelihood function against Box–Cox parameter λ giving transformation to approximate normality. Data used are annual flood sequences (50 years) for Rio Itajaí-Açú at Rio do Sul and Ibirama.

example, the flood records from the gauging stations on the Rio Itajaí-Açú at Rio do Sul and on the Rio Hercílio at Ibirama, we find the plots of $\ln L_{max}(\lambda)$ against λ as shown in Figure 7.3. Using the values of $\ln L_{max}$, shown in Figure 7.3, to fit a quadratic curve relating $\ln L_{max}$ to λ, it is found that the two quadratics are

$$\ln L_{max} = -198.5104 + 4.3898\lambda - 3.2012\lambda^2 \tag{7.9}$$

and

$$\ln L_{max} = -281.429 - 2.042\lambda - 8.436\lambda^2 \tag{7.10}$$

for Rio do Sul and Ibirama, respectively, these curves having their maxima where $\hat{\lambda} = 0.6856$ and $\hat{\lambda} = -0.1210$. Denoting the general quadratic by $\ln L_{max} = b_0 + b_1\lambda + b_2\lambda^2$ with b_2 negative, the estimate of λ is $-b_1/(2b_2)$ and the 95% confidence limits for the λ which maximises $\ln L_{max}$ are given by

$$(-b_1 \pm \sqrt{(-4b_2\chi^2)})/(2b_2) \tag{7.11}$$

where χ^2 is the tabulated value with one degree of freedom. For Rio do Sul and Ibirama, we find that the 95% confidence intervals are $(-0.4097, 1.781)$ and $(-0.7958, 0.5537)$, respectively. The limits for Rio do Sul are particularly wide, as is evident from the low curvature of the plot of $\ln L$ shown in Figure 7.3; the Rio do Sul limits even include the case $\lambda = 1$, the value of λ corresponding to no transformation. For both Ibirama and Rio do Sul, the 95% limits include the value $\lambda = 0$, corresponding to the log transformation. Indeed, we can make normal $(Q\text{-}Q)$ plots for the annual floods at these sites, after the raw data have been log-transformed, together with the envelopes as described in Chapter 3. Figure 7.4 shows these $Q\text{-}Q$ plots for Ibirama and Rio do Sul; plots for the other two sites, Apiuna and Indaial, that are used in this chapter are also shown.

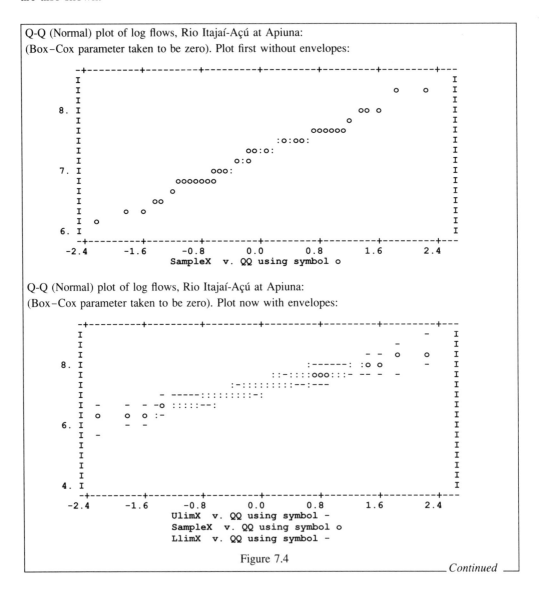

Figure 7.4

Continued

__ Figure 7.4 *(continued)* __

Q-Q (Normal) plot of log flows, Rio Itajaí-Açú at Rio do Sul:
(Box–Cox parameter taken to be zero). Plot first without envelopes:

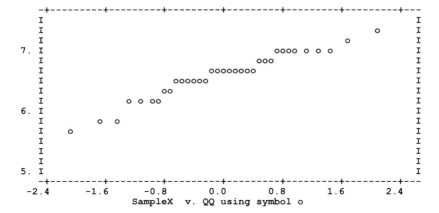

Q-Q (Normal) plot of log flows, Rio Itajaí-Açú at Rio do Sul:
(Box–Cox parameter taken to be zero). Plot now with envelopes:

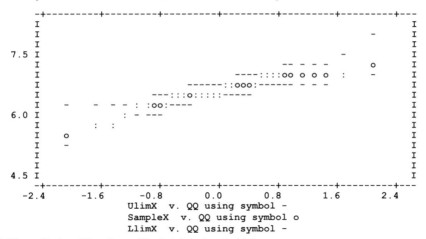

Q-Q (Normal) plot of log flows, Rio Itajaí-Açú at Indaial:
(Box–Cox parameter taken to be zero). Plot first without envelopes:

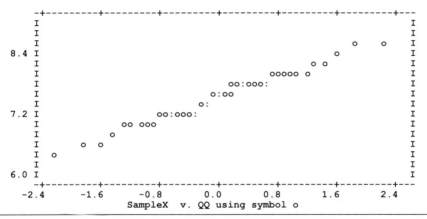

_____ Figure 7.4 *(continued)* _____

Q-Q (Normal) plot of log flows, Rio Itajaí-Açú at Indaial:

(Box–Cox parameter taken to be zero). Plot now with envelopes:

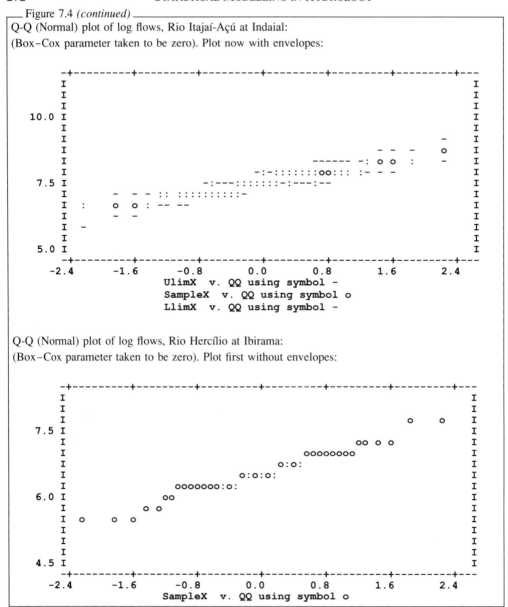

Q-Q (Normal) plot of log flows, Rio Hercílio at Ibirama:

(Box–Cox parameter taken to be zero). Plot first without envelopes:

—— Figure 7.4 *(continued)* ————————————————————————
Q-Q (Normal) plot of log flows, Rio Hercílio at Ibirama:
(Box–Cox parameter taken to be zero). Plot now with envelopes:

```
  -+---------+---------+---------+---------+---------+---------+---
 I                                                              -    I
 I                                                                   I
 I                                                         -    o    I
7.5 I                                                    - :         I
 I                                                               -   I
 I                                          -------:o o o  -    -    I
 I                                   -:-::oooooooo        -         I
 I                            --:-:o:o:----------  -              I
 I                     ---:-::o:::-:--                              I
 I             -::::oo::::-:-:                                      I
6.0 I       -  - - -:o  ----                                        I
 I  -         o  :----                                             I
 I  o    o  : -                                                   I
 I  -                                                             I
 I                                                               I
 I -                                                              I
4.5 I                                                              I
  -+---------+---------+---------+---------+---------+---------+---
 -2.4      -1.6      -0.8       0.0       0.8       1.6       2.4
```
UlimX   v. QQ using symbol -
SampleX v. QQ using symbol o
LlimX   v. QQ using symbol -

Figure 7.4 This program performs *Q-Q* plots for the values in cross-correlated sequences $\{X_t\}, \{Y_t\}, \ldots$ after joint transformation to near-normality by means of transformation to logarithms $X_t = \ln(x_t)$, $Y_t = \ln(y_t), \ldots$ where $\{x_t\}, \{y_t\}, \ldots$ are the original data. For each sequence, 95% envelopes are calculated.

A more general form of the Box–Cox transformation involves two parameters, λ and α, the transformation being

$$y = \begin{cases} ((x+\alpha)^{\lambda} - 1)/\lambda & \lambda \neq 0 \\ \ln(x+\alpha) & \lambda = 0 \end{cases}$$

Estimation of the two parameters α and λ, which most nearly transform a single data series $\{x_t\}$ to normality, requires a two-dimensional plot of $\ln L$ as a function of α and λ, where $\ln L$ is given by

$$\ln L = -(n/2) \ln \left[\sum_t^n ((x_t + \alpha)^{\lambda} - 1)/(\lambda - \mu)^2/n \right] + (\lambda - 1) \sum_t^n \ln x_t \qquad (7.12)$$

where $\mu = \Sigma[((x_t + \alpha)^{\lambda} - 1)/\lambda]/n$.

Joint estimation of λ and α is not always straightforward. Using, for example, the data from Rio do Sul, a plot of $\ln L$ as a function of α and λ shows that the log-likelihood surface is almost featureless without any well-defined maximum.

The above discussion has considered only the transformation to normality of single annual flood sequences, considered separately. In the multivariate case, it will usually be adequate to transform each flood sequence separately to the near-normal scale, although strictly this does not guarantee that the multivariate distribution of the transformed sequences is multivariate normal. If, using the methods described below, we find that (say) two separately transformed flood sequences do not appear to be bivariate normal after transformation, we must then search for two Box–Cox parameters, say λ_x and λ_y, which *jointly* transform the bivariate flood sequence $\{x_t, y_t\}$ $(t = 1, 2, \ldots, N)$ to bivariate normality. We would then look for two transformations, say,

$$U = \begin{cases} (x^{\lambda_x} - 1)/\lambda_x & \lambda_x \neq 0 \\ \ln x & \lambda_x = 0 \end{cases}$$

$$V = \begin{cases} (y^{\lambda_y} - 1)/\lambda_y & \lambda_y \neq 0 \\ \ln y & \lambda_y = 0 \end{cases}$$

lamdax	1	2	3	4	5	6	7
1	−2.00	−2.00	−2.00	−2.00	−2.00	−2.00	−2.00
2	−1.33	−1.33	−1.33	−1.33	−1.33	−1.33	−1.33
3	−0.67	−0.67	−0.67	−0.67	−0.67	−0.67	−0.67
4	0.00	0.00	0.00	0.00	0.00	0.00	0.00
5	0.67	0.67	0.67	0.67	0.67	0.67	0.67
6	1.33	1.33	1.33	1.33	1.33	1.33	1.33
7	2.00	2.00	2.00	2.00	2.00	2.00	2.00

lamday	1	2	3	4	5	6	7
1	−2.00	−1.33	−0.67	0.00	0.67	1.33	2.00
2	−2.00	−1.33	−0.67	0.00	0.67	1.33	2.00
3	−2.00	−1.33	−0.67	0.00	0.67	1.33	2.00
4	−2.00	−1.33	−0.67	0.00	0.67	1.33	2.00
5	−2.00	−1.33	−0.67	0.00	0.67	1.33	2.00
6	−2.00	−1.33	−0.67	0.00	0.67	1.33	2.00
7	−2.00	−1.33	−0.67	0.00	0.67	1.33	2.00

logL	1	2	3	4	5	6	7
1	−435.8902	−424.2011	−415.9522	−411.8146	−411.7025	−414.9273	−420.6949
2	−425.1129	−413.0952	−404.4612	−400.0267	−399.8150	−403.1220	−409.0701
3	−417.4631	−405.0905	−395.9913	−391.1220	−390.6765	−393.9838	−400.0815
4	−413.2984	−400.6698	−391.1661	−385.8292	−385.0353	−388.2147	−394.3621
5	−412.5181	−399.8114	−390.1079	−384.4459	−383.3099	−386.2585	−392.3165
6	−414.6362	−402.0159	−392.3438	−386.5926	−385.2491	−387.9598	−393.8297
7	−419.0729	−406.6369	−397.1537	−391.5093	−390.1225	−392.6662	−398.3300

77 CONTOUR logL

Contour plot of logL at intervals of   5.258

** Scaled values at grid points **

−79.7015	−77.3364	−75.5328	−74.4593	−74.1956	−74.6793	−75.7565
−78.8577	−76.4575	−74.6180	−73.5242	−73.2687	−73.7843	−74.9006
−78.4549	−76.0383	−74.1928	−73.1160	−72.8999	−73.4607	−74.6128
−78.6033	−76.2015	−74.3941	−73.3791	−73.2281	−73.8327	−75.0019
−79.3954	−77.0423	−75.3117	−74.3857	−74.3009	−74.9299	−76.0896
−80.8502	−78.5646	−76.9226	−76.0792	−76.0389	−76.6679	−77.7991
−82.8999	−80.6768	−79.1080	−78.3211	−78.2998	−78.9131	−80.0100

Figure 7.5   Calculation and contour plot of bivariate log-likelihood (using annual flood sequences from Rio Itajaí-Açú at Rio do Sul and Ibirama) for Box–Cox parameters λ_x, λ_y giving approximate bivariate normality (y-sequence is Rio do Sul, x-sequence is Ibirama). The contour plot shows that the values of λ_x, λ_y which maximise the bivariate log-likelihood are close to the values of λ_x, λ_y obtained by maximising the univariate log-likelihoods.

_____ Continued _____

— Figure 7.5 *(continued)* —

```
       0.0000      0.1667     0.3333     0.5000     0.6667     0.8333     1.0000
         '           '          '          '          '          '          '
  1.000-000      22222    4444444                                  4444444-
       000       2222     4444444                   666             44444
        00       2222      444444          6666666666666             4444
         0      22222      444444       6666666666666666666            44
         0       2222     444444       666666666666666666666666          4
                22222     444444      6666666666666666666666666666        4
  0.833-  2222            44444       66666666666666666666666666666        -
         22222            44444     6666666666666666666666666666666666
         2222             44444     66666666666666666666666666666666666
         2222            44444     66666666666666    666666666666666
         2222            44444    666666666666         66666666666666
         2222            44444    6666666666            666666666666
  0.667- 2222            44444    6666666666              666666666666      -
         2222            44444    6666666666              6666666666666
         2222            44444    6666666666              666666666666
         2222            44444    6666666666              66666666666666
         2222            44444    666666666666           6666666666666
         2222            44444    666666666666666666666666666666666666
         2222            44444     6666666666666666666666666666666666
  0.500- 2222            44444     666666666666666666666666666666666      4-
         2222            44444     66666666666666666666666666666666        4
         2222           444444      6666666666666666666666666666          444
         2222           444444      666666666666666666666666              4444
       0  2222          444444        6666666666666666                 444444
       0  22222         444444                                        4444444
  0.333-00   22222     4444444                         444444444  -
       000    22222     44444444                        4444444444
      0000     22222      444444444                    44444444444
      0000      22222          44444444444444444444444444444444444
      0000       222222        4444444444444444444444444444             22
       0000       222222        44444444444444444444444               2222
  0.167- 0000     2222222           444444                        2222222-
       8       00000      22222222                           222222222
       88      00000      222222222                         2222222222
      8888      00000         2222222222222222222222222222222222
      8888      000000         222222222222222222222222222               00
        8888     000000         2222222222222                          00000
  0.000-  88888     0000000                             000000000  -
         '           '          '          '          '          '          '
```

The joint maximum-likelihood estimation of λ_x and λ_y is an iterative calculation, for which initial estimates are required. These can be the estimates of λ_x, λ_y which transform the individual sequences to near-normality, taken separately. That is, we find λ_x which transforms $\{x_t\}$ to near-normality, and similarly we find λ_y which transforms $\{y_t\}$ to near-normality, and use these estimates to start the bivariate calculation. As an example, we consider the calculation of λ_x, λ_y which transform the flood sequences for Ibirama and Apiuna to near-bivariate normality. Treating each sequence separately, we find that λ_x and λ_y are both very close to zero, suggesting log transformations of each sequence. The log-likelihood for the bivariate sequence of annual floods at Ibirama and Apiuna is

$$\ln L(\lambda_x, \lambda_y) = -(N/2)\ln|\mathbf{S}(\lambda_x, \lambda_y)| + (\lambda_x - 1)\sum_{t=1}^{N}\ln x_t + (\lambda_y - 1)\sum_{t=1}^{N}\ln y_t \qquad (7.13)$$

where $\mathbf{S}(\lambda_x, \lambda_y)$ is the matrix of variances and covariances of the two sequences $\{x_t, y_t\}$, $t = 1, 2, \ldots, N$. To search for the maximum of this log-likelihood function, λ_x for the Ibirama sequence was varied from -0.7 to 0.5 in steps of 0.2 and λ_y for the Apiuna sequence from -0.7 to 0.7 in steps of 0.2. Contours of the log-likelihood function for λ_x and λ_y are shown in Figure 7.5, from which it is seen that the maximum is close to $\lambda_x = \lambda_y = 0$, the values given by taking the logarithms of each flood sequence separately.

```
 6   PROCEDURE 'GAMMAPLOT'
 7   PARAMETER NAME='V','N','ROSE','GAMABSC','MAT';MODE=p
 8   VARIATE [NVALUES=ROSE] W[1...N]
 9   VARIATE [NVALUES=ROSE] T
10      &    [VALUES=1...ROSE] I
11   CALCULATE W[1...N]=V[1...N]
12      &     W[1...N]=W[1...N]-MEAN(W[1...N])
13      &        T=(I=0.5)/ROSE
14      & GAMABSC=CED(T;N)
15   MATRIX [ROWS=ROSE;COLUMNS=N] DEVS
16      &   [ROWS=N; COLUMNS=ROSE] DEVSDASH
17   SYMMETRICMATRIX [ROWS=N] SIG,SIGINV
18   EQUATE !P(V[1...N]); NEWSTRUCTURES=DEVSDASH
19   CALCULATE DEVS=TRANSPOSE(DEVSDASH)
20      &        SIG=DEVSDASH*+DEVS
21      &     SIGINV=INVERSE(SIG)
22      &        MAT=DEVS*+SIGINV*+DEVSDASH
23   ENDPROCEDURE
```

Now apply the GAMMAPLOT procedure to the flood records for Rio do Sul, Ibirama, Apiuna and Indaial:

```
38   MATRIX [ROWS=32;COLUMNS=32] Mat
39   VARIATE [NVALUES=32] Distance,Gamabsc
40   GAMMAPLOT V; N=4; ROSE=32; Gamabsc; Mat
41   EQUATE [OLDFORMAT=!((1,-32)31)] Mat; NEWSTRUCTURES=Distance
42   CALCULATE Distance=SORT(Distance)
43   GRAPH [NROWS=16;NCOLUMNS=70] Distance;Gamabsc
```

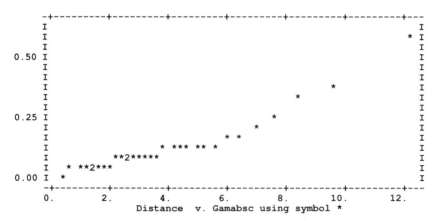

The resulting cumulative plot shows evidence of curvilinearity. We now transform the four flood sequences to logs, to see whether the cumulative plot is more nearly linear.

```
53   CALCULATE V[1...4]=LOG(V[1...4])
54   GAMMAPLOT V; N=4; ROSE=32; Gamabsc; Mat
55   EQUATE [OLDFORMAT=!((1,-32)31)] Mat; NEWSTRUCTURES=Distance
```

Figure 7.6 A GENSTAT procedure to perform a 'gamma plot' as a visual check for multivariate normality. After the procedure listing, it is used to produce gamma plots for the four annual flood records from Rio do Sul, Ibirama, Apiuna and Indaial on the Rio Itajaí-Açú. The first plot is obtained using the raw (untransformed) data; the second is obtained after transforming all four sequences to logarithms.

_____ Continued _____

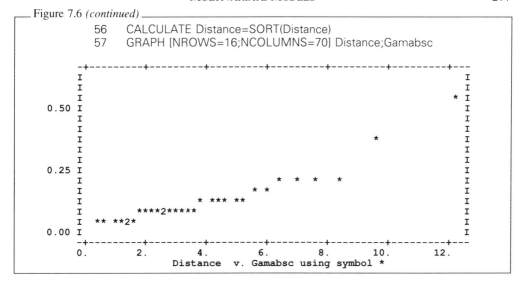

Figure 7.6 *(continued)*

```
      56    CALCULATE Distance=SORT(Distance)
      57    GRAPH [NROWS=16;NCOLUMNS=70] Distance;Gamabsc
```

An approximate assessment of the agreement of transformed sequences with bivariate normality can be made as follows. Denote the transformed sequences by $\{\mathbf{U}_t, \mathbf{V}_t\}$, $t = 1, 2, \ldots, N$.

(i)   Calculate the quantities d_t^2, given by

$$d_t^2 = (\mathbf{U}_t - \overline{\mathbf{U}})^{\mathrm{T}} \mathbf{S}^{-1} (\mathbf{U}_t - \overline{\mathbf{U}}) \qquad t = 1, 2, \ldots, N \tag{7.14}$$

In this expression $\mathbf{S}$ is the variance-covariance matrix calculated from the N vectors $[\mathbf{U}_t, \mathbf{V}_t]^{\mathrm{T}}$, and $\overline{\mathbf{U}}, \overline{\mathbf{V}}$ are the means of these vectors.

(ii) Place the quantities d_t^2 in ascending order, say $d_{(t)}^2$.

(iii) Plot the pairs $d_{(t)}^2$, $\chi_2^2(t - 1/2)/N$, where $\chi_2^2(t - 1/2)/N$ is the $100(t - 1/2)/N$ percentile of the χ^2 distribution with, for the bivariate case, two degrees of freedom. Where, more generally, we are dealing with G sequences of annual floods, the matrix $\mathbf{S}$ will have dimensions $G \times G$, and the χ^2 values will be $\chi_G^2(t - 1/2)/N$ percentiles of the χ^2 distribution with G degrees of freedom.

Figure 7.6 shows a GENSTAT procedure for such a plot applied to annual flood data from four flow gauging sites (Rio do Sul, Ibirama, Apiuna, Indaial) in the basin of the Rio Itajaí-Açú. The curvature of the plot shown in the output indicates that the raw data—as expected—cannot be considered multivariate normal. We therefore seek four Box–Cox parameters which jointly transform the four flood sequences to approximate multivariate normality. For this purpose we seek $\boldsymbol{\lambda} = [\lambda_1, \lambda_2, \lambda_3, \lambda_4]^{\mathrm{T}}$ which together maximise the log-likelihood

$$\ln L = -(n/2) \ln |\mathbf{S}(\lambda)| + \sum_{i=1}^{G} (\lambda_i - 1) \sum_{j=1}^{N} (\ln x_{ij}) \tag{7.15}$$

MATLAB is particularly appropriate for such calculations, and Figure 7.7 shows the edited output from a program which effects this maximisation using the Nelder–Mead algorithm for minimising a function. We see that the estimate of the vector λ is $\hat{\boldsymbol{\lambda}} = [0.4184, 0.7060, 0.4030, 0.0146]^{\mathrm{T}}$. Figure 7.8 shows the gamma plot, as described above, after transforming the original four annual flood sequences. Much of the curvature evident in Figure 7.6 has now disappeared, so we have gone a long way towards securing a transformation to multivariate normality; however, there remains one point which appears to lie away from the linear trend. The year giving rise to this point can be readily traced by identifying which year gave rise to the largest ordinate plotted in the graph; a check on the validity of the original record is called for.

The gamma plot procedure, described above, requires the use in equation (7.14) of the estimated means $\overline{U}$, $\overline{V}$ and estimated variance-covariance matrix S, instead of their true values. If the sample size is large — a rough rule being that the number of years of record, N, should be greater than ten times the number of gauging sites, G — this substitution causes no complications. For smaller samples, some modifications must be made. We now give the modified steps necessary when $N < 10G$ (supposing N years' data from each of G gauging sites):

(i) Calculate the *jackknifed* means $\overline{x}_{(i)}$ obtained by omitting the data from year i and calculating the means of the remaining $(N - 1)$ data values for each of the G gauge sites. There will

```
     ≫            lamda =[0.5943;0.3736;0.4651;0.1201];
     ≫            [lamda,cnt] =NELDER('bmult',lamda,0.00001)
  lamda =
      0.4184
      0.7060
      0.4030
      0.0146
```

Figure 7.7   Edited output from a MATLAB program, using the Nelder–Mead algorithm to find four Box–Cox parameters, $\lambda_1, \lambda_2, \lambda_3, \lambda_4$ which jointly maximise the log-likelihood function given in equation (7.15), where S is the variance-covariance matrix of the variables after Box–Cox transformation, using parameters $\lambda_1, \lambda_2, \lambda_3, \lambda_4$. The first line gives the starting value of the vector λ, obtained by maximising the log-likelihood for each of the four variables individually. Data used are the 32-year annual flood sequences from gauging stations Rio do Sul, Ibirama, Apiuna and Indaial, in the basin of the Rio Itajaí-Açú.

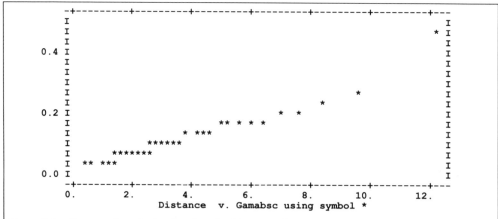

Figure 7.8   Gamma plot calculated, using the 32 years of common annual flood records from Rio do Sul, Ibirama, Apiuna and Indaial on the Rio Itajaí-Açú, southern Brazil, after transformation using the jointly optimised Box–Cox parameters λ, found in Figure 7.7. The plot should be near-linear if the joint Box–Cox transformation achieves multivariate normality.

obviously be different values $\bar{\mathbf{x}}_{(i)}$ for each of the N years of record. In the bivariate case considered in (7.14), $\bar{\mathbf{x}}_{(i)}$ would have just two components, $\bar{U}_{(i)}$, $\bar{V}_{(i)}$; in the general case of G gauging stations, $\bar{\mathbf{x}}_{(i)}$ will have G components.

(ii) Calculate the jackknifed covariance matrices, one for each year of record, using the expression

$$\mathbf{S}_{(i)}^{-1} = a\mathbf{S}^{-1} + ab\mathbf{S}^{-1}(\mathbf{x}_i - \bar{\mathbf{x}})^{\mathrm{T}}(\mathbf{x}_i - \bar{\mathbf{x}})S^{-1}/D \qquad (7.16)$$

where

$$D = 1 - b(\mathbf{x}_i - \bar{\mathbf{x}})\mathbf{S}^{-1}(\mathbf{x}_i - \bar{\mathbf{x}})^{\mathrm{T}} \qquad (7.17)$$

$a = (N - 2)/(N - 1)$, and $b = N/(N - 1)^2$. Here, $\mathbf{x}_i$ is the ith row of the $N \times G$ matrix of data values, and the dimensions of $\mathbf{S}^{-1}$, $\mathbf{S}_{(i)}^{-1}$ are $G \times G$; $\mathbf{x}_i$ is ith row of the $N \times G$ data matrix, and $\bar{\mathbf{x}}$ is the $1 \times G$ vector of means of the N data from each of the G sites.

(iii) Plot the quantities $(\mathbf{x}_i - \bar{\mathbf{x}}_{(i)})\mathbf{S}_{(i)}^{-1}(\mathbf{x}_i - \bar{\mathbf{x}}_{(i)})^{\mathrm{T}}$ against $G(N^2 - 1)/(N(N - G))$ times the quantiles of an F-distribution with G and $N - G$ degrees of freedom.

Figure 7.9 shows a GENSTAT procedure for this calculation, together with the output obtained when it is used with data from the four sites Rio do Sul, Ibirama, Apiuna and Indaial. The first

---

The GENSTAT procedure FPLOT for this calculation is given first:

```
PROCEDURE 'FPLOT'
PARAMETER NAME='V','N','ROSE','FABSC','FORD','FORD2';MODE=p
MATRIX [ROWS=1;COLUMNS=ROSE] S
    &    [ROWS=1;COLUMNS=N] VBAR,VI,VBARI,XI,XII
    &    [ROWS=ROSE;COLUMNS=N] W,W1
    &    [ROWS=ROSE;COLUMNS=ROSE] PICK,PICKMINUS1
    &    [ROWS=ROSE;COLUMNS=1] EYE
    &    [ROWS=N;COLUMNS=1]TOP
SYMMETRICMATRIX [ROWS=N] SIG,SIGINV,SGINVM1
VARIATE [NVALUES=ROSE] T
    &    [VALUES=1...ROSE] I
SCALAR U,A,B,AA,BB,CC,ROSEM1
CALCULATE ROSEM1=ROSE-1
    &              S=1
    &              EYE=1
    &              PICK=0
    &    PICK$[1;1]=1
    &              VBAR=S*+V/ROSE
    &              W=V-EYE*+VBAR
    &              SIG=(TRANSPOSE(W)*+W)/ROSEM1
    &         SIGINV=INVERSE(SIG)
CALCULATE          T=(I-0.5)/ROSE
    &              A=N
    &              B=ROSE-N
```

Figure 7.9   Edited output from a program to check multivariate normality using jackknifed means and covariance matrices. The procedure given here is valid when the number of data units is less than about 10 times the number of variates. The plot is illustrated using flood data from Rio do Sul, Ibirama, Apiuna and Indaial on the Rio Itajaí-Açú; the number of years having data at all four sites is 32, and the number of flood sequences (variates) is four, so this visual test is likely to be more appropriate than that of Figure 7.8.

Continued

Figure 7.9 *(continued)*

```
            &     FABSC=FED(T;A;B)
CALCULATE VI=PICK*+W
    &          U=QPRODUCT(VI;SIGINV)
    &       FORD=U*TRANSPOSE(PICK)
FOR [NTIMES=ROSEM1]
CALCULATE PICK=CIRCULATE(PICK;1)
        &     VI=PICK*+W
        &        U=QPRODUCT(VI;SIGINV)
        &     FORD=FORD+U*TRANSPOSE(PICK)
ENDFOR
CALCULATE               PICK=1
        &       PICK$[1;1]=0
        &            W1=V
        &       PICKMINUS1=0
        & PICKMINUS1$[1;1]=1
        &             VBARI=PICK*+V/(ROSE-1)
        &              W1=W1-TRANSPOSE(PICKMINUS1)*+VBARI
        &              AA=(ROSE-2)/(ROSE-1)
        &              BB=ROSE/(ROSE-1)**2
        &              XI=PICKMINUS1*+W
        &              TOP=SIGINV*+TRANSPOSE(XI)
        &              CC=PICKMINUS1*+FORD
        &          SGINVM1=AA*SIGINV+(AA*BB*TOP*+TRANSPOSE(TOP))\
                              /(1-BB*CC)
        &              XII=PICKMINUS1*+W1
        &               U=XII*+SGINVM1*+TRANSPOSE(XII)
        &             FORD2=U*TRANSPOSE(PICKMINUS1)
FOR [NTIMES=ROSEM1]
    CALCULATE  PICK=CIRCULATE(PICK;1)
        &     PICKMINUS1=CIRCULATE(PICKMINUS1;1)
        &             VBARI=PICK*+V/(ROSE-1)
        &              W1=W1-TRANSPOSE(PICKMINUS1)*+VBARI
        &              XI=PICKMINUS1*+W
        &              TOP=SIGINV*+TRANSPOSE(XI)
        &              CC=PICKMINUS1*+FORD
        &          SGINVM1=AA*SIGINV+(AA*BB*TOP*+TRANSPOSE(TOP))\
                              /(1-BB*CC)
        &              XII=PICKMINUS1*+W1
        &               U=XII*+SGINVM1*+TRANSPOSE(XII)
        &             FORD2=FORD2+U*TRANSPOSE(PICKMINUS1)
ENDFOR
"FORD2 now contains the jack-knifed distances. They now need to be put
in increasing order of magnitude using SORT."
''
''
CALCULATE FORD2=SORT(FORD2)
   &             FORD=SORT(FORD)
 ''
 ''
ENDPROCEDURE
''
''
MATRIX [ROWS=32;COLUMNS=4]V,VV
OPEN 'fig7-7.dat';CHANNEL=2;FILETYPE=input
```

Figure 7.9 *(continued)*

```
READ [CHANNEL=2] V
CLOSE 'fig7-7.dat';CHANNEL=2; FILETYPE=input
MATRIX [ROWS=32;COLUMNS=1]Ford,Ford2,Fabsc
FPLOT V;N=4;ROSE=32;Fabsc;Ford;Ford2
GRAPH [NROWS=16;NCOLUMNS=70] Ford;Fabsc
GRAPH[ NROWS=16;NCOLUMNS=70] Ford2;Fabsc
CALCULATE VV=LOG(V)
FPLOT VV;N=4;ROSE=32;Fabsc;Ford;Ford2
GRAPH [NROWS=16;NCOLUMNS=70] Ford2;Fabsc
```

The following graph shows the 'jack-knifed' plot for the raw data (four annual flood sequences from Rio do Sul, Ibirama, Apiuna and Indaial):

```
105 GRAPH [NROWS=16;NCOLUMNS=70] Ford2;Fabsc
```

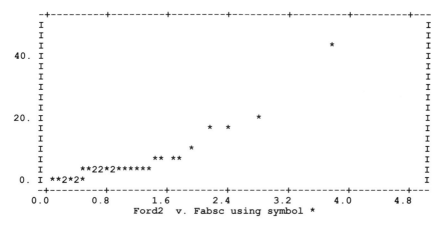

The following graph shows the probability plot after transforming all four annual flood sequences by taking logarithms:

```
106    CALCULATE VV=LOG(V)
107    FPLOT VV;N=4;ROSE=32;Fabsc;Ford;Ford2
108    GRAPH [NROWS=16;NCOLUMNS=70] Ford2;Fabsc
```

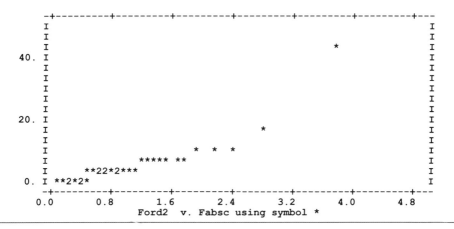

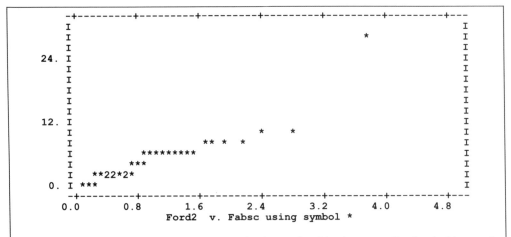

Figure 7.10   This edited output also presents a visual test of multivariate normality, but in this case the original four 32-year annual flood sequences have been transformed using the four Box–Cox parameters found in Figure 7.7, which jointly maximise the four-variate log-likelihood function.

plot was obtained applying the method to the untransformed data, and the second to the data after transforming to logarithms (for Rio do Sul and Ibirama at least, the log transformation is consistent with the univariate Box–Cox parameters, as the discussion of Figure 7.3 showed). Figure 7.10 shows the same plot after transformation of data using the four Box-Cox parameters $\hat{\lambda} = [0.4184, 0.7060, 0.4030, 0.0146]^{\mathrm{T}}$; the problem noted with the gamma plot, associated with the year giving rise to the largest ordinate, remains.

7.2.6   Likelihood-based Confidence Limits for Quantiles: a Shorter Series $\{y_t\}$ Extended by Means of a Longer Series $\{x_t\}$, Correlated with It

In the above sections we have discussed at some length the procedures for transforming annual flood sequences (or, more generally, any multivariate hydrological sequences) to approximate multivariate normality. We now proceed to illustrate the relevance of such transformations to the calculation of likelihood-based confidence intervals floods with T-year return period, at a site with short record $\{y_t\}$, using additional information at a site with long record, denoted by $\{x_t\}$. We assume that Box–Cox (or other) transformations have been obtained such that $\{x_t\}$, $\{y_t\}$ can be taken as bivariate normal. The method to be described does not require that the bivariate distribution be necessarily bivariate normal; however the calculations are merely simpler if bivariate normality can be assumed, and this is why we have discussed transformations to normality at some length. But given any appropriate bivariate distribution, the same procedures would apply, although it would, of course, be necessary to use a different likelihood function in what follows.

We illustrate the calculation using the annual flood sequence $\{y_t\}$ at Rio do Sul on the Rio Itajaí-Açú, and the longer annual flood sequence $\{x_t\}$ for its tributary the Rio Hercílio at Ibirama. For illustrative purposes, we assume that transformation of each sequence to logarithms is adequate to achieve bivariate normality, so that five parameters are required: μ_x, μ_y, σ_x, σ_y, and ρ; the 100-year flood at Rio do Sul is then $\exp(\mu_y + 2.33\sigma_y)$, the value 2.33 being the 99th percentile of the standard normal $N(0, 1)$ distribution (that is, a probability $1/T = 1/100 =$

0.01 in the upper tail). Although the annual flood sequence at Rio do Sul is shorter than that at Ibirama, a gap in the Ibirama record occurred when data were available from Rio do Sul, and our analysis takes this into account. It would be a simple matter, also, to modify the calculation given below for the case where data at one or other of the two sites is censored (and, since the data are bivariate normal which is relatively easy to integrate numerically, where both records are censored in the same year or years).

Apart from a constant, the log-likelihood for the data is

$$\ln L = -(n_x + n_{xy}) \ln \sigma_x - (n_y + n_{xy}) \ln \sigma_y - (1/2) n_{xy} \ln(1 - \rho^2)$$

$$- (1/2) \sum_{i=1}^{n_x} X_i^2 - (1/2) \sum_{i=1}^{n_y} Y_i^2 - 1/(2(1 - \rho^2)) \sum_{i=1}^{n_{xy}} (X_i^2 - 2\rho X_i Y_i + Y_i^2) \qquad (7.18)$$

where

$$X = (\ln x - \mu_x)/\sigma_x$$

$$Y = (\ln y - \mu_y)/\sigma_y$$

and n_x is the number of years for which only the x-sequence has data, n_y is the number of years for which only the y-sequence has data, and n_{xy} is the number of years for which both sequences have data. By differentiating $\ln L$ with respect to μ_x, μ_y, σ_x, σ_y and ρ, it is found that the maximum-likelihood estimates of these five parameters are $\hat{\mu}_x = 6.5335$, $\hat{\mu}_y = 6.6081$, $\hat{\sigma}_x = 0.5071$, $\hat{\sigma}_y = 0.4313$ and $\hat{\rho} = 0.7378$. The maximum-likelihood estimate of X_0 is therefore $\exp(6.6081 + 2.33 \times 0.4313) = 2024.4$ m^3 s^{-1}. The 95% confidence region for μ_y, σ_y — the parameters for the site with the shorter record, Rio do Sul — is given by all values of these parameters which satisfy

$$2\{\ln L(\hat{\mu}_x, \hat{\mu}_y, \hat{\sigma}_x, \hat{\sigma}_y, \hat{\rho}) - \ln L(\tilde{\mu}_x, \mu_y, \tilde{\sigma}_x, \sigma_y, \tilde{\rho})\} < \chi^2_{2 \text{ df}} \qquad (7.19)$$

where $\tilde{\mu}_x$, $\tilde{\sigma}_x$, $\tilde{\rho}_x$ are the maximum-likelihood estimates of μ_x, σ_x, ρ, given the values of μ_y, σ_y. For each point-pair (μ_y, σ_y) in this region, there corresponds an estimate of the 100-year flood at Rio do Sul, namely $X_0 = \exp(\mu_y + 2.33\sigma_y)$, and the largest and smallest values of these values, say X_L and X_U, give the 95% confidence interval for X_0. One way of calculating X_L, X_U would be by searching over a grid of points in the (μ_y, σ_y) plane. However, instead of plotting X_0 over the whole region, we can search for the maximum and minimum values of X_0 over the region for which the inequality in expression (7.19) becomes an equality: namely

$$2\{\ln L(\hat{\mu}_x, \hat{\mu}_y, \hat{\sigma}_x, \hat{\sigma}_y, \hat{\rho}) - \ln L(\mu_x, \mu_y, \sigma_x, \sigma_y, \rho)\} - \chi^2_{2 \text{ df}} = 0 \qquad (7.20)$$

That is, we can search for the maximum and minimum values of $X_0 = \exp(\mu_y + 2.33\sigma_y)$ along this curve. Figure 7.11 shows the output from a MATLAB program which finds these values, using the Nelder–Mead algorithm; we see that the 95% confidence limit for X_0 is (1727.6, 2860.3) m^3 s^{-1}, the maximum-likelihood estimate of X_0 being $\hat{X}_0 = 2024.4$ m^3 s^{-1}. As a check that the values of $[\mu_x, \mu_y, \sigma_x, \sigma_y, \rho]$ for which X_0 has its maximum and minimum values do in fact lie on the curve (7.20), we can substitute the two sets of five values in equation (7.20) to verify that the left-hand side is close to zero. We find that X_0 has its minimum value, X_L, at [6.4199, 6.4437, 0.4861, 0.4009, 0.6467] and its maximum value, X_U, at [0.6304, 0.7416, 0.5106, 0.4833, 0.7188]. Substituting these values in the left-hand side of equation (7.20), we obtain numerical values of 0.0073 and 0.000 17, respectively. We conclude that the limits X_L and X_U do in fact lie close to the curve defined by equation (7.20).

Calculation of the LOWER 95% confidence interval for X_0 (100-year flood at Rio do Sul). The output shows a value of 1727.6 m^3 s^{-1} for the LOWER 95% confidence limit):

```
>>          theta =[6.5335;6.6081;0.5071;0.4313;0.7378];
>>          [theta,cnt] =NELDER('optmin',theta,0.001,1)
cnt =
      0
initial
v =
   6.5335e+000   6.5335e+000   6.5335e+000   5.8802e+000   7.1869e+000   6.5335e+000
   6.6081e+000   6.6081e+000   6.6081e+000   5.9473e+000   6.6081e+000   7.2689e+000
   5.0710e−001   5.5781e−001   5.0710e−001   4.5639e−001   5.0710e−001   5.0710e−001
   4.3130e−001   4.3130e−001   4.7443e−001   3.8817e−001   4.3130e−001   4.3130e−001
   8.1158e−001   7.3780e−001   7.3780e−001   6.6402e−001   7.3780e−001   7.3780e−001
f =
   1.6820e+005   2.0452e+005   2.3628e+005   1.4471e+007   3.8557e+007   1.1229e+008
cnt =
      203
contract
v =
   6.4052e+000   6.4053e+000   6.4055e+000   6.4055e+000   6.4051e+000   6.4052e+000
   6.4132e+000   6.4134e+000   6.4133e+000   6.4133e+000   6.4132e+000   6.4132e+000
   5.3319e−001   5.3354e−001   5.3359e−001   5.3329e−001   5.3367e−001   5.3336e−001
   4.4694e−001   4.4685e−001   4.4688e−001   4.4688e−001   4.4693e−001   4.4692e−001
   8.2180e−001   8.2194e−001   8.2177e−001   8.2180e−001   8.2181e−001   8.2185e−001
f =
   1.7279e+003   1.7279e+003   1.7279e+003   1.7279e+003   1.7279e+003   1.7279e+003
test =
   1.5376e−003
theta =
   6.4052
   6.4132
   0.5332
   0.4469
   0.8218
cnt =
      203
>>          exp(6.4132+2.33*0.4469)
ans =
   1.7276e+003
```

Calculation of the UPPER 95% confidence interval for X_0 (100-year flood at Rio do Sul). Only output from the initial and final iterations is shown. The output shows a value of 2860.3 m^3 s^{-1} for the UPPER 95% confidence limit):

```
>>          theta =[6.5335;6.6081;0.5071;0.4313;0.7378];
>>          [theta,cnt] =NELDER('optmax',theta,0.001,1)
cnt =
      0
initial
```

Figure 7.11   Calculation of the likelihood-based confidence limits of the 100-year flood at Rio do Sul in the basin of the Rio Itajaí-Açú, using information from the longer flood record available for the Rio Hercílio at Ibirama. Calculation of the confidence limit is effected by Nelder–Mead minimisation of the function 'optmin' shown in the file FIG7-11.PRT. Only the outputs from the first and last iterations are shown. The likelihood-based confidence interval for the 100-year flood is (1727.6, 2860.3) m^3 s^{-1} (the ML estimate is 2024.4 m^3 s^{-1}).

_____ Continued _____

___ Figure 7.11 *(continued)* ___

```
v =
    6.5335e+000    6.5335e+000    6.5335e+000    6.5335e+000    7.1869e+000    5.8802e+000
    6.6081e+000    6.6081e+000    6.6081e+000    7.2689e+000    6.6081e+000    5.9473e+000
    5.0710e−001    5.0710e−001    5.5781e−001    5.0710e−001    5.0710e−001    4.5639e−001
    4.3130e−001    4.7443e−001    4.3130e−001    4.3130e−001    4.3130e−001    3.8817e−001
    8.1158e−001    7.3780e−001    7.3780e−001    7.3780e−001    7.3780e−001    6.6402e−001
f =
    4.1042e−002    4.7158e−002    4.9904e−002    7.3078e+000    9.4083e+000    1.6187e+001
cnt =
       271
reflect
v =
    6.6094e+000    6.6099e+000    6.6090e+000    6.6089e+000    6.6090e+000    6.6093e+000
    6.7867e+000    6.7868e+000    6.7866e+000    6.7866e+000    6.7866e+000    6.7867e+000
    4.5292e−001    4.5289e−001    4.5293e−001    4.5295e−001    4.5297e−001    4.5296e−001
    5.0296e−001    5.0290e−001    5.0301e−001    5.0301e−001    5.0298e−001    5.0293e−001
    7.8640e−001    7.8652e−001    7.8632e−001    7.8628e−001    7.8628e−001    7.8636e−001
f =
    3.4966e−004    3.4966e−004    3.4966e−004    3.4966e−004    3.4966e−004    3.4966e−004
test =
    1.7429e−003
theta =
    6.6094
    6.7867
    0.4529
    0.5030
    0.7864
cnt =
          271
≫            exp(6.7867+2.33∗0.5030)
ans =
    2.8603e+003
≫            newg(theta)
ans =
    0.0055
```

It is of interest to compare the 95% confidence interval for the 100-year flood at Rio do Sul, calculated by the above procedure, with the 95% confidence interval for the same event, calculated using only the annual flood sequence from Rio do Sul itself: that is, without using the longer record from Ibirama. Using the Nelder–Mead algorithm of MATLAB, we can easily compute the maximum-likelihood estimate of the 100-year flood as $\exp(\hat{\mu}_y + 2.33\hat{\sigma}_y) = \exp(6.5908 + 2.33 \times 0.3979) = 1840.7$ m^3 s^{-1}, a value rather less than the estimate of 2024.4 m^3 s^{-1} obtained when the Ibirama data are also used, probably because the Rio do Sul record does not include the two extreme floods which occurred in 1983 and 1984; information transferred to Rio do Sul, by virtue of the correlation between flood peaks at Rio do Sul and those at Ibirama, adjusts the estimate of the 100-year flood upwards. However, the 95% confidence interval, calculated from the Rio do Sul data only, is (1418.2, 2764.9) m^3 s^{-1}, compared with (1727.6, 2860.3) m^3 s^{-1} when both Rio do Sul and Ibirama data are used. The shorter confidence interval in the latter case is a measure of the information introduced by using the longer (Ibirama) flood record.

The above discussion has concentrated on the derivation of likelihood-based confidence intervals for floods with T-year return period, with $T = 100$ in the numerical examples. We can, of course, calculate any $100(1 - \alpha)\%$ confidence intervals for any function of the distribution parameters μ_y, σ_y in exactly the same way. In particular, we can calculate 95% likelihood-based confidence limits for the mean annual flood at Rio do Sul. When the Nelder–Mead algorithm is used to calculate such limits, the maximum-likelihood estimate of the mean annual flood is (when the longer Ibirama record is also included in the analysis) 813.3 m^3 s^{-1}, and the 95% confidence limits are (650.3, 1061.9) m^3 s^{-1}.

We may speculate that evaluation of the widths of confidence limits, calculated from a short record at the site of interest and then by also using longer records from neighbouring gauging stations, provides a means of determining which longer records are worth using, and what is the gain from using them. If the 95% confidence interval for the 100-year flood at Rio do Sul is shortened substantially by using a longer flood record from a station nearby, then there is a clear gain in information from using it; if, however, the confidence interval is lengthened, then there is a dilution of information resulting from introduction of the longer record. The possibility of assessing gains and losses of information does not appear to have been discussed in the hydrological literature, and research is required to establish the circumstances under which the approach is likely to be worthwhile.

7.2.7   A Note on the Bivariate Gamma Distribution

The emphasis of the above discussion has been on transformations to multivariate normality, because much the greater part of multivariate analysis has been developed for normal distributions only. We mention here, however, that some multivariate analogues of the distributions discussed in Chapter 3 exist. The three-parameter log-normal distribution has an obvious multivariate extension; also of interest is the multivariate gamma which has the bivariate and trivariate forms

$$f_{XY}(x, y) = \lambda^{k_1+k_2} y^{k_1-1}(x - y)^{k_2-1} \exp(-\lambda x)/[\Gamma(k_1)\Gamma(k_2)] \tag{7.21}$$

and

$$f_{XYZ}(x, y, z) = \lambda^{k_1+k_2+k_3}(x - y)^{k_1-1}(y - z)^{k_2-1}z^{k_3-1} \exp(-\lambda x)/[\Gamma(k_1)\Gamma(k_2)\Gamma(k_3)] \tag{7.22}$$

where, for equation (7.21), $0 < y < x < \infty$, and, for equation (7.22), $0 < z < y < x < \infty$. For equation (7.21), the means of y and x are k_1/λ and $(k_1 + k_2)/\lambda$, respectively, with variances k_1/λ^2 and $(k_1 + k_2)/\lambda^2$. The correlation between x and y is $\sqrt{(k_1/(k_1 + k_2))}$ and therefore always positive. Generalisations corresponding to the three-parameter univariate gamma are also possible. The inequalities restrict the usefulness of these distributions to particular instances; we could use equation (7.21), for example, to extend records of annual runoff from a river basin, given longer records of annual rainfall, since we could safely assume that annual runoff y is less than annual rainfall x. Similarly, equation (7.21) may be appropriate for extending records of annual runoff y from a headwater basin, given a longer record of annual runoff x at a downstream site; or for extending annual flood records in headwater catchments, given longer flood records downstream. By dividing the joint distribution for x and y, given by (7.21), by the marginal distribution for x, we obtain the conditional distribution of y, given x, as

$$\Gamma(k_1 + k_2)(y/x)^{k_1-1}(1 - y/x)^{k_2-1}(1/x)/[\Gamma(k_1)\Gamma(k_2)] \tag{7.23}$$

which is a beta distribution with conditional mean

$$E[Y|x] = k_1 x/(k_1 + k_2) \tag{7.24}$$

and conditional variance

$$\text{var}[Y|x] = k_1 k_2 x^2/(k_1 + k_2) \tag{7.25}$$

so that the variance of y is proportional to the square of x. The beta distribution of expression (7.23) is a particularly 'flexible' distribution of two parameters; the rectangular distribution ($k_1 = 1$, $k_2 = 1$) and triangular ($k_1 = 2$, $k_2 = 1$) are special forms, and the distribution is symmetric, skewed to the right or left according as $k_1 = k_2$, $k_1 < k_2$, or $k_1 > k_2$. If $k_1 > 1$ and $k_2 > 1$, the distribution is unimodal, and in general skewed. So the distribution (7.21) gives rise to a wide variety of conditional distributions, subject always to the physical constraint that $y < x$. Where hydrological variables can safely be assumed to satisfy it, the bivariate gamma is particularly useful.

The trivariate form (7.23) is equally flexible, subject to the constraint that $x < y < z$. The conditional distribution of z given x and y is a beta distribution in z/y with parameters k_1 and k_2; that of y given x and z is a beta distribution in $(y - z)/(x - z)$ with parameters k_2 and k_3; and that of x given y and z is a gamma distribution in $(x - y)$ with parameters k_3 and λ. The expected values of all three conditional distributions are linear functions in y, x and z, and y, respectively.

Estimation of the parameters of equations (7.21) and (7.22) can be derived by maximum likelihood, although solution of the equations $\partial \ln L/\partial k_1 = 0$, $\partial \ln L/\partial k_2 = 0$, $\partial \ln L/\partial \lambda = 0$ (subject, of course, to graphical verification that the maximum has indeed been found) for expression (7.21) requires the digamma and trigamma functions.

7.3 MULTI-SITE MODELLING OF MONTHLY FLOWS: SOME GENERAL OBSERVATIONS

Hitherto in this chapter we have considered some estimation problems for the case where the hydrological records available at several sites were cross-correlated, but without any dependence amongst data within the same sequence. That is, if we consider a data matrix with G rows, corresponding to data from G sites, and N columns, corresponding to data from N years, our discussion hitherto has dealt with cases where data values in the same column were correlated, whilst those in the same row were not. These earlier discussions considered in particular the analysis of annual flood records (assumed statistically independent from year to year, but possibly correlated between different sites in the same year); and of annual runoff, provided that the G sites measured runoff from catchment areas without any extensive storage which would result in one year's precipitation emerging as runoff in subsequent years.

Where we are concerned with modelling runoff during time intervals shorter than a year (a month, say), we must expect there to be *serial* dependence amongst the data, as well as the cross-correlation between sites, considered hitherto. By way of notation, suppose that $x_t^{(1)}$, $x_t^{(2)}, \ldots$ is the runoff in month t at gauging sites $1, 2, \ldots$; the discussion of this section deals with the modelling of sequences in which there is dependence of $x_t^{(1)}$ on $x_t^{(2)}$ as before, but also dependence of $x_t^{(1)}$, $x_t^{(2)}$ on $x_{t-1}^{(1)}$, $x_{t-1}^{(2)}$. We shall denote by $\mathbf{x}_t$ in bold type the $G \times 1$ vector of observations from the G sites in time interval t, and we shall generally refer to the time interval as a month, although largely the same approach would apply where the time interval is, say, ten days or longer. We shall refer to the data as 'monthly runoff sequences', leaving the reader to make the necessary mental adjustments where hydrological variables other than runoff are to be modelled, possibly over different time intervals.

It is important to emphasise that, when modelling hydrological sequences with serial dependence at several sites, the modeller should not immediately plunge into a multivariate model expressing the vector $\mathbf{x}_t$ in terms of $\mathbf{x}_{t-1}, \mathbf{x}_{t-2}, \ldots$ with its consequent problems of estimating entire matrices of coefficients. It will be sound practice to begin by modelling the monthly runoff sequence at each site separately in the first instance; that is, we adopt a candidate model for each site separately, estimate the model parameters, and examine the structure of the model residuals, modifying the model structure in the light of results of the diagnostic tests discussed in earlier chapters. If, for example, we have used as a candidate model one that expresses $x_t^{(i)}$ in terms of $x_{t-1}^{(i)}$ at site i, then our diagnostic tests would almost certainly include a plot of model residuals against $x_{t-2}^{(i)}$ to assess whether this variable also should be included in the model. However, in addition to the diagnostic tests already discussed, it will be necessary to plot residuals from the candidate model for site i against x_{t-2}^{j}, $j \neq i$, to determine whether these lagged terms from other sites should also be included. If, by means of such tests, we can obtain sequences of residuals $\{e_t^{(i)}\}$ which are approximately normally distributed and have negligible cross-correlations, then we shall have 'decoupled' the sequences: a highly desirable thing to do, since it means that we can model each site independently of the others.

We can express these ideas more formally as follows. Suppose we use the symbol $x_t^{(1)}$ to denote the event $x_t^{(1)} < X_t^{(1)} < x_t^{(1)} + \Delta x_t^{(1)}$; then with $P[E]$ denoting probability of the event E, and considering two sites 1, 2 only, we shall look for ways of describing

$$P[x_t^{(1)}, x_t^{(2)}, x_{t-1}^{(1)}, x_{t-1}^{(2)}, \ldots]$$

and clearly this description will be much simplified if

$$P[x_t^{(1)}, x_t^{(2)}, x_{t-1}^{(1)}, x_{t-1}^{(2)}] = P[x_t^{(1)}, x_t^{(2)}, x_{t-1}^{(1)}, x_{t-1}^{(2)} | x_{t-2}^{(1)}, x_{t-2}^{(2)}, \ldots] \qquad (7.26)$$

since we can then restrict consideration to the joint modelling of $x_t^{(1)}, x_t^{(2)}, x_{t-1}^{(1)}, x_{t-1}^{(2)}$. But we can also write

$$P[x_t^{(1)}, x_t^{(2)}, x_{t-1}^{(1)}, x_{t-1}^{(2)}] = P[x_t^{(1)}, x_t^{(2)}, | x_{t-1}^{(1)}, x_{t-1}^{(2)}] P[x_{t-1}^{(1)}, x_{t-1}^{(2)}]$$

$$= P[x_t^{(1)} | x_t^{(2)}, x_{t-1}^{(1)}, x_{t-1}^{(2)}] P[x_t^{(2)} | x_{t-1}^{(1)}, x_{t-1}^{(2)}] P[x_{t-1}^{(1)}, x_{t-1}^{(2)}] \qquad (7.27)$$

suggesting that the description of $x_t^{(1)}, x_t^{(2)}, x_{t-1}^{(1)}, x_{t-1}^{(2)}$ in terms of their probabilities can be broken down into components: in the first of which we deal with $x_{t-1}^{(1)}, x_{t-1}^{(2)}$ alone, in the second of which we deal with $x_t^{(2)}$ given $x_{t-1}^{(1)}, x_{t-1}^{(2)}$, and in the third of which we deal with $x_t^{(1)}$ given $x_t^{(2)}, x_{t-1}^{(1)}, x_{t-1}^{(2)}$. And the components can be dealt with independently, since the probabilities in the above relations multiply. In simple terms, if we find that we do not need to include $x_{t-2}^{(1)}$ and $x_{t-2}^{(2)}$ in models of $x_t^{(1)}$ and $x_t^{(2)}$, we can model $x_t^{(2)}$ in terms of $x_{t-1}^{(1)}$ and $x_{t-1}^{(2)}$, and then model $x_t^{(1)}$ in terms of $x_t^{(2)}, x_{t-1}^{(1)}$ and $x_{t-1}^{(2)}$.

A comment is necessary concerning the relation between the kinds of model described in the remainder of this chapter, and the time-series models which have received much attention in the hydrological literature in recent years. These time-series models came to hydrology through the admirable text *Time Series Analysis: Forecasting and Control* by Box and Jenkins (1970), following which there was an explosion of interest in the hydrological applications of models denoted by acronyms such as ARMA, ARIMA and ARIMAX. These models are not discussed in any depth in the present book for the following reason.

Box–Jenkins models deal with time series which either have a stationary correlational structure, or which can be simply transformed into other series having this property: for example, by differencing the original series. It seems to us unlikely, in general, that many hydrological series will have a stationary correlational structure, even if they can be transformed into series with stationary mean and variance. As an example, consider the runoff in two successive periods (let us say, of ten days) during the dry season, and then from two successive periods during the wet season, for a basin in which wet and dry seasons are pronounced. Denote the two runoff data values by x_1, x_2, and their correlation by r_{12}. We would expect r_{12} to be very large during the dry season, recession period, because most of the information about x_2 would be contained in x_1. We would expect r_{12} to be considerably smaller during the wet season periods, reflecting the period-to-period variability in rainfall input and consequent rapid runoff during that season. These correlations will not be altered by transformations to attain stationarity of mean and variance of x_1 and x_2; it would therefore be inappropriate to assume that the correlation between x_t and x_{t-1} is constant for all t, as the time-series models assume. The correlational structure clearly differs between wet and dry seasons in this example. A more extreme case occurs in catchment areas of semi-arid regions, where flow may be totally absent in some months of the year; wet season flow will be of variable duration, giving a fragmentary record inappropriate for the use of time-series models of the autoregressive moving-average type.

In this book, therefore, we prefer to work with models in which we express March runoff in terms of runoff during February and, if necessary, January; April runoff in terms of runoff in preceding months March, February, ... The model for April runoff need not necessarily be of the same structure as the model for March runoff: it may have more terms or fewer, and it may include more or fewer lagged variables from neighbouring flow gauging sites. In summary, we prefer to let each month's data, at each site, 'speak for itself', and we shall not impose any rigid model, for example of the form $\mathbf{x}_t - \boldsymbol{\mu} = \mathbf{A}(\mathbf{x}_{t-1} - \boldsymbol{\mu}) + \mathbf{B}\boldsymbol{\varepsilon}_t$, including the same number of terms at each site and in each season.

7.3.1   The Multi-site Thomas–Fiering Model: Multivariate Multiple Regression

However, since we have to start somewhere, let us begin by generalising the simple Thomas–Fiering model to the multi-site case. We have seen that the univariate Thomas–Fiering model, as commonly used, is often unsatisfactory because assumptions implicit in the model are unlikely to be valid; the same is true of the multi-site Thomas–Fiering model, but it serves as a point of departure for discussion of alternative models which may be more appropriate.

We define the runoff model for G sites, and consider the estimation of its parameters, for just one month, denoted by t; t may be the month of June. There will be models similar to that described below for each month of the year. If we have N complete years of runoff record from each of the G sites, we can arrange the data from the N months (June) as a data matrix with N rows and G columns. (In earlier sections of this chapter, we took G to be the number of rows in the data matrix, with P — the number of years — being the number of columns; in the present section, we take G as the number of columns, in order to be consistent with the notation commonly used in regression analysis, and used previously in Chapter 4.) Call this matrix $\mathbf{Y}$. Suppose that for month t (June) we have p explanatory variables for each of the G sites: these may include, for example, the runoffs at the G sites in month $t-1$, perhaps also the runoffs from the G sites in month $t-2$, and so on. These we can arrange as an $N \times (p+1)$ matrix of explanatory-variable data, in which the first column consists entirely of ones, and the second, third, ... , $(p + 1)$th columns are the explanatory variables. Call this matrix $\mathbf{X}$. To relate $\mathbf{Y}$ to $\mathbf{X}$, we need a $(p + 1) \times G$ matrix of coefficients $\mathbf{B}$, and an $N \times G$ matrix of residuals ε. Then, for month t (June), we can write

$$\mathbf{Y} = \mathbf{XB} + \varepsilon \qquad (7.28)$$

where the $N \times G$ matrix of residuals is assumed to be such that, denoting by $\varepsilon_{(i)}$, $\varepsilon_{(k)}$ its ith and kth columns, $E[\varepsilon_{(i)}] = 0$; $\mathrm{cov}(\varepsilon_{(i)}, \varepsilon_{(k)}) = E[\varepsilon_{(i)}\varepsilon_{(k)}^{\mathrm{T}}] = \sigma_{ik}\mathbf{I}$ where $i, k = 1, \ldots, G$ and $\mathbf{I}$ is the $N \times N$ identity matrix. The $G \times G$ covariance matrix $\mathbf{\Sigma} = [\sigma_{ik}]$ is the covariance matrix of the flows at the G sites. The estimate of the matrix $\mathbf{B}$ is analogous to that obtained in the multiple regression case, namely

$$\hat{\mathbf{B}} = (\mathbf{X}^{\mathrm{T}}\mathbf{X})^{-1}\mathbf{X}^{\mathrm{T}}\mathbf{Y} \qquad (7.29)$$

The estimated residuals are given by the matrix

$$\hat{\varepsilon} = \mathbf{Y} - \hat{\mathbf{Y}} = [\mathbf{I} - \mathbf{X}(\mathbf{X}^{\mathrm{T}}\mathbf{X})\mathbf{X}^{\mathrm{T}}]\mathbf{Y} \qquad (7.30)$$

and it can be shown that when the matrix $\mathbf{X}$ is of full rank p (that is, when no coefficients $l_1, l_2, \ldots, l_p$, not all zero, can be found such that $l_1\mathbf{x}_{(1)} + l_2\mathbf{x}_{(2)} + \cdots + l_p\mathbf{x}_{(p)} = \mathbf{0}$), and when the errors ε have a G-variate normal distribution with covariance matrix $\mathbf{\Sigma}$, then the estimate of $\mathbf{\Sigma}$ is

$$\hat{\mathbf{\Sigma}} = \hat{\varepsilon}^{\mathrm{T}}\hat{\varepsilon}/N = (\mathbf{Y} - \mathbf{X}\hat{\mathbf{B}})^{\mathrm{T}}(\mathbf{Y} - \mathbf{X}\hat{\mathbf{B}})/N \qquad (7.31)$$

The rank of the matrix $\mathbf{X}$ can be determined very easily in MATLAB by simply entering the instruction 'rank($\mathbf{X}$)'.

When using the multivariate multiple regression for simulation purposes (that is, for the generation of artificial sequences) it will of course be necessary to have fitted the model (7.28) to each month's flow record. Given the estimates of the 12 matrices $\mathbf{B}$ and $\mathbf{\Sigma}$, say $\hat{\mathbf{B}}$ and $\hat{\mathbf{\Sigma}}$, the procedure for generating the monthly flows at the G sites in, say, June, consists basically of two steps. First, a G-dimensional pseudo-random variable is generated from the normal distribution with zero mean and covariance matrix $\hat{\mathbf{\Sigma}}$, the covariance matrix for that month. Denote this pseudo-random variable by $\mathbf{e}$. Then, using the matrix of explanatory variables $\mathbf{X}$ appropriate for June — constructed from the monthly flows previously generated for the month of May, and perhaps one or more previous months — calculate the generated flow for June as $\mathbf{Y} = \mathbf{X}\hat{\mathbf{B}} + \mathbf{e}$. The generation of the G-variate pseudo-random variable with covariance matrix $\hat{\mathbf{\Sigma}}$ is achieved as follows.

(i) Calculate the eigenvalues and eigenvectors of $\hat{\mathbf{\Sigma}}$. This is easily achieved, notably with both GENSTAT and MATLAB, or with other programs. Denote the eigenvalues by $\lambda_1, \lambda_2, \ldots, \lambda_G$ and the eigenvectors by $\mathbf{w}_1, \mathbf{w}_2, \ldots, \mathbf{w}_G$.

(ii) Construct the matrix $\mathbf{W}$ having the eigenvectors $\mathbf{w}_1, \mathbf{w}_2, \ldots, \mathbf{w}_G$ as its columns.

(iii) Generate G pseudo-random variables $u_1, u_2, \ldots, u_G$ from the $N(0,1)$ distribution, and transform them to normally distributed variates with variances $\lambda_1, \lambda_2, \ldots, \lambda_G$ by calculating $u_i^* = u_i \sqrt{(\lambda_i)}$.

(iv) Calculate the $G \times 1$ vector $\mathbf{W}^{\mathrm{T}}\mathbf{u}^*$, which has zero mean and the desired covariance matrix $\hat{\mathbf{\Sigma}}$.

To illustrate the mechanics, we give an abbreviated example using 50 years of monthly flow records from three gauging stations lying within the basin of the Rio Itajaí-Açú: Ibirama (drainage area 3314 km^2), Apiuna (drainage area 9242 km^2), and Indaial (drainage area 11 151 km^2). The file STN4.M contains data from the three months April, May and June (each a 46×7 array) and Figure 7.12 shows the output from the following short MATLAB program:

```
% The file 'fig7-12.out' is the output file.
% stn4.m is the file with 50 years' of monthly flow
% data from each of the seven sites.
diary fig7-12.out
stn4
Apr=[data(:,2) data(:,5) data(:,8)]
May=[data(:,3) data(:,6) data(:,9)]
Jun=[data(:,4) data(:,7) data(:,10)]
[m,n]=size(May);
May=[ones(m,1) May];
R=rank(May)
if R>0,
   XTY=May'*Jun
   XTX=May'*May
      B=inv(XTX)*XTY
      e=Jun-May*B
      sigma=cov(e)
      correlations=corr(e)
      [W,lamda]=eig(sigma)
else
      'matrix of explanatory variables is not of full rank'
end
```

Having constructed the 50×3 matrices **Apr**, **May**, **Jun**, the expressions necessary for the calculation of equations (7.29) and (7.30) are obtained. The matrix **X** is constructed from the May monthly flows, by adding an initial column of ones to allow for the intercepts in the regression. We find that the rank of the matrix **X** (monthly flows for May, in this example) is then 4 (3 for the three sites plus 1 for the column of ones), so the matrix of regression coefficients **B** is calculated and printed. Values in the first row are the three intercepts in the regressions fitted for the three sites; the remaining three rows multiply the May flows in the product **XB**. The residuals $\hat{\varepsilon}$, their covariance matrix $\hat{\Sigma}$, and the correlation matrix follow; last of all are the matrix **W** giving the eigenvectors, and the matrix **lamda** giving the three eigenvalues 94, 62 and 20 980. Clearly the latter eigenvalue is much greater than the others. Values in the matrices are printed to four figures; more figures can be obtained if required (by inserting the MATLAB instruction 'format long').

To generate the artificial sequences, the following MATLAB instructions generate a 50×7 matrix of pseudo-random errors **e** for substitution in **XB** + **e**, sufficient for the generation of 50 artificial June flows (such as might be appropriate for generating the June flows of one 50-year artificial series: in a real simulation problem, many such sequences would be generated):

Apr =

9.3100	30.5000	45.5000
10.6000	35.3000	51.8000
52.3000	166.0000	252.0000
⋮		
41.8000	131.0000	182.0000

May =

4.8400	19.1000	27.8000
18.0000	49.5000	72.5000
21.9000	67.7000	120.0000
⋮		
51.4000	161.0000	227.0000

Jun =

12.6000	43.2000	56.5000
73.3000	237.0000	287.0000
18.9000	43.7000	68.2000
⋮		
105.0000	327.0000	396.0000

R =
   4

XTY =
1.0e+006 *

0.0026	0.0073	0.0096
0.1648	0.4466	0.5721
0.4549	1.2503	1.6005
0.6011	1.6474	2.1145

XTX =
1.0e+006 *

0.0001	0.0021	0.0060	0.0081
0.0021	0.1890	0.4867	0.6380
0.0060	0.4867	1.3123	1.7095
0.0081	0.6380	1.7095	2.2413

B =

17.5294	52.4165	78.1378
−0.4370	−1.7808	−2.1686
−0.0360	0.4184	0.2622
0.3568	0.7337	1.0789

e =

−12.0461	−28.9877	−46.1442
39.5498	142.7290	156.6954
−29.4390	−86.0944	−109.6679
⋮		
34.7324	132.1889	142.1958

Figure 7.12 Edited output from MATLAB program (see file FIG7-12.PRT) to fit multivariate multiple regression to June flows ($= Y$), in terms of May flows ($= X$), for 50 years' data from three gauging stations on the Rio Itajaí-Açú and tributaries: Ibirama, Apiuna, Indaial. Data are in the MATLAB file STN4.M on the diskette distributed by The MathWorks. For matrices with 50 rows, only the first three and the last three rows are shown.

_____ Continued _____

Figure 7.12 *(continued)*

```
sigma =
  1.0e+004 *
        0.1132          0.2807          0.3630
        0.2807          0.7567          0.9614
        0.3630          0.9614          1.2437

correlations =
        1.0000          0.9591          0.9676
        0.9591          1.0000          0.9910
        0.9676          0.9910          1.0000

W =
       -0.5173          0.8256          0.2253
        0.7422          0.3019          0.5983
       -0.4259         -0.4767          0.7690

lamda =
  1.0e+004 *
        0.0094               0               0
             0          0.0062               0
             0               0          2.0980
```

```
rand('normal');        U=rand(50,7);
for i=1:3,
    ll = sqrt(lamda(i,i));
    for j=1:50,
        U(j,i)=ll*U(j,i);
    end;
end;
for i=1:50,
x=U(i,:)';        y=W'*x;
    for j=1:3,
        V(i,j)=y(j);
    end;
end;
```

Our discussion of multivariate multiple regression has been in the context of hydrological simulation. However, the method is applicable in many other contexts. For example, we may wish to relate both of the two variables y_1, ln(discharge), and y_2, ln(suspended sediment concentration), to explanatory variables describing basin climatological, geological, and vegetation characteristics, perhaps to combine y_1 and y_2 subsequently to give ln(suspended sediment yield) = ln(discharge $\times$ suspended sediment concentration) = $y_1 + y_2$. In this case, it may be sensible to use multivariate regression of the response variable $[y_1, y_2]^T$ on the explanatory variables. We return to this topic in Section 7.3.6 below.

Note, however, that care is necessary in the use of multivariate response variables; if, say, y_1 and y_2 were sediment yield (tonnes per square kilometre) and specific yield (thousands of cubic metres of runoff per square kilometre), then the calculation of sediment yield will already have utilised runoff data, so that a spurious correlation will have been introduced between y_1 and y_2 in this case.

7.3.2   Criticism of the Model

Section 7.3.1 illustrated only the straightforward mechanics of the calculation. It is important to emphasise, however, that we need to look very critically at the model before using it. In the uncritical fitting procedure used above, we have fitted in all $4 \times 3 = 12$ regression coefficients, the elements of the matrix **B**; and $3(3 + 1)/2 = 6$ variances and covariances. One important question to be asked, therefore, is whether all these parameters are required or whether an adequate model can be fitted using rather fewer. Likewise, our uncritical fitting of the model, and its use for generating artificial flow sequences, has omitted any steps to verify whether we are justified in treating the residuals in $\hat{\varepsilon}$ as multivariate normal; nor have we considered whether the covariance matrix of ε is homogeneous. In fact, the multivariate multiple regression as fitted above is a very poor model for these three flow records, not least because, when it is used to generate artificial flow sequences, a considerable number of the generated flows are negative.

A further criticism appears if we consider the nature of the drainage areas of which we are modelling the monthly flows. The three areas are of increasing size, the drainage area of Indaial incorporating the other two. The first column of regression coefficients in the matrix **B** are those for the component in which June flow at Ibirama is expressed in terms of May flow at Ibirama, at Apiuna and at Indaial, respectively: or, in obvious notation,

$$IbJun = 17.5294 - 0.4370 IbMay - 0.0360 ApMay + 0.3568 IndMay + \text{error terms} \quad (7.32)$$

However, the logic of using, as an explanatory variable for predicting June flow at Ibirama, the previous month's flow from larger, downstream catchment areas (Indaial, Apiuna) is questionable, so an obvious way of reducing the number of fitted regression coefficients from 12 to some smaller number is by looking at whether these terms can be omitted. In fact in the univariate analysis for Ibirama, $IbMay$ accounts for only 30% of the variation in $IbJun$; if $ApMay$ is included this percentage rises slightly to 34%, and is virtually unchanged if $IndMay$ is also included. As expected, $ApMay$ and $IndMay$ contribute nothing to explaining the variance of $IbJun$, so there is a reasonable case for using as the model for June flows at Ibirama, the simplified form

$$IbJun = \beta_0 + \beta_1 IbMay + \text{an error term} \quad (7.33)$$

When we come to the downstream catchment at Apiuna, we again find that June flow, $ApJun$, is not very strongly correlated with the previous month's flow $ApMay$, which accounts for about 37% of its variation. But it is hydrologically reasonable to look at the relation between $ApJun$ and $IbJun$, the June flows from the upstream and downstream catchment areas; not surprisingly, we find that $IbJun$ accounts for over 94% of the variation in $ApJun$, and the inclusion of further explanatory variables — even $ApMay$ — brings no significant improvement. So there are good reasons why $IbJun$ should be used as an explanatory variable for $ApJun$, and why $IndMay$ should not. Similarly, the univariate analysis of $IndJun$ shows that June flow at the upstream catchment area, $ApJun$, accounts for nearly 99% of the variance, other variates making no significant contribution.

Further univariate analyses suggest that a parsimonious model can be developed from the following form:

$$IbJun = \beta_0 + \beta_1 IbMay + \text{error term}$$

$$ApJun = \beta_0 + \beta_1 IbJun + \beta_2 ApMay + \text{error term} \quad (7.34)$$

$$IndJun = \beta_0 + \beta_1 IbJun + \beta_2 IndMay + \text{error term}$$

When these univariate models are fitted, the raw residuals *IbJun−IbJun* (fitted), *ApJun−ApJun* (fitted) and *IndJun−IndJun* (fitted), plot against each other as shown in Figure 7.13, and this output also shows the covariance matrix calculated from the raw residuals. The correlations between Ibirama residuals with those from Apiuna and Indaial are both small (0.0734, 0.1154) suggesting that — always provided the assumption of multivariate normality is valid — we may be able to 'decouple' the Ibirama model from the model for Apiuna and Indaial. If this decoupling is possible, the model for Ibirama can be developed independently of the model for Apiuna and Indaial, with a consequent reduction both in the number of model parameters and, hence, in the dimensionality of the parameter space in which maximum-likelihood estimates are to be sought.

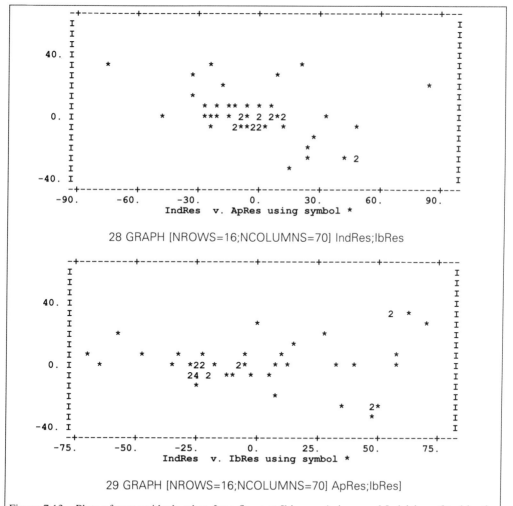

Figure 7.13   Plots of raw residuals when June flows at Ibirama, Apiuna, and Indaial are fitted by the models described in the text: residuals are from univariate models fitted at each site. Also shown is the covariance matrix calculated from these residuals.

_ Continued _

Figure 7.13 *(continued)*

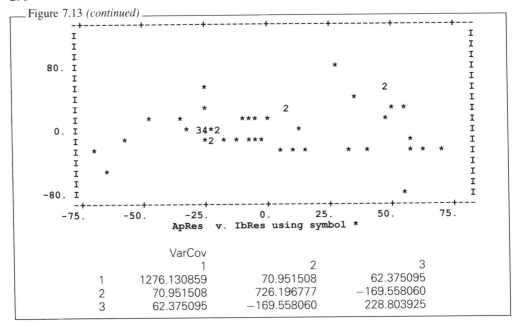

ApRes v. IbRes using symbol *

	VarCov		
	1	2	3
1	1276.130859	70.951508	62.375095
2	70.951508	726.196777	−169.558060
3	62.375095	−169.558060	228.803925

A very useful tool, therefore, in the modelling of multi-site sequences of monthly flow, is a test of whether a matrix of correlations, calculated from the residual matrix ε in equation (7.30), is consistent with the hypothesis that it is of diagonal form. This we present in Section 7.3.3.

7.3.3 Test that a Correlation Matrix is of Diagonal Form

The test which follows is strictly appropriate only where the residuals ε can be assumed multivariate normal, and it is based upon the following result from statistical theory. Given a correlation matrix $\mathbf{R}$ calculated from N multivariate normal observations of G variates (in the present context of modelling monthly flows N is the number of years and G is the number of gauging stations) then when the hypothesis H_0 (that the correlation matrix is diagonal) is true, the statistic

$$-n' \ln |\mathbf{R}| \qquad (7.35)$$

where $n' = N - (2G + 11)/6$, is approximately distributed as χ^2 with $G(G - 1)/2$ degrees of freedom when N is not too small. $|\mathbf{R}|$ is the determinant of the matrix $\mathbf{R}$, and is equal to the product of its eigenvalues. In GENSTAT it can be found by entering

CALCULATE y = det(R)

and in MATLAB by entering

y = det(R)

This test needs to be modified when it is applied to the correlation matrix calculated from the residuals $\hat{\varepsilon}$ derived from the records for Ibirama, Apiuna and Indaial, since these residuals are constrained by the fact that the sum of the elements in each of the three columns of $\hat{\varepsilon}$ is zero; we suggest that n' in expression (7.35) be replaced by $(N - 1) - (2G + 11)/6$. In the present case, $n' = 46.1667$ and $\chi^2 = 10.2361$ with three degrees of freedom, a value lying well into the upper 5% tail of the tabulated χ^2. We conclude that the full 3×3 matrix **R** is not consistent with the hypothesis of diagonal form; the univariate analyses therefore do not suggest that our suggested model for all three sites will decouple.

7.3.4 The Use of Transformations when Simulating Flow Sequences

Earlier in this chapter, we made extensive reference to the transformation of annual flood sequences: transformation does not greatly complicate the estimation of annual floods with T-year return periods, since the inverse transformation gives the estimate of the T-year flood on the original scale of measurement. Where, however, we are modelling sequences of monthly flow in order to generate synthetic flow sequences for simulation purposes, transformation must usually be avoided. This is because the statistical characteristics in the transformed scale, in which a model is fitted, are changed after inverse transformation. A simple example occurs where a log transformation of data is used to achieve normality; the mean $\hat{\mu}$ estimated on the log scale gives a negatively biased estimate of the mean on the original scale of measurement. To avoid such difficulties it is desirable to identify and fit models intended for simulation purposes in the scale of measurement.

7.3.5 Modelling Monthly Flow for the Apiuna and Indaial Sites (continued)

We now discuss in a little more detail the modelling of June flows at the two sites Apiuna and Indaial, using the suggested model (7.34) above. When the June flows at either site are plotted against, say, the explanatory variable June flow at Ibirama, two features stand out: first, the extreme divergence of the June flows in the year 1983 from the other June flows in the period of record; and second, the wider dispersion at large values of the explanatory variable. The different nature of the (*el Niño*) flow in May–June 1983 causes frequent messages, warning of large residuals and high leverage, whatever model is fitted, whether linear or generalised linear, so that it is tempting to omit from the analysis the year 1983, without which the data would be reasonably well behaved. Such a temptation should be strongly resisted. The divergence from the remainder of the high May–June flows of 1983 arises from a real and widely observed meteorological phenomenon, and is not the consequence of questionable data. There is therefore no case for omitting 1983 from the record.

Because June flows at Apiuna and Indaial cannot be decoupled, it is necessary to model the June flows at these sites taking account of the cross-correlation between them. Probably the simplest model with any degree of realism is

$$E[ApJun] = \gamma_0 + \gamma_1 IbJun + \gamma_2 ApMay$$
$$E[IndJun] = \beta_0 + \beta_1 IbJun + \beta_2 IndMay$$
(7.36)

with residuals normally distributed; however, we must take account of the way that dispersion increases at higher flows. Extending Aitkin's model for heterogeneous variance described in Chapter 6, we take

$$\text{var}(ApJun) = \exp(\lambda_{x0} + \lambda_{x1} IbJun)$$
$$\text{var}(IndJun) = \exp(\lambda_{y0} + \lambda_{y1} IbJun)$$
(7.37)

where the parameters $\lambda_{x0}, \lambda_{x1}, \lambda_{y0}, \lambda_{y1}$ are to fitted as well as the parameters γ_i and β_i ($i = 0, 1, 2$), and the correlation ρ, assumed constant. This model therefore requires 11 parameters to be fitted, and we have assumed only that the dispersion of *ApJun* and *InJun* increases only as one of the explanatory variables, namely *IbJun*. Direct maximisation of the log-likelihood function — for example, by the GENSTAT directive FITNONLINEAR or the MATLAB directive NELDER — is either difficult or not possible without severely modifying these standard routines. The maximum permitted number of parameters accommodated by NELDER is ten, but even with nine the search procedure will 'hunt': it approaches what appears to be an optimum, perhaps several times, but on finding that the convergence criterion (tolerance) cannot be met the simplex is expanded and the search moves off in another direction, eventually to diverge.

Fortunately, the blunt instrument of direct maximisation can be avoided. We can exploit the linearity of the relations (7.36). Let the variables *ApJun*, *IndJun*, *IbJun*, *ApMay* and *IndMay* be denoted by x, y, u, v, w, respectively. Writing

$$\sigma_x = \sqrt{(\exp(\lambda_{x0} + \lambda_{x1}u))}$$

$$= \exp(\lambda_{x0} + \lambda_{x1}u)/2$$

and with σ_y given by an analogous expression, then, for given ρ, σ_x and σ_y the equations for the β_i and γ_i are easily found to be of the form

$$\begin{bmatrix} \mathbf{A} & -\rho\mathbf{B} \\ -\rho\mathbf{D} & \mathbf{E} \end{bmatrix} \begin{bmatrix} \beta \\ \gamma \end{bmatrix} = \begin{bmatrix} \mathbf{C} \\ \mathbf{F} \end{bmatrix} \tag{7.38}$$

where

$$\mathbf{A} = \begin{bmatrix} -\beta_0\Sigma(1/\sigma_y) & -\beta_1\Sigma(u/\sigma_y) & -\beta_2\Sigma(w/\sigma_y) \\ -\beta_0\Sigma(u/\sigma_y) & -\beta_1\Sigma(u^2/\sigma_y) & -\beta_2\Sigma(uv/\sigma_y) \\ -\beta_0\Sigma(w/\sigma_y) & -\beta_1\Sigma(uw/\sigma_y) & -\beta_2\Sigma(w^2/\sigma_y) \end{bmatrix}$$

$$\mathbf{B} = \begin{bmatrix} -\gamma_0\Sigma(1/\sigma_x) & -\gamma_1\Sigma(u/\sigma_x) & -\gamma_2\Sigma(v/\sigma_x) \\ -\gamma_0\Sigma(u/\sigma_x) & -\gamma_1\Sigma(u^2/\sigma_x) & -\gamma_2\Sigma(uv/\sigma_x) \\ -\gamma_0\Sigma(w/\sigma_x) & -\gamma_1\Sigma(uw/\sigma_x) & -\gamma_2\Sigma(wv/\sigma_x) \end{bmatrix}$$

$$\mathbf{D} = \begin{bmatrix} -\beta_0\Sigma(1/\sigma_y) & -\beta_1\Sigma(u/\sigma_y) & -\beta_2\Sigma(w/\sigma_y) \\ -\beta_0\Sigma(u/\sigma_y) & -\beta_1\Sigma(u^2/\sigma_y) & -\beta_2\Sigma(uv/\sigma_y) \\ -\beta_0\Sigma(v/\sigma_y) & -\beta_1\Sigma(uv/\sigma_y) & -\beta_2\Sigma(vw/\sigma_y) \end{bmatrix}$$

$$\mathbf{E} = \begin{bmatrix} -\gamma_0\Sigma(1/\sigma_x) & -\gamma_1\Sigma(u/\sigma_x) & -\gamma_2\Sigma(v/\sigma_x) \\ -\gamma_0\Sigma(u/\sigma_x) & -\gamma_1\Sigma(u^2/\sigma_x) & -\gamma_2\Sigma(uv/\sigma_x) \\ -\gamma_0\Sigma(v/\sigma_x) & -\gamma_1\Sigma(uv/\sigma_x) & -\gamma_2\Sigma(v^2/\sigma_x) \end{bmatrix}$$

$$\mathbf{C} = \begin{bmatrix} -(\Sigma(y/\sigma_y) - \rho\Sigma(x/\sigma_x)) \\ -(\Sigma(uy/\sigma_y) - \rho\Sigma(ux/\sigma_x)) \\ -(\Sigma(wy/\sigma_y) - \rho\Sigma(wx/\sigma_x)) \end{bmatrix}$$

$$\mathbf{F} = \begin{bmatrix} -(\Sigma(x/\sigma_x) - \rho\Sigma(y/\sigma_y)) \\ -(\Sigma(ux/\sigma_x) - \rho\Sigma(uy/\sigma_y)) \\ -(\Sigma(vx/\sigma_x) - \rho\Sigma(vy/\sigma_y)) \end{bmatrix}$$

Equations (7.38) can be solved for the β_1 and γ_i, and the x and y residuals calculated. The 50*IbJun* flows are then placed in increasing order of magnitude, divided into five groups of 10, and the variances of the *ApJun* residuals are calculated within the five groups. By equations (7.37), the logarithms of these variances are related linearly to λ_{x0} and λ_{x1}, which can then be estimated by linear regression. The process is repeated for the y residuals, leading to estimates

Iter	b0	b1	b2	c0	c1
1	−13.936654	3.498309	0.055498	−5.799549	2.613497
2	−8.057239	3.551253	0.169565	−5.490576	2.722896
3	−7.562183	3.169824	0.245339	−5.504190	2.716131
4	−7.860626	3.359300	0.206370	−5.542428	2.758868
⋮	⋮	⋮	⋮	⋮	⋮
19	−7.745454	3.298446	0.219262	−5.530632	2.745648
20	−7.745455	3.298447	0.219262	−5.530632	2.745649

c2	10x	11x	10y	11y	ro
0.073212	3.986122	0.033064	4.623472	0.028708	0.879220
0.102947	4.089027	0.032790	4.915574	0.030829	0.873737
0.104167	4.088454	0.032790	5.166280	0.029332	0.827591
0.092894	4.086072	0.032784	5.024742	0.030180	0.852403
⋮	⋮	⋮	⋮	⋮	⋮
0.096380	4.086432	0.032789	5.069576	0.029929	0.844718
0.096380	4.086432	0.032789	5.069576	0.029929	0.844718

Figure 7.14   Convergence of estimates β_i ($i = 0, 1, 2$), denoted b_i; γ_i, denoted c_i; and $\lambda_{j,k}$ ($j = x, y$; $k = 0, 1$), denoted $l_{j,k}$, using the iterative calculation of estimates in the model (7.35).

of λ_{y0} and λ_{y1}. Finally, the correlation between the x- and y-residuals gives an estimate of ρ. The calculation is repeated until convergence is obtained. For the June flows at Apiuna and Indaial, Figure 7.14 shows that satisfactory convergence had been achieved by 20 iterations.

In this example we have discussed only how June flows might be fitted, and it would, of course, be necessary to identify appropriate models for the other months also. We have not considered whether two λ-terms, in each of the variance expressions (7.37), are adequate or indeed necessary, and to test such hypotheses likelihood ratio tests would be needed; nor have we included λ-parameters for the other explanatory variables *IndMay* and *ApMay*. Again, we have not considered whether three β- and three γ-coefficients are adequate or necessary for the description of the expected values of Apiuna and Indaial flows. Likelihood-ratio tests would be appropriate here also. The point of the example, however, is to demonstrate that to identify and fit a model for simulating monthly flows at two sites or more is not simply a question of calculating an estimate of the matrix **B** in equation (7.28) using equation (7.29). Fitting a multi-site, parsimonious, simulation model to monthly flow data requires careful and critical application of the statistical procedures described in this and earlier chapters.

7.3.6   Multi-site Models for Monthly Flows from Basins with Intermittent Flows

Where monthly flow sequences contain seasons in which no measurable streamflow occurs, the difficulties are further exacerbated. Consider the simplest case of two flow gauging sites: a monthly flow model will require two components, one of which will simulate the occurrence or non-occurrence of measurable flow, and the other component will model the magnitudes of flow when it occurs. Denoting by 0, 1 the events 'no flow occurs', 'flow occurs', the states for the two basins in any particular month may be denoted by the four pairs 00, 01, 10, 11: the pairs 00 and 11, for example, denoting the events 'no flow from either basin' and 'flow from both basins'. Between any one month and its successor, the transitions representing possible changes of state can be arranged as a 4×4 transition matrix, in which the probabilities are the probabilities of passing from state ij $(i, j = 0, 1)$ in month m, to state kl $(k, l = 0, 1)$ in month $m + 1$:

		State in month $(m + 1)$			
		00	01	10	11
	00	$p_{00,00}$	$p_{00,01}$	$p_{00,10}$	$p_{00,11}$
State in	01	$p_{01,00}$	$p_{01,01}$	$p_{01,10}$	$p_{01,11}$
month m	10	$p_{10,00}$	$p_{10,01}$	$p_{10,10}$	$p_{10,11}$
	11	$p_{11,00}$	$p_{11,01}$	$p_{11,10}$	$p_{11,11}$

where the probabilities in each row sum to one. There are 12 such matrices, corresponding to months $m = 1, 2, \ldots, 12$; the probability $p_{00,01}$ in month m, for example, is estimated as the proportion of times in the N (say) years of record in which both sites were dry in month m, but flow was measured at the second site in month $m + 1$. The 12 transition matrices will be larger, with more rows and columns if there are more than two sites (for three sites, there are 8 states $000, \ldots, 111$) and/or if the occurrence of measurable flow in any month depends upon the presence or absence of flow in two (or more) preceding months.

Having modelled the occurrences of months with flow, the second stage is to model the magnitudes of flow in those months. With two sites and flows occurring at both, an approach along the lines of the last section (7.3.5) will be appropriate; univariate models will be required for those months in which flow is observed at one site only, and particular care will be necessary where flows corresponding to transitions from states 01 to 11, 10 to 11, 11 to 01 and 11 to 10 are to be modelled.

Clearly the principal difficulties arise from the need to identify different models for the different transitions within each month. These difficulties are diminished if decoupling is possible, and close attention is needed to establish whether parameters can be omitted or combined.

7.3.7 Estimating Suspended Sediment Load: the Case of Two Variates at One Site

In the present chapter, several response variables have hitherto been measured at different sites: annual floods, or monthly streamflows, at several gauging stations. A different case occurs where two or more variables are measured at the same site: for example, concentration c_t of suspended sediment may be measured at irregular intervals throughout the year, and combined with discharge q_t to form an estimate of annual sediment yield, of the form $\Sigma c_t q_t$ or, more generally, $S = \Sigma w_t c_t q_t$, where the $\{w_t\}$ are weights. The mean and variance of S are generally the quantities of interest; in some cases, it is possible to infer some additional characteristics of its probability distribution—see, for example, Ferguson (1986, 1987). Walling and Webb (1981, 1988) have made extensive empirical studies of the properties of S, with particular reference to the biases and variances of alternative estimators. Little attention has been paid to the problems of how best to utilise the longer records $\{q_t\}$ commonly available at the gauging site, and how to transfer information in records of sediment concentration $\{c_t\}$ between sites.

7.4 SOME OTHER POSSIBLE MULTIVARIATE DISTRIBUTIONS

Throughout this chapter, whenever assumptions have been made about the multivariate distribution from which data have been sampled, we have usually taken this distribution to be either multivariate normal, or of a form which could be transformed to multivariate normal by means of a Box–Cox transformation. Other options are possible. For example, we may have observed that the annual floods at each of two sites are Gumbel-distributed, and wish to use a bivariate distribution with Gumbel marginals for the purpose of information transfer. A general result is that, given two marginal cumulative distribution functions $F(x)$ and $F(y)$ (not necessarily Gumbel, or indeed not necessarily of the same family) then a bivariate cumulative distribution function having these marginal cdfs is $G(x, y)$, where

$$G(x, y) = F(x)F(y)\{1 + \theta(1 - F(x))(1 - F(y))\} \tag{7.39}$$

where the parameter θ lies in the interval $-1 \leq \theta \leq 1$. Similar generalisations are possible for distributions of three or more variates.

The probability density function corresponding to equation (7.39) is

$$g(x, y) = f(x)f(y)\{1 + \theta(1 - 2F(x))(1 - 2F(y))\} \tag{7.40}$$

Although they have obvious relevance to the problem of information transfer, distributions of this form do not appear to have attracted the attention of hydrologists, perhaps because the functional form of equation (7.40) imposes certain constraints on conditional expectations $E[Y|x]$ and $E[X|y]$. Thus a consequence of equation (7.40) is that, when $F(x)$ and $F(y)$ are both Gumbel cdfs given by

$$F(x; \alpha, u) = \exp(-\exp(-\alpha(x - u)))$$

$$F(y; \beta, v) = \exp(-\exp(-\beta(y - v)))$$

the conditional expectation is not linear, but has the form

$$E[Y|x] = (v + \gamma/\beta - \theta\gamma \ln 2/\beta) + (2\theta\gamma \ln 2/\beta)F(x; \alpha, u) \tag{7.41}$$

where γ is Euler's constant, with a similar expression for $E[X|y]$. The parameter θ behaves like a correlation coefficient in that when $\theta = 0$, the expected value of Y reduces to its univariate form, $E[Y|x] = E[Y] = v + \gamma/\beta$. Further research is required to establish whether such distributions are useful for information transfer in situations where transformation to normality is inappropriate or unsatisfactory.

REFERENCES

Atkinson, A. C. (1987). *Plots, Transformations and Regression: An Introduction to Graphical Methods of Diagnostic Regression Analysis.* Clarendon Press, Oxford.

Box, G. E. P. and Jenkins, G. M. (1970). *Time Series Analysis: Forecasting and Control.* Holden Day, San Francisco.

Beale, E. M. L. and Little, R. J. A. (1975). Missing values in multivariate analysis. *J. Roy. Statist. Soc. B*, **37**, 129–45.

Dempster, A. P., Laird, N. M. and Rubin, D. B. (1977). Maximum likelihood from incomplete data via the EM algorithm. *J. Roy. Statist. Soc B*, **79**, 1–38.

Ferguson, R. I. (1986). River loads underestimated by rating curves. *Water Resour. Res.*, **22**, 74–6.

Ferguson, R. I. (1987). Accuracy and precision of methods for estimating river loads. *Earth Surf. Processes and Landforms*, **12**, 95–104.

Walling, D. E. and Webb, B. W. (1981). The reliability of suspended sediment load data. In *Erosion and Sediment Transport Measurement* (Proc. Florence Symp. June 1981) IAHS Pub. No. 133, 177–94. IAHS Press, Wallingford.

Walling, D. E. and Webb, B. W. (1988). The reliability of rating curve estimates of suspended sediment yield: some further comments. In *Erosion and Sediment Transport Measurement* (Proc. Porto Alegre Symp. Dec. 1988) IAHS Pub. No. 174, 337–50. IAHS Press, Wallingford.

EXERCISES AND EXTENSIONS

7.1. The data below, also given in file EX7-1.DAT on the diskette distributed by The MathWorks contain annual flood data over a 55-year period from 23 gauging stations. Using the data from R. Benedito at Timbó and R. Itajaí-Mirím at Brusque:

(a) Explore the alternative method for using data from incomplete years, suggested in Section 7.2.

(b) Explore which Box–Cox transformations transform the two sequences to approximate joint multivariate normality.

Gauging site	Basin area	No. of years	
		Missing:	Incomplete:
R Taió at Taió	1 575	5	0
Pouso Redondo	130	23	0
R Trombudo at Trombudo Central	55	31	4
R Adaggo at Barração	163	33	3
R Itajaí at Barração	364	33	4
R Itajaí do Sul at Saltinho	483	44	6
R Itajaí do Sul at Jararaca	728	29	2
R Itajaí-Açú at Rio do Sul	5 100	16	4
R Itajaí-Açú at Rio do Sul Novo	5 100	48	1
R Hercílio at Barra da Prata	1 420	47	2
R Hercílio at Ibirama	3 314	5	2
R Neisse Central at Neisse Central	195	30	1
R Itajaí-Açú at Apiuna	9 242	4	1
R Itajaí-Açú at Warnow	9 714	51	1
R Benedito at Benedito Novo	692	4	2
R Benedito at Timbó	1 342	7	4
R Itajaí-Açú at Indaial	11 151	4	5
R do Testo at Rio do Testo	105	22	0
R Itajaí-Açú at Itoupava Seca	11 719	40	1
R Garcia at Garcia	127	22	1
R Luiz Alves at Luiz Alves	204	14	5
R Itajaí-Mirím at Botuvera	859	48	2
R Itajaí-Mirím at Brusque	1 240	5	6

Chapter 8

Rainfall-runoff models

8.1 THE NATURE OF 'LUMPED CONCEPTUAL MODELS' OF RIVER BASIN RESPONSE TO RAINFALL

All but one of the models discussed so far have consisted of an equation in which the response variable (vector-valued in Chapter 7) was expressed explicitly in terms of a systematic component and an error, assumed random. The exception was the Green–Ampt equation of Chapter 5, for which no explicit equation of this form was possible. In all other cases, the systematic component in the model description 'response variable = systematic component + error' consisted of a fairly simple algebraic expression, involving explanatory variables and constants.

A very large class of models exists, however, in which it is impractical to write the systematic component of the model explicitly in terms of the explanatory variables and constants. This class includes the *lumped conceptual* models relating daily runoff from a river basin (or runoff in intervals shorter than a day, for small basins) to mean daily rainfall averaged over basin area and, often, mean areal estimates of potential evaporation. The models attempt to describe three basic processes within any river basin, namely

(i)   *storage* of water within soil, vegetation, aquifers and water bodies;

(ii) *loss*, from storage, of water to the atmosphere through evaporation, or by lateral flow across the basin's topographic boundaries;

(iii) *routing* of water, whether over the surface or through soil and aquifers, from within the basin to the catchment outfall.

The basis of such models is the equation of continuity: that is, the water balance equation, $P = Q + AE + \Delta S$, expressing rainfall 'input', P, to the basin in terms of the 'outputs' runoff Q, evaporation AE, and change of storage ΔS, all measured in units of length. The output Q being of primary interest, the water balance equation is written as $Q = P - AE - DS$. However, all four terms in this equation present difficulties.

In the case of runoff Q, this is almost always estimated using measured water levels in the stream draining the river basin, and a rating curve relating water level (in units of length, [L]) to discharge (measured in terms of volume per unit time, $[L^3 T^{-1}]$). The number of points available for fitting the rating curve, and the reliability of water-level and discharge measurements defining the points, will determine the accuracy and the precision of the rating curve; and because large discharges occur less frequently than smaller discharges, and are more difficult to measure, rating curves are generally considerably less reliable at higher water levels than lower. Furthermore, use of a rating curve to estimate discharges $q_1, q_2, q_3, \ldots, q_t$ given water levels $Z_1, Z_2, Z_3, \ldots, Z_t$, automatically introduces a correlation amongst the $\{q_t\}$, for the same reason that predicted values $\hat{y}_t$ obtained from a linear regression $E[Y] = \beta_0 + \beta_1 x$ are correlated: namely, because each q_t is calculated from the same data values used in fitting the rating curve. Indeed, using the methods of earlier chapters, it would be possible to calculate a prediction interval for each q_t (which, strictly, should be written $\hat{q}_t$, since it is estimated), together with an estimate of the covariance between any pair $\hat{q}_t$ and $\hat{q}_{t+k}$.

There is a further difficulty in the estimation of runoff $\{Q_t\}$. To obtain Q_t, with units [L], from q_t, with units $[L^3][T^{-1}]$, it is necessary to multiply by a factor with units of time [T], and divide by the area of the basin, in units $[L^2]$. But basin area can be extremely difficult to measure, particularly in terrain of low relief covered with dense vegetation precluding the use of areal survey or satellite imagery. Much of the Amazon basin, for example, consists of valleys, often quite deeply incised, separated by plateaus of very gentle relief, covered by forest with trees of height 25–35 m; defining where drainage boundaries occur is a matter of very great practical difficulty, but of very great importance if reliable water budgets are to be calculated.

Much has been written about the efficiency of rainfall catch by an individual rain gauge, and about the estimation of mean areal rainfall, P, using a network of such instruments. The possibilities for consistent bias (for example, by siting gauges in places of easy access) and large sampling errors in the estimation of P, are considerable. There are also great practical difficulties in measuring rainfall above trees, and in keeping gauge orifices clear of debris. In consequence of these factors, mean areal rainfall may be estimated with low accuracy.

Although instruments are available for the direct measurement of evaporation, AE, at a point, their use in a network to provide areal estimates under all weather conditions is still some way off. Particularly in developing countries, estimates of AE must be derived from routine climatological measurements collected at a site not necessarily representative of the basin area as a whole. In practice, routine climatological data are used to calculate the potential evapotranspiration, ET, of a 'short green crop plentifully supplied with water', and plausible assumptions are introduced to allow an estimate of AE to be derived from ET.

Change of storage ΔS, whether above the rooting depth of plants or below, also presents formidable difficulties of measurement, and these difficulties are aggravated where soil and aquifer conditions permit flow of water into, or away from, the basin area as defined at the soil surface. As with the estimation of AE, ΔS is estimated by introducing assumptions expressing the fact that ΔS increases during rainfall, and is decreased by evaporation loss and runoff during dry periods.

The difficulties in the estimation of Q, P, AE and ΔS, sketched briefly above, suggest that the water balance equation $Q = P - AE - \Delta S$ should be written:

$$\text{estimated } Q = \text{ estimated } \hat{P} - \text{ estimated } AE - \text{ estimated } \Delta S + \text{ error}$$

or symbolically

$$q_t = f(\{P_t\}; \{EO_t\}; \ldots ; \theta) + \varepsilon_t$$

(where we now change the notation so that q_t, with units [L], replaces the Q_t used previously). The systematic component of the model, denoted by $f(.)$, contains as explanatory variables rainfall P_t, together with climatological variables from which, say, an estimate of potential evaporation EO can be calculated, and perhaps other variables which describe soil moisture status. As with models previously discussed, the systematic component $f(.)$ contains parameters, θ, which must be estimated from data. Since $f(.)$ is founded upon the concept of continuity, models of the above form are often called *conceptual models*.

Since the concept of storage is a principal characteristic of conceptual models, it is natural to represent it — whether it be in the soil, in the vegetation canopy, or in aquifers — by one or more 'reservoirs' or 'storages', which are replenished during wet periods and which continue to release water when rain ceases. For each storage within the model, hypothetical rules are postulated which express how release from storage is related to storage volume. Thus for a linear storage, rate of release $q(t)$ is taken to be proportional to storage volume $S(t)$, $q(t) = \lambda S(t)$; this, together with the continuity equation $\mathrm{d}S/\mathrm{d}t = i(t) - q(t)$, where $i(t)$ is rate of inflow to storage, and the initial rate of release, q_0, defines the rate of release, since $q(t) = \lambda \exp(-\lambda t) \int_0^t \exp(\lambda u) i(u) \mathrm{d}u$. If $i(t) = i$, a constant, with $q(t) = q_0$ at $t = 0$, we find that $q(t) = i + (q_0 - i) \exp(-\lambda t)$, a form reminiscent of the Horton infiltration law discussed in Chapter 5. The quantity λ is a parameter to be estimated. In a more general case, where an input $i(t)$ is routed through a cascade of linear reservoirs not necessarily all with the same parameter λ, a linear differential equation results, which has the general solution

$$q(t) = \int_{u=0}^{\infty} h(u) i(t-u) \mathrm{d}u \qquad (8.1)$$

the function $h(.)$ being the instantaneous unit hydrograph. For a real river basin, it must be emphasised that these storages are 'conceptual' and need not have an existence in reality.

The storages of the preceding paragraph were *linear* storages. In practice, rather few conceptual storages within a rainfall-runoff model can be taken as linear. Consider the 'storage' representing precipitation retained on a canopy surface: experience shows that when standing under a tree, we do not get wet immediately it begins to rain. The canopy must become sufficiently wet before 'overflow' from 'canopy storage' reaches the soil surface. Thus the conceptual canopy storage has a 'threshold', say μ, such that when $i(t) = i$, a constant, and when the storage is initially empty, $q = 0$ for $t < \mu/i$. For larger values of t, we have, with the same assumptions,

$$q(t) = i[1 - \exp(-\lambda(t - \mu/i))] \qquad t \geq \mu/i \qquad (8.2)$$

The effect of the threshold, μ, is to introduce a point of a non-differentiability in the function $q(t)$ since, for $t < \mu/i$, $\mathrm{d}q/\mathrm{d}t = 0$ and for $t > \mu/i$, $\mathrm{d}q/\mathrm{d}t = i\lambda \exp[-\lambda(t - \mu/i)]$ which tends to the value $i\lambda$ when $t \to \mu/i$ from above. Hence the derivative of $q(t)$ has different values according to whether $t \to \mu/i$ from below ($t < \mu/i$) or above ($t > \mu/i$), a fact which may be expected to complicate matters if we need to differentiate with respect to a parameter θ, to find the maximum of a likelihood function, or the minimum of a least-squares measure of goodness of model fit. In addition, we see that the threshold has introduced a second parameter, μ, which must also be estimated; and this parameter enters in a complicated manner, first in so far as the time axis is divided into two parts by $t = \mu/i$, so that the two parts themselves depend upon μ, and second because μ enters as a multiplier of λ in the term $\exp(\lambda\mu/i)$, when $t > \mu/i$.

Suppose now that, for a storage representing the behaviour of a plant canopy, subject to an idealised, constant rainfall i beginning at time $t = 0$, we take measurements of the water penetrating the canopy during discrete intervals of time $[0, T), [T, 2T), \ldots, [kT, (k+1)T), \ldots$. Denote the measured values of accumulated 'throughflow' by $Q_1, Q_2, Q_3, \ldots, Q_t, \ldots$. Then,

provided T is small relative to the unknown μ/i, the measured quantities Q_t, which we indicate by $Q_{\text{obs},t}$, should estimate the following quantities (using the symbol $Q_{\text{fit},t}$ to indicate fitted
values, obtained from the model, and $[kT]$ for the largest multiple of T not exceeding μ/i):

$$Q_{\text{fit},t} = \begin{cases} 0 & t < [kT] \\ i\{[(k+1)T] - \mu/i\} - \{\exp(-\lambda\mu/i)/\lambda\}\cdot \\ \quad \{\exp(-\lambda\mu/i) - \exp(-\lambda[(k+1)T])\} & [kT] \le t \le [(k+1)T] \\ iT - i\{\exp(\lambda\mu/i)/\lambda\}\cdot \\ \quad \{\exp(-\lambda kT) - \exp(-\lambda(k+1)T)\} & t > [(k+1)T] \end{cases} \quad (8.3)$$

So for this very elementary system of a single storage, with a threshold, representing plant
canopy interception response to constant rainfall, we might seek to estimate the unknown
parameters λ, μ by:

(i)   choosing some criterion measuring the lack of agreement between $Q_{\text{obs},t}$ and $Q_{\text{fit},t}$ for
      $t = T, 2T, 3T, \ldots$;
(ii) choosing numerical values for λ, μ which 'optimise' this criterion, perhaps by minimising
      the lack of agreement between $Q_{\text{obs},t}$ and $Q_{\text{fit},t}$.

Our discussion so far has addressed the modelling of just one 'process' (interception) acting
within a river basin. In practice, it is usually necessary to describe each of many hypothesised
processes — interception; infiltration; surface runoff; evaporation from plant canopy storage;
evapotranspiration from soil, and leaf stomata; percolation to groundwater; release of water
from aquifers, as baseflow; channel flow — in terms of storages, fluxes between storages, fluxes
from storage to the atmosphere as evaporation, and fluxes from storage to the basin outfall. In
contrast with the example discussed above, where we assumed measurements of throughfall,
there are commonly no measurements of variables describing within-basin behaviour. The
measurements that are usually available for the formulation, fitting and testing of a conceptual
model are those discussed earlier, namely:

(i)   daily rainfall, from a network of gauges;
(ii)   a continuous trace (apart from periods of instrumental failure) of water level at the
       catchment outfall, together with a rating curve by which water level can be converted to
       discharge, and hence to runoff per unit of basin area;
(iii) standard meteorological measurements of wet and dry bulb temperatures, sunshine hours
       and wind-run, from which an estimate of atmospheric evaporation demand — commonly
       Penman potential evapotranspiration — can be estimated.

Because of the multiplicity of storages necessary to describe the component ΔS in river basin
behaviour, and because of the thresholds in the hypothetical storages, the systematic component
$f(.)$ cannot, in general, be written down easily in an explicit algebraic form. It is therefore not
possible, in general, to write down a goodness-of-fit criterion (such as least squares) and use
algebra to differentiate it with respect to the parameters θ. On the other hand, it is relatively
easy to describe how the model operates using a diagram.

Our discussion so far has concentrated on the systematic component $f(.)$. We also need
to consider the 'error' ε_t, describing the discrepancy between observed runoff per unit area,
and the runoff per unit area as computed from the systematic component of the model. It
is convenient to think of this error ε_t as a random variable, and it would exist even if the

systematic component were fully correct as a description of basin response to rainfall (which, of course, it never is) and the parameters θ were known exactly. The errors would exist because rain gauges do not give a perfect measure of mean areal rainfall; because basin area is never known exactly, so that runoff per unit area is always in error; and because rating curves never give instantaneous discharges that are error-free.

Having written that the errors ε_t can conveniently be regarded as random variables, we postpone for the time being a discussion of their statistical characteristics, returning to this topic in a later section (8.6).

8.2 EXAMPLES OF LUMPED CONCEPTUAL MODELS OF RIVER BASIN RESPONSE TO RAINFALL

The easiest way to acquire an understanding of the systematic components of lumped conceptual models of the rainfall-runoff relation is to study some examples of models that have been used in practice. Of the enormous number that have been described, we give four. Two are venerable in that they laid the basis for much subsequent work on lumped conceptual modelling; the other two are more recent and are extensively used. Despite much further effort by hydrologists to understand hydrological processes, and despite the enormous increase in computational power and its ready availability, the general structure of rainfall-runoff models has not greatly changed over the 20 years spanned by the lumped models now to be described.

8.2.1 The Dawdy–O'Donnell Model

The first model, described some 25 years ago by David Dawdy and Terence O'Donnell, was proposed when computers were first beginning to be used in hydrological computation; and despite its age, few lumped models from later periods have brought significant further developments. The model is shown in Figure 8.1.

In the model, a river basin is represented by four interconnected reservoirs with volumes at any instant denoted by R, S, M and G. The 'surface storage', R, is augmented by rainfall P; it is depleted by evaporation E_R, by infiltration F and, when the volume R exceeds a threshold R^*, by flow into stream channels Q_I. The 'channel storage', S, is augmented by this channel flow Q_I; and is depleted by the surface runoff emerging at the gauging station Q_S. The 'soil moisture storage', M, is augmented by infiltration F (entering it from the surface storage) and by capillary rise C; it is depleted by transpiration E_M and, when the volume M exceeds a threshold M^*, by deep percolation D. The 'groundwater storage', G, is augmented by deep percolation D (entering it from the soil moisture storage); it is depleted by capillary rise C, and by the baseflow B which emerges at the gauging station. If, and whilst, G exceeds G^*, M is absorbed into G, C and D no longer operate, but E_M and F now act on G.

There are nine parameters which control the functioning of the model. At the beginning of each interval, the volume in R lies between zero and R^*, the first parameter; P is added to R; and E_R, if any, is given first call on the sum. Next, F is calculated according to certain criteria based on a Horton-type equation, considering the rate of supply available from surface storage and the potential rate of infiltration at the start of the interval. This involves maximum and minimum infiltration rates f_o and f_c, and an exponent k (three more parameters). In preparation for the next interval, a potential ratio f_i is calculated for the end of the current interval. Then Q_I is determined by the excess, if any, over R^* left in storage after E_R and F have been extracted.

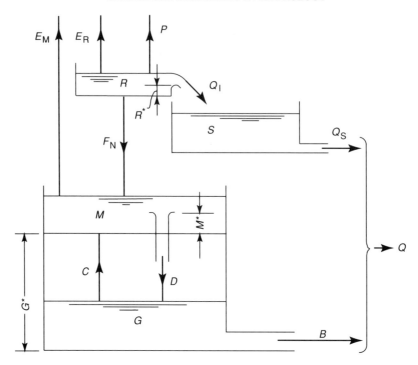

Figure 8.1   Diagram illustrating the Dawdy–O'Donnell conceptual model (see text for description of model function).

The channel storage S is assumed to be a linear storage with constant K_S, the fifth parameter. Then, Q_S is a function of the volume in S at the beginning of the interval, of the inflow Q_1, and of K_S. A water balance calculation gives the volume left in S ready for the start of the next interval.

At the beginning of any time interval, M lies between zero and M^*, the sixth parameter. Either E_M is removed or F is added, for one of the two will be zero depending on whether or not E_R satisfied E_P, the potential evapotranspiration. One of several alternatives is now followed depending on whether or not G, at the start of the time interval, is greater than G^*, the seventh parameter, and, if not, whether or not the quantity in M is now greater than M^*. If G is less than G^*, D is set equal to the excess, if any, over M^* now in M; C is zero if D exists, otherwise it is determined as a function of demand in M, of supply in G, and of a maximum rate of rise c_{max}, the eighth parameter.

Also M is left in at M^* if D exists, or is augmented by C, if not. If G, at the beginning of the time interval, is greater than G^*, F, if any, acts on G directly in place of D, and C similarly in place of E_M. In this alternative, M remains at M^*.

Then, G is assumed to be a linear storage with storage constant K_G (the ninth parameter); B is then a function of the volume in G, at the start of the time interval, of the inflow D or abstraction C, and of K_S. Again, a water balance calculation yields the volume left in G ready for the start of the next time interval.

In addition to the nine parameters listed above, the initial volume in each of the four reservoirs must be specified. To estimate these quantities would increase the number of parameters from 9 to 13. To avoid this complication, it is common to start the model-fitting procedure at the end of a long dry period, if one can be found, and to assume that the volumes then in storage are all zero, and that the potential infiltration rate has recovered to its maximum value f_0.

Three of the parameters (R^*, M^*, G^*) in the Dawdy–O'Donnell model are 'threshold' parameters, which give rise to complications in the model-fitting procedure, as described in Section 8.1.

8.2.2   The Nash–Sutcliffe Layer Model

Dating from about the same period as the Dawdy–O'Donnell model, and equally classical in its seminal influence, the model proposed in 1969 by Nash and Sutcliffe is shown in Figure 8.2. This model assumes that the basin is analogous to a vertical stack of horizontal soil layers,

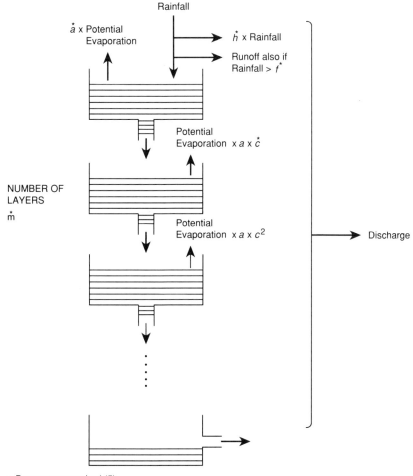

Figure 8.2   Diagram illustrating the Nash–Sutcliffe 'layer' model (see text for description of model function).

each of which can contain a certain amount of water at field capacity. Evaporation from the top layer takes place at the potential rate, and from the second layer only upon exhaustion of the first, and then at the potential rate multiplied by a parameter c, the value of which is less than unity. On exhaustion of the second layer, evaporation occurs from the third layer at the potential rate multiplied by c^2, and so on. A constant evaporation potential applied to the basin would reduce the soil moisture in a roughly exponential manner.

When rainfall exceeds evaporation, a proportion (h) of the excess contributes to generated runoff, and of the remainder anything in excess of a threshold value (f) also contributes to generated runoff. The remaining rainfall excess is used to restore the storages in the several layers to field capacity, beginning with the first and proceeding downwards until the rainfall is exhausted, or until all the layers are at field capacity. Any final excess also contributes to generated runoff.

Nash and Sutcliffe calculate the potential evaporation using the Penman formula with an albedo of 0.25; to allow for systematic error, the potential evaporation is multiplied by a factor, a, before it is compared with rainfall. The capacity of each soil layer, except the lowest, is taken as 1 inch (25.4 mm) and the number of soil layers, m, is a parameter to be estimated. To allow for fractional values of m, as the estimation procedure requires, a related parameter, z, is specified for use instead of m: this is defined as the total storage at field capacity, and m is redefined as z rounded upwards to a whole number.

Thus, the Nash–Sutcliffe model contains five parameters: c, z, a, f and h, with $0 < c < 1$; $z > 0$; $a > 0$; $f > 0$; $0 < h < 1$.

8.2.3 The Institute of Hydrology Lumped Model

The Institute of Hydrology lumped model, incorporating the results of extensive work which followed the Dawdy–O'Donnell and Nash–Sutcliffe models, has been described by Blackie and Eeles (1985). The version of the model that they describe is designed to produce hourly estimates of streamflow from hourly catchment rainfall and hourly potential evaporation derived from meteorological data using the Penman formula. The structure of the model is shown in Figure 8.3. It consists of four storages representing, notionally, the interception by the vegetation and surface litter, the surface runoff storage, the soil moisture storage, and the groundwater storage. The model contains 15 parameters to be estimated, and it represents catchment behaviour as follows.

Incoming rainfall $RAIN$ enters the interception storage until its contents CS reach the storage capacity SS. Any rainfall excess $ERAIN$, reaching the notional soil surface, is divided into two components, one of which is 'surface runoff' and the other infiltrates to the soil moisture storage. The volume assigned to surface runoff, $ROFF$, is determined by the expression $ROFF = ROP \times ERAIN$, where ROP ('runoff proportion') is a function of the soil moisture deficit, DC, and the rainfall intensity. In fact ROP is given by

$$RO = RC[\exp(-RS \times DC) + \exp(RR \times ERAIN) - 1] \tag{8.4}$$

where RC, RS and RR are parameters to be estimated. After $ROFF$ has been calculated, the remaining component of excess rainfall, $ERAL - ROFF$, infiltrates to the soil moisture storage to reduce the soil moisture deficit DC.

If soil moisture storage is less than field capacity (that is, if DC is positive) no drainage to groundwater occurs. If DC is negative, water in soil moisture storage drains to groundwater at a rate GPR, given by $GPR = -A \times DC$.

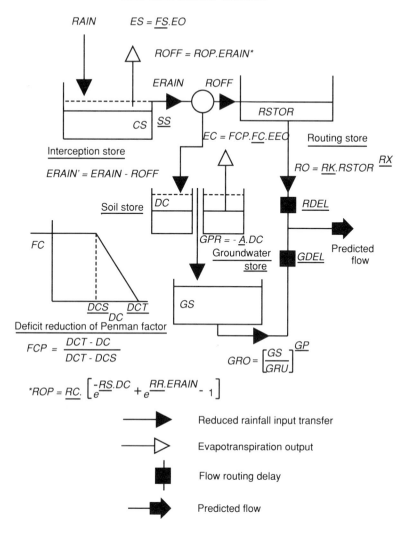

Figure 8.3   The Institute of Hydrology lumped conceptual model (following Blackie and Eeles, 1985: see text for description of model function).

The interception storage is depleted by evaporation at a rate ES, given by $ES = FS \times EO$, where EO is Penman potential evaporation for the hour interval. But ES cannot exceed the storage content CS, so that when the interception storage becomes empty (that is, when $ES = FS \times EO$ is greater than CS), the residual evaporative demand, $EEO = EO - CS/FS$ is abstracted from the soil moisture storage. This storage is depleted by transpiration at a rate EC determined by $EC = FCP \times FC \times EEO$, where FC is a function of the deficit, DC; it is given by

$$FCP = \begin{cases} 1 & DC < DCS \\ (DCT - DC)/(DCT - DCS) & DCT > DC > DCS \end{cases} \qquad (8.5)$$

where *DCS* and *DCT* represent, respectively, the soil moisture deficits at which transpiration begins to be constrained and finally ceases. Thus total evaporation, relative to Penman *EO* and to soil moisture storage, is determined by the four parameters *FS*, *FC*, *DCS* and *DCT*.

The surface runoff store is treated as a non-linear reservoir, giving the volume contribution to flow as

$$RO = RK \times RSTOR^{RX} \tag{8.6}$$

where *RSTOR* is the reservoir content at the start of the interval. This in turn is delayed by *RDEL* time intervals.

The groundwater store contributes to baseflow as a non-linear reservoir. In each time interval the volume output, *GRO*, from the store content, *GS*, is given by

$$GRO = (GS/GSU)^{GSP} \tag{8.7}$$

where *GSU* and *GSP* are parameters to be estimated. Arrival of this output at the catchment outfall is delayed by *GDEL* time intervals, where *GDEL* is a parameter. Thus, total runoff in time interval t is $FLOW(t) = RO(t - RDEL) + GRO(t - GDEL)$.

The 15 parameters whose values must be estimated are: *SS* and *FS* for the interception storage; *RC*, *RS*, *RR*, *RK*, *RX* and *RDEL* for the surface runoff storage; *FC*, *DCS*, *DCT* and *A* for the soil moisture storage; and *GSU*, *GSP* and *GDEL* for the groundwater storage. In addition, the initial contents of the stores have to be fixed at the start of each model run. Whenever possible, runs are started at a point preceded by several dry days so that *CS*, the interception storage content, can be assumed to be zero, so that *DC* is a positive soil moisture deficit, and so that the contents of the surface runoff storage, *RSTOR*, are close to zero. *GSU* is computed from the initial observed flow, assumed to consist only of baseflow under these conditions. This leaves just the initial value of soil moisture deficit to be estimated from field observation, or to be treated as a sixteenth parameter.

The following code gives a MATLAB version of the part of the Institute of Hydrology model which is concerned with the calculation of the flow sequence (*FLOWHAT*), given input sequences of rainfall and Penman *EO*, and trial values of the model parameters. When the model is fitted to the flow record which resulted from the rainfall and *EO* sequences used as inputs, it is necessary to define a measure of goodness of fit between the observed flow sequence, and the calculated sequence *FLOWHAT*; and estimates of model parameters are calculated which match the observed and fitted flow sequences as nearly as possible, judged by the numerical value of a goodness-of-fit criterion, discussed later in Section 8.6.

```
function FLOWHAT=ihmodel(prmters15);
%
% This function calculates the vector of fitted flow values
% calculated by the IH model (see J R Blackie and C W O Eeles,
% 'Lumped catchment models', Chapter 11 of 'Hydrological
% Forecasting' edited by M G Anderson and T P Burt, 1985, John
% Wiley). The model contains 15 parameters:
%
% SS and FS describing the interception store;
% RC, RS, RR, RK, RX, RDEL for surface runoff;
% FC, DCS, DCT and A for the soil moisture store;
```

```
% GSU, GSP and GDEL for the groundwater store.
%
%
% In addition, the initial contents of the four stores
% (interception store; routing store; soil store; and groundwater
% store) must be specified at the start of each model run.
%
% The output of the function is the vector of calculated
% flows, 'FLOW', corresponding to the 15 parameter values.
% 'NTIMES' is the number of values in the observed data sequences
% for RAIN and EO. (We use Blackie and Eeles's notation
% throughout).
%
% It is assumed that the vectors RAIN and EO (of length NTIMES)
% are read from files named RAIN.MAT and EO.MAT, with the rain
% and EO values for each time interval on a new line; and that
% the 4 initial values for store contents are read from a file
% STORINIT.MAT, in which the contents of interception store,
% routing store, soil store and groundwater store are on separate
% lines (i.e. <ENTER> is pressed after the entry of each)..
%
%
%
load RAIN;                          % Read in the rainfall sequence.
load EO;                            % Read in the potential evaporation
NTIMES=length(RAIN);
FLOWHAT=ones(NTIMES,1)-1;           % Initialises the vector FLOWHAT.
%
%
load STORINIT                       % This file contains initial storage
                                    % contents.
   CS=STORINIT(1);                  % Set storage contents to their
RSTOR=STORINIT(2);                  %         initial values.
   DC=STORINIT(3);
   GS=STORINIT(4);
%
% Now identify the 15 parameters contained in the vector
% 'prmters15'.
%
%
%
SS=prmters15(1);                    % SS and FS are interception
                                    %               parameters.
FS=prmters15(2);
%
%
```

```
RC=prmters15(3);                          % RC, RS, RR, RK, RX and RDEL are
                                          % the routing store parameters.
RS=prmters15(4);
RR=prmters15(5);
RK=prmters15(6);
RX=prmters15(7);
RDEL=prmters15(8);
%
%
FC=prmters15(9);                          % FC, DCS, DCT and A are soil store
                                          %                      parameters.
DCS=prmters15(10);
DCT=prmters15(11);
A=prmters15(12);
%
%
GSU=prmters15(13);                        % GSU, GSP and GDEL are groundwater
                                          % store parameters.
GSP=prmters15(14);
GDEL=prmters15(15);
%
%
%
for I=1:NTIMES,
%
    ERAIN=0;                              % Interception store operation
                                          %                      follows.
    CS=CS+RAIN(I);                        % ERAIN is rain minus
                                          % interception loss.
        if CS >= SS,                      % SS is interception store
                                          % maximum depth.
          ERAIN=CS-SS;
              CS=SS;
        end;
%
%
    if ERAIN >= 0,
            ROP=RC*(exp(-RS*DC)+exp(RR*ERAIN)-1);
            ROFF=ROP*ERAIN;               % RC, RS, RR are parameters.
            RSTOR=RSTOR+ROFF;             % ROFF goes to the routing store.
            DC=DC-(ERAIN-ROFF);           % The difference ERAIN-ROFF
                                          % infiltrates
                                          %    and reduces soil moisture
                                          deficit DC.
    end;
%
```

```
%
% If soil moisture deficit DC is positive, there is no loss to
% groundwater; if DC is negative, GPR is the drainage to
% groundwater:
%
%
    if DC > 0,
          GPR=0;
    else
          GPR=−A∗DC;                          % Note: this program assumes A is
                                              % positive: hence the negative DC
          DC=DC+GPR;                          % becomes more positive.
          GS=GS+GPR;
    end;
%
%
% Now deal with evaporation loss:
%
%
    ES=FS∗EO(I);                              % ES is taken from the
                                              % interception
    if ES <= CS,                              % store, if there is enough
                                              % there;
       CS=CS−ES;                              % if not, the interception store
                                              % is set to zero, and the excess
    else                                      % evaporation demand is passed
       EEO=EO(I)−CS/FS;                       % to the soil store.
         CS=0;
                if DC <= DCS,
                      FCP=1;
              elseif DCT >= DC & DC > DCS,
                      FCP=(DCT-DC)/(DCT-DCS);
                  else
                      FCP=0;                  % FCP is a function of the
                                              % deficit DC. This
                  end;                        % function is defined by
       EC=FCP∗FC∗EEO;                         % parameters DCT and DCS, where
       DC=DC+EC;                              % DCT>DC>DCS. Hence 4 parameters
    end;                                      % determine operation of the
                                              % soil moisture store: DCT,DCS,
                                              % FC and FS.
%
%
% Now deal with the routing from the routing store:
%
%
```

```
    v=NTIMES-RDEL;                          % Check whether the routed
                                            % component arrives at the gauge
    if I <= v,                              % site before the end of
        RO=RK*(RSTOR^RX);                   % the record. If so, add
        J=I+RDEL;                           % it to FLOWHAT and reduce
                                            % contents of the routing store
        if RO>RSTOR,                        % accordingly.
                RO=RSTOR;
                FLOWHAT(J)=FLOWHAT(J)+RO;
                RSTOR=0;
        else
                FLOWHAT(J)=FLOWHAT(J)+RO;
                RSTOR=RSTOR-RO;
        end;
    end;
%
%
%
% Now do the same for the flow component coming from groundwater
% store:
%
%
    w=NTIMES-GDEL;
    if I <= w,
        GRO=(GS/GSU)^GSP;                   % GSU, GSP, GDEL are parameters
                                            % controlling operation of the
        J=I+GDEL;                           % groundwater store.
        if GRO>GS,
                GRO=GS;
                FLOWHAT(J)=FLOWHAT(J)+GRO;
                GS=0;
        else
                FLOWHAT(J)=FLOWHAT(J)+GRO;
                    GS=GS-GRO;
        end;
    end;
%
%
end                                         % end of loop over NTIMES.
```

Thus, following a call on the function *ihmodel* by means of

FLOWHAT=ihmodel(prmters15)

the vector *FLOWHAT* contains the flows predicted by the Institute of Hydrology model, for the
particular set of parameters used in the vector of parameters (**prmters**15) and the 'input' variables *RAIN* and *EO*. Clearly in searching for the 'best-fitting' model, the vector (**prmters**15)

will be modified perhaps many times, until the values in *FLOWHAT* agree satisfactorily with the observed flow sequence.

8.2.4   The IPH2 Model

Like the Institute of Hydrology lumped model of Section 8.2.3, the Instituto de Pesquisas Hidráulicas IPH2 model has been developed over a number of years with contributions from people who have used it extensively in South America. The model has essentially three components: the first describing losses from interception and evaporation; the second defining the separation of infiltration from rainfall excess; and the third defining the routing of surface runoff and groundwater contributions to flow at the catchment outfall. We deal with these in turn.

Potential evaporation is subtracted from rainfall, when rainfall exceeds it. Otherwise, the residual potential evaporation is taken from the interception storage, or, if this is empty, from soil moisture storage according to a linear relation $ES_t = EP_t S_t / S_{max}$, where ES_t is the evaporation taken from soil storage, EP_t is potential evaporation, S_t is soil moisture storage at time t, and S_{max} is the maximum depth of soil moisture storage, a parameter.

When rainfall exceeds potential evaporation, the excess is retained in the interception storage up to its maximum capacity R_{max}, a parameter. Any excess over R_{max} is passed to the soil moisture storage, except for a proportion (defined by a constant of proportionality *IMP*, or 'impervious area') which contributes directly to surface runoff.

Infiltration I_t is calculated from the Horton equation of Chapter 5, in the form

$$I_t = I_b + (I_0 - I_b)h^t \tag{8.8}$$

where I_t is infiltration capacity at time t, I_b is the minimum infiltration capacity occurring as $t \to \infty$, I_0 is the initial infiltration capacity at time $t = 0$, and $h = \exp(-k)$, with k a parameter related to soil type. Downward percolation from the upper soil layer is defined by

$$T_t = I_b(1 - h^t) \tag{8.9}$$

so that, by applying continuity $dS/dt = I_t - T_t$ to the upper soil layer and substituting for I_t and T_t, we have

$$S_t = S_0 + I_0(h^t - 1)/(\ln h) \tag{8.10}$$

whence we see that

$$S_t = a_i + b_i I_t = a_t + b_t T_t \tag{8.11}$$

where

$$a_i = -I_0^2/[(I_0 - I_b)\ln h]$$
$$b_i = I_0/[(I_0 - I_b)\ln h] \tag{8.12}$$
$$a_t = 0$$
$$b_t = -I_0/[I_b \ln h]$$

and S_0 has been set equal to zero. The relations between S_t, I_t and T_t are shown in Figure 8.4.

When the effective precipitation (rainfall less the abstractions listed above) is greater than the infiltration capacity I_t, I_{t+1} is calculated from the Horton equation (8.8), S_{t+1} is calculated

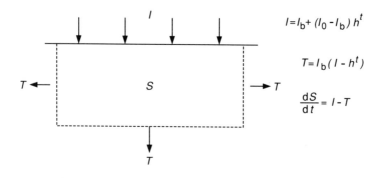

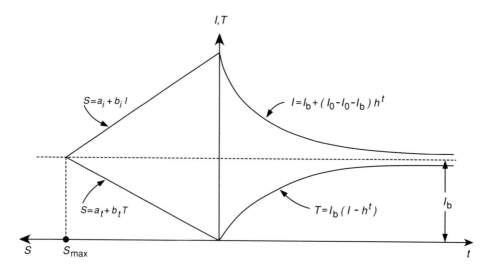

$$S = a_i + b_i I;$$

$$a_i = -I_0^2/[(I_0 - I_b)\ln h]$$

$$b_i = I_0/[(I_0 - I_b)\ln h]$$

$$S = a_t + b_t T;$$

$$a_t = 0$$

$$b_t = -I_0/[I_b \ln h]$$

Figure 8.4　Relationships between storage S, infiltration I, and percolation T for IPH2 model.

from equation (8.11), and T_{t+1} also from equation (8.11). The volumes V_e, V_p going to surface runoff and to percolation are then given by

$$V_e = (P_t - I_b)\Delta t - (I_t - I_b)(h^{\Delta t} - 1)/\ln h \qquad (8.13)$$

$$V_p = I_b \Delta t + (T_{t+1} - T_t)\ln h \qquad (8.14)$$

There remains the description of routing of the generated surface and subsurface flows to the catchment outfall. In the case of surface runoff, this is achieved by routing the volume V_e through a linear reservoir with parameter K_{sup}; and in the case of subsurface flow, the volume V_p is routed through a linear reservoir with parameter K_{sub}. Summation of the outputs from the two notional reservoirs gives the modelled streamflow at the catchment flow gauging site, as a function of the parameters S_{max}, R_{max}, I_b, I_0, h, K_{sub}, K_{sup}. In applications, S_{max} is commonly set equal to $-I_0/\ln h$, obtained by using equation (8.11) in the form $S_{max} = a_i + b_i I_b$. This reduces the seven parameters to six. The parameter K_{sub} is found by fitting an exponential curve to hydrograph recessions, giving a further reduction to five, of which R_{max} has been found to have little influence. The four most sensitive parameters are thus I_0, I_b, h and K_{sup}.

8.2.5   The Sugawara Tank Model

As a final illustration of a lumped conceptual model of river basin behaviour, we cite the tank model of Sugawara *et al.* (1974) shown in Figure 8.5. The generalised structure is analogous to the corresponding storages in the real river basin; a vertical line of storages (termed 'tanks') represents interception, soil moisture and groundwater storages at different depths in the profile,

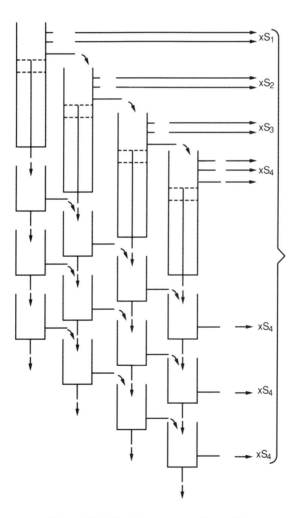

Figure 8.5   The Sugawara 'tank' model.

whilst the lateral disposition of tanks represents zones of the basin at different depths in the profile. Each tank has four interconnections, representing lateral and vertical inputs and outputs.

Clearly the number of parameters needed to specify the Sugawara model could be very large, and estimating them would require a high degree of subjectivity. Nevertheless, the generality of the model form gives a certain satisfaction, and it has led to new approaches, more economic in terms of parameters, such as the model developed by Moore and Clarke (1981, 1982, 1983) described below.

8.3 'LUMPED' AND 'DISTRIBUTED' MODELS

Despite the lengthy period over which these models have been developed, it will be seen that they differ more in matters of detail than in broad concept. All the models in Sections 8.2.1 to 8.2.4 work with spatially averaged values of rainfall, evaporation, runoff and infiltration: no account is taken of topographical features of hill-slopes, or of channel geometry, or of spatial variation in interception and infiltration processes. An early attempt to take account of spatial variability in hydrological processes, particularly those that give rise to rapid runoff during and immediately following rain, is the model of Beven and Kirkby (1979).

The Beven–Kirkby model differs from the lumped models of Sections 8.2.1 to 8.2.4 in two respects: first, it combines the *distributed* effects of contributing areas within an otherwise lumped conceptual model; and second, the model parameters are estimated from measurements taken in the field. Both constitute significant departures from the lumped-modelling concept, and subsequent work has developed the distributed approach still further. The structure of the model is shown in Figure 8.6; it consists essentially of the following components:

(i) A variable contributing area (see Figure 8.6) related to subsurface soil water storage. Rain falling upon the contributing area, A_C, immediately becomes overland flow.

(ii) A surface interception and depression storage, S_1, with maximum depth S_D, which must be filled before infiltration from it can take place. Water evaporates from this storage, at the potential rate, until it becomes empty. (This lumped storage represents processes assumed to be spatially averaged.)

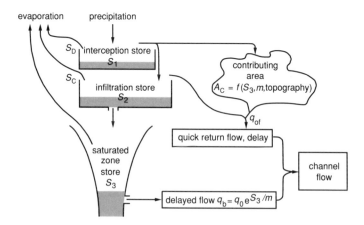

Figure 8.6   The Beven–Kirkby model (see text for description of model function).

(iii)   A near-surface infiltration storage S_2 . Water drains from this storage to another storage, S_3, in the part of the basin complementary to A_C, at a constant rate i_0. Once the interception-depression storage S_1 is filled, water enters S_2 at the rainfall rate i, unless $i > i_{max} = i_0 + b/S_2$. In this case, the excess $i - i_{max}$ is routed to the basin outfall by a surface route. Water is lost from S_2 as evaporation, at a rate depending on the depth within the storage S_2; denoting by e_r the potential evapotranspiration which remains to be satisfied after the storage S_1 becomes empty, the loss e_a from the storage S_2 is given by $e_a = e_r S_2 / S_C$, where S_C is the maximum depth of the storage S_2.

(iv)   A non-linear, subsurface, saturated, soil water store with instantaneous depth S_3, with behaviour defined by $q_b = q_0 \exp(S_3/m)$, where q_b is the flow reaching the channel from the store, q_0 is the flow when $S_3 = 0$, and m is a constant. S_3 is taken to be zero when the average soil water storage, over the entire basin, is just saturated. Positive values of S_3 constitute a moisture surplus, negative values of S_3 a state of deficit.

(v)   A saturated area A_C, where A_C is a function of S_3, m and topography. Effects of topography are introduced as follows: for any point in the catchment, let a be the area drained per unit length of the contour passing through that point, and let β be the slope angle there. Beven and Kirkby argue that the saturated area A_C is that area for which

$$\ln(a/\tan \beta) > S_T/m - S_3/m + \lambda \tag{8.15}$$

where S_T is the local maximum storage, assumed constant over the basin, and λ is given by

$$\lambda = (1/A) \int_0^A \ln(a/\tan \beta) dA' \tag{8.16}$$

The authors show how λ may be obtained from maps, and S_T from field measurements.

(vi)   A channel routing relationship of the form

$$c(t) = CHA \cdot Q(t)^{CHB} \tag{8.17}$$

where $c(t)$ is an average kinematic flow velocity, assumed to be spatially constant and equal to water velocities measured by tracer experiments; and $Q(t)$ is discharge from the entire basin at time t. The constants CHA, CHB are derived from the use of fitting procedures demonstrated in earlier chapters of this book; it is assumed that, with $c(t)$, $Q(t)$ measured, equation (8.17) can be written $\ln c(t) = \ln CHA + CHB[\ln Q(t)] + \varepsilon_t$, and that $\ln CHA$ and CHB can be calculated by linear regression. Likewise, Beven and Kirkby estimated the other parameters of the model using data from a sprinkling infiltrometer, leading to infiltration curves expressing cumulative infiltration as a function of time.

The Beven–Kirkby model, in so far as it introduced a simplistic description of topographical influences on rapid runoff, prepared the way for a surge of activity in distributed modelling which accompanied the development of computers with parallel processing facilities. The SHE (Système Hydrologique Européen) model (Abbott *et al.*, 1986), developed jointly by hydraulic and hydrological institutes in England, France and Denmark during the 1980s, demonstrates the complexities of modelling, on a basin-wide scale, processes which act at time scales differing by several orders of magnitude. The model was developed in the expectation that the costs of instrumentation for the collection of data in the field would fall rapidly, and that the ready availability of field data would gradually eliminate the need to estimate model parameters by the use of statistical procedures. It can be argued, with the benefit of 15 years of hindsight, that this expectation was unlikely to be realised. An optimist might claim that the SHE model remains in advance of its time; a pessimist, on the other hand, might claim that the vision of instrumentation that is at once cheap, rugged, vandal-proof, and capable of being distributed in networks of a density adequate for the spatial scale at which processes operate, is never likely to be achieved, even in developed countries.

In recent years, also, there has been a stocktaking of what is being, and can be, achieved by the use of distributed models, even with the most powerful computer support. Grayson *et al.* (1992) expressed the doubts as follows, in a paper discussing future directions for physically based, distributed-parameter models (that is, those in which parameters describing processes are allowed to vary spatially over the river basin):

> The attraction of these models is their potential to provide information about the flow characteristics at points within the catchments, but current representations in process-based models are often too crude to enable accurate, a priori application to predictive problems. The difficulties relate to both the perception of model capabilities and the fundamental assumptions and algorithms used in the models. In addition, the scale of measurement for many parameters is often not compatible with their use in hydrologic models. The most appropriate uses of process-based, distributed-parameter models are to assist in the analysis of data, to test hypotheses in conjunction with field studies, to improve our understanding of processes and their interactions and to identify areas of poor understanding in our process descriptions. The misperception that model complexity is positively correlated with confidence in the results is exacerbated by the lack of full and frank discussion of a model's capability/limitations and reticence to publish poor results ... Model development is often not carried out in conjunction with field programs designed to test complex models, so the link with reality is lost.

Allowing for a degree of overemphasis in the views expressed by Grayson *et al.*, there remains much that is true in what they have written. Developments in computing power have proceeded far more rapidly than developments in instrumentation for measuring and recording hydrological variables at many points within a catchment area. Sophisticated distributed-parameter models, formulated so as to combine the latest findings from field studies of individual processes, are often 'reduced' — for want of a better word — to assuming that certain parameters are constant over space, because knowledge of how they vary spatially does not exist. When this becomes necessary, sophistication becomes pointless, because the model becomes effectively a lumped-parameter model.

So, on the one hand, we have lumped models of catchment behaviour, which present difficulties, often severe, as discussed in subsequent sections; and, on the other, we have distributed models of a sophistication often unwarranted by the current availability of data on spatial variability of hydrological processes. However, the distinction between lumped and distributed models is not only one of lesser or greater sophistication, but also intimately bound up with the purposes for which such models are to be used; answers to certain questions can only be attempted by using models with spatially distributed parameters. This topic is discussed in Section 8.4.

8.4 WHEN CAN LUMPED MODELS BE USED, AND WHEN MUST DISTRIBUTED MODELS BE USED?

Before embarking on work that involves development of a model expressing mean daily runoff (or runoff over some shorter period) in terms of rainfall and some index of evaporation, it is essential to hold the objective clearly in view. Blackie and Eeles (1985) give a good account of the uses of lumped catchment models, under five headings:

  (i)   quality control and infilling of missing data;

 (ii)   extensions of historic flow records;

(iii)   generation of synthetic data runs for civil engineering design work and other applications;

 (iv)   water resource assessment;

  (v)   water resources management including real-time forecasting.

Using the Institute of Hydrology lumped model described in Section 8.2.3 above, Blackie and Eeles state that it 'should be capable of making acceptable estimates of the following outputs for time-intervals of 1 hour or any multiple thereof: streamflow; evapotranspiration; soil moisture deficits and storage changes; changes in groundwater storage'.

In a paper which appeared in the same collection as Blackie and Eeles (1985), Beven (1985) quotes Beven and O'Connell (1982) as defining the role of distributed models in hydrology. The four areas offering the greatest potential are given as:

  (i)   forecasting the effects of land-use change;

 (ii)   forecasting the effects of spatially variable inputs and outputs;

(iii)   forecasting the movements of pollutants and sediments;

 (iv)   forecasting the hydrological response of ungauged catchments where no data are available for calibration of a lumped model.

Despite the views expressed by Grayson *et al.* above, distributed-parameter models remain the most objective approach to answering questions such as: how the pattern of runoff will be changed if the area above 500 m in a catchment is planted to forest; or, if an accidental spillage of toxic chemical occurs at the bridge crossing this stream, how far the river channel network will be affected, how long its effects will last, and to what extent groundwater will be affected. Lumped-parameter models have little to offer for such questions.

Important as such questions are, it must also be recognised that, when we attempt to answer them, we are stepping beyond the boundaries of science, if this is defined as the construction of hypotheses to be accepted or rejected by comparison of prediction with observation. There will not be — or so we hope — field observations with which we can compare (say) the movement of toxic chemical predicted by a distributed-parameter model. But hydrological modelling is replete with instances where quantitative estimates are calculated which cannot be subject to assessment by scientific method: estimation of the 100-year flood is another such example. Philip (1975) comments perceptively on what Weinberg (1972) has termed 'trans-science', defined as 'questions which can be asked of science and yet cannot be answered by science'. Philip writes:

> For a genuinely stationary catchment, the task of performing an accurate detailed physical characterization and using this in reliable predictive calculations is scientifically feasible; but it would almost always demand for its proper performance the expenditure of resources out of all proportion to the benefits. Weinberg would recognize this as a trans-scientific situation, *par excellence*. The problem of a catchment subject to precisely known changes involves further complications, and is even more trans-scientific: and when changes on the catchment are subject to the random forces of the market-place or the even more random whims of cabinet ministers, the incertitudes multiply further and trans-science has broken out all over.

Trans-science also enters the statistical aspects of catchment prediction. In fact Weinberg specifically sees problems involving the probability of improbable but crucial events as lying squarely in trans-science. We simply cannot afford to be scientific about such questions and sit on our bottoms for the next 1000 or 10 000 years while we get the data in. And I venture to suggest that the question of stationarity (which bedevils the current world-wide climate-change argument) also inhabits the republic of trans-science.

All this may seem pessimistic, defeatist even. What, you may ask, is the point of scientific hydrology if the problems it seeks to solve are ultimately trans-scientific? In my opinion, the answer is that it remains our obligation to ensure that our methods are as scientific and objective as possible. Let us work towards a situation where the trans-scientific judgements which practical hydrologists are forced to make are informed and sustained by a truly scientific hydrology.

8.5   A SIMPLE LUMPED-PARAMETER MODEL USING ARTIFICIAL DATA

In this section we return to lumped-parameter models — those for which model inputs, outputs and the parameters of hydrological processes are all spatially averaged — and illustrate some aspects of their use with a short sequence of artificial data on mean areal rainfall, potential evapotranspiration, and runoff.

Suppose that we model a basin as a single linear reservoir, with area equal to the basin area and of unlimited depth. That is, we assume that instantaneous storage $S(t)$ in the reservoir is linearly related to the instantaneous discharge leaving the reservoir.

Let S_i be the depths of water in storage at times $i = 0, 1, \ldots, t$; suppose that the reservoir receives rainfall $P_{01}, P_{12}, \ldots, P_{t,t+1}, \ldots$ in the time intervals between $t = 0$ and $t = 1$; between $t = 1$ and $t = 2$; .... Suppose that evaporation and other losses are known in the same intervals, denoted by $E_{01}, E_{12}, \ldots$. Because the reservoir is linear, the discharge is proportional to storage, so we write $q(t) = \lambda S(t)$ where λ is some constant to be estimated; or, for the discrete time intervals being considered, mean discharge $q_{t,t+1} = \lambda S_{t+1}$. (In this section, we use a double subscript for q to assist understanding.) To simplify matters still further, assume that all of the discharge generated in the interval $(t, t + 1]$ passes the basin's flow gauging station in the same interval: that is, no part of it is delayed in its passage until the intervals $(t + 1, t + 2], (t + 2, t + 3], \ldots$.

The parameters of the very simple model are therefore the initial storage S_0, and the constant for the linear reservoir, λ. We assume that a sequence of runoff measurements $q_{01}, q_{12}, q_{23}, \ldots$ are available as a basin 'response variable' from which λ and S_0 are to be estimated.

From water balance considerations, we have (subject to the provision given below)

$$S_1 = S_0 + P_{01} - E_{01}$$

giving

$$q_{01} = \lambda[S_0 + P_{01} - E_{01}]$$

Also

$$S_2 = S_1 - q_{01} + P_{12} - E_{12}$$
$$= (1 - \lambda)[S_0 + P_{01} - E_{01}] + (P_{12} - E_{12})$$

giving

$$q_{12} = \lambda(1 - \lambda)[S_0 + P_{01} - E_{01}] + \lambda(P_{12} - E_{12}) \qquad (8.18)$$

Similarly,

$$S_3 = S_2 - q_{12} + P_{23} - E_{23}$$

giving

$$q_{23} = [S_0 + P_{01} - E_{01}]\{(1 - \lambda) - \lambda(1 - \lambda)\} + (1 - \lambda)(P_{12} - E_{12}) + (P_{23} - E_{23}) \quad (8.19)$$

and so on. Thus we have expressed q_{01}, q_{12}, q_{23} in terms of λ and S_0, and it is clear that whilst we could continue to write explicit expressions for $q_{34}, q_{45}, \ldots, q_{t,t+1}$ in this manner, they will rapidly become very lengthy. We also note that, already, q_{23} is non-linear in the parameters because of terms involving $\lambda^2 S_0$, λS_0. Furthermore, in writing down q_{01}, q_{12} and q_{23}, we have assumed that the reservoir never becomes empty in any of the three periods; to check that this is so, a logical test is necessary, of the form

$$q_{01} = \begin{cases} \lambda[S_0 + P_{01} - E_{01}] & \text{if } S_0 + P_{01} - E_{01} \geq 0 \\ 0 & \text{otherwise} \end{cases} \quad (8.20)$$

with similar tests for $q_{12}, q_{23}, \ldots$.

But the complications are just beginning. Evaporation $E_{t,t+1}$ is measured only in very exceptional circumstances, so the actual evaporation must be estimated by, say, assuming that it is a function of storage S_t, and additional parameters are needed to describe the form of this function. The runoff generated in the interval $(t, t+1]$ is unlikely to emerge in its entirety from the basin in this interval, so more parameters are needed to describe how runoff is distributed over the successive time intervals $(t+1, t+2], (t+2, t+3], \ldots$. The simple example discussed above assumed that the linear reservoir had unlimited depth; in practice, we might prefer to assume that the depth is limited and equal to D, say, with any excess occurring during $(t, t+1]$ contributing to the runoff generated in that interval. Thus D becomes another parameter to be estimated. Finally, we may consider that a single linear reservoir is a gross simplification of how a river basin behaves, and wish to introduce other reservoirs, not necessarily linear, to represent storage within vegetation, within successive soil layers, and in deep storage. Thus the number of storages and, if we are not careful, the number of parameters, can proliferate rapidly.

Even where the number of parameters has been kept within limits, much research effort by hydrologists has been addressed to the difficult problem of estimating the parameters by minimising some measure of the discrepancies between the observed runoff sequence $q_{01}, q_{12}, q_{23}, \ldots$ and the 'fitted' sequence given by the model. Commonly, this measure has been the sum of squared differences, and it is this measure that we shall use for the time being.

The data in Figure 8.7 show a sequence of net inputs (pme: daily precipitation minus daily evaporation) to a linear reservoir with initial contents S_0. The outputs, q_{obs}, are shown for a period of 20 days; these were generated as described earlier in this section, with values $S_0 = 100$ mm and λ — the factor which multiplies S_t to give the 'observed' outflow, q_{obs} — set at 0.05.

To begin with, we assume that q_{obs} is free from error, and we explore the sum of squares surface RSS, regarded as a function of S_0 and λ, to find the point at which RSS is a minimum; this would be one approach if the parameters were unknown. We set up a 7×7 grid, centred on $S_0 = 100$, $\lambda = 0.05$, and therefore expect to find the minimum value of RSS at the grid centre. Note that, as well as having eliminated the errors of measurement in q_{obs}, we have also eliminated another kind of error entirely, namely, the error introduced by approximating a complex reality by a much-simplified representation of it. When any real-world sequence, q_{obs}, is to be modelled, both kinds of error — errors of measurement and model error — will be present, although inseparable.

Figure 8.8 shows a lineprinter contour plot of the RSS surface. We see that there is a pronounced curvilinearity in the relation between λ and S_0, for any fixed value of RSS. This

P_ET	q	S
−1	5.000	100.00
−1	4.700	94.00
−1	4.415	88.30
9	4.644	92.89
−1	4.362	87.24
−1	4.094	81.88
−1	3.839	76.78
−1	3.597	71.95
−1	3.367	67.35
19	4.149	82.98
9	4.392	87.83
−1	4.122	82.44
−1	3.866	77.32
−1	3.623	72.45
−1	3.391	67.83
29	4.672	93.44
−1	4.388	87.77
−1	4.119	82.38
−1	3.863	77.26
−1	3.620	72.40

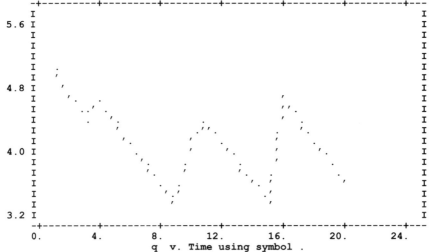

Figure 8.7 Artificial sequences of 20 values of rainfall minus ET, $P - ET$, and runoff, q, obtained by routing $P - ET$ through a linear reservoir with parameter $\lambda = 0.05$. Initial storage in the linear reservoir is taken as 100 mm.

is a common feature of the RSS surface for many rainfall-runoff models, and it occurs for the following reason. If, in the 'explicit' expressions for q(calculated), we neglect terms in λ^2, we have terms of the form q(calculated) = (a function of the input data) times (λS_0) plus (a function of the input data). When this enters the calculation of RSS, terms in $\lambda^2 S_0^2$ are obtained, and the curves of constant RSS become rectangular hyperbolas of the form λS_0 = constant. Where this occurs, searching the RSS surface to find the values of λ and S_0 at which the surface has its minimum commonly presents computational difficulties. These are the consequence of a lack

RSS 1	2	3	4	5	6	7
1 133.432144	116.801720	101.291100	86.900269	73.629257	61.478035	50.446602
2 78.139610	60.980896	45.967236	33.098614	22.375031	13.796481	7.362971
3 42.330250	27.330545	15.615085	7.183845	2.036814	0.174005	1.595417
4 20.109526	8.937181	2.234010	0.000001	2.235164	8.939499	20.112997
5 7.531932	1.229261	0.591561	5.618842	16.311098	32.668365	54.690548
6 1.950162	1.180574	7.266864	20.209042	40.007080	66.661034	100.170868
7 1.580590	6.782138	20.020319	41.295120	70.606514	107.954567	153.339264

Contour plot of RSS at intervals of 15.334

∗∗ Scaled values at grid points ∗∗

0.1031	0.4423	1.3056	2.6931	4.6046	7.0402	10.0000
0.1272	0.0770	0.4739	1.3179	2.6091	4.3473	6.5326
0.4912	0.0802	0.0386	0.3664	1.0637	2.1305	3.5666
1.3114	0.5828	0.1457	0.0000	0.1458	0.5830	1.3117
2.7606	1.7824	1.0183	0.4685	0.1328	0.0113	0.1040
5.0959	3.9769	2.9977	2.1585	1.4592	0.8997	0.4802
8.7018	7.6172	6.6057	5.6672	4.8017	4.0093	3.2899

```
        0.0000     0.1667     0.3333     0.5000     0.6667     0.8333     1.0000
           '          '          '          '          '          '          '
  1.000-00000000000000000000          222222      4444      6666      888    0-
         0000000000000000000000        222222     44444     6666      888
         0000000000000000000000000      2222222    44444     6666     888
         00000000000000000000000000      222222    44444     6666     8
         0000000000000000000000000000     222222    44444    6666
         00000000000000000000000000000    2222222    44444    6666
  0.833-000000000000000000000000000000     2222222    44444   666-
         00000000000000000000000000000000   2222222   44444
         0000000000000000000000000000000000   2222222   44444
         0000000000000000000000000000000000    2222222    444444
         00000000000000000000000000000000000    2222222    444
         0000000000000000000000000000000000000   2222222    4
  0.667-0000000000000000000000000000000000000000  22222222   -
         0000000000000000000000000000000000000000  222222222
         00000000000000000000000000000000000000000000  222222
         000000000000000000000000000000000000000000000000  222
         0000000000000000000000000   0000000000000000000000
         0000000000000000000000    000000000000000000000000000
  0.500-    0000000000000000000000000000000000000000000000000000    -
          0000000000000000000000000000   000000000000000000000000000
          000000000000000000000000000   0000000000000000000000000000
           000000000000000000000000000   000000000000000000000000000
   22      0000000000000000000000000   0000000000000000000000000000000
   22222       000000000000000000000000000000000000000000000000000000
  0.333-22222222         000000000000000000000000000000000000000000000-
         2222222222          000000000000000000000000000000000000000000
          22222222222          0000000000000000000000000000000000000000
           2222222222          000000000000000000000000000000000000
   44       22222222222          000000000000000000000000000
   444444       22222222222          00000000000000000000
  0.167- 444444444          222222222222222          000000000000-
           444444444          222222222222          00000
   66        4444444444          22222222222222
   6666666       44444444444          2222222222222222
      666666666          44444444444          22222222222222222
   8        666666666          444444444444          222222
  0.000-8888888          6666666666          4444444444444          -
           '          '          '          '          '          '          '
```

Figure 8.8 Sum-of-squares function evaluated over a 7 × 7 grid of points, centred on the true values $S_0 = 100$ mm, $\lambda = 0.05$, for the artificial data of Figure 8.6. Also shown is the contour plot of this surface (note the narrow valley indicative of a functional relation between S_0 and λ); and the GENSTAT program for calculating the sum-of-squares surface, using the artificial data of Figure 8.7.

Continued

_ Figure 8.8 *(continued)* _

```
        VARIATE  [VALUES=3(−1),9,5(−1),19,9,4(−1),29,4(−1)]P_ET
            &    [VALUES=5.000,4.700,4.415,4.644,4.362,4.094,3.839,3.597,\
                          3.367,4.149,4.392,4.122,3.866,3.623,3.391,4.672,\
                          4.388,4.119,3.863,3.620] qobs
            &    [NVALUES=20] qfit,S,dev
        SCALAR SO, lamda,u,SSQ
        SCALAR lamdaO,SOO
        MATRIX [ROWS=7;COLUMNS=7] RSS
        CALCULATE    lamdaO=0.05
                 & SOO=100
        FOR i=1...7
            CALCULATE ii=i−4
                &       lamda=lamdaO+0.01∗ii
                    FOR j=1...7
                        CALCULATE    jj=j−4
                            &        SO=SOO+10∗jj
                            &        qfit$[1]=lamda∗SO
                            &        S$[1]=SO
                        FOR t=2...20
                            CALCULATE  u=t−1
                                &        S$[t]=S$[u]+P_ET$[t]− qfit$[u]
                                &        test=S$[t].GE.O
                                &        qfit$[t]=lamda∗S$[t]∗test
                        ENDFOR
            CALCULATE        dev=qfit−qobs
                &            SSQ=SUM(dev∗∗2)
                &            RSS$[i;j]=SSQ
            ENDFOR
        ENDFOR
        PRINT RSS; DECIMALS=6
        CONTOUR RSS
```

of definition in the model: we could get very similar observed output sequences, q_{obs}, by taking either a larger value for the initial reservoir contents S_0, together with a smaller value of λ, the factor which multiplies it to give q_{obs}; or a smaller value of S_0 and a larger value for λ. Thus S_0 and λ are by their nature inversely related. The problem can be evaded (and usually is, in practice) by fixing one of the parameters — say, S_0 — but our hope that the model bears some approximation to reality is then diluted further. Fixing S_0 at, say, 200 mm, or indeed any other positive value, then simplifies the search for the value of λ which minimises the *RSS* surface, as Figure 8.8 will make clear: with S_0 fixed, we need only calculate *RSS* at points on the line S_0 = constant, and the 'dimensionality' of the problem has been reduced from two to one.

   Now let us look at what happens if, as in real life, the 'data' (artificial in this example) are subject to error. We introduce errors of magnitude 10% only in the values of q_{obs} on those days when rain fell (that is, when precipitation minus evaporation is positive, in the present case). Values of q_{obs} for four days are affected (days 4, 10, 11 and 16). The signs of the 10% errors, chosen by tossing a coin, were +, −, + and +. In practice, of course, mean areal rainfall and mean areal evaporation will also be subject to random errors, and also to biases in the sense that, for example, a mountainous catchment area will have most of its rain gauges and meteorological stations at lower altitudes which are likely to be unrepresentative of the catchment area as a whole.

RSS

	1	2	3	4	5	6	7
1	147.041107	129.765793	113.610291	98.574570	84.658676	71.862556	60.186234
2	88.965179	70.927101	55.034054	41.286068	29.683102	20.225176	12.912289
3	50.744713	34.678082	21.895693	12.397532	6.183575	3.253843	3.608328
4	26.430605	14.043423	6.125415	2.676568	3.696893	9.186391	19.145052
5	12.030144	4.398079	2.430961	6.128834	15.491686	30.519545	51.212322
6	4.855232	2.669599	7.339846	18.865976	37.247974	62.485870	94.579674
7	3.087087	6.809238	18.568022	38.363430	66.195435	102.064087	145.969406

Contour plot of RSS at intervals of 14.461

** Scaled values at grid points **

0.2135	0.4709	1.2840	2.6529	4.5775	7.0579	10.0940
0.3357	0.1846	0.5076	1.3046	2.5758	4.3210	6.5403
0.8319	0.3041	0.1681	0.4238	1.0713	2.1105	3.5414
1.8277	0.9711	0.4236	0.1851	0.2556	0.6353	1.3239
3.5091	2.3980	1.5141	0.8573	0.4276	0.2250	0.2495
6.1521	4.9047	3.8057	2.8550	2.0526	1.3986	0.8929
10.1681	8.9735	7.8563	6.8166	5.8543	4.9694	4.1620

```
           0.0000      0.1667      0.3333      0.5000      0.6667      0.8333      1.0000
           ,           ,           ,           ,           ,           ,           ,
     1.000-00000000000000000000         2222222      4444        666      888     0-
           00000000000000000000         222222      44444       6666      888
           000000000000000000000         2222222      44444      6666      888
           000000000000000000000000       2222222      44444     6666      8
           00000000000000000000000000      222222      44444     6666
           0000000000000000000000000000     222222      44444     6666
     0.833-00000000000000000000000000000     2222222     44444      666-
           0000000000000000000000000000000    2222222     44444
           00000000000000000000000000000000    2222222    44444
           000000000000000000000000000000000    2222222     44444
           0000000000000000000000000000000000    2222222      444
           00000000000000000000000000000000000    2222222
     0.667-000000000000000000000000000000000000     2222222      -
           0000000000000000000000000000000000000000    22222222
           0000000000000000000000000000000000000000000   222222
           00000000000000000000000000000000000000000000    222
           000000000000000000000000000000000000000000000
           000000000000000000000000000000000000000000
     0.500-           0000000000000000000000000000000000000000000      -
           2           000000000000000000000000000000000000000000000
           222         00000000000000000000000000000000000000000000000
           222222      0000000000000000000000000000000000000000000000
           222222222   00000000000000000000000000000000000000000000
           222222222    00000000000000000000000000000000000000000
     0.333-    2222222222   000000000000000000000000000000000000000-
               222222222    0000000000000000000000000000000000
           444      22222222222    000000000000000000000000000000
           444444     222222222222    0000000000000000000000000
           44444444      2222222222222   00000000000000000000
             444444444       222222222222    000000000
     0.167-66     444444444       222222222222     000-
           666666     444444444        2222222222222
           66666666       4444444444      2222222222222
             666666666       44444444444      222222222222
           888888     666666666         444444444444      2222
             88888888      6666666666      444444444444
     0.000-00     888888888        6666666666      44444444444-
           ,           ,           ,           ,           ,           ,           ,
```

Figure 8.9   Sum-of-squares surface (plot of *RSS*) for the artificial data of Figure 8.7, but with random 10% errors $(+,-,+,+)$ added to the four peak flows in the observed flow sequence, q_{obs} (compare with Figure 8.8). Also shown is the GENSTAT program used to obtain the surface.

Continued

Figure 8.9 *(continued)*

```
           "Same artificial data as in Figure 8.7, but with random 10% errors
           added to the peak flows in the sequence qobs."
           ''
           ''
           ''
           VARIATE  [VALUES=3(−1),9,5(−1),19,9,4(−1),29,4(−1)]P_ET
               &    [VALUES=5.000,4.700,4.415,5.108,4.362,4.094,3.839,3.597,\
                        3.367,3.735,5.831,4.122,3.866,3.623,3.391,5.139,\
                        4.388,4.119,3.863,3.620] qobs
               &    [NVALUES=20] qfit,S,dev
           SCALAR SO, lamda,u,SSQ
           SCALAR lamdaO,SOO
           MATRIX [ROWS=7;COLUMNS=7] RSS
           CALCULATE     lamdaO=0.05
                     & SOO=100
           FOR i=1...7
                 CALCULATE ii=i−4
                     &     lamda=lamdaO+0.01*ii
                       FOR j=1...7
                           CALCULATE    jj=j−4
                              &             SO=SOO+10*jj
                              &         qfit$[1]=lamda*SO
                              &         S$[1]=SO
                            FOR t=2...20
                                CALCULATE   u=t−1
                                  &             S$[t]=S$[u]+P_ET$[t]− qfit$[u]
                                  &             test=S$[t].GE.O
                                  &         qfit$[t]=lamda*S$[t]*test
                            ENDFOR
                  CALCULATE      dev=qfit−qobs
                      &          SSQ=SUM(dev**2)
                      &          RSS$[i;j]=SSQ
                  ENDFOR
           ENDFOR
           PRINT RSS; DECIMALS=6
           CONTOUR RSS
```

Having introduced these random errors to q_{obs}, let us now look at their effects on the *RSS* surface and on the calculation of S_0 and λ. Figure 8.9 shows that, whilst the general shape of the *RSS* surface is left broadly unchanged, the narrow trench defining the relation between λ and S_0 is no longer apparent. The 'lower' parts of the *RSS* surface have therefore become a broad and featureless plain, like a rectangular hyperbola drawn with a very broad brush. Even if we fix the value of S_0 at some arbitrary value, therefore, the value of λ which minimises the *RSS* surface will be rather poorly defined.

The model used in this artificial example is grossly simplified by comparison with those in common use by hydrologists. Nevertheless, it serves to illustrate some important points. First, calculation of the estimates of parameter values — those that minimise the *RSS* surface, in the case of the example, or which optimise whatever other fitting criterion that might be adopted — is not straightforward. Although our discussion has considered only the simplest of cases, the difficulties increase rapidly as the number of parameters increases.

Second, when the difficulties are evaded by fixing the values of some of the parameters, the physical resemblance between the model and reality, which was the basis for its adoption in the first place, may be substantially diminished.

8.6 MEASURES OF DISCREPANCY BETWEEN OBSERVATIONS AND FITTED VALUES

A commonly used measure of the degree of agreement between a sequence of observed flows q_t (observed), and a series q_t (calculated) of flows obtained when we assume particular numerical values for the parameters of a conceptual hydrological model, is the sum of squares criterion

$$\sum (q_t(\text{observed}) - q_t(\text{calculated}))^2 \qquad (8.21)$$

We revert to our usual practice of denoting the parameters by the vector $\theta = [\theta_1, \theta_2, \ldots]^{\text{T}}$, and denote by $q_t^* = q_t(\theta^*)$ the flows calculated for specific numerical values of θ. Fitting the model can then be regarded as a search to determine the parameter values $\hat{\theta}$ which minimise the criterion (8.21).

If we were prepared to assume that the discrepancies $\varepsilon_t = [q_t(\text{observed}) - q_t(\text{fitted})]$ were random variables with zero means and constant variances, and that they were normally and independently distributed, the likelihood procedures outlined in earlier chapters would lead immediately to the criterion (8.21) as the log-likelihood from which θ could be estimated. On the other hand, we might be prepared to abandon statistical considerations, and minimise a criterion like expression (8.21) without introducing any notion of random variation in the q_t. This would be both unwise (since we could not, for example, make probabilistic statements about the reliability of forecasts obtained from the model) and illogical (where, for example, we are to use the model to extend records of annual floods, for the purpose of calculating the flood with T-year return period). More is to be lost than gained by discarding the concept that the ε_t are random variables.

Assuming, then, that the ε_t are to be treated as random variables, requirements that they are normal, mutually independent, and homogeneous in variance are unlikely to be satisfied. The variance of ε_t during periods of recession, for example, is likely to be very much less than the variance of ε_t during periods when discharge is rising; the ε_t will also be strongly intercorrelated. The criterion (8.21), regarded as a log-likelihood, is therefore inappropriate as it stands, and we must modify it if we are to make use of likelihood theory and other statistical methods which assume statistical independence of observations of the response variable.

One difficulty with the criterion (8.21) is that it gives equal weight to each observation q_t in the record $\{q_t\}$. If we intend to use the model to extend flood records, it would be more appropriate to give particular weight to the differences $\varepsilon_t = q_t(\text{observed}) - q_t(\text{calculated})$ for values of t where $q_t(\text{observed})$ has a local maximum: that is, where $q_t(\text{observed})$ is greater than both $q_{t-1}(\text{observed})$ and $q_{t+1}(\text{observed})$. Instead of using the criterion (8.21), we should then be using

$$\sum w_t [q_t(\text{observed}) - q_t(\text{calculated})]^2 \qquad (8.22)$$

where w_t is unity if $q_t(\text{observed}) > q_{t-1}(\text{observed})$ and $q_t(\text{observed}) > q_{t+1}(\text{observed})$, and is zero otherwise. The weights w_t are easily calculated in GENSTAT using the following instructions:

```
> CALCULATE qm1 = SHIFT(q;1): & qp1 = SHIFT(q;−1)
> & w = (q−qm1.GT.0).AND.(q−qp1.GT.0)
```

which puts 1s where q_t has its maximum values. Substituting LT instead of GT in the second line would put 1s where q_t has its minima, whilst the instructions

```
>CALCULATE qm1 = SHIFT(q;1): & qp1 = SHIFT(q;−1)
> & dqm1 = q−qm1: & dqp1 = q−qp1
> & w1=(dqm1.GT.0).AND.(dqp1.GT.0)
> & w2=(dqm1.LT.0).AND.(dqp1.LT.0)
> & w = w1+w2
```

puts 1s where q_t has its local maxima or minima. We can in fact generalise the expression (8.22) still further; if S_v represents summation over sequences of v values, so that $S_v q_t$ (observed), for example, is the sum of v successive flow measurements q_t, then a generalisation of (8.22) is

$$\sum w_{v,t}[S_v q_t(\text{observed}) - S_v q_t(\text{calculated})]^2$$

We might wish to use this criterion if, for example, we wanted to fit the model using monthly total flows, instead of mean daily flows.

If we have confidence in the structure of a rainfall-runoff model, and believe that it constitutes a good description of the physical behaviour of a river basin, there is no intrinsic reason why the model should not be fitted to high flows when we are concerned with problems of flood estimation, and to low flows when we are concerned with droughts. There is, of course, no reason to expect estimates of the model parameters to be the same in both cases. Also, if by fitting the model to high flows we find that the assumption of statistical independence of the residuals ε_t becomes more tenable, we can obtain from likelihood theory values for the variances and covariances of the estimates of model parameters, using the second derivatives at the point where the log-likelihood has its maximum, as described in Chapter 3. However, care is needed if the shape of the log-likelihood function is very far from being a quadratic surface near this point, since the likelihood will not then be adequately summarised by the coordinates of the point at which it has its maximum (these coordinates being estimates of the model parameters) and the curvature at that point.

Some of these ideas are illustrated using the Institute of Hydrology lumped model, introduced earlier in Section 8.2.3. Blackie and Eeles (1985) describe a fitting procedure for the 15 model parameters which proceeds through four stages, as follows. In the first stage, all parameters except the interception parameters are held constant, and the interception parameters alone are optimised: whether by trial and error, or by processes of 'automatic' optimisation as described in Section 8.7, or by some combination of the two, in which the model user 'steers' the automatic optimisation in some way, by limiting the range over which the search for the best-fitting model parameter is allowed to vary. Thus at the end of this first stage, improved estimates of the interception parameters have been found, whilst the others remain to be fitted. In the second stage, all parameters except the soil parameters are held constant, and the soil parameters are optimised by a similar procedure to that used in the first stage. Similarly in the third and fourth stages, all parameters except the groundwater and routing parameters, respectively, are held fixed, and the groundwater and routing parameters are optimised in the third and fourth stages, respectively. Thus at the end of the fourth stage, all 15 parameters have been subject to an optimisation procedure. The calculation now recommences at the first stage, and the iterations proceed until the estimates of all the parameters have converged to some optimum value which, it is hoped, is a global optimum. This kind of segmented iterative procedure is necessary because 15 parameters is a very large number to estimate simultaneously.

The following MATLAB code, in the file OPTIHLM.M in the computer material provided with this book, shows how this calculation can be effected. At each stage, a group of parameters (interception; soil; groundwater; routing) is optimised (see Section 8.7) using the MATLAB directive NELDER; each call on NELDER requires a function to be specified (*fitint*, *fitsoil*, *fitrout*, *fitgw*), and it is these functions that are minimised with respect to their parameters. The four functions are given in the files FITINT.M, FITSOIL.M, FITROUT.M and FITGW.M on the diskette, and they are listed at the end of OPTIHLM.M below.

```
% Listing of MATLAB file OPTIHLM.M
%
load flow;                          % Read in the flow sequence.
N = length(flow);
Time = ones(N,1);
Time = cumsum(Time);                % 'Time' is now 1, 2, ..., N.
tol = 0.001;                        % 'tol' is the tolerance for
                                    % convergence in
nits = 5;                           % the Nelder minimisation. 'nits' is
                                    % the number of iterations permitted.
        load intprm;                % 'intpm' is the set of interception
                                    % parameters.
        load soilprm;               % Now load the soil, routing, and
                                    % groundwater
        load routprm;               % parameters, and put them into one
                                    % vector named
        load gwprm;                 % 'prmter15'. Use 'prmter15' to
                                    % calculate the fitted flows,
                                    % 'FLOWHAT' as demonstrated earlier in
                                    % this chapter.
                                    %
for i = 1:nits,                     % The number of iterations.
                                    %
        intprm = nelder('fitint',intprm,tol,1);
        save intprm;                % This saves the optimised
                                    % interception parameters.
                                    %
                                    %
        prmter15=[intprm' routprm' soilprm' gwprm' storinit']';%This
                                    % line updates the file 'prmter15' by
                                    % replacing former values of 'intprm'
                                    % by their newly optimised values
                                    % —assuming that they have converged
                                    % satisfactorily.
```

```
flowhat = ihmodel(prmter15);        % This calculates the
                                    % predicted flows using the updated
                                    % parameter estimates.
plot(Time,flowhat,'-',Time,flow,':');   % This plots them.
pause
                                    % Observed and fitted flows, 'flow' and
                                    % 'flowhat', are plotted on the same
                                    % graph with full and dotted lines,
                                    % respectively.
                                    % The graph is shown until the
                                    % modeller
                                    % strikes any key to continue by
                                    % breaking
                                    % the 'pause'.
%
%
keyboard                            % 'keyboard' transfers control to the
                                    % keyboard,
                                    % so that the modeller can, if he
                                    % wishes, change the
                                    % values of any of the parameter
                                    % estimates. The program can be set
                                    % running again
                                    % by entering 'CONTROL-Z'.
%
%
% Having used nelder to optimise the interception
% parameters, we
% now do the same for the soil parameters.
%.
%
soilprm = nelder('fitsoil',soilprm,tol,1) % Optimises soil
                                    % moisture parameters.
save soilprm;                       % Saves the optimised values.
prmter15=[intprm' routprm' soilprm' gwprm' storinit]';
                                    % Updates the vector 'prmter(15)'.
flowhat = ihmodel(prmter15); % Recomputes predicted flows.
plot(Time,flowhat,'-',Time,flow,':'); % Plots them.
pause
                                    % Observed and fitted flows, 'flow' and
                                    % 'flowhat', are plotted on the same
                                    % graph with
                                    % full and dotted lines, respectively.
                                    % The graph is shown until the
                                    % modeller strikes
                                    % any key to continue by breaking
```

```
                                              % the 'pause'.
%
%
keyboard
%
%
%
% Having used nelder to optimise the soil parameters, we
% now do the same for the routing parameters.
%
%
   routprm = nelder('fitrout',routprm,tol,1)
   save routprm;
   prmter15=[intprm' routprm' soilprm' gwprm' storinit]';
   flowhat = ihmodel(prmter15);
   plot(Time,flowhat,'-',Time,flow,':');
   pause
                                   % Observed and fitted flows, 'flow' and
                                   % 'flowhat', are plotted on the same
                                   % graph with full
                                   % and dotted lines, respectively.
                                   % The graph is shown until the
                                   % modeller strikes any
                                   % key to continue by breaking
                                   % the 'pause'.
%
%
keyboard
%
%
% Having used nelder to optimise the routing parameters, we
% now do the same for the groundwater parameters.
%
%
   gwprm = nelder('fitgw',gwprm,tol,1)
   save routprm;
   prmter15=[intprm' routprm' soilprm' gwprm' storinit]';
   flowhat = ihmodel(prmter15);
   plot(Time,flowhat,'-',Time,flow,':');
   pause
                                   % Observed and fitted flows, 'flow' and
                                   % 'flowhat', are plotted on the same
                                   % graph with full
                                   % and dotted lines, respectively.
                                   % The graph is shown until the
                                   % modeller strikes any
```

```
                                        % key to continue by breaking
                                        % the 'pause'.
% Having used nelder to optimise the groundwater parameters,
% we now begin the loop again
% to reoptimise the interception parameters.
%
%
%
%
keyboard
%
%
end
```

The following are the listings of the four MATLAB files FITINT.M, FITROUT.M, FITSOIL.M and FITGW.M which calculate the function to be minimised when the optimising directive nelder is called. By selecting appropriate values for the vector **weights**, the parameters can be fitted to peak flows only, or to minimum flows only, or to both, or indeed to any values in the observed flow sequence that are the most appropriate for the model purpose.

```
% Listing of MATLAB file FITINT.M
function y = fitint(intprm);
load flow;
load routprm;
load soilprm;
load gwprm;
load storinit;
load weights;
prmter15=[intprm' routprm' soilprm' gwprm' storinit']';
q = ihmodel(prmter15);
y = sum(weights.*(flow−q).*(flow−q));

% Listing of MATLAB file FITROUT.M
function y = fitrout(routprm);
load flow;
load intprm;
load soilprm;
load gwprm;
load storinit;
load weights;
prmter15=[intprm' routprm' soilprm' gwprm' storinit']';
q = ihmodel(prmter15);
y = sum(weights.*(flow−q).*(flow−q));

% Listing of MATLAB file FITSOIL.M
function y = fitsoil(soilprm);
load flow;
```

```
load intprm;
load routprm;
load gwprm;
load storinit;
load weights;
prmter15=[intprm' routprm' soilprm' gwprm' storinit']';
q = IHMODEL(prmter15);
y = sum(weights.*(flow−q).*(flow−q));

% Listing of MATLAB file FITGW.M
function y = fitgw(gwprm);
load flow;
load intprm;
load routprm;
load soil prm;
load storinit;
load weights;
prmter15=[intprm' routprm' soilprm' gwprm' storinit']';
q = IHMODEL(prmter15);
y = sum(weights.*(flow−q).*(flow−q));
```

8.7 'AUTOMATIC' OPTIMISATION OF PARAMETERS

In Section 8.5 we sought to locate the minimum of the RSS surface by simple grid search, a procedure made possible by the simple form of the two-parameter model. With more than two parameters, grid searches rapidly become too unwieldly for practical use. One alternative is a trial-and-error search which, for lengthy data sets, is time-consuming unless the user has access to very powerful computing facilities. Another is to use a computational procedure which makes a search for the minimum according to some programmed rules. Several procedures of this kind exist; the FITNONLINEAR directive of GENSTAT uses a modified Newton–Raphson method (although the method used by the directive can be changed to a Gauss–Newton procedure with ease), and the Nelder–Mead algorithm, mentioned above and illustrated below, is another. The algorithm due to Rosenbrock (1960) has been widely used by hydrologists to fit rainfall-runoff models. These methods are neither as robust nor as automatic as, for example, the methods used to fit the linear models of Chapter 4, or the generalised linear models of Chapter 6. Minimisation of an RSS function (or, equivalently, maximization of a likelihood function; the general term 'optimisation' covers both cases) is easiest with few parameters, when the functions are approximately quadratic, when the parameters are not interrelated, and when initial parameter estimates are good. Convergence of the calculation to an optimum is not guaranteed to succeed, and indeed is likely to fail with complicated functions unless careful thought is given to the way in which the function is parametrised, to starting values for the parameters, to initial steplengths for the search process, and to upper and lower bounds to exclude values of the parameters which would cause the calculation to break down.

To illustrate the use of one automatic search procedure to find the minimum of an RSS function, we return to the artificial data of Section 8.5. The use of the Nelder–Mead algorithm requires:

(i) A convergence criterion, which is used to test whether the minimum has been attained. This is some small positive number; in the case of the example, it was set, rather arbitrarily, at 0.001. Smaller values require more iterations to attain convergence to the minimum (when indeed it is possible to attain it), larger values less.

(ii) The number of iterations at which the calculation should stop, if the minimum has not been attained by then. This was set to 30.

(iii) Three parameters, α, β and γ. For the example, these were set equal to 1, 0.5 and 2, respectively.

(iv) Initial values of the parameters. Two sets were tried: first, $S_0 = 110$, $\lambda = 0.06$; and second, $S_0 = 120$, $\lambda = 0.08$.

(v) A subroutine specifying the form of the function to be minimised. Where a function, such as a likelihood, is to be maximised, the subroutine must define the negative of this function.

For the example, a subroutine ('m-file', in MATLAB terminology) was composed to define the calculation of RSS, and the MATLAB directive NELDER was used to minimise it. Starting from $S_0 = 110$, $\lambda = 0.06$, the minimum of RSS was given as 3.3009×10^{-4} after 20 iterations, where the fitted values were $\hat{S}_0 = 100.2674$ and $\hat{\lambda} = 0.0498$. Starting at $S_0 = 120$, $\lambda = 0.08$, the minimum was given after 26 iterations as 9.8354×10^{-4}, with values $\hat{S}_0 = 100.1484$, $\hat{\lambda} = 0.0498$. Calculation times were of the order of a minute on a 386 desk-top computer. This would increase substantially with a larger number of parameters, longer data sequences, and a more stringent criterion for convergence.

When the 10% errors were introduced, and starting from $S_0 = 110$, $\lambda = 0.06$, convergence was obtained after 19 iterations, with $\hat{S}_0 = 92.1569$, $\hat{\lambda} = 0.0561$, the minimum of RSS being 0.6795. Starting from $S_0 = 120$, $\lambda = 0.08$, the minimum was attained after 27 iterations, at $\hat{S}_0 = 91.0547$, $\hat{\lambda} = 0.0569$, the minimum being 0.6794. Clearly the introduction of these modest errors to the output variable q_{obs} has led to an appreciable change in the estimate of S_0 from its 'true' value of 100.

For models which are less oversimplified than that used in the above trivial example, the difficulties are formidable, and have been well illustrated in papers by Pickup (1977) and Johnston and Pilgrim (1976). These authors used the Boughton (1965) model, but it can be fairly supposed that the difficulties are typical of all but the most exceptional rainfall-runoff models, and are not a particular characteristic of the model used.

Pickup used a 12-parameter version of the Boughton model and began by assuming that all 12 parameters were known exactly; the model was then used to transform an artificial sequence of precipitation measurements to streamflow (presumably, synthetic sequences of daily potential evaporation were also used, although this was not stated). The generated streamflow sequence, together with the artificial precipitation and evaporation sequences used to derive it, were then used to fit the model by minimisation of a sum of squares function with respect to the 12 parameters, now assumed to be unknown. Four alternative algorithms were used to minimise the RSS function: the simplex method of Nelder and Mead, the Rosenbrock (1960) algorithm, the method described by Powell (1965), and the Davidon method as described by Fletcher and Powell (1963). Clearly all four algorithms should have given estimates of the model parameters close to the true (known) values, if not identical with them. However, Pickup found that no algorithm gave the true set of parameter values, and even for the better methods the estimates commonly differed from their true values by a factor of 2 or more, nor was the sense in which they differed consistent from algorithm to algorithm in any way. Pickup's experience is not unique; Johnston and Pilgrim (1976) gave results which were equally disturbing. Using a nine-parameter version of the Boughton model, their study

> highlighted the complexities involved in finding the optimum set of values for the parameters of rainfall-runoff models. A true optimum set of values was not found in over two years of full-time work concentrated mainly on the Lidsdale No. 2 catchment, although many apparently optimum sets were readily obtained .... . Although the Boughton model was used, it is thought that most of the findings are applicable to all rainfall-runoff models.

In the face of such difficulties, it may be tempting to abandon the attempt to estimate parameters by a computer optimization algorithm and to use a trial-and-error fitting procedure instead. However, it would be illusory to suppose that similar difficulties would be so avoided. It is the model that is commonly at fault, not the method used to fit it; the difficulties are introduced by the use of thresholds to describe the behaviour of hypothetical storages.

8.8 SPLIT-RECORD TESTS

When a rainfall-runoff model, of the kind here discussed, is fitted to the observed runoff sequence from a river basin, the value RSS_{min} of the RSS function at its minimum is often small relative to the total sum of squared deviations of q_{obs} about its mean. Thus for the artificial example with the errors added, the value of RSS_{min} is about 0.679 (smaller, indeed, than the sum of squared deviations for the introduced errors, which is 0.7983), whereas the sum of squared deviations for q_{obs} is 6.0462, about ten times as great. Indeed, the ratio $1 - RSS_{min}/$(total sum of squares for q_{obs}), expressed as a percentage, is one of many measures of 'goodness of fit' that have been proposed by hydrological modellers. Its value, usually denoted by R^2, for the artificial data is 88.74%, which might lead us to think that we have a quite a good model.

However, it frequently happens that if we use a fitted rainfall-runoff model to estimate the values of q_{obs} from future observations of rainfall and evaporation variables (or for other rainfall and evaporation data that were not included at the time of fitting), we find that the values predicted by the model are very far removed from what actually occurred. Furthermore, an R^2 calculated from these additional data — using, of course, the values of q calculated from the fitted model to obtain RSS — is very likely to be a good deal less than the R^2 computed at the model-fitting stage. When this occurs, the conclusion must be that whilst the fitted model represents quite well the characteristics of the flow record q_{obs} over the period of record, its ability to represent other periods of record from the same gauging site is not nearly so good. The value of the model is then clearly much diminished.

To explore this phenomenon, it is common to divide the period of record into two halves. The first is a 'fitting' period, the data from which are used to estimate the model parameters, and the second is a 'test' period, its data being used to explore model performance. Subsequently we can also interchange the two periods, the data used for testing now being used for fitting and conversely. The ideal would be to obtain high R^2-values both at the fitting and testing stages, with similar estimates of the model parameters from both halves of the record, so that we could confidently combine them, perhaps by weighting them inversely as their variances.

To give an example, we return to the artificial data, with the 10% errors added. If we divide these already limited data in half, and use the first ten values of the two variables *pem* and q_{obs}

to fit the model, we find after application of the Nelder–Mead algorithm that the minimum of the RSS function, $RSS = 0.3713$, was reached after 17 iterations, and that the estimates of S_0 and λ were $\hat{S}_0 = 93.5430$, $\hat{\lambda} = 0.0545$, giving an R^2-value of 88.39%, a fairly good result. However, if, using these values of $\hat{S}_0$ and $\hat{\lambda}$, we 'predict' the values of q_{obs} for the second half of the full record, we find an R^2-value of -12.6%, a truly dreadful result. The meaning of the negative sign is that the predictions given by the model are considerably worse than if we had just used their mean value. Despite the fact that we might have hoped naively that the way in which a linear reservoir functions had some physical resemblance to the way in which a river basin behaves, our 'physically based' model has proved totally inadequate.

There is a good deal in the literature on the use of split-record tests, but we may speculate that still more may be gained from modelling subsets of the data in particular ways. Suppose, for example, that we are developing a rainfall-runoff model fitted to peak flows, of which the record contains $2N$, say. Let these peak flows be numbered sequentially $q_1, q_2, \ldots, q_{2N}$. Now suppose that we start by fitting the model to the set $q_1, q_2, \ldots, q_N$, as we would for an ordinary split-record test. The next step is different: we drop q_1, and fit to $q_2, q_3, \ldots, q_N$, and then add q_{N+1}, fitting to $q_2, q_3, \ldots, q_{N+1}$. Next, we drop q_2, fitting to $q_3, \ldots, q_{N+1}$, and then add q_{N+2}, fitting to $q_3, q_4, \ldots, q_{N+3}$. Finally, of course, we end up fitting to $q_{N+1}, q_{N+2}, \ldots, q_{2N}$, the other 'half' of the split-record test. By subtracting and adding the peak flows, one by one, in this manner, we may be able to identify peak flows which are particularly influential in their effect on model fit. This influence may be because the influential peak flows are in gross error, or because the model is inappropriate for these particular data. One example might be where certain floods are produced by snowmelt; if the model contains no snowmelt component, it cannot be expected to describe them satisfactorily, and a test like the above may thereby help to identify where the model is inadequate. The technique described above is similar in nature to the calculation of deletion residuals in the linear regression case, in which influential points are identified by omitting them one at a time from the calculation; the present case would, of course, be much more demanding computationally.

8.9   SOME GENERAL OBSERVATIONS ON RAINFALL-RUNOFF MODELLING

Earlier sections have considered in some detail aspects of modelling sequences of mean daily flow, or of flows averaged over other short time intervals Δt. (We must be aware, however, that the word 'mean' in this context, although standard hydrological usage, may be misleading. Often, one measurement only of water level is taken in each Δt, and is converted to an estimated discharge by stage–discharge curve, another source of error. The estimated discharge is then taken to be constant over Δt, an assumption whose validity will depend upon the size of Δt and the responsiveness of the basin being modelled.) To simplify matters, take $\Delta t = 1$ day.

The essence of the approach is to look for physically realistic relations between the response variable q_t and the explanatory variables, rainfall P_t, and some appropriate estimate or index of evaporation, EO_t. This defines the systematic component of the model. The difference between the observed flow sequence, regarded as response variable, and the flow sequence predicted by the model when the explanatory variables are 'fed into' it, constitutes a sequence of deviations, ε_t; in this book we consider the ε_t to be random variables, and some of them — provided that they are not too near together in time — may for practical purposes be considered statistically independent. Where it is safe to assume statistical independence — for example, where models are fitted to peak flows not too near to each other in time — we immediately have access to a large body of statistical methods, based on likelihood, which would otherwise be unavailable. Likelihood theory, for example, allows us to calculate confidence regions for model parameters; and by sampling from this confidence region, we can calculate empirically a probability distribution for future values $\hat{q}_{t+k}$, predicted by the model.

The models discussed are also *lumped* in the sense that they take no account of spatial variability in hydrological processes, although it would be quite possible to do so, in theory, by dividing the basin into sub-areas that we would hope were homogeneous, and modelling the components separately. The cost, in terms of additional computation, would be very considerable if we wished to fit different parameters to different sub-areas. Models which allow for spatial variability are *distributed* models.

We have seen that rainfall-runoff models derived from water balance considerations contain parameters that enter the model non-linearly, and that these non-linearities lead to computations that are time-consuming and not guaranteed to succeed. A consequence is that, when these difficulties are met, there is a temptation to tinker with the model, or with its computer coding, until it is fitted 'successfully'; no one likes to see his model fail. However, the questions that must constantly be asked are, why one needs a model that must be fitted to daily flows and whether one is convinced that the physical 'reality' built into the model is a reality and not an illusion.

8.10   A GENERALISED DISTRIBUTED-PARAMETER MODEL

We have written at some length about the difficulties encountered in the fitting of conceptual rainfall-runoff models. Moore and Clarke (1981, 1982, 1983) attempted a new approach to rainfall-runoff modelling which avoided some of the difficulties implicit in models then existing. In their approach, a simple store (of the type with volumes M, R, S, G in the Dawdy–O'Donnell model of Section 8.2.1) was replaced by a statistical population of stores: that is, by a population of stores, each of which has maximum depth s, with the parameter s having a probability distribution $p(s; \theta)$ defined on the interval $[0, \infty)$. In this way a basin may be regarded as consisting of a population of storage elements, or narrow vertical tubes, of varying depths, each tube being closed at the bottom and open at the top. Rain falling on each storage element is stored until the element becomes full, any excess becoming runoff. The contents of the element are depleted by evaporation at the potential rate EO until empty, after which evaporation ceases until the storage element is replenished by more rain. Since the depth s of the storage element varies over the basin area according to $p(s; \theta)$, shallow elements (with small s) will be rapidly depleted by evaporation, and rapidly filled to capacity during periods of rainfall, when they will contribute to generated runoff; conversely, deep storage elements (with large s) will become empty much less frequently, and will contribute to basin evaporative demand even during prolonged dry spells (although they will also contribute much less frequently to generated runoff). For any particular alternation of wet and dry periods, therefore, a time-variant proportion of the storage elements within the basin area will be contributing to runoff whilst it is raining, and another time-variant proportion of the storage elements will be meeting basin evaporative demand when rain has stopped. The longer the period of rain, the greater the proportion of storage elements generating runoff; and the longer the dry period, the smaller is the proportion of storage elements still evaporating at the potential rate (and hence the smaller will be the rate of *actual* evaporation, averaged over the whole basin area). Thus, we no longer need to estimate parameters controlling how a small number of storages operate, but need instead to estimate the parameters θ of a probability distribution $p(s; \theta)$: so that the estimation problems of Chapters 3 to 6, and the estimation problems of this particular rainfall-runoff model, become one and the same. Furthermore, the same parameters θ control the mechanism by which actual evaporation diminishes as the basin becomes drier, and by which the proportion of basin area contributing to runoff increases as it becomes wetter.

The above description shows how basin runoff is generated; the generated runoff must now be routed to the basin outfall. For this purpose, the distribution of storage depths $p(s; \theta)$ is generalised to the form $p(s, t; \theta, \phi)$, where t is the time taken for runoff generated by the storage element with depth s, to reach the basin outfall. Thus the depths s of storage elements, and the routing times t, have a bivariate probability distribution with parameters θ, ϕ. A simple case occurs where the bivariate distribution $p(s, t; \theta, \phi)$ factorises to the product $p(s; \theta)p(t; \phi)$, so that runoff generation and routing processes become decoupled.

The Moore–Clarke model succeeded in simplifying the nature of the sum of squares surface obtained when fitting the model, in so far as contours commonly became much more nearly elliptical. It also provided good fits to hydrographs using only two parameters: one of which was the parameter of an exponential distribution of storage depths $p(s; \theta_1) = \theta_1 \exp(-\theta_1 s)$, and the other the parameter of an exponential distribution of routing times $p(t; \theta_2) = \theta \exp(-\theta_2 t)$. Like all other models, however, it performed less well during 'test' periods than in 'calibration' periods. The model was subsequently adapted (Moore and Clarke, 1983) to the joint modelling of runoff and sediment yield, taking account of the accumulation of sediment, its removal by runoff, and the exhaustion of sediment supply during the course of a storm.

There is a logical link between the Moore–Clarke model, and the Sugawara model described in Section 8.2.5. The diagram illustrating that model shows four storages (tanks) in each of a series of layers. The Moore–Clarke model increases this number to infinity, subject to three conditions:

(i)   that each tank has just one threshold;

(ii) that all input, increasing tank level to above its threshold, is transferred to the next layer of tanks, where this is postulated; if it is not, the excess is routed to the basin outfall to arrive there at times specified by the distribution $p(t; \phi)$, where t is time since the excess was generated;

(iii) that each tank has a different threshold denoted by the variable s, with s having a distribution $p(s; \theta)$ over the infinite population of tanks in each layer.

In the case of a single, infinite, layer of tanks, Hosking and Clarke (1990) derived analytical results relating runoff to rainfall, assuming:

(i)   that the occurrence/non-occurrence of rainfall was represented by a first-order, two-state Markov chain with transition probabilities α, β (α is the probability that the $(i + 1)$th time interval is wet given that the ith time interval is dry; β is the probability that the $(i + 1)$th time interval is dry given that the ith interval is wet);

(ii) that the depth of rainfall, in wet intervals, has cumulative density function (cdf) denoted by $W(x)$, with probability density $dW/dx = w(x)$, and is serially independent;

(iii) that potential evapotranspiration has cdf $D(x)$, probability density $d(x)$, and is serially independent.

With these assumptions, Hosking and Clarke (1990) showed that the cdfs for the depth of water retained in a store of typical depth s, when the ith time interval is dry (index D) and wet (index W), respectively, are

$$F_{i+1}^{D}(x) = \int_{0}^{x} \{(1 - \alpha) F_i^{D}(y) + \alpha F_i^{W}(y)\} d(x - y) dy$$

$$1 - F_{i+1}^{W}(x) = \int_{x}^{s} [\beta\{1 - F_i^{D}(y)\} + (1 - \beta) F_i^{D}(y)] w(x - y) dy$$

The cdfs for quantity of water in storage, within the whole basin, are obtained by multiplying these quantities by $Ap(s; \theta)$ and integrating, over s, from zero to infinity. Further, they showed that if R_i is the depth of runoff generated by a typical store of depth s in the ith time interval, given that the ith time interval is wet ($J_i = 1$), the cdf of R_i is

$$P[R_i < x | J_i = 1] = \int_{0}^{s} [\beta\{1 - F_{i-1}^{D}(z)\}] + (1 - \beta)\{1 - F_{i-1}^{W}(z)\}] w(x + z) dz + W(x)$$

Finally, resting on the assumption that discharge Q_i at the basin outfall could be modelled by routing the generated runoff through a linear reservoir, an analytical expression was derived for the cdf of the discharge Q_i.

Hosking and Clarke (1990) pointed out, also, that it was straightforward to modify their analysis to the case where the cdfs of rainfall depth $W(x)$, and of potential evaporation depth $D(x)$, are time dependent. Thus, a rainfall model of the Stern–Coe type (see Chapter 6) could be used to represent input to the model, with a similar model to represent potential evaporation during dry periods between storm events.

The analytical results obtained by Hosking and Clarke referred only to the case of a single population of storages: that is, to one row of storages in a diagram like that of Sugawara *et al.* (1974). It would be vastly more difficult to derive similar analytical relations where the overspill from one population of storages enters a second, possibly with diversion of some proportion of the overspill to rapid runoff. The difficulty arises principally because the overspill is serially dependent; in this more complicated case, no alternative to Monte Carlo simulation has yet been found.

8.11 SOME OBSERVATIONS ON MODELS FOR SEDIMENT YIELD AND WATER QUALITY

The difficulties encountered when modelling rainfall-runoff relationships are magnified when we begin to model sediment yield from river basins, whether as bedload or suspended material. Basin sediment yield depends both upon the availability of material to be transported, and upon the presence of water with sufficient energy to transport it. Hence, in addition to modelling the rainfall-runoff processes, it is necessary to include additional parameters to model the availability of erodable material, in time, and the rate at which it can be removed when it is available. As before, it is necessary to ask ourselves whether a model of (say) daily sediment yield, with rainfall and flow as explanatory variables, is really essential for the purpose at hand.

The same remarks hold for the modelling of transport of nutrients at a basin scale. At a more restricted level — for example, a water treatment plant operating under relatively controlled conditions — modelling becomes less difficult, unless biological processes influence the source and sink terms of the model.

8.12   THE GENERALISED LIKELIHOOD UNCERTAINTY ESTIMATION PROCEDURE

In a series of papers — for example, Binley and Beven (1991), Beven and Binley (1992) — a method has been described which is intended to serve as a general framework for evaluating the uncertainty in model predictions. The generalised likelihood uncertainty estimation (GLUE) procedure is based upon multiple simulations using parameter values selected at random from some parameter space. This space, and the weights given to the various parameters within it, summarise the modeller's beliefs about the possible ranges of the various parameters, based upon his or others' experience in similar modelling situations. It is, the authors state, a variation of Monte Carlo simulation, which requires the modeller to define the distributions from which to select parameter values, together with any expected intercorrelation between parameter values, where the parameters may also include those specifying the initial conditions for a simulation (such as initial storage contents).

The GLUE approach is an interesting and, in some respects, new approach to modelling rainfall-runoff relationships of river basins. It is important to observe, however, that the term 'likelihood' as used by Beven and Binley bears no relation to the likelihood function as it is used throughout this book or, indeed, more widely in common statistical usage. This is not a criticism of the GLUE approach; provided that a word is used consistently, and provided that its use is defined for the benefit of others, we can choose any words that we wish to explain ideas. However, some of the statements by Beven and Binley (1992) do less than full justice to existing likelihood methods. We find, for example, the following in their paper (we highlight part of the quote):

> We use the term likelihood in a very general sense, as a fuzzy, belief, or possibilistic measure of how well the model conforms to the observed behaviour of the system, and not in the restricted sense of maximum likelihood theory *which is developed under specific assumptions of zero mean, normally distributed errors ... Our experience with physically-based distributed models suggests that the errors associated with even optimal parameter sets are neither zero mean or normally distributed.*

Earlier chapters of this book have demonstrated that it is by no means necessary to assume that random components are normally distributed.

The GLUE procedure has more serious shortcomings, however, than the merely semantic. First, GLUE requires that each parameter of a model be given a prior distribution, from which values are drawn at random. In the absence of knowledge about a form for the prior distribution, a uniform distribution is adopted. This leads to conceptual problems which are briefly mentioned by the authors, but which are not satisfactorily resolved. Take, for example, the Horton infiltration law; we saw in Section 8.4 that the IPH2 model expresses this either in the form $I_t = I_b + (I_0 - I_b)h^t$ or in the form $I_t = I_b + (I_0 - I_b)\exp(-kt)$, where $h = \exp(-k)$. In the absence of knowledge of h, the GLUE procedure would require a uniform distribution for h on the interval $[0,1]$. But expressed in terms of k, this is equivalent to an exponential distribution $\exp(-k)$ over the interval $[0, \infty)$, which is very far from being a distribution of 'equal ignorance' concerning values of k. Beven and Binley recognise the problem, but state:

> In practice, the assumption of a uniform reference prior is unlikely to prove critical, because as soon as information is added in terms of comparisons between observed and predicted responses then, if this information has value, the distribution of calculated likelihood values should dominate the uniform prior distribution when uncertainty estimates are recalculated.

Our interpretation of this is that, given some data, we shall no longer need to assume a uniform prior distribution for h. This may be true, but it does not remove the problem that our knowledge — or lack of knowledge — about the alternative parameters h and k should be identical.

A second difficulty is the following. Any two parameters, θ_1 and θ_2, say, may or may not be interrelated, like the λ and S_0 of the artificial example discussed earlier. If they are interrelated, some form of bivariate prior distribution $f(\theta_1, \theta_2)$ must be assumed. GLUE then requires that values θ_1, θ_2 be sampled from this distribution. But it is by no means clear how to choose the distribution $f(\theta_1, \theta_2)$ such that, for example, the marginal distribution of θ_1 is rectangular and the marginal distribution of θ_2 is normal with specified mean and variance. Concerning this problem, Binley and Beven (1991) state only that 'It is necessary to define distributions from which to select the parameter values, *together with any expected intercorrelation between the parameter values*'. How this is to be achieved is far from clear, and a proper statistical solution is likely to be very elusive.

Third, the degree of subjectivity — whether in the choice of distributions for the parameter values, or in the range over which parameter values are allowed to vary, or in the choice of the Beven–Binley likelihood criterion — in the GLUE procedure is substantial. In the 'conventional' approach to model fitting of preceding chapters, there are established procedures for ascertaining whether model assumptions — concerning whether random components have some specified distribution, for example, or whether a parameter in the model can be set to zero without significant loss of information — are satisfied or not. It is not clear that corresponding procedures exist for the GLUE procedure.

The above comments are not the only criticisms that can be levelled against GLUE. Nevertheless, it represents a bold attempt to introduce some much-needed new thinking into a field of modelling that is in grave danger of becoming intellectually sterile. At present, however, GLUE remains a research tool having many difficulties to be solved before its use can be recommended.

REFERENCES

Abbott, M. B., Bathurst, J. C., Cunge, J. A., O'Connell, P. E. and Rasmussen, J. (1986). An introduction to the European Hydrological System — Système Hydrologique Européen, 'SHE', 1: History and philosophy of a physically-based, distributed modelling system. *J. Hydrology*, **87**, 45–59.

Beven, K. J. (1985). Distributed models. In Anderson, M. G. and Burt, T. P. (eds), *Hydrological Forecasting*. Wiley, Chichester, 405–35.

Beven, K. J. and Binley, A. M. (1992). The future of distributed models: model calibration and uncertainty prediction. *Hydrological Processes*, Special Issue on the Future of Distributed Models, 1–20.

Beven, K. J. and Kirkby, M. J. (1979). A physically based, variable contributing area model of basin hydrology. *Hydrological Sciences Bulletin*, **24** (1), 43–69.

Beven, K. J. and O'Connell, P. E. (1982). *On the Role of Distributed Models in Hydrology*. Report No. 81. Institute of Hydrology, Wallingford.

Binley, A. M. and Beven, K. J. (1991). Physically-based modelling of catchment hydrology: a likelihood approach to reducing predictive uncertainty. In Farmer, D. G. and Rycraft, M. J. (eds), *Computer Modelling in the Environmental Sciences*, IMA Series No. 28. Clarendon Press, Oxford.

Blackie, J. R. and Eeles, C. W. O. (1985). In Anderson, M. G. and Burt, T. P. (eds) *Hydrological Forecasting*. Wiley, Chichester, 311–45.

Boughton, W. F. (1965). *A New Simulation Technique for Estimating Catchment Yield*. Report No. 78. Water Research Laboratory, University of New South Wales, Manley Vale, Australia.

Dawdy, D. and O'Connell, T. (1965). Mathematical models of catchment behaviour. *J. Hydraul. Div. Am.*, Soc. Civ. Eng., **91** HY4, 123–37.

Fletcher, R. and Powell, M. J. D. (1963). A rapidly convergent descent method for minimisation. *Computer J.*, **6**, 163–8.

Grayson, R. B., Moore, I. D. and McMahon, T. A. (1992). Physically based hydrologic modelling 2. Is the concept realistic? *Water Resour. Res.*, **28**, 2659–66.

Hosking, J. R. M. and Clarke, R. T. (1990). Rainfall-runoff relations derived from the probability theory of storage. *Water Resources Res.*, **26**, 1455–63.

Johnston, P. R. and Pilgrim, D. H. (1976). Parameter optimisation for watershed models. *Water Resour. Res.*, **12**, 477–86.

Moore, R. J. and Clarke, R. T. (1981). A distribution function approach to rainfall-runoff modelling. *Water Resour. Research*, **17**, 1367–82.

Moore, R. J. and Clarke, R. T. (1982). A distribution function approach to modelling basin soil moisture deficit and streamflow. In Singh, V. P. (ed.), *Statistical Analysis of Rainfall and Runoff*. Water Resources Publications, Fort Collins, CO, 173–90.

Moore, R. J. and Clarke, R. T. (1983). A distribution function approach to modellling basin sediment yield. *Journal of Hydrology*, **65**, 239–57.

Nash, J. E. and Sutcliffe, J. V. (1969). *Flood Wave Formation*. WMO Publication, No. 228, TP 122. World Meteorological Organisation, Geneva.

Philip, J. R. (1975). Some remarks on science and catchment prediction. In Chapman, T. G. and Dunin, F. X. (eds) *Prediction in Catchment Hydrology*. Australian Academy of Science, Canberra.

Pickup, G. (1977). Testing the efficiencies of algorithms and strategies for automatic calibration of rainfall-runoff models. *Hydrol. Sci. Bull.*, **22**, 257–74.

Powell, M. J. D. (1965). A method for minimising the sum of squares of non-linear functions without calculating derivatives. *Computer J.*, **7**, 155–62.

Rosenbrock, H. H. (1960). An automatic method for function minimisation. *Computer J.*, **3**, 175–84.

Sugawara, M., Ozaki, E., Watanabe, I. and Katouyama, Y. (1974). *Tank Model and Its Application to Bird Creek, Wollombi Brook, Bikin River, Kitsu River, Sanaja River and Nam Mune*. Research Note No. 11. National Research Centre for Disaster Prevention, Tokyo.

Weinberg, A. M. (1972). Science and trans-science. *Minerva*, **10**, 209–22.

EXERCISES AND EXTENSIONS

8.1. The data given below (and in file EX8-1.DAT on the diskette distributed through The MathWorks) are three-hourly values of mean areal rainfall (mm), Penman potential evaporation *EO* (mm), and flow (mm) over a 55-day period, from a 1.5 km² first-order basin of the Rio Potiribu, Brazil. The data are used by kind permission of Sr Eduardo Mediondo.

Use the MATLAB coding given in the text to prepare a program to estimate the Institute of Hydrology model parameters by trial and error. Graph the observed and fitted hydrographs after each trial, paying particular attention to the residuals.

Rain	0.00	0.00	0.00	0.00	0.00
EO	0.1	0.1	0.3	1.0	1.5
Flow	0.069	0.069	0.067	0.065	0.065
Rain	0.00	0.00	0.00	0.00	0.00
EO	0.5	0.1	0.1	0.1	0.1
Flow	0.065	0.065	0.065	0.065	0.065
Rain	0.00	0.00	0.00	0.00	0.00
EO	0.3	1.0	1.5	0.5	0.1
Flow	0.066	0.066	0.066	0.066	0.066
Rain	0.00	0.00	0.00	0.00	0.00
EO	0.1	0.1	0.1	0.3	1.0
Flow	0.067	0.067	0.066	0.064	0.064
Rain	0.00	0.00	0.30	2.20	2.00
EO	1.5	0.5	0.1	0.1	0.1
Flow	0.064	0.064	0.064	0.064	0.064

Rain	4.80	0.00	0.00	0.00	0.00
EO	0.1	0.3	1.0	1.5	0.5
Flow	0.066	0.071	0.067	0.065	0.065
Rain	0.00	0.00	0.00	0.00	0.00
EO	0.1	0.1	0.1	0.1	0.3
Flow	0.066	0.066	0.066	0.063	0.063
Rain	0.00	0.00	0.00	0.00	0.00
EO	1.0	1.5	0.5	0.1	0.1
Flow	0.063	0.063	0.064	0.064	0.064
Rain	0.00	0.00	0.00	0.00	0.00
EO	0.1	0.1	0.3	1.0	1.5
Flow	0.064	0.064	0.064	0.064	0.063
Rain	0.00	0.00	0.00	0.00	0.00
EO	0.5	0.1	0.1	0.1	0.1
Flow	0.059	0.058	0.058	0.057	0.057
Rain	0.00	0.00	0.00	0.00	3.10
EO	0.3	1.0	1.5	0.5	0.1
Flow	0.056	0.056	0.056	0.056	0.068
Rain	6.20	0.10	3.10	0.60	0.00
EO	0.1	0.1	0.1	0.3	1.0
Flow	0.089	0.073	0.076	0.064	0.067
Rain	0.00	0.00	0.00	0.00	0.00
EO	1.5	0.5	0.1	0.1	0.1
Flow	0.058	0.055	0.055	0.056	0.055
Rain	0.00	0.00	0.00	0.00	0.00
EO	0.1	0.3	1.0	1.5	0.5
Flow	0.055	0.054	0.054	0.054	0.054
Rain	0.00	0.00	0.00	0.00	0.00
EO	0.1	0.1	0.1	0.1	0.3
Flow	0.054	0.055	0.055	0.055	0.055
Rain	0.00	0.00	0.00	0.00	0.00
EO	1.0	1.5	0.5	0.1	0.1
Flow	0.056	0.056	0.056	0.056	0.056
Rain	0.00	0.00	0.00	0.00	0.00
EO	0.1	0.1	0.3	1.0	1.5
Flow	0.056	0.056	0.056	0.056	0.056
Rain	0.00	7.50	11.60	1.20	0.00
EO	0.5	0.1	0.1	0.1	0.1
Flow	0.057	0.057	0.062	0.120	0.107
Rain	0.00	0.00	0.00	0.00	0.00
EO	0.3	1.0	1.5	0.5	0.1
Flow	0.072	0.065	0.064	0.064	0.064
Rain	0.00	0.00	0.00	0.00	0.00
EO	0.1	0.1	0.1	0.3	1.0
Flow	0.064	0.064	0.064	0.064	0.064
Rain	0.00	0.00	0.00	0.00	0.00
EO	1.5	0.5	0.1	0.1	0.1
Flow	0.064	0.064	0.064	0.064	0.064
Rain	0.00	0.00	0.00	0.00	0.00
EO	0.1	0.3	1.0	1.5	0.5
Flow	0.064	0.063	0.061	0.061	0.061

(*continued*)

Rain	0.00	0.00	0.00	0.00	0.00
EO	0.1	0.1	0.1	0.1	0.3
Flow	0.062	0.062	0.062	0.062	0.062
Rain	0.00	0.00	0.00	0.00	0.00
EO	1.0	1.5	0.5	0.1	0.1
Flow	0.062	0.062	0.062	0.063	0.063
Rain	5.10	22.30	0.00	0.00	0.00
EO	0.1	0.1	0.3	1.0	1.5
Flow	0.063	0.070	0.256	0.211	0.094
Rain	0.00	0.00	0.00	0.00	0.00
EO	0.5	0.1	0.1	0.1	0.1
Flow	0.075	0.070	0.069	0.069	0.069
Rain	0.00	0.10	0.00	0.60	0.10
EO	0.3	1.0	1.5	0.5	0.1
Flow	0.069	0.069	0.069	0.063	0.060
Rain	0.00	0.00	0.20	5.10	1.40
EO	0.1	0.1	0.1	0.3	1.0
Flow	0.060	0.060	0.061	0.061	0.072
Rain	2.10	4.90	0.10	0.00	0.10
EO	1.5	0.5	0.1	0.1	0.1
Flow	0.074	0.063	0.073	0.074	0.064
Rain	1.60	0.00	0.00	0.00	0.00
EO	0.1	0.3	1.0	1.5	0.5
Flow	0.063	0.063	0.063	0.061	0.060
Rain	0.00	0.00	0.00	0.00	0.00
EO	0.1	0.1	0.1	0.1	0.3
Flow	0.060	0.060	0.060	0.060	0.060
Rain	0.00	0.00	0.00	0.00	0.00
EO	1.0	1.5	0.5	0.1	0.1
Flow	0.060	0.060	0.060	0.060	0.061
Rain	0.00	0.00	0.00	0.00	0.00
EO	0.1	0.1	0.3	1.0	1.5
Flow	0.061	0.061	0.061	0.060	0.058
Rain	0.00	0.00	0.00	0.00	0.00
EO	0.5	0.1	0.1	0.1	0.1
Flow	0.058	0.058	0.058	0.058	0.058
Rain	0.00	0.00	0.00	0.00	0.00
EO	0.3	1.0	1.5	0.5	0.1
Flow	0.058	0.058	0.059	0.059	0.059
Rain	0.00	0.00	0.00	0.00	0.00
EO	0.1	0.1	0.1	0.3	1.0
Flow	0.059	0.059	0.059	0.059	0.058
Rain	0.00	0.00	0.00	0.00	0.00
EO	1.5	0.5	0.1	0.1	0.1
Flow	0.058	0.058	0.058	0.058	0.058
Rain	0.00	0.00	0.00	0.00	0.00
EO	0.1	0.3	1.0	1.5	0.5
Flow	0.058	0.058	0.058	0.058	0.058
Rain	0.00	2.20	5.80	0.10	1.40
EO	0.1	0.1	0.1	0.1	0.3
Flow	0.058	0.058	0.058	0.057	0.056

Rain	0.20	0.00	0.00	0.00	0.00
EO	1.0	1.5	0.5	0.1	0.1
Flow	0.056	0.056	0.056	0.056	0.056
Rain	0.00	3.80	1.50	0.00	0.00
EO	0.1	0.1	0.3	1.0	1.5
Flow	0.056	0.056	0.056	0.056	0.056
Rain	0.00	0.00	0.00	0.00	0.00
EO	0.5	0.1	0.1	0.1	0.1
Flow	0.056	0.056	0.056	0.056	0.054
Rain	0.00	0.00	0.00	0.00	0.00
EO	0.3	1.0	1.5	0.5	0.1
Flow	0.054	0.054	0.055	0.055	0.055
Rain	0.00	0.00	0.00	0.00	0.00
EO	0.1	0.1	0.1	0.3	1.0
Flow	0.055	0.055	0.055	0.055	0.055
Rain	0.00	0.00	0.00	0.00	0.00
EO	1.5	0.5	0.1	0.1	0.1
Flow	0.056	0.056	0.056	0.056	0.053
Rain	0.00	0.00	0.00	0.00	0.00
EO	0.1	0.3	1.0	1.5	0.5
Flow	0.053	0.053	0.053	0.054	0.054
Rain	0.00	0.00	0.00	0.00	0.20
EO	0.1	0.1	0.1	0.1	0.3
Flow	0.054	0.054	0.054	0.055	0.055
Rain	0.00	0.00	2.50	1.40	9.20
EO	1.0	1.5	0.5	0.1	0.1
Flow	0.055	0.055	0.056	0.056	0.053
Rain	4.50	0.00	0.00	0.00	0.00
EO	0.1	0.1	0.3	1.0	1.5
Flow	0.080	0.129	0.066	0.056	0.055
Rain	0.00	0.00	0.00	0.00	0.00
EO	0.5	0.1	0.1	0.1	0.1
Flow	0.055	0.054	0.054	0.055	0.055
Rain	0.00	0.00	0.00	0.00	0.00
EO	0.3	1.0	1.5	0.5	0.1
Flow	0.055	0.055	0.056	0.056	0.055
Rain	0.00	0.00	0.00	0.00	0.00
EO	0.1	0.1	0.1	0.3	1.0
Flow	0.054	0.054	0.054	0.055	0.055
Rain	0.00	0.00	0.00	0.00	0.00
EO	1.5	0.5	0.1	0.1	0.1
Flow	0.055	0.055	0.055	0.055	0.055
Rain	0.00	0.00	0.00	0.00	0.00
EO	0.1	0.3	1.0	1.5	0.5
Flow	0.056	0.056	0.056	0.056	0.054
Rain	0.00	0.00	0.00	0.00	0.00
EO	0.1	0.1	0.1	0.1	0.3
Flow	0.053	0.053	0.053	0.053	0.053
Rain	0.00	0.00	0.00	0.00	0.00
EO	1.0	1.5	0.5	0.1	0.1
Flow	0.053	0.053	0.053	0.054	0.054

(*continued*)

Rain	0.00	0.00	0.00	0.00	0.00
EO	0.1	0.1	0.3	1.0	1.5
Flow	0.054	0.054	0.054	0.054	0.053
Rain	0.00	0.00	0.00	0.00	0.00
EO	0.5	0.1	0.1	0.1	0.1
Flow	0.053	0.053	0.053	0.053	0.053
Rain	0.00	0.00	0.00	0.00	0.00
EO	0.3	1.0	1.5	0.5	0.1
Flow	0.053	0.053	0.053	0.054	0.054
Rain	0.00	0.00	0.00	0.00	0.00
EO	0.1	0.1	0.1	0.3	1.0
Flow	0.054	0.054	0.054	0.054	0.054
Rain	0.00	0.00	0.10	14.90	1.40
EO	1.5	0.5	0.1	0.1	0.1
Flow	0.052	0.052	0.052	0.051	0.054
Rain	0.20	0.00	0.00	0.00	0.00
EO	0.1	0.3	1.0	1.5	0.5
Flow	0.054	0.054	0.054	0.054	0.054
Rain	0.00	0.00	0.00	4.10	7.90
EO	0.1	0.1	0.1	0.1	0.3
Flow	0.053	0.053	0.053	0.053	0.052
Rain	11.20	0.90	1.50	5.90	1.30
EO	1.0	1.5	0.5	0.1	0.1
Flow	0.055	0.111	0.082	0.055	0.060
Rain	1.20	0.20	0.00	0.00	0.00
EO	0.1	0.1	0.3	1.0	1.5
Flow	0.060	0.053	0.053	0.054	0.054
Rain	0.00	0.00	0.00	0.00	0.00
EO	0.5	0.1	0.1	0.1	0.1
Flow	0.054	0.054	0.054	0.054	0.054
Rain	0.00	0.00	0.00	0.00	0.00
EO	0.3	1.0	1.5	0.5	0.1
Flow	0.054	0.054	0.054	0.054	0.054
Rain	0.00	0.00	0.00	0.00	0.00
EO	0.1	0.1	0.1	0.3	1.0
Flow	0.054	0.054	0.054	0.054	0.054
Rain	0.00	0.00	0.00	0.00	0.00
EO	1.5	0.5	0.1	0.1	0.1
Flow	0.054	0.053	0.053	0.053	0.053
Rain	0.00	0.00	0.00	0.00	0.00
EO	0.1	0.3	1.0	1.5	0.5
Flow	0.052	0.052	0.052	0.052	0.052
Rain	0.00	0.00	0.00	0.00	0.00
EO	0.1	0.1	0.1	0.1	0.3
Flow	0.051	0.051	0.051	0.053	0.054
Rain	0.00	0.00	0.00	0.00	0.00
EO	1.0	1.5	0.5	0.1	0.1
Flow	0.054	0.054	0.054	0.054	0.054
Rain	0.00	0.00	0.00	0.00	0.00
EO	0.1	0.1	0.3	1.0	1.5
Flow	0.053	0.053	0.053	0.053	0.053

Rain	0.00	0.00	0.00	0.00	0.00
EO	0.5	0.1	0.1	0.1	0.1
Flow	0.053	0.053	0.053	0.054	0.054
Rain	0.00	0.00	0.00	0.00	0.00
EO	0.3	1.0	1.5	0.5	0.1
Flow	0.054	0.054	0.054	0.054	0.054
Rain	0.00	0.00	0.00	0.00	0.00
EO	0.1	0.1	0.1	0.3	1.0
Flow	0.054	0.053	0.052	0.052	0.052
Rain	0.00	0.00	0.00	0.00	0.00
EO	1.5	0.5	0.1	0.1	0.1
Flow	0.052	0.052	0.052	0.052	0.052
Rain	0.00	0.00	0.00	0.20	0.10
EO	0.1	0.3	1.0	1.5	0.5
Flow	0.052	0.052	0.052	0.052	0.052
Rain	0.00	0.00	0.50	0.40	0.00
EO	0.1	0.1	0.1	0.1	0.3
Flow	0.052	0.052	0.052	0.050	0.049
Rain	0.00	0.00	0.50	3.30	38.00
EO	1.0	1.5	0.5	0.1	0.1
Flow	0.049	0.049	0.049	0.049	0.153
Rain	9.90	0.00	0.00	0.00	0.00
EO	0.1	0.1	0.3	1.0	1.5
Flow	0.473	0.335	0.123	0.077	0.067
Rain	0.20	8.20	0.10	0.00	0.00
EO	0.5	0.1	0.1	0.1	0.1
Flow	0.062	0.077	0.097	0.067	0.060
Rain	0.00	0.00	0.00	0.00	0.00
EO	0.3	1.0	1.5	0.5	0.1
Flow	0.057	0.056	0.056	0.056	0.054
Rain	0.00	0.00	0.00	0.00	0.00
EO	0.1	0.1	0.1	0.3	1.0
Flow	0.054	0.054	0.054	0.054	0.054
Rain	0.00	0.00	0.00	0.00	0.00
EO	1.5	0.5	0.1	0.1	0.1
Flow	0.054	0.054	0.054	0.054	0.054
Rain	0.00	0.00	0.00	0.00	0.00
EO	0.1	0.3	1.0	1.5	0.5
Flow	0.054	0.054	0.054	0.054	0.054
Rain	0.00	0.00	0.00	0.00	0.00
EO	0.1	0.1	0.1	0.1	0.3
Flow	0.052	0.051	0.051	0.050	0.050
Rain	0.00	0.00	0.00	0.00	0.00
EO	1.0	1.5	0.5	0.1	0.1
Flow	0.050	0.050	0.050	0.050	0.050

Chapter 9

The modelling of spatial processes

9.1 THE NATURE OF SPATIAL VARIATION

This book has sought to develop topics in the statistical modelling of hydrological data sequences by introducing increasing levels of complexity. We began with the very simplest of models in which a single time-dependent variable of interest, y_t, has a constant systematic component, with the random components ε_t independently distributed and sampled from a probability distribution of specified form but containing unknown parameters. Additional variables were then introduced in order to explain the statistical behaviour of y_t; at first these models were linear in the parameters to be estimated, but subsequently non-linear forms were introduced as the level of modelling complexity increased. At all stages throughout this process (except for one short diversion when we considered linear regressions subject to serially correlated random components), we have assumed that random components were statistically independent at a single site, although we allowed for the possibility of cross-correlation between sites in Chapter 7.

This assumption of independent errors ε is commonly not satisfied in several areas of hydrology, notably where measurements are taken near-simultaneously at a number of locations for the purpose of exploring the spatial variability of a hydrological process. This includes a very large number of cases, of which the following are a small selection:

(i) Measurements of volume of water per unit volume of soil (moisture volume fraction, commonly denoted by θ but not to be confused with our use of θ to represent unknown parameters in a probability density) at a series of depths throughout the soil profile, at a number of sites on a grid, with regular or irregular spacing. Commonly, soil moisture suction φ will also be observed, and we may be interested in how the pair of variables (θ, φ) vary jointly through the three-dimensional space defined by three coordinates (x, y, z). Alternatively, if the sites that are sampled are a series of depths (10 cm, 20 cm, 30 cm, ...) in each of a number of access tubes in a line running down a hillslope, the points at which θ and φ are measured can be considered two-dimensional, lying in the vertical plane defined by coordinates (x, z).

(ii) Measurements of precipitation at a dense network of gauges distributed over a relatively small area, for the purpose of studying rainfall as a spatial process. Equally, the plan position indicator of a weather radar will show representations of the spatial distribution of an index associated with precipitable water. This index will vary spatially in a manner in which the assumption of independently varying random components is likely to be an unacceptable oversimplification. Appropriate statistical modelling of the 'rainfall surface' may result in more accurate forecasts of heavy rainfall and flood flows.

(iii) The numbers of fractures per unit of aquifer area, their directions, and the numbers of fracture intersections, are important for the modelling of groundwater flow.

(iv) Saturated hydraulic conductivity varies spatially over quite small areas, as shown by the data in Table 9.1 (used by kind permission of Eduardo Mediondo). Using the inverted well method, saturated hydraulic conductivity was measured at each of an array of points, most of which took the form of a cross but with a smaller number of diagonal points, within a small first-order drainage basin within the catchment area of the Rio Potiribu in the basaltic region of southern Brazil. The 'vertical' line of points (those with easting of 200 m) run vertically down the slope (see Figure 9.1); points at the 'lower' end of this vertical lie near the catchment outfall, whilst those at the 'top' end lie on or near the interfluve. Calculated in this direction, the mean saturated hydraulic conductivity is 0.272 m day^{-1} with a standard deviation of ± 0.216 m day^{-1}. Using data from points lying in the 'horizontal' arm of the grid, the mean saturated hydraulic conductivity is very similar, 0.255 m day^{-1}, but the standard deviation is much less at ± 0.074 m day^{-1}.

Table 9.1   Saturated hydraulic conductivity (measured by inverted well method) at 53 points in the form of a cross, in a 12.5 ha sub-basin ('Anfiteatro') of the Rio Potiribú, southern Brazil.

Easting	200	200	200	200	200	200
Northing	10	20	30	40	50	60
K	0.201	0.082	0.129	0.096	0.189	0.232
Easting	200	200	200	200	200	200
Northing	70	80	90	100	110	120
K	0.173	0.182	0.128	0.120	0.115	0.336
Easting	200	200	200	200	200	200
Northing	130	140	150	160	170	180
K	0.036	0.173	0.235	0.826	0.363	0.353
Easting	200	200	200	200	200	200
Northing	190	200	210	220	230	240
K	0.324	0.448	0.169	0.210	0.814	0.700
Easting	100	110	120	130	140	150
Northing	140	140	140	140	140	140
K	0.345	0.335	0.252	0.394	0.250	0.259
Easting	160	170	180	190	200	210
Northing	140	140	140	140	140	140
K	0.211	0.187	0.277	0.288	0.173	0.288
Easting	220	230	240	250	260	270
Northing	140	140	140	140	140	140
K	0.237	0.234	0.259	0.446	0.184	0.192
Easting	280	290	300	172	186	214
Northing	140	140	140	112	126	154
K	0.230	0.144	0.254	0.249	0.216	0.220
Easting	228	172	186	214	228	
Northing	168	168	154	126	112	
K	0.223	0.425	0.187	0.164	0.133	

Figure 9.2 shows the correlograms (plots of correlations between saturated hydraulic conductivity K at distance s, and the same sequence displaced by h intervals of 10 m, where h is the 'lag') both in the north–south direction (upper part of the figure) and in the east–west direction (lower part of the figure). These were calculated using the Genstat directive

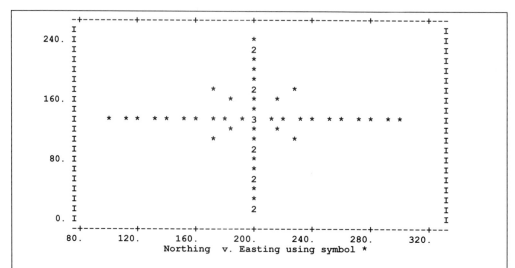

Figure 9.1   Dispositions of grid-points for the sampling of saturated hydraulic conductivity K, first-order basin in the catchment area of the Rio Potiribu, southern Brazil. Spacing between sampling points in the horizontal and vertical is 10 m. 'Vertical' points lie in a north–south direction, running from interfluve (top of diagram) to basin outfall (bottom of diagram).

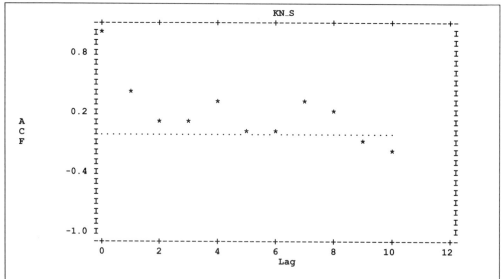

Figure 9.2   Plot of autocorrelation function of saturated hydraulic conductivity K, first-order basin ('Amfiteatro') of the Rio Potiribu: correlogram calculated in the north–south direction.

Figure 9.2 *(continued)*

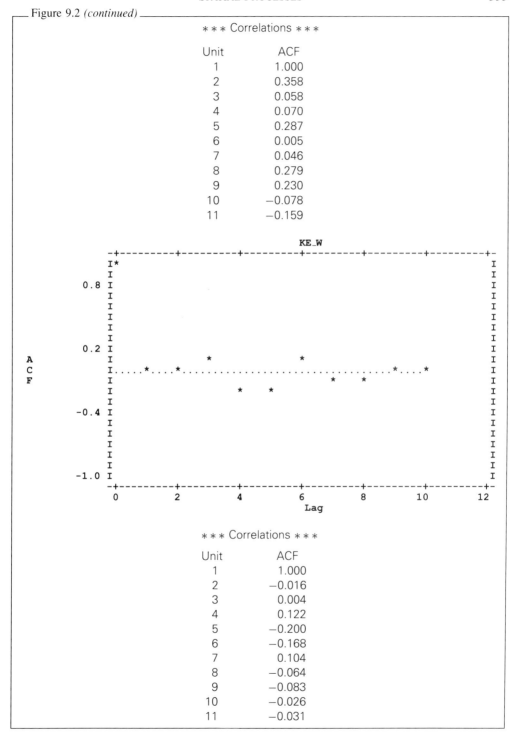

*** Correlations ***

Unit	ACF
1	1.000
2	0.358
3	0.058
4	0.070
5	0.287
6	0.005
7	0.046
8	0.279
9	0.230
10	−0.078
11	−0.159

KE_W

*** Correlations ***

Unit	ACF
1	1.000
2	−0.016
3	0.004
4	0.122
5	−0.200
6	−0.168
7	0.104
8	−0.064
9	−0.083
10	−0.026
11	−0.031

```
>CORRELATE [PRINT=autocorrelations; GRAPH=autocorrelations; \
           MAXLAG=10] SERIES = K
```

It is seen that the lag-1 correlation in the north–south direction is considerably larger than that in the east–west direction, although an approximate test of significance (see, for example, Box and Jenkins, 1970) suggests that none of the correlations is significantly greater than zero. Nevertheless, the sample of data is small for the use of such a test, and the visual impression remains strong that 'downslope' measurements exhibit different variance and correlation characteristics than 'across-slope' measurements.

9.2   THE VARIOGRAM

Where the assumption of statistical independence is inappropriate, as is commonly the case for spatial data, the starting point for the development of methods for drawing inferences (for example, by estimating areal mean values, their variances and approximate confidence limits) must begin with the specification of a model of the statistical process from which the data are a *realization*. We take this starting point to be the definition of *intrinsic stationarity*:

$$E[Z(s+h) - Z(s)] = 0 \qquad (9.1a)$$

$$\text{var}[Z(s+h) - Z(s)] = 2\gamma(h) \qquad (9.1b)$$

The quantity $2\gamma(h)$ on the right-hand side of equation (9.1b) is the crucial parameter for the study of spatial data consisting of measurements $\{Z_i; i = 1, \ldots, N\}$ made at the points denoted by $\{s_i; i = 1, \ldots, N\}$. In two-dimensional space, s_i is determined by the two coordinates (x_i, y_i), and h is the length of the vector joining the points $s = (x, y)$ and $s+h = (x+h_x, y+h_y)$. The extension to three dimensions is straightforward, the point s now being defined by three coordinates (x, y, z).

Reference to texts in spatial statistics (see, for example, Cressie, 1993) shows that, when the process $Z(s)$ has *second-order stationarity* (that is, when $E[Z] = \mu$ and $\text{cov}[Z(s_1) - Z(s_2)] = C(s_1 - s_2)$ depends only on the difference $s_1 - s_2$) then the variogram and the covariance function $C(.)$ are related by

$$\text{var}[Z(s+h) - Z(s)] = \text{var}[Z(s+h)] + \text{var}[Z(s)] - 2\,\text{cov}[Z(s+h), Z(s)]$$

which, when $\text{var}[Z(s+h)] = \text{var}[Z(s)]$, gives

$$2\gamma(h) = 2[C(0) - C(h)] \qquad (9.2)$$

but the definition of the variogram given in equations (9.1) does not require the more restrictive assumption of second-order stationarity.

Essentially, we regard the variogram $2\gamma(h)$ as a parameter of the statistical process which generated the observations Z_i. Typical problems in spatial analysis concern the estimation of the variogram, fitting a model with a small number of parameters to the variogram calculated from data, and using the variogram for interpolation and for the calculation of approximate confidence intervals for interpolated values.

9.3   CALCULATION OF THE VARIOGRAM FOR POTIRIBU DATA ON SATURATED HYDRAULIC CONDUCTIVITY K

In general, spatial data are not collected on a regular grid. Depending upon the type of data being analysed, we may require to consider variograms in two or three dimensions. For example, when modelling rainfall characteristics over an area without topographic extremes, it will commonly be sufficient to calculate a variogram in two dimensions; but when modelling the distribution of a pollutant throughout a soil matrix, we shall almost certainly require a three-dimensional variogram. Furthermore, we shall often want to calculate the variogram in several directions, to explore the extent of *anisotropy*: that is, whether the variogram is a function of angular direction θ, as well as absolute distance h. Figure 9.3 gives a GENSTAT procedure which calculates the variogram, for data from a two-dimensional irregular grid, in any direction,

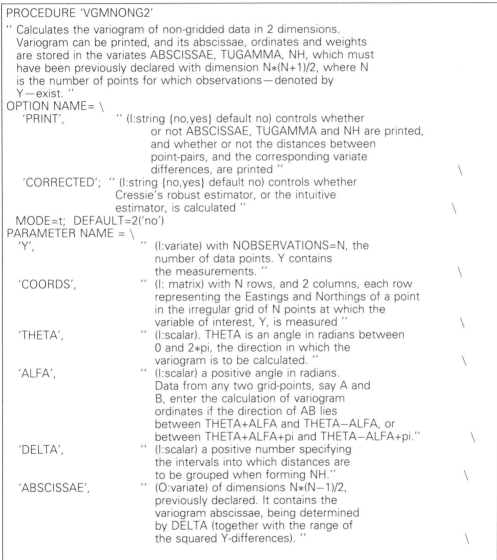

```
PROCEDURE 'VGMNONG2'
'' Calculates the variogram of non-gridded data in 2 dimensions.
   Variogram can be printed, and its abscissae, ordinates and weights
   are stored in the variates ABSCISSAE, TUGAMMA, NH, which must
   have been previously declared with dimension N*(N+1)/2, where N
   is the number of points for which observations—denoted by
   Y—exist. ''
OPTION NAME= \
   'PRINT',            '' (I:string {no,yes} default no) controls whether
                          or not ABSCISSAE, TUGAMMA and NH are printed,
                          and whether or not the distances between
                          point-pairs, and the corresponding variate
                          differences, are printed ''                        \
   'CORRECTED';  '' (I:string {no,yes} default no) controls whether
                          Cressie's robust estimator, or the intuitive
                          estimator, is calculated ''                        \
   MODE=t;  DEFAULT=2('no')
PARAMETER NAME = \
   'Y',               '' (I:variate) with NOBSERVATIONS=N, the
                          number of data points. Y contains
                          the measurements. ''                               \
   'COORDS',          '' (I: matrix) with N rows, and 2 columns, each row
                          representing the Eastings and Northings of a point
                          in the irregular grid of N points at which the
                          variable of interest, Y, is measured ''           \
   'THETA',           '' (I:scalar). THETA is an angle in radians between
                          0 and 2*pi, the direction in which the
                          variogram is to be calculated. ''                  \
   'ALFA',            '' (I:scalar) a positive angle in radians.
                          Data from any two grid-points, say A and
                          B, enter the calculation of variogram
                          ordinates if the direction of AB lies
                          between THETA+ALFA and THETA−ALFA, or
                          between THETA+ALFA+pi and THETA−ALFA+pi.''         \
   'DELTA',           '' (I:scalar) a positive number specifying
                          the intervals into which distances are
                          to be grouped when forming NH.''                   \
   'ABSCISSAE',       '' (O:variate) of dimensions N*(N−1)/2,
                          previously declared. It contains the
                          variogram abscissae, being determined
                          by DELTA (together with the range of
                          the squared Y-differences). ''                     \
```

Figure 9.3   GENSTAT procedure for calculation of variogram in two dimensions. Comments in the procedure give details of procedure inputs and outputs.

Continued

___ Figure 9.3 *(continued)* ___

```
'TUGAMMA',           ''  (O:variate) of dimensions N*(N−1)/2,
                         previously declared. It contains the
                         ordinates of the variogram. ''                    \
      'NH';          ''  (O:variate) of dimensions N*(N−1)/2,
                         previously declared. It contains the
                         numbers of Y-differences used in
                         calculating each variogram ordinate . ''          \
  MODE=p
CALCULATE E=0
CHECKARGUMENTS [ERROR=E] \
STRUCTURE=PRINT,CORRECTED,Y,COORDS,THETA,ALFA,DELTA,ABSCISSAE,TUGAMMA,
                                                                   NH; \
VALUES=2(!T(no,yes)),8(*);                         \
SET=10('yes');                                     \
DECLARED=4('yes'),3('no'),3('yes');                \
TYPE=2('text'),'variate','matrix',3('scalar'),3('variate');     \
PRESENT=7('yes'),3('no')
EXIT [CONTROL=procedure] E
SCALAR Nrows, Ncols, NvalsY, Nabsc, NDimTug
CALCULATE    Nrows=NROWS(COORDS)
      &          Ncols=NCOLUMNS(COORDS)
      &          NvalsY=NVALUES(Y)
      &          Nabsc=NVALUES(TUGAMMA)
      &          NDimTug=0.5*Nrows*(Nrows−1)
IF Nrows.NE.NvalsY
    PRINT 'Data error: numbers of data-points and \
variate values are not equal.'
    EXIT [CONTROL=procedure]
ELSIF Nrows.LE.2
    PRINT ' Insufficient data for variogram calculation.'
    EXIT [CONTROL=procedure]
ELSIF Nrows.NE.NvalsY
    PRINT 'Number of gridpoints differs from number of data \
values. '
    EXIT [CONTROL=procedure]
ELSIF NDimTug.NE.Nabsc
    PRINT 'Length of TUGAMMA should be 0.5*N*(N−1). '
    EXIT [CONTROL=procedure]
ELSE
    VARIATE [NVALUES=Nrows] YDiffs, SelAngle, Dist, Mask, Angle
          &                      EDiffs, NDiffs
    VARIATE [NVALUES=Nrows] x[1...#Ncols]
    VARIATE [NVALUES=Nrows] Temp[1...#Ncols]
      SCALAR Max, Min, Ulimit, Llimit, DifVec, DifVecPl, \
              NEm1, Count, Rest, SumElements, LthminCount, \
              Sestimator, Sprint,Big,Small,piover2,phi, \
              Bigppi,Smallppi,pi
    MATRIX [ROWS=Ncols;COLUMNS=Nrows] Mdash
    CALCULATE Mdash=TRANSPOSE(COORDS)
    EQUATE OLDSTRUCTURES=Mdash; NEWSTRUCTURES=x
    DELETE Mdash
    CALCULATE NEm1=Nrows−1
      &      DifVec=0.5*Nrows*NEm1
      &    DifVecPl=DifVec+Nrows
```

Figure 9.3 *(continued)*

```
   VARIATE [NVALUES=DifVec] Distances, FinalY, Choose
   VARIATE [NVALUES=DifVecPl] TuDistances,TuFinalY
   VARIATE [VALUES=1...#DifVec] Order
   IF CORRECTED .EQS. 'no'
            CALCULATE Sestimator=1
   ELSE
            CALCULATE Sestimator=0
   ENDIF
   IF PRINT .EQS. 'yes'
            CALCULATE Sprint=1
   ELSE
            CALCULATE Sprint=0
   ENDIF
'' Calculate the upper and lower angles between which the angle defined
   by each point-pair must lie if it is to enter calculation of the
   variogram in the direction THETA. Also rotate the axes so that
   the vertical axis lies in the direction of THETA.''
   CALCULATE    pi =3.14159265
      &     piover2 =pi/2
      &         phi =piover2−THETA
            CALCULATE    Temp[1] =x[1]*COS(phi)+x[2]*SIN(phi)
                 &       Temp[2] =−x[1]*SIN(phi)+x[2]*COS(phi)
                 &          x[1] =Temp[1]
                 &          x[2] =Temp[2]
          DELETE Temp
   CALCULATE       Big =piover2+ALFA
      &          Small =piover2−ALFA
      &         Bigppi =Big+pi
      &       Smallppi =Small+pi
      &          Count =0
      &      Distances =0
      &         FinalY =0
'' Now calculate the differences between Eastings and Northings, and
   between the data values Y, for each pair of points: and calculate
   the distance and angle determined by each point pair. The variate
   SelAngle selects those point-pairs giving an angle with the direction
   THETA which lies in the specified range.''
      FOR I = 1... NEm1
            CALCULATE    EDiffs =DIFFERENCE(x[1];−I)
                 &       NDiffs =DIFFERENCE(x[2];−I)
                 &       YDiffs =DIFFERENCE(Y;−I)
                 &         Dist =SQRT(EDiffs**2+NDiffs**2)
                 &        Angle =EDiffs/Dist
                 &        Angle =ARCCOS(Angle)
                 &     SelAngle =((Angle.LE.Big).AND.(Angle.GE.Small)).OR. \
                                   ((Angle.LE.Bigppi).AND.(Angle.GE.Smallppi))
                 &         Mask =2*SelAngle−1
                 & SumElements =SUM(SelAngle)
            EXIT [REPEAT=yes] SumElements .EQ. 0
'' Now pick out the point pairs for which SelAngle=1, and combine
   these differences with others previously selected.''
      CALCULATE Rest=Nrows−SumElements
         & LthminCount=DifVec−Count
      EQUATE [OLDFORMAT=Mask] OLDSTRUCTURES=Dist; \
                                NEWSTRUCTURES=Dist
```

_____ Figure 9.3 *(continued)* _____

```
     EQUATE [OLDFORMAT=Mask] OLDSTRUCTURES=YDiffs; \
                             NEWSTRUCTURES=YDiffs
     EQUATE OLDSTRUCTURES=!P(Distances,Dist); NEWSTRUCTURES=TuDistances
     EQUATE OLDSTRUCTURES=!P(FinalY,YDiffs);   NEWSTRUCTURES=TuFinalY
     EQUATE [OLDFORMAT=!(#Count(1),#LthminCount(−1), \
                   #SumElements(1), #Rest(−1))] \
                   OLDSTRUCTURES=TuDistances; NEWSTRUCTURES=Distances
     EQUATE [OLDFORMAT=!(#Count(1),#LthminCount(−1), \
                   #SumElements(1), #Rest(−1))] \
                   OLDSTRUCTURES=TuFinalY;     NEWSTRUCTURES=FinalY
     CALCULATE Count=Count+SumElements
     ENDFOR
'' Now group the distances into the appropriate class intervals defined
   by DELTA; pick out those which lie in each class using the variate
   named Choose; count them, to give the vector of variogram weights NH;
   and calculate the variogram ordinates using the method specified
   by the option CORRECTED.''
     RESTRICT Distances, FinalY, Order, Choose; CONDITION=Order.LE.Count
     PRINT 'Data used to calculate variogram are'
     IF Sprint
          PRINT 'Distances between point-pairs, and variate differences:-'
          PRINT [ORIENTATION=across] Distances, FinalY
     ENDIF
     CALCULATE     Max =MAX(Distances)
          &        Min =MIN(Distances)
          &        Nintervals =ROUND((Max−Min)/DELTA)+1.0
     FOR J=1...Nintervals
          CALCULATE      Llimit =(J−1)*DELTA
          &              Ulimit =Llimit+DELTA
          & ABSCISSAE$[J] =Ulimit
          & Choose =(Distances .GT. Llimit) .AND. (Distances .LE. Ulimit)
          &         NH$[J] =SUM(Choose)
             IF Sestimator
                IF NH$[J].GT.0
                     CALCULATE TUGAMMA$[J]=SUM(Choose*FinalY*FinalY)/NH$[J]
                ENDIF
             ELSE
                IF NH$[J].GT.0
                     CALCULATE TUGAMMA$[J]=(SUM(Choose*SQRT(ABS(FinalY))))/ \
                          NH$[J])**4/(0.457 + 0.494/NH$[J] )
                ENDIF
             ENDIF
     ENDFOR
     RESTRICT Distances, FinalY, Order, Choose
  IF Sprint
     PRINT 'Abscissae, ordinates, and numbers of differences \
from which ordinates were calculated, are'
     RESTRICT ABSCISSAE, TUGAMMA, NH; CONDITION=NH>0
     PRINT ABSCISSAE,TUGAMMA,NH; DECIMALS=1,4,0
     RESTRICT ABSCISSAE,TUGAMMA,NH
     ENDIF
ENDIF
ENDPROCEDURE
''
''
```

and Figure 9.4 gives a procedure doing the same thing in three dimensions. Both procedures provide two possible estimates of the variogram: the first is the *moments estimator*, given by

$$2\hat{\gamma}(h) = \sum_h [Z(s+h) - Z(s)]^2 / N(h) \qquad (9.3a)$$

where $N(h)$ is the number of differences, in the direction in which the variogram is being calculated, of size h in absolute magnitude; and the second is a *bias-corrected estimator* given

---

```
PROCEDURE 'VGMNONG3'
'' Calculates the variogram of non-gridded data in 3 dimensions.
  Variogram can be printed, and its abscissae, ordinates and weights
  are stored in the variates ABSCISSAE, TUGAMMA, NH, which must
  have been previously declared with dimension N*(N+1)/2, where N
  is the number of points for which observations—denoted by
  Y—exist. ''
OPTION NAME= \
   'PRINT',          '' (I:string {no,yes} default no) controls whether
                        or not ABSCISSAE, TUGAMMA and NH are printed '' \
   'CORRECTED'; '' (I:string {no, yes} default no) controls whether
                     Cressie's robust estimator, or the intuitive
                     estimator, is calculated ''                           \
   MODE=t; DEFAULT=2('no')
PARAMETER NAME = \
   'Y',             '' (I:variate) with NOBSERVATIONS=N, the
                        number of data points. Y contains
                        the measurements. ''                               \
   'COORDS',        '' (I: matrix) with N rows, and 3
                        columns, giving the Eastings, Northings, and Height
                        of measurements Y made on a 3-dimensional grid '' \
   'DIRECTION',     '' (I:variate). DIRECTION contains the direction
                        cosines of the direction in which the
                        variogram is to be calculated. ''                  \
   'ALFA',          '' (I:scalar) a positive angle in radians.
                        Data from any two grid-points, say A and
                        B, enter the calculation of variogram
                        ordinates if the direction of AB lies
                        within an angle ALFA of the direction
                        specified by DIRECTION, or if the direction
                        of AB makes an angle with it lying between
                        pi and pi−ALFA. ''                                 \
   'DELTA',         '' (I:scalar) a positive number specifying
                        the intervals into which distances are
                        to be grouped when forming NH. ''                  \
   'ABSCISSAE',     '' (O:variate) of dimensions N*(N+1)/2,
                        previously declared. It contains the
                        variogram abscissae, being determined
                        by DELTA (together with the range of
                        the squared Y-differences). ''                     \
```

Figure 9.4   GENSTAT procedure for calculation of a variogram, using non-gridded data sampled in a three dimensions. Comments within the body of the procedure give details of input and output variables.

Continued

__ Figure 9.4 *(continued)* __

```
'TUGAMMA',          '' (O:variate) of dimensions N*(N+1)/2,
                    previously declared. It contains the
                    ordinates of the variogram.          '' \
  'NH';             '' (O:variate) of dimensions N*(N+1)/2,
                    previously declared. It contains the
                    numbers of Y-differences used in
                    calculating each variogram ordinate .'' \
  MODE=p
CALCULATE E=0
CHECKARGUMENTS [ERROR=E] PRINT,CORRECTED,Y,COORDS, \
                            DIRECTION,ALFA,DELTA,ABSCISSAE,TUGAMMA,NH; \
TYPE=2('text'),'variate','matrix','variate',2('scalar'),3('variate'); \
VALUES=2(!T(no,yes)),8(*); \
DECLARED=10('yes'); \
PRESENT=10('yes')
EXIT [CONTROL=procedure] E
SCALAR NvalsE, NvalsN, NvalsY, Nabsc, NDimTug
CALCULATE    Nrows=NROWS(COORDS)
       &         Ncols=NCOLUMNS(COORDS)
       &         NvalsY=NVALUES(Y)
       &         Nabsc=NVALUES(TUGAMMA)
       &      NDimTh=NVALUES(DIRECTION)
       &      NDimTug=0.5*Nrows*(Nrows−1)
IF Nrows.NE.NvalsY
   PRINT 'Data error: numbers of data-points and \
variate values are not equal.'
   EXIT [CONTROL=procedure]
ELSIF Nrows.LE.2
   PRINT 'Insufficient data for variogram calculation.'
   EXIT [CONTROL=procedure]
ELSIF Nrows.NE.NvalsY
   PRINT 'Number of gridpoints differs from number of data \
values. '
   EXIT [CONTROL=procedure]
ELSIF NDimTug.NE.Nabsc
   PRINT 'Length of TUGAMMA should be 0.5*N*(N−1). '
   EXIT [CONTROL=procedure]
ELSIF NDimTh.NE.3
   PRINT 'Vector specifying direction requires 3 cosines.'
   EXIT [CONTROL=procedure]
ELSE
     VARIATE [NVALUES=Nrows] YDiffs, SelAngle, Dist, Mask, Angle
          &                         EDiffs, NDiffs, HtDiffs
       SCALAR Max, Min, Ulimit, Llimit, DifVec, DifVecPl, \
             NEm1, Count, Rest, SumElements, LthminCount, \
             Sestimator, Sprint,phi,pimalfa,l,m,n
        VARIATE [NVALUES=Nrows] x[1...3]
       MATRIX [ROWS=Ncols;COLUMNS=Nrows] Mdash
       CALCULATE Mdash=TRANSPOSE(COORDS)
       EQUATE OLDSTRUCTURES=Mdash; NEWSTRUCTURES=x
       DELETE Mdash
       CALCULATE NEm1=Nrows−1
          &     DifVec=0.5*Nrows*NEm1
          & DifVecPl=DifVec+Nrows
```

Figure 9.4 *(continued)*

```
VARIATE [NVALUES=DifVec] Distances, FinalY, Choose
VARIATE [NVALUES=DifVecPl] TuDistances,TuFinalY
VARIATE [VALUES=1...#DifVec] Order
IF CORRECTED .EQS. 'no'
        CALCULATE Sestimator=1
ELSE
        CALCULATE Sestimator=0
ENDIF
IF PRINT .EQS. 'yes'
        CALCULATE Sprint=1
ELSE
        CALCULATE Sprint=0
ENDIF

CALCULATE        pi=3.14159265
     &        pimalfa=pi−ALFA
     &           Count=0
     &        Distances=0
     &           FinalY=0
     &               l=DIRECTION$[1]
     &               m=DIRECTION$[2]
     &               n=DIRECTION$[3]
     FOR I = 1...NEm1
       CALCULATE        EDiffs=DIFFERENCE(x[1];−I)
          &             NDiffs=DIFFERENCE(x[2];−I)
          &             HtDiffs=DIFFERENCE(x[3];−I)
          &             YDiffs=DIFFERENCE(Y;−I)
          &              Dist=SQRT(EDiffs**2+NDiffs**2+HtDiffs**2)
          &             Angle=l*EDiffs+m*NDiffs+n*HtDiffs
          &             Angle=Angle/Dist
          &             Angle=ARCCOS(Angle)
          &          SelAngle=(Angle.LE.ALFA).OR. \
                             ((Angle.LE.pi).AND.(Angle.GE.pimalfa))
          &             Mask=2*SelAngle−1
          &          SumElements=SUM(SelAngle)
       EXIT [REPEAT=yes] SumElements .EQ. 0
       CALCULATE Rest=Nrows−SumElements
          &          LthminCount=DifVec−Count
       EQUATE [OLDFORMAT=Mask] OLDSTRUCTURES=Dist; \
                             NEWSTRUCTURES=Dist
       EQUATE [OLDFORMAT=Mask] OLDSTRUCTURES=YDiffs; \
                             NEWSTRUCTURES=YDiffs
       EQUATE OLDSTRUCTURES=!P(Distances,Dist); NEWSTRUCTURES=TuDistances
       EQUATE OLDSTRUCTURES=!P(FinalY,YDiffs); NEWSTRUCTURES=TuFinalY
       EQUATE [OLDFORMAT=!(#Count(1),#LthminCount(−1),              \
                 #SumElements(1), #Rest(−1))]                       \
                 OLDSTRUCTURES=TuDistances; NEWSTRUCTURES=Distances
       EQUATE [OLDFORMAT=!(#Count(1),#LthminCount(−1),              \
                 #SumElements(1), #Rest(−1))]                       \
                 OLDSTRUCTURES=TuFinalY; NEWSTRUCTURES=FinalY
       CALCULATE Count=Count+SumElements
       ENDFOR
       RESTRICT Distances, FinalY, Order, Choose; CONDITION=Order.LE.Count
       PRINT 'Data used to calculate variogram are'
       PRINT [ORIENTATION=across] Distances, FinalY
```

Figure 9.4 *(continued)*

```
    CALCULATE Max=MAX(Distances)
         &          Min=MIN(Distances)
         & Nintervals=ROUND((Max−Min)/DELTA)+1.0
    FOR J=1...Nintervals
              CALCULATE Llimit=(J−1)*DELTA
              &              Ulimit=Llimit+DELTA
              & ABSCISSAE$[J]=Ulimit
              & Choose=(Distances .GT. Llimit) .AND. (Distances .LE. Ulimit)
              &          NH$[J]=SUM(Choose)
         IF Sestimator
              IF NH$[J].GT.0
                   CALCULATE TUGAMMA$[J]=SUM(Choose*FinalY*FinalY)/NH$[J]
              ENDIF
         ELSE
              IF NH$[J].GT.0
                   CALCULATE TUGAMMA$[J]=(SUM(Choose*SQRT(ABS(FinalY)))/ \
                                        NH$[J])**4/(0.457 + 0.494/NH$[J] )
              ENDIF
         ENDIF
    ENDFOR
    RESTRICT Distances, FinalY, Order, Choose
    IF Sprint
       PRINT 'Abscissae, ordinates, and numbers of differences \
from which ordinates were calculated, are'
         RESTRICT ABSCISSAE, TUGAMMA, NH; CONDITION=NH>0
         PRINT ABSCISSAE,TUGAMMA,NH; DECIMALS=1,4,0
         RESTRICT ABSCISSAE,TUGAMMA,NH
    ENDIF
ENDIF
ENDPROCEDURE
''
''
```

by Cressie (1993) in the form

$$2\tilde{\gamma}(h) = \{\sum |Z(s + h) - Z(s)|^{1/2} / N(h)\}^4 / [0.457 + 0.494/N(h)] \qquad (9.3b)$$

Figures 9.5 and 9.6 show the variograms for the Potiribú data, in the north–south and east–west directions, respectively, calculated using equation (9.3a). We can see that the variogram confirms the impression of anisotropy suggested by the correlograms of Figure 9.2. We note a suggestion of periodic fluctuations in the north–south variogram, and it has been suggested that these might be associated with the 'bunds' (small terraces for the prevention of soil erosion) running along the contours, which are perpendicular to this line of sampling points.

When the variogram of data collected on an irregular grid is to be calculated in a direction making an angle θ with a suitably chosen x-axis, it is necessary to define conditions which specify which pairs of grid-points are to be used in calculating the variogram. If, for example, the direction of the line joining points A and B of the grid is at right angles to the chosen direction θ, then the point-pair AB will not contribute to the variogram calculation (although it will, of course, contribute when we calculate the variogram in the direction perpendicular to θ).

Abscissae, ordinates, and numbers of differences from which ordinates were calculated, are

Abscissa	TuGamma	Number
10.0	0.0502	25
20.0	0.0843	24
30.0	0.0624	25
40.0	0.0467	22
50.0	0.0763	21
60.0	0.0727	22
70.0	0.0502	19
80.0	0.0584	18
90.0	0.1067	17
100.0	0.1149	16
110.0	0.0930	14
120.0	0.1044	13
130.0	0.1270	12
140.0	0.1548	10
150.0	0.1523	9
160.0	0.1177	8
170.0	0.1151	7
180.0	0.1289	6
190.0	0.1704	5
200.0	0.2129	4
210.0	0.2873	3
220.0	0.3788	2
230.0	0.2490	1

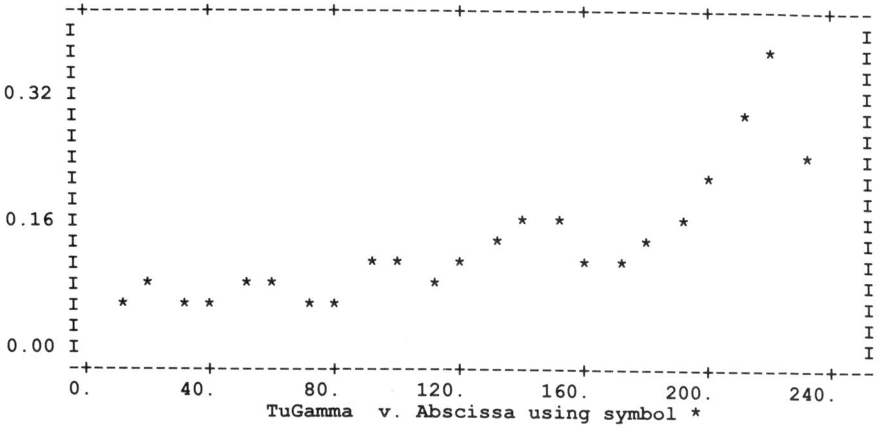

Figure 9.5 Variogram (moments estimator) calculated using the saturated hydraulic conductivity data from the first-order basin ('Amfiteatro') of the Rio Potiribú: variogram calculated in the north–south direction.

So it is common to specify an angle α, say, such that any point-pair AB will enter the variogram calculation if the direction AB makes an angle with the direction θ, lying between the limts $\theta\pm\alpha$ (or, between $(\theta+\pi)\pm\alpha$). Clearly, if the points all lie on a rectangular grid parallel to the x- and y-axes, the value of α can be set very small; thus with data collected on an $m \times n$ rectangular grid and α set, say, to $\pi/100$, there will be $n(1 + 2 + 3 + \cdots + (m - 1)) = mn(m - 1)/2$ differences of the form $Z(s + h) - Z(s)$ contributing to the variogram calculation in the

Abscissae, ordinates, and numbers of differences from which ordinates were calculated, are

Abscissa	TuGamma	Number
10.0	0.0109	22
20.0	0.0098	21
30.0	0.0078	22
40.0	0.0116	19
50.0	0.0156	18
60.0	0.0087	19
70.0	0.0117	16
80.0	0.0097	15
90.0	0.0108	14
100.0	0.0122	13
110.0	0.0096	10
120.0	0.0040	9
130.0	0.0141	8
140.0	0.0115	7
150.0	0.0125	6
160.0	0.0219	5
170.0	0.0164	4
180.0	0.0166	3
190.0	0.0235	2
200.0	0.0083	1

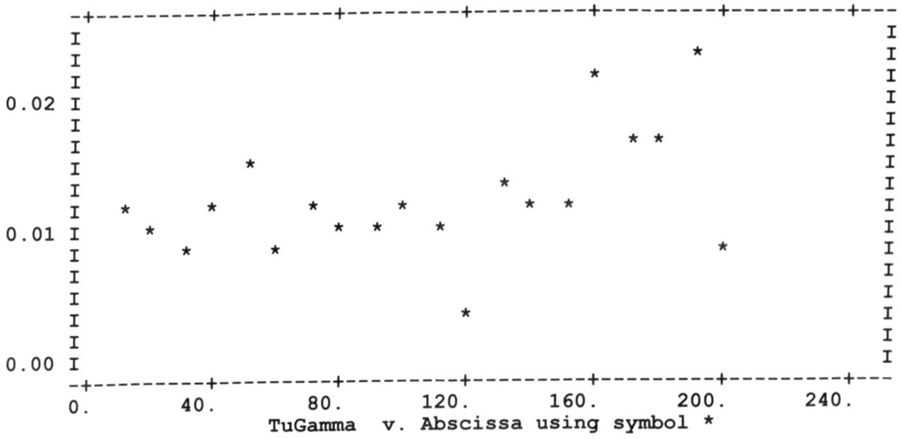

Figure 9.6   Variogram (moments estimator) calculated using the saturated hydraulic conductivity data from the first-order basin ('Amfiteatro') of the Rio Potiribú: variogram calculated in the east–west direction.

x-direction ($\theta = 0$), and $nm(n-1)/2$ differences contributing to the variogram calculated in the y-direction ($\theta = \pi/2$). For $0 < \theta < \pi/2$, the number of differences will vary considerably with θ and α, and it is common to set the value of α as high as $\pi/4$.

In three dimensions, the extension is straightforward; the direction θ must be specified by its direction cosines (say l_0, m_0, n_0) relative to the three axes defined by intersections of the planes $x = y = z = 0$; and a point-pair AB will contribute if the direction AB with direction cosines (l, m, n) satisfies the relation $\arccos(ll_0 + mm_0 + nn_0) < \alpha$.

9.4   SOME ISOTROPIC VARIOGRAM MODELS

Variograms calculated from data yield a table of values showing variogram ordinates, abscissae, and the numbers of point-pairs contributing to the estimation of each variogram ordinate; and, of course, these can be plotted, as for the Potiribu data. Frequently we wish to know whether the estimated variogram is consistent with a particular specified form deduced from theoretical considerations. A number of such forms have been proposed where isotropy can be assumed, including the following (all of which are valid in one, two or three dimensions): in terms of the semi-variogram $\gamma(h)$, Cressie (1993) lists the following:

(i)   Linear model:

$$\gamma(h; c_0, b_l) = \begin{cases} 0 & h = 0 \\ c_0 + b_l\,|h| & h \neq 0 \end{cases} \tag{9.4}$$

where c_0, b_l are non-negative parameters.

(ii)   Spherical model:

$$\gamma(h; c_0, c_s, a_s) = \begin{cases} 0 & h = 0 \\ c_0 + c_s\{(3/2)(|h|/a_s) - (1/2)(|h|/a_s)^3\} & 0 < |h| \leq a_s \\ c_0 + c_s & a_s < |h| \end{cases} \tag{9.5}$$

where c_0, c_s and a_s are non-negative parameters.

(iii)   Exponential model:

$$\gamma(h; c_0, c_e, a_e) = \begin{cases} 0 & h = 0 \\ c_0 + c_e\{1 - \exp(-|h|/a_e)\} & h \neq 0 \end{cases} \tag{9.6}$$

where c_0, c_e, a_e are non-negative parameters.

(iv)   Rational quadratic model:

$$\gamma(h; c_0, c_r, a_r) = \begin{cases} 0 & h = 0 \\ c_0 + c_r\,|h|^2/(1 + |h|^2/a_r) & h \neq 0 \end{cases} \tag{9.7}$$

where c_0, c_r and a_r are non-negative parameters.

(v)   Power function model:

$$\gamma(h; c_0, b_p, \lambda) = \begin{cases} 0 & h = 0 \\ c_0 + b_p\,|h|^\lambda & h \neq 0 \end{cases} \tag{9.8}$$

where c_0, b_p are non-negative parameters, and the parameter λ satisfies $0 \leq \lambda < 2$.

For interpolation purposes, it is convenient to represent the calculated variogram using ($2\times$) a theoretical semivariogram such as those listed in (i) to (v) above. Whilst generalised least-squares estimates of semivariogram parameters — such as c_0, c_s and a_s in the spherical model — can be obtained by minimizing

$$[2\gamma - 2\gamma(c_0, c_s, a_s)]^\mathrm{T} \mathbf{V}^{-1}[2\gamma - 2\gamma(c_0, c_s, a_s)] \tag{9.9}$$

with respect to the parameters c_0, c_s, a_s, Cressie (1993) suggests the simpler alternative of approximate weighted least squares, in which estimates of the parameters are found by minimising

$$\sum_{j=1}^{K} N(h_j)\{2\gamma(h_j)/2\gamma(h_j; c_0, c_s, a_s) - 1\}^2 \tag{9.10}$$

where expression (9.10) refers to the parameters c_0, c_s, a_s of the spherical model, and $2\gamma(h_j; c_0, c_s, a_s)$ to the form (9.5): for other models, the parameters and the theoretical variogram will of course be those appropriate to the model used. The GENSTAT procedure given in Figure 9.7 minimises the expression (9.10) for each of the variogram forms given in (i) to (v) above. The minimisation is achieved using the GENSTAT directive NONLINEARFIT in order to incorporate the inequality constraints on the parameters; thus, the constraint $c_0 \geq 0$ is achieved by replacing c_0 by a new parameter c_0^* where $c_0 = \exp(c_0^*)$ and minimising with respect to c_0^* which can vary over the unrestricted range $(-\infty < c_0^* < \infty)$. As with all uses of minimisation algorithms when the surface to be minimised is not necessarily even approximately quadratic, there is no guarantee that the iterative calculation will converge, nor, if it does converge, that it converges to a global minimum.

Following Journel and Huijbregts (1978), Cressie (1993) recommends that fitting should use only the variogram ordinates up to half the maximum lags, and only those for which $N(h_j) \geq 30$. To illustrate the variogram fitting procedure, we set these recommendations aside

```
PROCEDURE 'VGMFIT'
OPTION NAME='SEMIVARIOGRAM'; MODE=t; DEFAULT='exponential'
PARAMETER NAME='NH',              \
                'TUGAMMA',         \
                'H',               \
                'EMATES'; MODE=p
SCALAR Scase, Rows
CALCULATE Rows=NVALUES(H)
VARIATE [NVALUES=#Rows] Gamma
IF SEMIVARIOGRAM .EQS. 'exponential'
          CALCULATE Scase=1
ELSIF SEMIVARIOGRAM .EQS. 'linear'
          CALCULATE Scase=2
ELSIF SEMIVARIOGRAM .EQS. 'spherical'
          CALCULATE Scase=3
ELSIF SEMIVARIOGRAM .EQS. 'rationalquadratic'
          CALCULATE Scase=4
ELSIF SEMIVARIOGRAM .EQS. 'wave'
          CALCULATE Scase=5
ELSIF SEMIVARIOGRAM .EQS. 'power'
          CALCULATE Scase=6
ELSE
          EXIT [CONTROL=procedure]
ENDIF
CALCULATE Gamma=TUGAMMA/2
CASE Scase                            "Exponential case"
SCALAR ce,co,ae,ece,eco,eae,f,f1,f2,f3
VARIATE [NVALUES=#Rows] f4, f5
```

Figure 9.7   GENSTAT procedure for estimating parameters of a theoretical semi-variogram, given variogram abscissae and ordinates computed from data.

_____ Continued _____

—— Figure 9.7 *(continued)* ——

```
CALCULATE co=Gamma$[1]
      &     ce=Gamma$[#Rows]
      &     ae=1/LOG(2)
      &     eco=LOG(co)
      &     ece=LOG(ce)
      &     eae=LOG(LOG(2))
EXPRESSION [VALUE=(f1=EXP(eae))]                        e[1]
      &     [VALUE=(f2=EXP(eco))]                       e[2]
      &     [VALUE=(f3=EXP(ecc))]                       e[3]
      &     [VALUE=(f4=(1-EXP(-H/f1)))   ]              e[4]
      &     [VALUE=(f5=(f2+f3*f4))]                     e[5]
      &     [VALUE=(f=SUM(NH*(f5/Gamma-1)**2))]         e[6]
  MODEL [FUNCTION=f]
RCYCLE eco, ece, eae
FITNONLINEAR [PRINT=model, summary, estimates, correlations,\
                    fittedvalues, monitoring; CALCULATION=e]
CALCULATE EMATES$[1]=EXP(eco)
      &     EMATES$[2]=EXP(ece)
      &     EMATES$[3]=EXP(eae)
PRINT 'Estimates of co, ce and ae are'
PRINT EMATES

OR                                        "Linear case"
SCALAR co, bl, eco, ebl, f
VARIATE [NVALUES=#Rows] f2
CALCULATE co=MIN(Gamma)
      &     eco=LOG(co)
      &     bl=(Gamma$[#Rows]-Gamma$[1])/(H$[#Rows]- H$[1])
        IF bl.LE.0
            CALCULATE ebl=0
        ELSE
            CALCULATE ebl=LOG(bl)
        ENDIF
EXPRESSION [VALUE=(f2=EXP(eco)+EXP(ebl)*H)   ]  e[1]
      &     [VALUE=(f=SUM(NH*(f2/Gamma-1)**2))] e[2]
MODEL [FUNCTION=f]
RCYCLE eco, ebl
FITNONLINEAR [PRINT=model, summary, estimates, correlations,\
                    fittedvalues, monitoring; CALCULATION=e]
CALCULATE EMATES$[1]=eco
      &     EMATES$[2]=ebl

OR                                        "Spherical case"
SCALAR co,cs,as,eco,ecs,eas,f1,f2,f3,f
VARIATE [NVALUES=#Rows] f4, f5
CALCULATE co=MIN(Gamma)
      &     cs=MAX(Gamma)-co
      &     as=1/(0.66667*(Gamma$[2]-co)/(cs*H$[2]))
      &     eco=LOG(co)
      &     ecs=LOG(cs)
        IF as.GT.0
            CALCULATE eas=LOG(as)
        ELSE
            CALCULATE eas=1
        ENDIF
```

─── Figure 9.7 *(continued)* ───

```
EXPRESSION [VALUE=(f1=EXP(eas)) ]                          e[1]
     &        [VALUE=(f2=EXP(eco)) ]                        e[2]
     &        [VALUE=(f3=EXP(ecs)) ]                        e[3]
     &        [VALUE=(f4=H.LE.f1) ]                         e[4]
     &        [VALUE=(f5=f4*(f2+f3*(1.5*H/f1 \
              −0.5*(H/f1)**3))+(1−f4)*(f2+f3)) ]            e[5]
     &        [VALUE=(f=SUM(NH*(f5/Gamma−1)**2))]           e[6]
MODEL [FUNCTION=f]
RCYCLE eco, ecs, eas
FITNONLINEAR [PRINT=model, summary, estimates, correlations,\
                    fittedvalues, monitoring; CALCULATION=e]
CALCULATE EMATES$[1]=EXP(eco)
     &       EMATES$[2]=EXP(ecs)
     &       EMATES$[3]=EXP(eas)
PRINT 'Estimates of parameters co, cs and as are'
PRINT EMATES

OR                                    "Rational quadratic case"
SCALAR co,cr,ar,eco,ecr,ear, f,f1,f2,f3
VARIATE [NVALUES=#Rows] f4
CALCULATE co=Gamma$[1]
     &        cr=(Gamma$[2]−co)*Gamma$[2]/H$[2]**2/co
       IF cr.GT.0
          CALCULATE ar=Gamma$[#Rows]/cr
       ELSE
          CALCULATE ar=2
       ENDIF
CALCULATE eco=LOG(co)
       IF cr.GT.0
             CALCULATE ecr=LOG(cr)
       ELSE
             CALCULATE ecr=1
       ENDIF
CALCULATE ear=LOG(ar)
EXPRESSION [VALUE=(f1=EXP(eco)) ]                          e[1]
     &        [VALUE=(f2=EXP(ecr)) ]                        e[2]
     &        [VALUE=(f3=EXP(ear)) ]                        e[3]
     &        [VALUE=(f4=f1+f2*H**2/(1+H**2/f3)) ]          e[4]
     &        [VALUE=(f=SUM(NH*(f4/Gamma−1)**2))]           e[5]
MODEL [FUNCTION=f]
RCYCLE eco,ecr,ear
FITNONLINEAR [PRINT=model, summary, estimates, correlations,\
                    fittedvalues, monitoring; CALCULATION=e]
CALCULATE EMATES$[1]=EXP(eco)
     &       EMATES$[2]=EXP(ecr)
     &       EMATES$[3]=EXP(ear)
PRINT 'Estimates of co, cr and ar are'
PRINT EMATES

OR                                            "Wave case"
SCALAR co, cw, aw, eco, ecw, eaw, f, f1, f2, f3
VARIATE [NVALUES=#Rows] f4
```

_ Figure 9.7 *(continued)* _

```
CALCULATE co=Gamma$[1]
       &      cw=Gamma$[#Rows]
       &      aw=0.5*(co+cw)
       &      eco=LOG(co)
       &      ecw=LOG(cw)
       &      eaw=LOG(aw)
EXPRESSION [VALUE=(f1=EXP(eco)) ]                    e[1]
       &        [VALUE=(f2=EXP(ecw)) ]               e[2]
       &        [VALUE=(f3=EXP(eaw)) ]               e[3]
       &        [VALUE=(f4=f1+f2*(1−f3*SIN(H/f3)/H)) ]   e[4]
       &        [VALUE=(f=SUM(NH*(f4/Gamma−1)**2))]  e[5]
MODEL [FUNCTION=f]
RCYCLE eco, ecw, eaw
FITNONLINEAR [PRINT=model, summary, estimates, correlations,\
                    fittedvalues, monitoring; CALCULATION=e]
CALCULATE EMATES$[1]=EXP(eco)
       &        EMATES$[2]=EXP(ecw)
       &        EMATES$[3]=EXP(eaw)
PRINT 'Estimates of parameters co, cw and aw are'
PRINT EMATES

OR                                      ''Power law case''
SCALAR co, bp,lamda, eco, ebp, elamda,f
VARIATE [NVALUES=#Rows] f1,f2
CALCULATE co=Gamma$[1]
       &      bp=(Gamma$[2]−co)/H$[2]
       &      eco=LOG(co)
       & elamda=0
         IF bp.GT.0
              CALCULATE ebp=LOG(bp)
         ELSE
              CALCULATE ebp=1
         ENDIF
EXPRESSION [VALUE=(f1=H**(2/(1+EXP(−elamda))))]     e[1]
       &        [VALUE=(f2=EXP(eco)+EXP(ebp)*f1)   ]  e[2]
       &        [VALUE=(f=SUM(NH*(f2/Gamma−1)**2))]  e[3]
MODEL [FUNCTION=f]
RCYCLE eco, ebp, elamda
FITNONLINEAR [PRINT=model, summary, estimates, correlations,\
                    fittedvalues, monitoring; CALCULATION=e]
CALCULATE EMATES$[1]=EXP(eco)
       &        EMATES$[2]=EXP(ebp)
       &        EMATES$[3]=2/(1+EXP(−elamda))
PRINT 'Estimates of parameters co, bp and lamda are'
PRINT EMATES

OR
PRINT 'Impermissible choice of semivariogram form. \
Check parameter settings.'
EXIT [CONTROL=procedure]
ENDCASE
ENDPROCEDURE
''
''
```

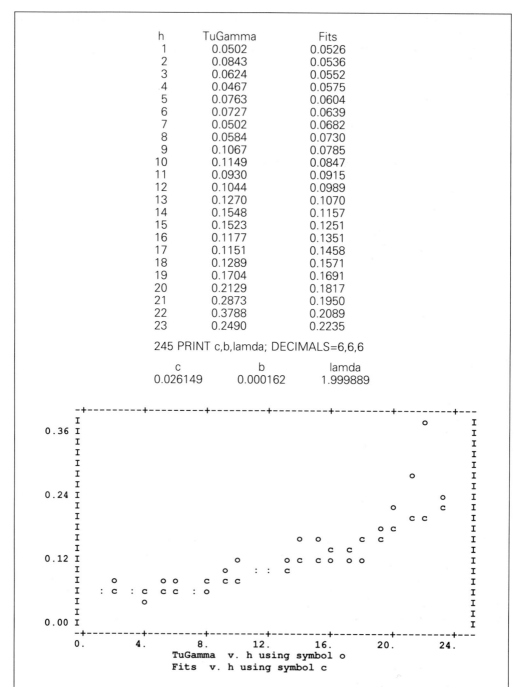

h	TuGamma	Fits
1	0.0502	0.0526
2	0.0843	0.0536
3	0.0624	0.0552
4	0.0467	0.0575
5	0.0763	0.0604
6	0.0727	0.0639
7	0.0502	0.0682
8	0.0584	0.0730
9	0.1067	0.0785
10	0.1149	0.0847
11	0.0930	0.0915
12	0.1044	0.0989
13	0.1270	0.1070
14	0.1548	0.1157
15	0.1523	0.1251
16	0.1177	0.1351
17	0.1151	0.1458
18	0.1289	0.1571
19	0.1704	0.1691
20	0.2129	0.1817
21	0.2873	0.1950
22	0.3788	0.2089
23	0.2490	0.2235

245 PRINT c,b,lamda; DECIMALS=6,6,6

c	b	lamda
0.026149	0.000162	1.999889

TuGamma v. h using symbol o
Fits v. h using symbol c

Figure 9.8 Output from GENSTAT procedure for fitting a power-law semi-variogram to Potiribú data on saturated hydraulic conductivity (note: estimates of co, bp and lamda had not converged by 20 iterations). Estimates of these parameters are given in the text. Also shown is a plot of observed and fitted variogram ordinates, denoted by the symbols 'o' and 'c' respectively.

and use the north–south Potiribu variogram to fit the power law in equation (9.8). In fact, the calculation had not converged after 20 iterations, the final estimates of the power-law parameters being $\hat{c}_0 = 0.026\,149$, $\hat{b}_p = 0.000\,162$, $\hat{\lambda} = 1.999\,889$, the estimate $\hat{\lambda}$ being very close to the permissible limit for this parameter given in equation (9.8). Figure 9.8 gives the observed and fitted variogram ordinates calculated from these estimates, together with a plot showing a good measure of agreement. We recall, however, that the variogram ordinates used in estimating c_0, b_p and λ did not satisfy the criteria recommended by Cressie (1993) set out above.

The Potiribú data, as we have seen, suggest a degree of anisotropy between the north–south and east–west directions. To represent the anisotropy in a single variogram form useful for the interpolation purposes discussed in the following section, it would be more appropriate to fit a variogram of the form

$$2\gamma(h, \theta; c_0, b_p, \lambda) = c_0 + b_p\{|h|/(1 + |\tan\theta|)\}^\lambda$$

where $\theta = 0$ is taken to lie along the north–south direction.

9.5 OPTIMAL INTERPOLATION USING THE VARIOGRAM

Fitting a parametric form to the calculated variogram is not, of course, an end in itself, but is a step which facilitates the interpolation of the variate Z at points within the area of the grid at which no measurements of Z were taken. The interpolated estimate commonly used is the minimum mean-square error estimate; in the case where the variate Z can be assumed to be of the form

$$Z(s) = \mu + \varepsilon(s) \tag{9.11}$$

where μ is an unknown constant, we have the case known as *ordinary kriging* in which we seek an estimate of the interpolated Z at the point s, of the form

$$\hat{Z}(\mathbf{s}) = \sum_{i=1}^{N} \lambda_i Z_i \tag{9.12}$$

where $Z_i (i = 1, \ldots, N)$ are the measured values of Z at the N grid-points, and the λ_i are coefficients which sum to unity and which minimise the mean-square error between the prediction $Z(\mathbf{s})$ and the true value of Z at the interpolated point. It can be shown (Cressie, 1993) that the $N \times 1$ vector λ of coefficients is given by

$$\lambda^{\mathrm{T}} = \{\gamma + \mathbf{1}(1 - \mathbf{1}^{\mathrm{T}}\Gamma^{-1}\gamma)/(\mathbf{1}^{\mathrm{T}}\Gamma^{-1}\mathbf{1})\}\Gamma^{-1} \tag{9.13}$$

where γ is an $N \times 1$ vector of values calculated using the fitted variogram and the coordinates $\mathbf{s} = (x, y)$ of the point at which interpolation is required; Γ is an $N \times N$ matrix using the fitted variogram and the coordinates $\mathbf{s}_i = (x_i, y_i)$ of the N grid-points; and $\mathbf{1}$ is the $N \times 1$ vector with every element equal to one. Specifically,

$$\gamma = [\gamma(\mathbf{s} - \mathbf{s}_1), \gamma(\mathbf{s} - \mathbf{s}_2), \ldots, \gamma(\mathbf{s} - \mathbf{s}_N)]^{\mathrm{T}} \tag{9.14}$$

and

$$\Gamma = \begin{bmatrix} 0 & \gamma(\mathbf{s}_1 - \mathbf{s}_2) & \gamma(\mathbf{s}_1 - \mathbf{s}_3) \ldots \gamma(\mathbf{s}_1 - \mathbf{s}_N) \\ \gamma(\mathbf{s}_2 - \mathbf{s}_1) & 0 & \gamma(\mathbf{s}_2 - \mathbf{s}_3) \ldots \gamma(\mathbf{s}_2 - \mathbf{s}_N) \\ \ldots & \ldots & \ldots \qquad\qquad \ldots \\ \gamma(\mathbf{s}_N - \mathbf{s}_1) & \gamma(\mathbf{s}_N - \mathbf{s}_2) & \gamma(\mathbf{s}_N - \mathbf{s}_3) \ldots 0 \end{bmatrix} \tag{9.15}$$

It can also be shown (see Cressie, 1993) that the variance of the estimate $\hat{Z}(\mathbf{s})$ is given by

$$\text{var}[\hat{Z}(\mathbf{s})] = \boldsymbol{\gamma}^{\mathrm{T}}\boldsymbol{\Gamma}^{-1}\boldsymbol{\gamma} - (\mathbf{1}^{\mathrm{T}}\boldsymbol{\Gamma}^{-1}\boldsymbol{\gamma} - 1)^2/(\mathbf{1}^{\mathrm{T}}\boldsymbol{\Gamma}^{-1}\mathbf{1}) \tag{9.16}$$

so that an approximate 95% prediction interval for $Z(\mathbf{s})$ is given by

$$[\hat{Z}(\mathbf{s}) - 1.96\sqrt{(\text{var}[\hat{Z}(\mathbf{s})])}, \hat{Z}(\mathbf{s}) + 1.96\sqrt{(\text{var}[\hat{Z}(\mathbf{s})])}] \tag{9.17}$$

The GENSTAT procedure listed in Figure 9.9 effects the interpolation (9.12), together with the calculation of the variances (9.16), for any number K of points, with (x, y) coordinates given as a $K \times 2$ array. The form of the variogram $2\gamma(h)$ must be specified, together with the (x, y) coordinates of the N grid-points as a $N \times 2$ array, and the values Z_i of the grid-point observations. The program will also function if interpolation is required in three dimensions; in this case, the array defining the coordinates of the K points must be $K \times 3$, and of the N gridpoints $N \times 3$.

```
PROCEDURE 'ORDKRIG'
"Calculates minimum mean square, unbiassed predictors in 2 or 3 dimensions"
OPTION NAME='PRINT','VARIOGRAM','DIMENSIONS';\
MODE=t; DEFAULT='no','exponential','two'
PARAMETER NAME='DATAVALUES','DATACOORDS','INTPOLATEDCOORDINATES',\
'VGMPARAMETERS','INTERPOLANDS','VARINTERPOLANDS';MODE=p
SCALAR VGMType, NoParams
CALCULATE NoParams=NVALUES(VGMPARAMETERS)
TEXT [ 'Number of variogram parameters does not agree with \
the type of variogram specified'] ErrorMessage
FOR Setting=VARIOGRAM
    IF     Setting .EQS. 'exponential'
           IF NoParams.EQ.3
                  CALCULATE VGMType=1
           ELSE
                  PRINT ErrorMessage
                  EXIT [CONTROL=procedure]
           ENDIF
    ELSIF Setting .EQS. 'linear'
           IF NoParams.EQ.2
                  CALCULATE VGMType=2
           ELSE
                  PRINT ErrorMessage
                  EXIT [CONTROL=procedure]
           ENDIF
    ELSIF Setting .EQS. 'spherical'
           IF NoParams.EQ.3
                  CALCULATE VGMType=3
           ELSE
                  PRINT ErrorMessage
                  EXIT [CONTROL=procedure]
           ENDIF
    ELSIF Setting .EQS.'rationalquadratic'
```

Figure 9.9 GENSTAT procedure for calculating estimates by spatial interpolation, using ordinary kriging, within a two- or three-dimensional set of points, not necessarily on a regular grid.

_____ Continued ___

Figure 9.9 *(continued)*

```
        IF NoParams.EQ.3
              CALCULATE VGMType=4
        ELSE
                    PRINT ErrorMessage
                    EXIT [CONTROL=procedure]
        ENDIF
   ELSIF Setting.EQS.'power'
        IF NoParams.EQ.3
              CALCULATE VGMType=5
        ELSE
                    PRINT ErrorMessage
                    EXIT [CONTROL=procedure]
        ENDIF
   ELSE
                    PRINT 'Variogram type not recognised.'
                    EXIT [CONTROL=procedure]
   ENDIF
ENDFOR
SCALAR Nrows,NDims, Nrowsm1, SumTerms, Longer, Npoints, NDimsm1
CALCULATE Nrows=NROWS(DATACOORDS)
   &       NDims=NCOLUMNS(DATACOORDS)
   &       Nrowsm1=Nrows−1
   &       SumTerms =0.5*Nrows*Nrowsm1
   &       Longer=SumTerms+Nrows
   &       Npoints=NROWS(INTPOLATEDCOORDINATES)
MATRIX [ROWS=#Nrows;COLUMNS=#Nrows] BigGamma
MATRIX [ROWS=#Nrows;COLUMNS=1] LitGamma
MATRIX [ROWS=1; COLUMNS=#Nrows] lamdadash
VARIATE [NVALUES=Nrows]x[1...#NDims]
VARIATE [NVALUES=Nrows] Distance
VARIATE [NVALUES=1] PointCoords[1...#NDims]
VARIATE [NVALUES=#Nrows] Ones
CALCULATE NDimsm1=−(NDims−1)
EQUATE [OLDFORMAT=!((1,#NDimsm1)#Nrows,−1)] DATACOORDS; x
SCALAR c0, ce, ae, D, IP1, H, Denom, x1, x2, x3, y1, y2, y3, Dist, \
       Factor, Factor2
CALCULATE BigGamma=0
   &       c0=0.5*VGMPARAMETERS$[1]
   &       ce=0.5*VGMPARAMETERS$[2]
''
''
FOR I=1...Nrowsm1
     CALCULATE x1=DATACOORDS$[I;1]
         &      x2=DATACOORDS$[I;2]
     IF NDims==3
         CALCULATE x3=DATACOORDS$[I;3]
     ENDIF
     CALCULATE IP1=I+1
     FOR J=IP1...Nrows
         CALCULATE y1=DATACOORDS$[J;1]
             &      y2=DATACOORDS$[J;2]
         IF NDims==3
                    CALCULATE y3=DATACOORDS$[J;3]
         ENDIF
```

Figure 9.9 (continued)

```
              CALCULATE Dist=(x1−y1)**2+(x2−y2)**2
              IF NDims==3
                  CALCULATE Dist=Dist+(x3−y3)**2
              ENDIF
              CALCULATE Dist = SQRT(Dist)
''
''
''
''

      CASE VGMType "Exponential variogram"
              CALCULATE  ae=VGMPARAMETERS$[3]
                      CALCULATE H=Dist
                      & D=c0+ce*(1−EXP(H/ae))
                      & BigGamma$[I;J]=D
                      & BigGamma$[J;I]=D
      OR
              "Linear variogram"
                      CALCULATE H=Dist
                      & D=c0+ce*H
                      & BigGamma$[I;J]=D
                      & BigGamma$[J;I]=D
      OR
              "Spherical variogram"
              CALCULATE ae=VGMPARAMETERS$[3]
                      CALCULATE H=Dist
                        IF H .EQ. 0
                              CALCULATE D=0
                        ELSIF (H .GT. 0) .AND. (H .LT. ae)
                        CALCULATE    D=c0+ce*(1.5*(H/ae) \
                                        −0.5*(H/ae)**3)
                        ELSE
                        CALCULATE D=c0+ce
                        ENDIF
                      CALCULATE BigGamma$[I;J]=D
                        &          BigGamma$[J;I]=D
      OR
              "Rational quadratic variogram"
              CALCULATE ae=VGMPARAMETERS$[3]
                      CALCULATE H=Dist
                        &      D=co+ce*H**2/(1+H**2/ae)
                        &      BigGamma$[I;J]=D
                        &      BigGamma&[J;I]=D
      OR
              "Power-law variogram"
              CALCULATE ae=VGMPARAMETERS$[3]
                      CALCULATE H=Dist
                        &      D=c0+ce*H**ae
                        &      BigGamma$[I;J]=D
                        &      BigGamma&[J;I]=D
      ENDCASE
   ENDFOR
ENDFOR
```

Figure 9.9 *(continued)*

```
"This ends the calculation of the matrix BigGamma which multiplies
    the vector of kriging coefficients, lamda. Now calculate
    LitGamma, the RHS of the equations for lamda. LitGamma depends
    upon the coordinates of the interpolated point: hence we need a
    FOR directive, running through the number of INTPOLATEDCO\
    ORDINATES, which was calculated above as Npoints. "
CALCULATE BgGamInv=INVERSE(BigGamma)
DELETE BigGamma
CALCULATE Ones=1
FOR I=1...Npoints
    CALCULATE     x1=INTPOLATEDCOORDINATES$[I;1]
        &         x2=INTPOLATEDCOORDINATES$[I;2]
        &   Distance=(x1*Ones−x[1])**2+(x2*Ones−x[2])**2
    IF NDims==3
        CALCULATE x3=INTPOLATEDCOORDINATES$[I;3]
        &   Distance=Distance+(x3*Ones−x[3])**2
ENDIF
CALCULATE Distance=SQRT(Distance)
CASE VGMType
    "Exponential variogram"
            CALCULATE ae=VGMPARAMETERS$[3]
            FOR J=1...Nrows
                CALCULATE H=Distance$[J]
                & D=c0+ce*(1−EXP(H/ae))
                & LitGamma$[J;1]=D
            ENDFOR
    OR
            "Linear variogram"
            FOR J=1...Nrows
                CALCULATE H=Distance$[J]
                & D=c0+ce*H
                & LitGamma$[J;1]=D
            ENDFOR
    OR
            "Spherical variogram"
            CALCULATE ae=VGMPARAMETERS$[3]
            FOR J=1...Nrows
                CALCULATE H=Distance$[J]
                &   D=c0+ce*(1.5*(H/ae) \
                    −0.5*(H/ae)**3)
                &   LitGamma$[J;1]=D
            ENDFOR
    OR
            "Rational quadratic variogram"
            CALCULATE ae=VGMPARAMETERS$[3]
            FOR J=1...Nrows
                CALCULATE H=Distance$[J]
                &   D=co+ce*H**2/(1+H**2/ae)
                &   LitGamma$[I;1]=D
            ENDFOR
    OR
            "Power-law variogram"
            CALCULATE ae=VGMPARAMETERS$[3]
            FOR J=1...Nrows
```

___ Figure 9.9 *(continued)* ___

```
                    CALCULATE H=Distance$[J]
                       &    D=c0+ce*H**ae
                       &    LitGamma$[J;1]=D
                    ENDFOR
           ENDCASE
  '' We can now calculate the coefficients lamda, appropriate
     for the I-th point to be interpolated; these coefficients
     multiply the data in DATAVALUES to give the mimimum mean-square
     unbiassed estimates of the variable of interest.''
     ''
     ''
           CALCULATE    Denom=TRANSPOSE(Ones)*+BgGamInv*+Ones
                &       Factor=TRANSPOSE(Ones)*+BgGamInv*+LitGamma
                &       Factor=1-Factor
                &       lamdadash=TRANSPOSE(LitGamma+Ones*Factor/Denom)*+BgGamInv
                & INTERPOLANDS$[I]=lamdadash*+DATAVALUES
                &       Factor2=TRANSPOSE(LitGamma)*+BgGamInv*+LitGamma
                & VARINTERPOLANDS$[I]=Factor2−Factor**2/Denom
  ENDFOR
  IF PRINT.EQS.'yes'
        PRINT INTERPOLANDS,VARINTERPOLANDS
  ENDIF
  ENDPROCEDURE

  MATRIX[ROWS=12;COLUMNS=2]DatCoords
  READ DatCoords
  1 1   1 2   1 4   1 7   2 1   2 2   2 4   2 7   3 1   3 2   3 4   3 7 :
  MATRIX[ROWS=4;COLUMNS=2]PtCoords
  READ PtCoords
  1.5 1.5 2.5 1.5 1.5 3 2.5 3 :
  VARIATE [NVALUES=4] Interpol,VarInterpol
  VARIATE [VALUES=0.10,0.16,4.387] VgmP
  VARIATE [VALUES=3,6,8,4,9,8,1,4,6,2,7,4]Data
  ORDKRIG [Print=yes]DATAVALUES=Data;DATACOORDS=DatCoords;\
          INTPOLATEDCOORDINATES=PtCoords;VGMPARAMETERS=VgmP;\
          INTERPOLANDS=Interpol;VARINTERPOLANDS=VarInterpol
```

9.6   UNIVERSAL KRIGING

The ordinary kriging interpolation procedure described in the last section assumes that $Z(\mathbf{s})$ has the form given by equation (9.11) in which μ is a constant. This would clearly be inappropriate if, say, we were interpolating moisture volume fraction within a three-dimensional soil matrix; it would also be appropriate if we were interpolating rainfall in a two-dimensional region with very variable topography. In such circumstances the interpolation method known as *universal kriging* is likely to be more appropriate; here, the model given by equation (9.11) is generalised to be of the form

$$Z(\mathbf{s}) = \sum_{j=1}^{K+1} f_{j-1}(\mathbf{s})\beta_{j-1} + \varepsilon(\mathbf{s}) \tag{9.18}$$

in which the constant μ has now been replaced by a linear combination of known functions (polynomials, for example). The $K+1$ coefficients β_j must be estimated. Given measurements of $Z(\mathbf{s})$ at N grid-points, we can use matrix notation to write equation (9.18) as:

$$\mathbf{Z} = \mathbf{X}\boldsymbol{\beta} + \boldsymbol{\varepsilon} \tag{9.19}$$

where $\mathbf{Z}$ is the $N \times 1$ vector of measurements $Z(s)$, $\mathbf{X}$ is the $N \times (K + 1)$ matrix of the functions f_{j-1} evaluated at the N grid-points, and $\boldsymbol{\beta}$ and $\boldsymbol{\varepsilon}$ are $(K + 1) \times 1$ and $N \times 1$ vectors, respectively. As in Section 9.5, we seek coefficients $\lambda_i (i = 1, \ldots, N)$ such that $\hat{\mathbf{Z}} = \boldsymbol{\lambda}^T \mathbf{Z}$ is a minimum mean-square error estimate of $Z(\mathbf{s})$ at the point with coordinates $\mathbf{s}$. If the variogram 2γ can be assumed known, it can be shown (see Cressie, 1993) that the vector $\boldsymbol{\lambda}$ is given by a generalisation of equation (9.13) of the form

$$\boldsymbol{\lambda}^T = \{\boldsymbol{\gamma} + \mathbf{X}(\mathbf{X}^T\boldsymbol{\Gamma}^{-1}\mathbf{X})^{-1}(\mathbf{x} - \mathbf{X}^T\boldsymbol{\Gamma}^{-1}\boldsymbol{\gamma})\}^T\boldsymbol{\Gamma}^{-1} \tag{9.20}$$

where $\mathbf{x}$ is the $(K + 1) \times 1$ vector of the functions $f_{j-1}(\mathbf{s})$ evaluated at the point $\mathbf{s}$ at which $Z(\mathbf{s})$ is to be estimated; $\boldsymbol{\gamma}, \boldsymbol{\Gamma}$ have the same meanings as in equations (9.14) and (9.15). The variance of $\hat{\mathbf{Z}}$ can be shown (Cressie, 1993) to be

$$\text{var}[\hat{\mathbf{Z}}(s)] = \boldsymbol{\gamma}^T\boldsymbol{\Gamma}^{-1}\boldsymbol{\gamma} - (\mathbf{x} - \mathbf{X}^T\boldsymbol{\Gamma}^{-1}\boldsymbol{\gamma})^T(\mathbf{X}^T\boldsymbol{\Gamma}^{-1}\mathbf{X})^{-1}(\mathbf{x} - \mathbf{X}^T\boldsymbol{\Gamma}^{-1}\boldsymbol{\gamma}) \tag{9.21}$$

from which approximate prediction intervals for the interpolated $Z(s)$ can be calculated using expression (9.17).

We have assumed in the above that the variogram 2γ is known, but of course it must be estimated from the data also. One method of estimating both the unknown coefficients $\boldsymbol{\beta}$ and the variogram 2γ, is the following:

(i)   Calculate ordinary least-squares estimates of $\boldsymbol{\beta}$:

$$\hat{\boldsymbol{\beta}} = (\mathbf{X}^T\mathbf{X})^{-1}\mathbf{X}^T\mathbf{Z}$$

(ii) Calculate the residuals $\mathbf{Z} - \mathbf{X}^T\hat{\boldsymbol{\beta}}$, and use them to fit a variogram model of the kind listed in Section 9.4.

(iii) Calculate a generalised least-squares estimate of the vector $\boldsymbol{\beta}$,

$$\hat{\boldsymbol{\beta}} = (\mathbf{X}^T\mathbf{V}^{-1}\mathbf{X})^{-1}\mathbf{X}^T\mathbf{V}^{-1}\mathbf{Z}$$

where $\mathbf{V}$ is the variance-covariance matrix of the vector $\mathbf{Z}$.

(iv) Repeat the steps (i) to (iii) until convergence.

Cressie (1993) points out that this procedure results in biased estimates of the variogram ordinates, but the bias is small where the h in $2\gamma(h)$ is small. The predictors $\boldsymbol{\lambda}^T\mathbf{Z}$ are unlikely to be greatly affected, although there is some evidence that the prediction variances are smaller than they should be.

The above short account of universal kriging describes the case in which the systematic part of the variation in $\mathbf{Z}$ is represented by the linear combination

$$\mu(\mathbf{s}) = \sum_{j=1}^{K+1} f_{j-1}(\mathbf{s})\beta_{j-1} \tag{9.22}$$

in which the functions f_{j-1} are typically polynomials, and both the number $K + 1$ and the coefficients β_{j-1} are to be determined in the course of the statistical analysis. Gambolati and

Volpi (1979) present the view that the systematic component μ should have a clear physical content and should be fully determined a priori. In their study of universal kriging to map groundwater contours in aquifers underlying Venice, they argue that $\mu(\mathbf{s})$ should include information related to what is believed to be the main trend, and that it is not necessary for it to be linear; and the function $\mu(s)$ used was

$$\mu(x, y) = A \ln[\{Bx^2 + Cxy + (y - d)^2\}/\{Bx^2 + Cxy + (y + d)^2\}] + z_0$$

where d is a geometric quantity drived from the available data, A, B and C are determined by least squares, and z_0 is a known potential at an upstream boundary. Whilst it is clearly good practice to utilise all available information to specify the form of $\mu(\mathbf{s})$, cases where an explicit form can be postulated by physical argument are likely to be the exception rather than the rule; and it is not clear what effect the least-squares estimation of B and C will have on the variogram estimate.

REFERENCES

Box, G. E. P. and Jenkins, G. M. (1970). *Time Series Analysis: Forecasting and Control*. Holden Day, San Francisco.

Cressie, N. A. C. (1993). *Statistics for Spatial Data* (revised edition). Wiley Interscience, New York.

Gambolati, G. and Volpi, G. (1979). Groundwater contour mapping in Venice by stochastic interpolators 1. Theory. *Water Resour. Res.* **15**, 281–90.

Journel, A. G. and Huijbregts, C. J. (1978). *Mining Geostatistics*. Academic Press, London.

Chapter 10

Some possible future developments in statistical modelling

10.1 OMISSIONS

The final chapter of a book such as this appears commonly to have two objectives: first, to mention, if only briefly, topics which have been omitted (one might claim, cynically, for the purpose of heading off critical reviews); and second, to hazard some guesses about future developments, possibly with some suggestions about research topics upon which work would be timely. Regarding the first of these objectives, the topics that have been omitted are many; indeed, it is not so much that certain individual topics have been omitted, as that whole areas of applied statistics with hydrological application have not been dealt with. The corpus of theory on time-series analysis is a case in point; over the past 20 years, methods of time-series analysis have been brought to bear on very many series of hydrological data. Their omission from this book has been deliberate, for three reasons. First, to deal properly with the subject would lengthen the book greatly. But second, and more to the point, the author is not fully convinced that some of the stationarity assumptions necessarily made in time-series analysis are appropriate for hydrology. To some extent, recent trends in the literature seem to support this view; papers on periodic ARMA models, for example, have moved towards the development of separate ARMA models for flows within distinct seasons. It is but a step from this to the separate modelling of each month's flow, using models of the kind discussed in Chapter 7. Lastly, the way in which statistical models are used in hydrology, for simulation of synthetic sequences as well as for prediction, requires us to consider properties of the probability distribution of Y_t, *conditional* on the values $y_{t-1}, \dots$. Regression methods therefore arise naturally in such applications. ARMA models, on the other hand, whilst constituting a particularly succinct way of describing the (assumed stationary) correlational structures of time series, do not have the naturalness of regression methods for hydrological application. This said, regression methods have been misapplied, often thoughtlessly and sometimes grossly, in hydrology.

For broadly the same reasons that time-series analysis was excluded, applications of the theory of stochastic processes have been given no place in this book. Textbooks on the subject show results of great elegance, and great satisfaction would be engendered by applying them to the difficult questions that the hydrologist is obliged to answer. However, he invariably finds himself burdened with frustrations similar to those of an unsuccessful marriage broker: on the one hand, a set of intractable hydrological records constitutes a bride indifferent to any suitor, often missing in her essential parts, of doubtful veracity and prone to mislead, and far, far too short; and, on the other hand, a set of elegant suitors, each in finest theoretical apparel, and each requiring an uncomfortably large marriage settlement by way of restrictive assumptions, before a union can be consummated. Often the most relevant statistical methods have come to light, it seems, through the efforts of hydrologists themselves, wrestling to make sense of their records, and not from seeking to unite elegant theory with ill-favoured data. An example that springs to mind is the Hurst coefficient.

10.2   SOME ASPECTS OF STATISTICAL MODELLING IN HYDROLOGY WHICH NEED FURTHER RESEARCH

We conclude by offering a brief selection of topics believed to be particularly deserving of greater research effort. The first (Section 10.2.1) is really an appeal for some new thinking on an old problem which is of extreme importance, particularly for developing countries; the second (Section 10.2.2) arises from the broadening horizons of hydrology and the increasing complexity of questions asked of hydrologists and their environmental scientist colleagues.

10.2.1   Regionalisation

At various points in this book we have touched upon the uses of statistical methods for transferring information from areas where hydrological records are longer or more reliable, to other areas where they are shorter, less reliable, or non-existent. Current methods of 'pooling' records tend to standardise the record at each site, prior to pooling over sites, commonly by dividing by an index variable. Where the record is of annual maxima, this index variable is usually taken as the mean of the annual maxima; a dimensionless cumulative probability curve ('growth curve') is then obtained when sites are pooled. Reed and Stewart (1993) summarise some of the problems to which this procedure gives rise: namely, that pooling by region causes discontinuities in the estimation of T-year extremes at the region boundaries; and that correlations among the annual maxima at different sites increase uncertainties in the growth curves.

The Use of Generalised Linear Models in Regionalisation

Generalised linear models (see Chapter 6) may afford an alternative to the index-variable method. Suppose that the data (which may be, for example, annual maxima or annual totals) from G sites and P years are arranged in a $G \times P$ table, following the format of Chapter 7, and let y_{ij} ($i = 1, \ldots, G; j = 1, \ldots, P$) be the (i, j)th data value in this table, which will usually have a high proportion of missing values. It would be reasonable to postulate that

$$E[Y_{ij}] = \mu + g_i + p_j$$

where μ represents a regional mean, g_i the effect of the ith gauge, and p_j the effect of the jth year. The component p_j therefore measures any tendency for all gauges in the same year to yield high (or low) data values, such as occur during *el Niño* events or in drought years. Introducing dummy variables x_i, z_j taking values zero or one, the expected value $E[Y_{ij}]$ can be written

$$E[Y_{ij}] = \mu + \sum_i g_i x_i + \sum_j p_j z_j \tag{10.1}$$

It would also be plausible to postulate a probability distribution for Y_{ij} coming from the exponential family: for annual floods, for example, a gamma distribution would be appropriate.

A link function would also need to be specified; the linear link (10.1) may be appropriate, but the log link function

$$\ln E[Y_{ij}] = \mu + \sum_i g_i x_i + \sum_j p_j z_j$$

ensures that $E[Y_{ij}]$ is always positive. Application of the generalised linear models fitting procedure yields estimates of μ, g_i and p_j.

Where the index-variable method has been used, efforts have been made to identify 'hydrologically homogeneous' regions; but homogeneity appears to this author to be an *ignis fatuus* inconsistent with the spatial variability commonly observed in many hydrological studies. The generalised linear model method would automatically allow for regional inhomogeneity, since the estimated effects $\hat{g}_i$ could be interpolated by the kriging methods discussed in Chapter 9. If desk-top computers are used, practical difficulties with the proposed method may arise if the number of parameters $G + P - 1$ is large, rendering necessary the use of more powerful computing facilities.

Parametrisation of Variance-covariance Matrices

Suppose that a multivariate model is to be used for simulation purposes. The errors in the model will have a variance-covariance matrix which must be estimated, and where the model is to simulate flows at G sites, the variance-covariance matrix of the errors will contain $G(G+1)/2$ parameters, often a substantial number.

However, it seems probable that the number of parameters necessary to describe the variance-covariance matrix of random components in the model could be reduced by using information about the system to be modelled. Where, say, two of the G gauging sites are on the same river, the correlation between errors for these sites is likely to be larger than between two sites in different drainage basins; and the correlation between the errors at two sites on the same river might be expected to be some function of the distance between them. Thus we could plausibly reduce the number of parameters in the variance-covariance matrix of errors by writing the correlation between errors at two sites in the same basin as a simple function of distance between the sites, say $\exp(-\lambda d)$; or by requiring that the correlation between errors at sites in widely separated basins be smaller than the correlation between errors at sites in adjacent basins. Thus, knowledge of the system being modelled would be used to define a particular structure for the variance-covariance matrix of errors, requiring fewer parameters than the full complement of $G(G+1)/2$. Likelihood ratio tests should serve to establish whether use of the restricted number of parameters results in a significant loss.

Regionalisation of Sediment Yields

Records of sediment concentration in streamflow are sparse, particularly in developing countries. In Brazil, even where sediment concentration is measured, samples are rarely collected more than four times a year, and at times that tend to be determined more by the prevailing workload than by the need to obtain samples representative of flow conditions. However, countries which depend heavily on hydropower generated from impounding reservoirs, need to know what sediment yields are to be expected, and how these yields are related to changes in land use. A field of study of great hydrological importance therefore concerns how very limited records of sediment concentration can be combined with longer records of discharge, to obtain reasonably unbiased estimates of sediment yield at a site; and, as a next step, how to regionalise these estimates to sites without sediment concentration records.

10.2.2 Regional Water Balance Studies

In recent years, a new kind of hydrological field study has come upon the scene, in which the exchanges of energy and water between the earth's surface and the atmosphere are studied on a much wider scale than hitherto. These field studies require close cooperation between hydrologists and meteorologists, since they often require the simultaneous collection of data by instruments at surface sites, by meteorological balloons and overflying aircraft, and by satellites. Besides their purely hydrological and meteorological aspects, the studies often include additional components looking at how elements such as carbon, and isotopes of oxygen, are cycled within the region. Element cycles of this kind are driven by the water cycle, so that the accuracy and precision with which water balance components are estimated determines to a large extent the outcome of the element cycling studies.

A common feature of regional water balance studies is that very large numbers of measurements are recorded, and their analysis presents two kinds of statistical problem which are encountered more frequently than in the past. The first concerns how to make the best use of data, when a high proportion of it may be redundant; and the second arises because instruments making the measurements may lose calibration or otherwise begin to perform badly. When instruments are making measurements many times a second, the volume of data is such that computational procedures must be largely automated, and it then becomes necessary to screen data, prior to their analysis, to identify suspect measurements and, if necessary, eliminate them. It is in such circumstances that the expert systems, mentioned briefly in Chapter 1, are likely to be especially useful.

Apart from problems arising from the volume and reliability of data, it is this author's view that the statistical analysis of data from large-scale, regional water balance studies requires much greater emphasis if their extremely high costs are to be justified. To fix ideas, we consider the LAMBADA Project (Large-scale Atmospheric Moisture Balance of Amazonia using Data Assimilation) which is being planned for the years 1996–98. This is intended to cover the 4.8 million square kilometres of the Amazon basin, and so may be considered the 'mother of all regional water balance studies'. A planning document for the project (LAMBADA-BATERISTA 1993) presents the straightforward equations whose components are to be measured. These are given as, for the atmospheric water balance,

$$\int_{z_s}^{z_t} dq/dt = \int_{z_s}^{z_t} \nabla \cdot \mathbf{v}q \, dz + E - P$$

where z_s, z_t are the heights of the surface (m) and the top of the atmosphere (m), respectively; q is water vapour concentration (kg m^{-3}) and $\mathbf{v}$ is the wind velocity vector (m s^{-1}); and E, P are evaporation rate (kg m^{-2} s^{-1} or mm s^{-1}) and precipitation rate (kg m^{-2} s^{-1} or mm s^{-1}). The equation given for the surface water balance is

$$\int_{z_b}^{z_s} dS/dt = P - E - R_o$$

where z_b is the depth of the hydrologically active soil layer (m), S is soil moisture (kg m^{-3}) and R_o is runoff (mm s^{-1}). The planning document proceeds to take space and time averages of these equations:

$$\langle E - P \rangle = \langle D \rangle - \langle \Delta S \rangle$$

for the land surface balance, where $\langle . \rangle$ denotes a basin-wide average, D is the downstream discharge from the basin, and ΔS is the change in surface and subsurface storage; and

$$\langle E - P \rangle = \langle \nabla . Q \rangle + \langle \Delta W \rangle$$

for the atmospheric water balance. The latter equation

states that the time-averaged, basin-averaged difference between evaporation and precipitation, i.e. the net source of vapour to the atmosphere, is equal to the time-space averaged vapour flux divergence from the overlying atmospheric column, corrected for any change in atmospheric vapor storage from the beginning to the end of the period.

Certain difficulties now present themselves. When considering the term $\langle \nabla \cdot . Q \rangle$, the report invites us to 'Consider the curve C which traces out the boundary of the drainage basin'. In 1863 H. W. Bates, describing one of his boat journeys during 14 years spent in the Amazon basin, wrote:

The fact of the tide being felt 530 miles up the Amazons, passing from the main stream to one of its affluents 380 miles from its mouth, and thence to a branch of the third degree, is a proof of the extreme flatness of the land which forms the lower part of the Amazonian valley. This uniformity of level is shown also in the broad lake-like expanses of water formed, near their mouths, by the principal affluents which cross the valley to join the main river.

This illuminates the difficulty in defining 'the curve C' defining any watershed boundary in the Amazon region.

Apart from the topographical difficulty of defining basin limits, there are considerable statistical problems in computing the water balance terms. The flux divergence $\nabla . Q$ will be calculated by establishing the wind vector, using radiosondes, at the nine vertices of a polygon approximating to the drainage basin boundary, each vertex being several hundred kilometres distant from those adjacent to it. Given the measured wind vector, at a given height, at each of the polygon vertices, various estimators of the flux across the polygon boundary can be proposed, and the question of how to choose between them is not trivial. Each estimator requires decisions concerning the degree of interpolation to be used when calculating the flux across each polygon side; concerning how variances are to be calculated for the estimate; concerning what to do about missing measurements; concerning the degree of interpolation to be used between altitudes; and concerning how estimates of flux across a polygon side for different altitudes are to be combined. These are statistical questions which must be answered at the planning stage; they cannot be postponed until the data have been collected.

Establishing a reliable water balance for the Amazon basin, and estimating how it may change if deforestation increases, is important not only for Brazil. This is not trans-science in the sense introduced in Chapter 8, namely, the seeking of answers that can feasibly be supplied by the scientific method, but at a cost out of all proportion to the benefit gained. The cost of LAMBADA will no doubt be considerable, probably far in excess of other regional water balance studies. But the benefit might be avoidance of the destruction of a natural resource important for the control of regional climate; statistical analysis of the data therefore needs to be planned with the utmost care.

REFERENCES

Bates, H. W. (1863, 1864). *The Naturalist in the River Amazons*. John Murray, London.

LAMBADA-BATERISTA (April, 1993). A preliminary science plan for a large-scale biosphere-atmosphere field experiment in the Amazon basin. Report from an International Workshop sponsored by WCRP and IGBP held at NASA/GSFC, Greenbelt, Maryland, 18–20 June, 1992. ISLSCP/IGPO. Washington, DC.

Reed, D. W. and Stewart, E. J. (1993). Inter-site and inter-duration dependence in rainfall extremes. SPRUCE II Conference Rothamsted Experimental Station, 13–16 September.

Appendix

Some results in probability and statistical theory

This Appendix gives definitions and results basic to the methods of this book. The reader interested in full mathematical rigour should consult a text such as that of Cramer (1946) and Cox and Hinckley (1974).

A.1 THE FUNDAMENTALS OF PROBABILITY THEORY

The theory of probability underlies all statistical modelling in hydrology. The fundamental concept is that of a *random variable*, defined over a *sample space*. Examples of hydrological (random) variables taking *discrete* values are: (i) the number of overbank floods occurring during a year (the variable, denoted by N, taking discrete values $0, 1, 2, \ldots$, so that the sample space consists of the non-negative integers on the real line); (ii) number of tips of a tipping-bucket gauge in a five-minute interval, a variable M defined on the integers $0, 1, 2, \ldots$. Examples of variables taking *continuous* values are (iii) annual maximum discharge (for which the random variable, say Q, takes values on the positive axis $Q \geq 0$; (iv) 30-minute maximum annual precipitation intensity I, also defined for $I \geq 0$. We adopt the common notation of representing random variables by capital letters $X, Y, Q, \ldots$, and observations of such variables by lower-case letters $x, y, q, \ldots$. Repeated observations of the random variable X are denoted by $x_1, x_2, x_3, \ldots$.

Events are defined on the sample space within which a random variable is defined. Examples are the events E: annual maximum discharge exceeds 4000 m^3 s^{-1}; 'A: Number of overbank floods N is zero'; or 'B: $Q \geq q$'. The capital letter gives a short-hand name for the event, which may be described in words, although it is more usual to describe the event simply by the statements '$N = 0$'; '$Q \geq q$'; '$I \leq i_0$'. Two events are *mutually exclusive* if they have no points of the sample space in common; thus when the random variable Q varies over a sample space $Q \geq 0$, the events 'A: $Q < 5$' and 'B: $Q > 10$' are mutually exclusive; the events 'A: $Q < 10$' and 'B: $Q > 7$' are not.

Associated with each event defined on a sample space is a *probability measure*. Probability can be regarded as a unit mass distributed, with varying density, over the entire sample space. For the random variable annual maximum discharge Q, varying over the sample space $Q \geq 0$, the unit mass of probability is distributed over the non-negative part of the real line; for the random variable N, the number of overbank floods in a year, the unit probability mass is distributed as particles placed at each of the points $0, 1, 2, \ldots$ on the real line. The *probability density* must obviously be greater than or equal to zero; to each and every event $A, B, N = 0, X \leq x, \ldots$ there will correspond a probability, denoted by symbols such as $P[A]$, $P[B]$, $P[N = 0]$, $P[X \leq x], \ldots$ representing the proportion of the unit mass lying within that part of the sample space to which the event corresponds.

A simple interpretation of probability is in terms of *limiting frequencies*. Suppose that, in a very large number of years, say Y, we observed the numbers of years $n_0, n_1, n_2, \ldots$ with zero, one, two, $\ldots$ overbank floods. Then, for large Y, the relative frequencies n_0/Y, n_1/Y, $n_2/Y, \ldots$ will approximate to the probabilities $P[N = 0]$, $P[N = 1]$, $P[N = 2], \ldots$. Thus one interpretation of $P[N = r]$ is

$$P[N = r] = \lim_{Y \to \infty} [n_r/Y]$$

In the case of a continuous random variable X, such as annual maximum discharge, the probability that X lies in any small interval $x < X \leq x + dx$ may be regarded as the limit as Y tends to infinity of the relative frequency n/Y, where n is the number of years out of Y that the annual maximum discharge lay within the interval $x < X \leq x + dx$. In symbols,

$$P[x < X \leq x + dx] = \lim_{Y \to \infty} [n/Y]$$

The *probability density function* (pdf) defines how the unit probability mass is distributed over the sample space. For a discrete random variable N, taking the values $0, 1, 2, \ldots$, the pdf can be represented by the set of values $\{p_r\}$ such that

$$P[N = r] = p_r \qquad r = 0, 1, 2, 3, \ldots$$

such that (i) $p_r \geq 0$; (ii) $p_r \leq 1$; (iii) $\Sigma p_r = 1$. In the case of a continuous random variable X, defined on, say $x \geq 0$, the pdf can be represented by the continuous function $f(x)$,

$$P[x < X \leq x + dx] = f(x)dx \qquad (x \geq 0)$$

such that (i) $f(x) \geq 0$; (ii) $\int f(x)dx = 1$, where integration is over the total sample space. Although hydrological variables commonly take non-negative values (exceptions are (a) net radiation; (b) change in storage ΔS within an experimental basin), there is no conceptual difficulty in thinking of random variables as being defined over the whole of the real line (for example, $-\infty < X < \infty$ in the case of a continuous random variable) with, if necessary, $f(x) = 0$ for negative values of x.

Discussion so far has been in terms of single events denoted by single symbols $A, B, \ldots$. We need also to consider the *union* of two events, denoted by $A \cup B$, and their *intersection*, denoted by $A \cap B$. The union $A \cup B$ is the event 'C: at least one of the events A, B occurs'; the intersection $A \cap B$ is the event 'D: both A and B occur'. An equivalent interpretation of the union $A \cup B$ is 'either A or B occurs, or they both occur'.

The *conditional probability* of A, given B, denoted by $P[A|B]$, is defined by

$$P[A|B] = P[A \cap B]/P[B]$$

which can also be written

$$P[A \cap B] = P[A|B]P[B]$$

Two events are said to be *statistically independent* if the probability of the event A is the same whether or not B has occurred: that is, $P[A|B] = P[A]$. The above definition of conditional probability then gives

$$P[A \cap B] = P[A]P[B]$$

This relation can be extended to give the condition for independence of more than two events. Thus events $A, B, C, \ldots$ are statistically independent if and only if

$$P[A \cap B \cap C \cap \ldots] = P[A]P[B]P[C] \ldots$$

This relation underlies many of the methods described in this book. For example, we assume that the annual maximum discharges $X_1, X_2, X_3, \ldots$ in a sequence of years $1, 2, 3, \ldots$ are statistically independent, so that

$$P[(x_1 < X_1 \le x_1 + dx_1) \cap (x_2 < X_2 \le x_2 + dx_2) \cap \ldots]$$
$$= P[x_1 < X_1 \le x_1 + dx_1]P[x_2 < X_2 \le x_2 + dx_2] \ldots$$

and if the random variables $X_1, X_2, X_3, \ldots$ all have the same pdf $f(x)$, the right-hand side becomes

$$f(x_1)f(x_2)f(x_3) \ldots dx_1 dx_2 dx_3 \ldots$$

so that the probabilities multiply.

The above intuitive ideas can be formalised in terms of the following:

(i)   There exists a sample space S.

(ii) Events $E_1, E_2, E_3, \ldots$ are defined on the sample space. These events include, as special cases, the 'null' event $\emptyset$, and the 'identity' event S.

(iii) A probability measure $P[.]$, is defined over S, such that

    (a)   for any event A, $P[A] \ge 0$;
    (b) $P[S] = 1$;
    (c)   for any two events A, B which are *mutually exclusive*, $P[A \cup B] = P[A] + P[B]$.

Parts (a), (b) and (c) of (iii) above together define the *axioms of probability*. Some results deduced from these axioms are given in the following section; these results form the basis for much of the application of probability theory in hydrology, as in other sciences.

A.2   DEDUCTIONS FROM THE AXIOMS OF PROBABILITY

A.2.1   The Probability of a Union of Two Events A and B that Are Not Mutually Exclusive

Recall that the union $A \cup B$ of two events A and B is defined as the event that either A or B occurs, or both occur. By reference to a Venn diagram (see Figure A.1) or otherwise, it can be shown that

$$P[A \cup B] = P[A] + P[B] - P[A \cap B] \tag{A.1}$$

and it can be seen that, for mutually exclusive events (for which $P[A \cap B] = 0$), equation (A.1) reduces to the axiom (iii)(c) of Section A.1. The relation can be generalised to give

$$P[A \cup B \cup C] = P[A] + P[B] + P[C] - P[A \cap B] - P[A \cap C] - P[B \cap C] + P[A \cap B \cap C]$$

and further generalisations are straightforward.

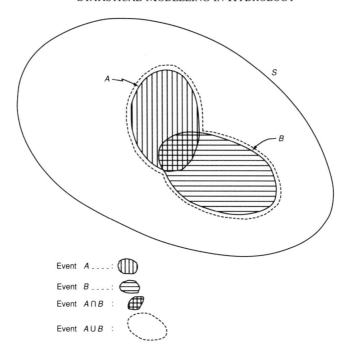

Figure A.1   Venn diagram of sample space S, showing events A, B, $A \cup B$ and $A \cap B$.

A.2.2   The Theorem of Total Probability

Suppose that the entire sample space S is covered by n mutually exclusive events $B_1, B_2, \ldots, B_n$: that is, $B_1 \cup B_2 \cup B_3 \cup \ldots \cup B_n = S$, with $B_1 \cap B_2 \cap B_3 \cap \ldots \cap B_n = \emptyset$, the 'null' event. For any event A, the events $A \cap B_1$, $A \cap B_2$, $\ldots$ are mutually exclusive. Hence we can use axiom (iii)(c) to write

$$P[A] = P[A \cap B_1] + P[A \cap B_2] + \cdots + P[A \cap B_n]$$

which, on writing $P[A \cap B_i] = P[A|B_i]P[B_i]$, gives the *theorem of total probability*:

$$P[A] = \sum_{i=1}^{n} P[A|B_i]P[B_i] \qquad (A.2)$$

To illustrate how this relation may be used, consider the maximum overbank discharge, Q_{max}, in a river reach, during a period $(0, T)$. The number, N, of overbank floods during the period can be $0, 1, 2, \ldots$ and the events 'B_0: no overbank floods', 'B_1: one overbank flood', 'B_2: two overbank floods', $\ldots$ are mutually exclusive. Hence the probability of the event 'A: $Q_{max} > q$' is given by

$$P[Q_{max} > q] = P[\text{maximum of 1 overbank discharge } > q | 1 \text{ overbank event}]P[B_1]$$

$$+ P[\text{maximum of 2 overbank discharges} > q | 2 \text{ overbank events}]P[B_2] + \cdots$$

Relations of this kind are especially useful, for example, in (i) the statistical model of flood occurrences known as the 'peaks over a threshold' (POT) model; (ii) models of bedload transport, where the events B_i are the events 'number of saltations of a bedload particle is i' ($i = 0, 1, 2, \ldots$) and the event A is 'distance travelled by particle is greater than x'.

A.3 SOME PARTICULAR PROBABILITY DENSITY FUNCTIONS

The idea of a density function, describing how a unit mass of probability is distributed over a sample space, was introduced in Section A.1. We now give some specific examples of *probability density functions* (pdfs), which have wide application in the water sciences. The section provides no more than a sketch of selected aspects of these pdfs, and the reader is advised to supplement his reading with more specialised texts, of which many exist.

The *uniform* distribution is defined by

$$f(x; a, b) = \begin{cases} 1/(b - a) & a \le x \le b \\ 0 & x < a \text{ or } x > b \end{cases}$$

The sample space S is therefore the segment of the real axis for which $a \le x \le b$, the random variable x being continuous in this region. The quantities a and b are *parameters* which describe the position and shape of the distribution; the particular uniform distribution for which $a = 0$, $b = 1$ plays a central role in the computer simulation of water resource systems and, more generally, in the mathematical procedures known as Monte Carlo.

The *Bernoulli* distribution is the pdf of a discrete random variable, X say, which takes the values $X = 1$ or $X = 0$ with probabilities π and $1 - \pi$. Thus

$$P[X = 1] = \pi$$

$$P[X = 0] = 1 - \pi$$

Examples are the occurrence or non-occurrence of rain on a day chosen at random; and the occurrence or non-occurrence of flow at a gauging station on an intermittent river, during a randomly chosen month.

The *binomial* distribution describes the behaviour of a discrete random variable X, taking integral values between 0 and n. If the outcome $X = 1$ in a Bernoulli trial is regarded as a 'success', the binomial distribution gives the probability of x successes in n independent trials of Bernoulli type with π constant. In symbols,

$$p(x; n, \pi) = \begin{cases} \binom{n}{x} \pi^x (1 - \pi)^{n-x} & x = 0, 1, 2, \ldots, n \\ 0 & \text{otherwise} \end{cases}$$

The *Poisson* distribution is shown in probability textbooks to be a limiting form of the binomial distribution. It describes the behaviour of a discrete random variable X, taking values $0, 1, 2, \ldots$ without upper limit, when (a) the number n of independent trials in a binomial distribution becomes very large, and (b) the probability π of a 'success' at each independent Bernoulli trial becomes very small, whilst (c) the product $n\pi$ (commonly denoted by λ) remains finite.

In symbols,

$$p(x; \lambda) = \begin{cases} \lambda^x \exp(-\lambda)/x! & x = 0, 1, 2, \ldots \\ 0 & \text{otherwise} \end{cases}$$

An important application occurs when events occur randomly in time such that (i) the probability of an event occurring in a very small interval of time $(t, t + \delta t)$ is proportional to δt:

say, equal to $\lambda\,\delta t$, where λ is a constant; (ii) the occurrences of events in non-overlapping intervals of time are statistically independent. The occurrences then follow a *Poisson process with parameter* λ; and for any interval of time $(0, T]$, the probability that x events occur is given by

$$p(x; \lambda) = \begin{cases} (\lambda T)^x \exp(-\lambda T)/x! & x = 0, 1, 2, 3, \ldots \\ 0 & \text{otherwise} \end{cases}$$

Events which, in certain circumstances, may be represented by a Poisson process, are the occurrence of floods exceeding a certain threshold, in the interval $(0, T]$; and the number of saltations made by a bedload particle in the interval $(0, T]$.

It is clear that, if we consider the probability of X_1 events in the interval $(0, T_1]$ and of X_2 events in the interval $(T_1, T_2]$, then the distribution of the sum $X = X_1 + X_2$ is

$$f(x; \lambda) = [\lambda(T_1 + T_2)]^x \exp[-\lambda(T_1 + T_2)]/x!$$

The sum of two Poisson variables X_1, X_2 is therefore also Poisson.

Closely related to the Poisson distribution is the *exponential* distribution. Where events occur in time according to a Poisson process, the intervals of time (now denoted by X; the double usage of X should not cause confusion) between successive events have the pdf

$$f(x; \lambda) = \begin{cases} \lambda \exp(-\lambda x) & 0 \le x < \infty \\ 0 & \text{otherwise} \end{cases}$$

The exponential distribution is the distribution of time intervals X between successive events in a Poisson process; the distribution of the sum of the two intervals, $X = X_1 + X_2$, defined by three successive events in a Poisson process, can also be calculated. This is found to be

$$f(x; \lambda) = \begin{cases} \lambda^2 x \exp(-\lambda x) & 0 \le x < \infty \\ 0 & \text{otherwise} \end{cases}$$

More generally, the sum of k independent random variables $X = X_1 + X_2 + \cdots + X_k$, where $X_1, X_2, \ldots, X_k$ have the same exponential distribution $\lambda \exp(-\lambda x)$, is the *gamma* distribution given by

$$f(x; k, \lambda) = \lambda(\lambda x)^{k-1} \exp(-\lambda x)/(k - 1)! \qquad 0 \le x < \infty$$

If we think of the gamma distribution in terms of the sum of k time intervals between any $k + 1$ successive events of a Poisson process, then k must be an integer. A more general form of the gamma distribution is permissible in which k is non-integral; provided that $k > 0$, we can write the more general form

$$f(x; k, \lambda) = \lambda(\lambda x)^{k-1} \exp(-\lambda x)/\Gamma(k) \qquad 0 \le x < \infty; k > 0$$

where $\Gamma(k)$ is the gamma function defined by

$$\Gamma(k) = \int_0^\infty u^{k-1} e^{-u} du \qquad k > 0$$

Thus defined, the gamma distribution has two parameters λ and k. Chapter 3 discusses a further generalisation having three parameters, λ, k and a, such that

$$f(x; \lambda, k, a) = \lambda[\lambda(x - a)]^{k-1} \exp(-\lambda(x - a))/\Gamma(k) \qquad a \le x < \infty$$

In the above development, we have considered at various points the distribution of sums of independent random variables with the same pdf. The *central limit theorem* states that, within very broad limits, the sum $X = X_1 + X_2 + \cdots + X_n$ of n independently distributed random variables $X_1, X_2, \ldots, X_n$, whether continuous or discrete, but having the same pdf, has approximately the *normal* or *Gaussian* distribution, given by

$$f(x; \mu, \sigma) = (1/\sqrt{(2\pi\sigma^2)})\exp[-(x - \mu)^2/(2\sigma^2)] \qquad -\infty < x < \infty$$

The larger the value n of random variables entering the sum X, the better the normal approximation to its distribution, which has position and 'spread' respectively determined by the two parameters μ and σ^2. These parameters are the *mean* and *variance* of the normal distribution, which is commonly denoted by the symbol $N(\mu, \sigma^2)$. The distribution plays a central role wherever statistical methods are used to draw conclusions from the analysis of data, for the following reasons. First, many real-world measurements are found to follow a normal distribution, or are such that some simple transformation of them follows a normal distribution. Second, because of the central limit theorem, the means of independent random variables, drawn from the same distribution, are for all practical purposes approximately normally distributed, whatever the underlying distribution.

Thus, when the number of trials n for a binomial distribution is large, the central limit theorem says that the number of successes r is *approximately* normally distributed with mean $\mu = n\pi$ and variance $\sigma^2 = n\pi(1 - \pi)$. If $X_1, X_2, X_3, \ldots, X_n$ are independent with the same Poisson distribution $p(x) = \lambda^x \exp(-\lambda)/x!$, the sum $X = X_1 + X_2 + \cdots + X_n$ has a distribution that is approximately normal with mean $\mu = n\lambda$ and variance $n\lambda$. Similar statements hold when $X_1, X_2, \ldots, X_n$ are independently distributed with exponential or gamma distributions. When $X_1, X_2, \ldots, X_n$ are independently distributed with the same *normal* distribution with mean μ and variance σ^2, then the sum $X_1 + X_2 + \cdots + X_n$ is *exactly* normally distributed with mean $n\mu$ and variance $n\sigma^2$.

Related to the normal distribution is the *log-normal* distribution, of a random variable X with positive values:

$$f(x; \mu, \sigma) = \{1/[x\sqrt{(2\pi\sigma^2)}]\}\exp[-(\ln x - \mu)^2/(2\sigma^2)] \qquad 0 \leq x < \infty$$

This asymmetric distribution has been extensively used to represent the distribution of annual maximum discharges at flow gauging stations; as with the gamma distribution, a three-parameter form is obtained by substituting x by $x - a$, when the range through which X may vary becomes $a \leq x < \infty$. By transforming from X to the new random variable $Y = \ln X$ in the two-parameter case (or $Y = \ln(X - a)$ in the three-parameter case), it can be seen that the variable Y is normally distributed with parameters μ and σ.

A.4 MOMENTS OF PROBABILITY DENSITY FUNCTIONS

Important characteristics of a pdf of a random variable X can be summarised by its *moments*.

The *first moment about the origin*, also called the *expected value*, or *mean*, of the random variable X, commonly denoted by the symbol $E[X]$ or, more succinctly, by μ, is given symbolically by

$$E[X] = \int_{-\infty}^{\infty} xf(x, \theta)\mathrm{d}x$$

when the random variable X is continuous, or by

$$E[X] = \sum xp(x, \theta)$$

when x is discrete; in both expressions θ are the parameters of the distribution. Thus for the uniform distribution with $f(x; a, b) = 1/(b - a)$, the expected value is

$$E[X] = \int_a^b x \, dx/(b - a)$$

or $(a + b)/2$, confirming intuition. For the binomial distribution, $E[X] = n\pi$; for the Poisson, $E[X] = \lambda$; for the exponential, $E[X] = 1/\lambda$; for the gamma with two parameters, $E[X] = k/\lambda$; and for the normal, $E[X] = \mu$.

The *second moment about the mean*, also called the *variance*, commonly denoted by the symbol var$[X]$ or σ^2, is defined by

$$\text{var}[X] = \int_{-\infty}^{\infty} (x - \mu)^2 f(x, \theta) dx$$

when X is a continuous random variable, or by

$$\text{var}[X] = \sum (x - \mu)^2 p(x, \theta)$$

when X is discrete. Using the 'expectation' symbol $E[.]$, we can easily see that

$$\text{var}[X] = E[(X - E[X])^2]$$

Thus, the variance of the uniform distribution $f(x; a, b) = 1/(b - a)$ is given by

$$\text{var}[X] = \int_a^b [x - (a + b)/2]^2 dx/(b - a)$$

$$= (b - a)^2/12$$

Similarly, for the Bernoulli distribution, var$[X] = \pi(1 - \pi)$; for the binomial distribution with X the number of successes in n trials, var$[X] = n\pi(1 - \pi)$; for the Poisson distribution, var$[X] = \lambda$, so that mean and variance are equal in this case; for the exponential distribution, var$[X] = 1/\lambda^2$; for the gamma distribution with two parameters, var$[X] = k/\lambda^2$; and for the normal, var$[X] = \sigma^2$.

The *third moment about the mean*, denoted by the symbol μ_3, is defined by

$$\mu_3 = \int_{-\infty}^{\infty} (x - \mu)^3 f(x, \theta) dx$$

when X is continuous, and by

$$\mu_3 = \sum (x - \mu)^3 p(x, \theta)$$

when X is discrete. A dimensionless measure of the skewness of a distribution is given by

$$g_3 = \mu_3/\sigma^3$$

where σ^2 is the variance discussed above.

Higher-order moments about the mean, such as the fourth moment about the mean,

$$\mu_4 = \int_{-\infty}^{\infty} (x - \mu)^4 f(x, \theta) dx$$

with obvious modification where X is a discrete random variable, are occasionally useful. The fourth moment about the mean can be used to form a dimensionless measure of the *pointedness* or *kurtosis* of a pdf:

$$g_4 = \mu_4/\sigma^4$$

where, σ^2 is, as before, the variance.

Given a sample $x_1, x_2, x_3, \ldots, x_t, \ldots, x_n$ of observations of the random variable X, sample moments can be calculated, analogous to the moments of the pdf. Thus the sample mean and variance are

$$\bar{x} = \sum x_t/n$$
$$s^2 = \sum (x_t - \bar{x})^2/n$$

The third and fourth sample moments are $\sum(x_t - \bar{x})^3/n$, $\sum(x_t - \bar{x})^4/n$. An observation x_t lying far from the sample mean $\bar{x}$ therefore makes large contributions to the calculation of third and fourth (and higher) moments, which therefore have 'large sampling errors'. Measures of asymmetry and kurtosis which are less subject to such sampling variability can be obtained by the use of l-moments, as defined in Chapter 3.

A.5 DISTRIBUTIONS OF FUNCTIONS OF RANDOM VARIABLES

In general, if a function $g(.)$ is a one-to-one transformation of the random variable X into another random variable $g(X)$, then $g(X)$ will itself have a probability distribution. More generally, any function $g(X_1, X_2, \ldots, X_n)$ of n random variables will also have a pdf. Given the function $g(.)$ and the pdf of $X_1, X_2, \ldots, X_n$, the pdf of $g(.)$ can sometimes be derived analytically; thus we get the following results:

(i) Where $X_1, X_2, \ldots, X_n$ are independently distributed $N(\mu, \sigma^2)$, the pdf of $\bar{X}$ is $N(\mu, \sigma^2/n)$.

(ii) Where $X_1, X_2, \ldots, X_n$ are independently distributed $N(0, 1)$, the pdf of $\chi^2 = X_1^2 + X_2^2 + \cdots + X_n^2$ is the chi-square distribution

$$f(\chi^2) = (\chi^2/2)^{n/2-1} \exp(-\chi^2/2)/[2\Gamma(n/2)]$$

where n is the number of degrees of freedom. The χ^2 distribution is a particular case of the gamma distribution with $\lambda = 1/2$, $k = n/2$.

(iii) Where $X_1, X_2, \ldots, X_n$ are independently distributed $N(\mu, \sigma^2)$, with $s^2 = \sum(x_t - \bar{x})/(n-1)$ the 'unbiased' sample variance, the statistic $(n-1)s^2/\sigma^2$ has the χ^2 distribution with $n-1$ degrees of freedom.

(iv) Where $X_1, X_2, \ldots, X_n$ are independently distributed $N(0, \sigma^2)$, with $s^2 = \sum(x_t - \bar{x})^2/(n-1)$, the statistic $t = \bar{x}/\sqrt{(s^2/n)}$ has the distribution

$$f(t) = [1/\sqrt{(\pi(n-1))}][\Gamma(n/2)/\Gamma((n-1)/2)][1 + t^2/(n-1)]^{-n/2}$$

which is the t-distribution with $n-1$ degrees of freedom.

(v) Given two independent samples, of n_1 and n_2 values, respectively, from $N(\mu, \sigma^2)$, with sample variances $s_1^2 = \Sigma(x_t - \bar{x})^2/(n_1 - 1)$, $s_2^2 = \Sigma(x_t - \bar{x})^2/(n_2 - 1)$, the ratio $F = s_1^2/s_2^2$ has the distribution

$$f(F) = cF^{(n_1 - 3)/2}[(n_2 - 1) + (n_1 - 1)F]^{-(n_1 + n_2 - 2)/2}$$

where the constant c is given by

$$c = [\Gamma((n_1 + n_2 - 2)/2)/\Gamma((n_1 - 1)/2)]\Gamma((n_2 - 1)/2)(n_1 - 1)^{(n_1 - 1)/2}(n_2 - 1)^{(n_2 - 1)/2}$$

The distribution $f(F)$ above is the basis of F-tests in the analysis of variance described in Chapter 4. Thus, in linear regression, the numerator and denominator in the expression (sum of squares due to the regression)/(residual mean square) have, when the null hypothesis $H_0 : \beta = 0$ is true, χ^2 distributions with 1 and $n - 2$ degrees of freedom, respectively. Hence, when H_0 is true, their ratio has the distribution $f(F)$ with 1 and $(n - 2)$ degrees of freedom.

A.6   MOMENTS OF FUNCTIONS OF RANDOM VARIABLES

Even where the distribution of a function $g(X)$ of a random variable X can be calculated, it is often necessary to calculate the moments of $g(X)$, particularly its expected value. Indeed the variance of X is a special case in which $g(X) = (X - \mu)^2$. More generally, the expected value of $g(X)$ is defined by

$$E[g(X)] = \int_{-\infty}^{\infty} g(x)f(x; \theta)dx$$

Approximately, $E[g(X)] = g(E[X]) = g(\mu)$. The variance of $g(X)$ can be similarly defined as

$$\text{var}[g(X)] = \int_{-\infty}^{\infty} (g(x) - E[g(x)])^2 f(x, \theta)dx$$

and, as a large-sample approximation,

$$\text{var}[g(X)] = [(\partial g/\partial x)^2]\text{var}[X]$$

A.7   MOMENTS OF BIVARIATE DISTRIBUTIONS

The concept of a (univariate) pdf $f(x, \theta)$ can be easily generalised to the pdf of two (or more) random variables. For two random variables X and Y, we can imagine the unit mass of probability distributed over the X-Y plane, with density $f(x, y; \theta)$. Two first moments about the origin can be defined as

$$\mu_x = \int_{-\infty}^{\infty} \int_{-\infty}^{\infty} xf(x, y; \theta)dxdy$$

$$\mu_y = \int_{-\infty}^{\infty} \int_{-\infty}^{\infty} yf(x, y; \theta)dxdy$$

Three second moments about the mean can now be defined. Two are of the form previously encountered, namely

$$\text{var}[X] = \int_{-\infty}^{\infty} \int_{-\infty}^{\infty} (x - \mu_x)^2 f(x, y; \theta)dxdy$$

$$\text{var}[Y] = \int_{-\infty}^{\infty} \int_{-\infty}^{\infty} (y - \mu_y)^2 f(x, y; \theta) \mathrm{d}x \mathrm{d}y$$

whilst the *covariance* of X and Y is

$$\text{cov}[X, Y] = \int_{-\infty}^{\infty} \int_{-\infty}^{\infty} (x - \mu_x)(y - \mu_y) f(x, y; \theta) \mathrm{d}x \mathrm{d}y$$

A dimensionless measure of the association between X and Y is the *correlation* of X and Y, denoted by $\text{corr}[X, Y]$, or more simply by the symbol ρ, and given by

$$\rho = \text{cov}[X, Y]/\sqrt{(\text{var}[X]\,\text{var}[Y])}$$

which can assume values, positive or negative, in the range $-1 \le \rho \le 1$.

Third moments for the bivariate case can be defined by considering the double integrals of $(x - \mu_x)^3$, $(x - \mu_x)^2(y - \mu_y)$, $(x - \mu_x)(y - \mu_y)^2$, and $(y - \mu_y)^3$. The first and last are simply the third moments of the (univariate) marginal distributions of X and Y (see Section A.8); the other two moments are rarely used. Fourth moments can also be defined, where similar comments apply.

A.8 CONDITIONAL DISTRIBUTIONS AND THEIR MEANS

For the bivariate distribution $f(x, y; \theta)$, the marginal distributions of X and Y are obtained by integrating over y and x, respectively. Denoting these distributions by $f_M(x; \theta_x)$ and $f_M(y; \theta_y)$, respectively,

$$f_M(x; \theta_x) = \int_{-\infty}^{\infty} f(x, y; \theta) \mathrm{d}y$$

$$f_M(y; \theta_y) = \int_{-\infty}^{\infty} f(x, y; \theta) \mathrm{d}x$$

The suffices x and y attached to the parameters θ on the left-hand side remind us that in general some of the parameters in θ may disappear when we integrate the function $f(x, y; \theta)$ shown on the right-hand side.

The marginal distributions, as defined above, can be used to obtain the *conditional pdf* of y, given x, and the conditional distribution of x, given y. Denoting these by $f(y|x; \theta)$ and $f(x|y; \theta)$ respectively, the probabilistic relation

$$P[A|B] = P[A \cap B]/P[B]$$

shows that

$$P[(y \le Y < y + \delta y)|(X = x)] = P[(y \le Y < y + \delta y) \cap (X = x)]/P[(X = x)]$$

and the conditional pdf of y given x is therefore given by

$$f(y|x; \theta) = f(x, y; \theta)/f_M(x; \theta_x)$$

the ratio of the bivariate distribution to the marginal distribution of x. Similarly the conditional pdf of x given y is

$$f(x|y; \theta) = f(x, y; \theta)/f_M(y; \theta_y)$$

Conditional expected values $E[Y|x]$, $E[X|y]$ and conditional variances $\text{var}[Y|x]$, $\text{var}[X|y]$ can be calculated for these conditional distributions. Thus

$$E[Y|x] = \int_{-\infty}^{\infty} yf(y|x; \theta)dy$$

$$E[X|y] = \int_{-\infty}^{\infty} xf(x|y; \theta)dx$$

The conditional variance $\text{var}[Y|x]$ is given by

$$\text{var}[Y|x] = E[(Y - E[Y|x])^2|x]$$

$$= \int_{-\infty}^{\infty} (y - E[Y|x])^2 f(y|x; \theta)dy$$

with a similar definition for $\text{var}[X|y]$.

At the expense of some complication in notation, the concepts of marginal and conditional distributions, and of their moments, can easily be generalised to the multivariate case $f(x, y, z, \ldots; \theta)$.

A.9 THE BIVARIATE NORMAL DISTRIBUTION

The concepts of Section A.8 take on a particular importance in the case of the *bivariate normal* distribution

$$f(x, y; \mu_x, \mu_y, \sigma_x, \sigma_y, \rho) = \frac{1}{2\pi\sigma_x\sigma_y\sqrt{(1 - \rho^2)}} \exp\left\{-\frac{1}{2(1 - \rho^2)}\left[\left(\frac{x - \mu_x}{\sigma_x}\right)^2\right.\right.$$

$$\left.\left.- 2\rho\left(\frac{x - \mu_x}{\sigma_x}\right)\left(\frac{y - \mu_y}{\sigma_y}\right) + \left(\frac{y - \mu_y}{\sigma_y}\right)^2\right]\right\}$$

The marginal distributions of x and y, obtained by integrating with respect to y and x over the ranges $(-\infty, \infty)$, are

$$f_M(x; \mu_x, \sigma_x) = (2\pi\sigma_x^2)^{-1/2} \exp[-(x - \mu_x)^2/(2\sigma_x^2)]$$

$$f_M(y; \mu_y, \sigma_y) = (2\pi\sigma_y^2)^{-1/2} \exp[-(y - \mu_y)^2/(2\sigma_y^2)]$$

Dividing the bivariate normal distribution $f(x, y; \mu_x, \mu_y, \sigma_x, \sigma_y, \rho)$ by these marginal distributions, we obtain the conditional distributions. The conditional pdf of Y given x is

$$f(y|x; \mu_x, \mu_y, \sigma_x, \sigma_y, \rho) = \{1/\sqrt{[2\pi\sigma_y^2(1 - \rho^2)]}\}$$

$$\times \exp\{-[y - \mu_y) - \rho\sigma_y(x - \mu_x)/\sigma_x]^2/[2(1 - \rho^2)\sigma_y^2]\}$$

It can be seen that this is a normal distribution with mean

$$E[Y|x] = \mu_y + \rho\sigma_y(x - \mu_x)/\sigma_x$$

and variance

$$\text{var}[Y|x] = \sigma_y^2(1 - \rho^2)$$

The expression for $E[Y|x]$ is of the general form $A + Bx$, showing that when $E[Y|x]$ is plotted against x, the result is a straight line. This result is, for example, the basis of the well-known Thomas–Fiering model, used in the simulation of water resource systems to derive artificial sequences of mean monthly flows.

A.10   THE MULTIVARIATE NORMAL DISTRIBUTION

Random variables $\mathbf{x} = X_1, X_2, \ldots, X_t, \ldots, X_n$ jointly normally distributed have the pdf

$$f(\mathbf{x}; \boldsymbol{\mu}, \boldsymbol{\Sigma}) = \{1/[(2\pi)^{n/2} |\boldsymbol{\Sigma}|^{1/2}]\} \exp[-(\mathbf{x} - \boldsymbol{\mu})^T \boldsymbol{\Sigma}^{-1}(\mathbf{x} - \boldsymbol{\mu})/2]$$

where $\boldsymbol{\mu}$ is the vector of means $[\mu_1, \mu_2, \ldots, \mu_t, \ldots, \mu_n]^T$ and $\boldsymbol{\Sigma}$ is the matrix of variances and covariances

$$\boldsymbol{\Sigma} = \begin{bmatrix} \sigma_1^2 & \rho_{12}\sigma_1\sigma_2 & \cdots & \rho_{1t}\sigma_1\sigma_t & \cdots & \rho_{1n}\sigma_1\sigma_n \\ \rho_{21}\sigma_1\sigma_2 & \sigma_2^2 & \cdots & \rho_{2t}\sigma_2\sigma_t & \cdots & \rho_{2n}\sigma_2\sigma_n \\ \cdots & & & \cdots & & \\ \cdots & & & \cdots & & \cdots & \sigma_n^2 \end{bmatrix}$$

Definitions of marginal and conditional distributions, and of moments about the origin and about means, follow easily from the bivariate normal case given in Section A.9. Particular results are the following:

(i)   All linear combinations of $X_1, X_2, \ldots, X_t, \ldots, X_n$ are normally distributed.

(ii)   All subsets of $X_1, X_2, \ldots, X_t, \ldots, X_n$ have (multivariate) normal distributions.

(iii) Zero covariance implies that the corresponding components are independently distributed.

(iv)   The conditional distributions of the components $X_1, X_2, \ldots, X_t, \ldots, X_n$ are (multi-variate) normal.

Corresponding to the univariate result which says that, when $X_1, X_2, \ldots, X_t, \ldots, X_n$ are distributed $N(\mu, \sigma^2)$, $[(X_1 - \mu)^2 + (X_2 - \mu)^2 + \cdots + (X_t - \mu)^2 + \cdots + (X_n - \mu)^2]/\sigma^2$ has a χ^2-distribution with n degrees of freedom, we have, for the multivariate normal case, the following:

(i) $(\mathbf{x} - \boldsymbol{\mu})^T \boldsymbol{\Sigma}^{-1}(\mathbf{x} - \boldsymbol{\mu})$ is distributed as χ^2 with n degrees of freedom.

(ii)   For the multivariate normal distribution $N(\boldsymbol{\mu}, \boldsymbol{\Sigma})$, the probability is $1 - \alpha$ that $\mathbf{x}$ lies within the solid ellipsoid $(\mathbf{x} - \boldsymbol{\mu})^T \boldsymbol{\Sigma}^{-1}(\mathbf{x} - \boldsymbol{\mu}) \leq \chi^2(\alpha)$, where $\chi^2(\alpha)$ is the quantile of the χ^2-distribution with n degrees of freedom, defining probability α in the distribution's upper tail.

A.11   MOMENTS OF FUNCTIONS OF TWO OR MORE RANDOM VARIABLES PUTTING $X = [X_1, X_2, \ldots, X_T, \ldots, X_N]^T$, $A = [A_1 A_2, \ldots, A_T, \ldots, A_N]^T$, SOME IMPORTANT RESULTS FOLLOW WHEN $G(X) = A^T X$:

(i) $E[\mathbf{a}^T \mathbf{x}] = \mathbf{a}^T \boldsymbol{\mu}$, where $\boldsymbol{\mu}$ is the $n \times 1$ vector of means of $X_1, X_2, \ldots, X_t, \ldots, X_n$.

(ii) $\text{var}[\mathbf{a}^T \mathbf{x}] = \mathbf{a}^T \boldsymbol{\Sigma} \mathbf{a}$, where $\boldsymbol{\Sigma}$ is the variance-covariance matrix of $\mathbf{x}$.

These results are true whatever the probability distribution of x. In the particular case $\boldsymbol{\mu} = [\mu, \mu, \ldots, \mu]^T$, $\boldsymbol{\Sigma} = \text{diag}(\sigma^2, \sigma^2, \ldots, \sigma^2, \ldots, \sigma^2)$, $\mathbf{a} = [1/n, 1/n, \ldots, 1/n]^T$, we have $E[\overline{\mathbf{x}}] = \mu$, $\text{var}[\overline{x}] = \sigma^2/n$.

Where $g(.)$ is any general function of the n random variables $X_1, X_2, \ldots, X_t, \ldots, X_n$ with means given by the vector $\boldsymbol{\mu} = [\mu_1, \mu_2, \ldots, \mu_t, \ldots, \mu_n]^T$ and variance-covariance matrix $\boldsymbol{\Sigma}$, then approximately

$$E[g(X_1, X_2, \ldots, X_t, \ldots, X_n)] = g(\mu_1, \mu_2, \ldots, \mu_t, \ldots, \mu_n)$$

and

$$\text{var}[g(X_1, X_2, \ldots, X_t, \ldots, X_n)] = [\mathbf{Dg}]^T \boldsymbol{\Sigma} [\mathbf{Dg}]$$

where

$$\mathbf{Dg} = [\partial g/\partial X_1, \partial g/\partial X_2, \ldots, \partial g/\partial X_t, \ldots, \partial g/\partial X_n]^T$$

and all partial derivatives $\partial g/\partial X_1, \partial g/\partial X_2, \ldots, \partial g/\partial X_t, \ldots, \partial g/\partial X_n$ are evaluated at $X_1 = \mu_1, X_2 = \mu_2, \ldots, X_n = \mu_n$.

A.12 MAXIMUM-LIKELIHOOD ESTIMATION OF μ, Σ FOR THE MULTIVARIATE NORMAL $N(\mu, \Sigma)$

We assume a sample of N vectors, sampled independently from $N(\boldsymbol{\mu}, \boldsymbol{\Sigma})$:

$$\mathbf{x}_1 = [X_{11}, X_{21}, \ldots, X_{t1}, \ldots, X_{n1}]^T$$
$$\mathbf{x}_2 = [X_{12}, X_{22}, \ldots, X_{t2}, \ldots, X_{n2}]^T$$
$$\cdots$$
$$\mathbf{x}_t = [X_{1t}, X_{2t}, \ldots, X_{tt}, \ldots, X_{nt}]^T$$
$$\cdots$$
$$\mathbf{x}_N = [X_{1N}, X_{2N}, \ldots, X_{tN}, \ldots, X_{nN}]^T$$

with mean $\overline{\mathbf{x}}$. Let $\mathbf{S} = \Sigma_{i=1}^{N}(\mathbf{x}_i - \overline{\mathbf{x}})(\mathbf{x}_i - \overline{\mathbf{x}})^T/(N-1)$. Then

(i) $\overline{\mathbf{x}}$ and $\mathbf{S}$ are independent.

(ii) $\overline{\mathbf{x}}$ and $\mathbf{S}$ are *sufficient statistics* for $\boldsymbol{\mu}$ and $\boldsymbol{\Sigma}$, containing all available information on these parameters, whatever the size N of sample.

(iii) $\overline{\mathbf{x}}$ is distributed as $N(\boldsymbol{\mu}, \boldsymbol{\Sigma}/N)$.

(iv) $(N-1)\mathbf{S}$ has the Wishart distribution with $N-1$ degrees of freedom (the Wishart distribution being analogous to the χ^2 distribution in the univariate case).

REFERENCES

Cox, D. R. and Hinckley, D. V. (1974). *Theoretical Statistics*. Chapman and Hall, London.
Cramer, H. (1946). *Mathematical Methods of Statistics*. Princeton University Press, Princeton, NJ.

Bibliography

GENERAL READING

Draper, N. R. and Smith, H. (1981). *Applied Regression Analysis* (2nd edition). Wiley, New York.
GENSTAT 5 Committee, Rothamsted Experimental Station (1992). *GENSTAT 5 Reference Manual.* Clarendon Press, Oxford.
Haan, C. T. (1977). *Statistical Methods in Hydrology.* Iowa State University Press, Ames.
Kendall, M. G. and Stuart, A. (1979). *The Advanced Theory of Statistics* (3 volumes). Oxford University Press, Oxford.
Linsley, R. K. Jr, Kohler, M. A. and Paulhus, J. L. H. (1982). *Hydrology for Engineers* (3rd edition). McGraw-Hill, New York.
MATLAB User's Guide (1993) The MathWorks, Inc, Natick, MA.
MATLAB Reference Guide (1993) The MathWorks, Inc, Natick, MA.
McCuen, R. H. (1985). *Statistical Methods for Engineers.* Prentice-Hall, Englewood Cliffs, NJ.
McCuen, R. H. (1989). *Hydrologic Analysis and Design.* Prentice-Hall, Englewood Cliffs, NJ.
Snedecor, G. W. and Cochran, W. G. (1980). *Statistical Methods* (7th edition) Iowa State University Press, Ames.

CHAPTER 1

Bastable, H. G., Shuttleworth, W. J., Dallarosa, R. L. G., Fisch, G. and Nobre, C. A. (1993). Observations of climate, albedo, and surface radiation over cleared and undisturbed Amazon forest. *Int. J. Climatology*, **13**, 783–96.
Chow, V. T., Maidment, D. R. and Mays, L. W. (1988). *Applied Hydrology.* McGraw-Hill, New York.
Loucks, D. P., Stedinger, J. R. and Haith, D. A. (1981). *Water Resource Systems Planning and Analysis.* Prentice-Hall, Englewood Cliffs, NJ.
Nachazel, K. (1993). *Estimation Theory in Hydrology and Water Systems.* Vol. 42 of Developments in Water Science. Elsevier Science Publishers, Amsterdam.
Sudler, C. E. (1927). Storage required for the regulation of streamflow. *Trans. ASCE*, **91**, 622–60.
SAS (1985). *SAS Users' Guide: Statistics* SAS Institute, Gary, North Carolina.
Wallis, J. R. (1988). The GIS/hydrology interface: the present and the future. *Environmental Software*, **3**, 171–73.
Wallis, J. R. (1991). The interface between GIS and hydrology. In Loucks, D. P. and da Costa, J. R. (eds), *Decision Support Systems.* NATO ASI Series Vol. G 26. Springer-Verlag, Berlin, 189–97.
Yevjevich, V. (1972). *Probability and Statistics in Hydrology.* Water Resources Publications, Fort Collins, CO.

CHAPTER 2

Aitkin, M., Anderson, D., Francis, B. and Hinde, J. (1989). *Statistical modelling in GLIM.* Oxford University Press, Oxford.
Anscombe, F. A. (1973). Graphs in statistical analysis. *Amer. Statist.*, **27**, 17–21.

Atkinson, A. C. (1987). *Plots, Transformations and Regression: An Introduction to Graphical Methods of Diagnostic Regression Analysis.* Oxford University Press, Oxford.

Chambers, J. M., Cleveland, W. S., Kleiner, B. and Tukey, P. A. (1983). *Graphical Methods for Data Analysis.* Duxbury Press, Boston, MA.

Cleveland, W. S. and McGill, R. (1984). The many faces of a scatterplot. *J. Amer. Statist. Assoc.*, **79**, 807–22.

Fennessey, N. and Vogel, R. M. (1990). Regional flow-duration curves for ungauged sites in Massachusetts. *J. Water Resour. Planning and Management*, **116**, 530–49.

Greenwood, J. A., Landwehr, J. M., Matalas, N. C. and Wallis, J. R. (1979). Probability weighted moments: definitions and relation to parameters of several distributions expressed in inverse form. *Water Resour. Res.*, **15**, 1049–54.

Hoaglin, D. C., Mosteller, F. and Tukey, J. W. (1983). *Understanding Robust and Exploratory Data Analysis.* Wiley, New York.

McGill, R., Tukey, J. W. and Larsen, W. A. (1978). Variations of box-plots. *Amer. Statist.*, **32**, 12–16.

Vellemann, P. F. and Hoaglin, D. C. (1981). *Applications, Basics and Computing of Exploratory Data Analysis.* Duxbury Press, Boston MA.

CHAPTER 3

Aitchison, J. and Brown, J. A. C. (1981). *The Lognormal Distribution.* Cambridge University Press, Cambridge.

Ashkar, F. and Bobee, B. (1988). Confidence intervals for flood events under a Pearson 3 or log Pearson 3 distribution. *Water Resour. Bull.*, **24**, 639–50.

Bobee, B. (1975). The log-Pearson Type 3 distribution and its application in hydrology. *Water Resour. Res.*, **11**, 681–89.

Bobee, B. and Robitaille, R. (1977). The use of the Pearson Type 3 and log Pearson Type 3 distributions revisited, *Water Resour. Res.*, **13**, 427–43.

Brakensiek, D. L. (1958). Fitting generalised log-normal distribution to hydrologic data. *Trans. Amer. Geophys. Union*, **39**, 469–73.

Burges, S. J., Lettenmaier, D. P., Bates, C. L. (1975). Properties of the three-parameter log-normal probability distribution. *Water Resour. Res.*, **11**, 229–35.

Chow, V. T. (1954). The log probability law and its engineering applications. *Proc. ASCE*, **80**, 1–25.

Chowdhury, J. U. and Stedinger, J. R. (1991). Confidence interval for design floods with estimated skew coefficient. *J. Hydraulic Eng. ASCE*, **117**, 811–30.

Cunnane, C. (1978). Unbiased plotting positions — a review. *J. Hydrology*, **37**, 205–22.

Cunnane, C. (1989). *Statistical Distributions for Flood Frequency Analysis.* Operational Hydrology Report 33. World Meteorological Organisation, Geneva.

Filliben, J. J. (1975). The probability plot correlation test for normality. *Technometrics*, **17**, 111–17.

Fontaine, T. A. and Potter, K. W. (1989). Estimating probabilistics of extreme rainfalls. *J. Hydraulic Eng. ASCE*, **115**, 1562–75.

Greenwood, J. A. and Durand, D. (1960). Aids for fitting the gamma distribution by maximum likelihood. *Technometrics*, **2**, 55–65.

Gumbel, E. J. (1954). Statistical theory of droughts. *Proc. Hydraulics Division ASCE*, **80**, 1–19.

Gumbel, E. J. (1958). *Statistics of Extreme Values.* Columbia University Press, New York.

Hirsch, R. M. and Stedinger, J. R. (1987). Plotting positions for historical flows and their precision. *Water Resour. Res.*, **23**, 715–27.

Houghton, J. C. (1978). Birth of a parent: the Wakeby distribution for modelling flood flows. *Water Resour. Res.*, **14**, 1105–9.

Hoshi, K., Stedinger, J. R. and Burges, S. (1984). Estimation of lognormal quantiles: Monte Carlo results and first-order approximations. *J. Hydrology*, **71**, 1–30.

Hosking, J. R. M. (1990). *l*-moments: analysis and estimation of distributions using linear combinations of order statistics. *J. Roy. Statist. Soc.*, *B*, **52**, 105–24.

Hosking, J. R. M. and Wallis, J. R. (1991). *Some Statistics Useful in Regional Flood Frequency Analysis.* Report RC 17096. IBM Research, Yorktown Heights, NY.

Hosking, J. R. M., Wallis, J. R. and Wood, E. F. (1985). Estimation of the generalized extreme-value distribution by the method of probability-weighted moments. *Technometrics*, **27**, 251–61.

Kelman, J. (1987). Cheias e aproveitamentos hidroelétricos. *Revista Brasileira de Engenharia*, **1**.

Kite, G. W. (1975). Confidence intervals for design events. *Water Resour. Res.*, **11**, 48–53.

Kite, G. W. (1977). *Frequency and Risk Analysis in Hydrology*. Water Resources Publications, Fort Collins, CO.

Landwehr, J. M., Matalas, N. C. and Wallis, J. R. (1979). Probability weighted moments compared with some traditional techniques in estimating Gumbel parameters and quantiles. *Water Resour. Res.*, **15**, 1055–64.

Landwehr, J. M., Matalas, N. C. and Wallis, J. R. (1979). Estimation of parameters and quantiles of Wakeby distributions. 2: Unknown lower bound. *Water Resour. Res.*, **15**, 1373–79.

Landwehr, J. M., Matalas, N. C. and Wallis, J. R. (1980). Quantile estimation with more or less floodlike distributions. *Water Resour. Res.*, **16**, 547–55.

Matalas, N. and Wallis, J. R. (1973). Eureka! it fits a Pearson type 3 distribution. *Water Resour. Res.*, **9**, 281–9.

National Research Council (1988). *Estimating Probabilities of Extreme Floods; Methods and Recommended Research*. National Academy Press, Washington, DC.

NERC (1975). *Flood Studies Report: Vol. 1: Hydrological studies*. Natural Environment Research Council, London.

Rossi, F., Fiorentino, M. and Versace, P. (1984). Two-component extreme value distribution for flood frequency analysis. *Water Resour. Res.*, **20**, 267–9.

Stedinger, J. R. (1980). Fitting log-normal distributions to hydrologic data. *Water Resour. Res.*, **16**, 481–90.

Stedinger, J. R. (1983). Design events with specified flood risk. *Water Resour. Res.*, **19**, 511–22.

Stedinger, J. (1983). Confidence intervals for design events. *J. Hydraul. Eng. ASCE*, **109**, 13–27.

CHAPTER 4

Atkinson, A. C. (1981). Two graphical displays for outlying and influential observations in regression. *Biometrika*, **68**, 13–20.

Atkinson, A. C. (1982). Regression diagnostics, transformations and constructed variables. *J. Roy. Statist. Soc., B*, **44**, 1–36.

Belsley, D. A., Kuh, E. and Welsch, R. E. (1980). *Regression Diagnostics*. Wiley, New York.

Bidwell, V. J. (1971). Regression analysis of non-linear catchment systems. *Water Resour. Res.*, **7**, 1118–26.

Cook, R. D. (1977). Detection of influential observations in linear regression. *Technometrics*, **19**, 15–16.

Cook, R. D. (1979). Influential observations in linear regression. *J. Amer. Statist. Ass.*, **74** (365), 169–74.

Cook, R. D. and Weisberg, S. (1982). *Residuals and Influence in Regression*. Chapman and Hall, London.

Cox, D. R. and Snell, E. J. (1968). A general definition of residuals. *J. Roy. Statist. Soc. B*, **2**, 248–75.

Hoerl, A. E. and Kennard, R. W. (1970). Ridge regression: biassed estimation for non-orthogonal problems. *Technometrics*, **12**, 55–67.

Matalas, N. C. and Gilroy, E. J. (1968). Some comments on regionalization in hydrologic studies. *Water Resour. Res.*, **4**, 1361–9.

Miller, A. J. (1984). Selection of subsets of regression variables (with discussion) *J. Roy. Statist. Soc. A*, **147**, 389–425.

Neter, J., Wasserman, W. and Kutner, M. H. (1985). *Applied Linear Statistical Models* (2nd edition). Irwin, Homewood, IL.

Montgomery, D. C. and Peck, E. A. (1982). *Introduction to Linear Regression Analysis*. Wiley, New York.

Stedinger, J. R. and Tasker, G. D. (1985). Regional hydrologic analysis I: Ordinary, weighted and generalised least squares compared. *Water Resour. Res.*, **21**, 1421–32.

Stedinger, J. R. and Tasker, G. D. (1986). Regional hydrologic analysis II: Model-error estimators, estimation of sigma, and log-Pearson type 3 distributions. *Water Resour. Res.*, **22**, 1487–99.

Tasker, G. D. and Stedinger, J. R. (1989). An operational GLS model for hydrologic regression. *J. Hydrology*, **111**, 361–75.

CHAPTER 5

Adams, R. S., Black, T. A. and Fleming, R. L. (1991). Evapotranspiration and surface conductance in a high elevation, grass-covered forest clearcut. *Agricultural and Forest Meteorol.*, **56**, 173–93.

Bates, D. M. and Watts, D. G. (1988). *Nonlinear Regression Analysis and Its Applications*. Wiley, New York.

Dolman, A. J., Gash, J. H. C., Roberts, J. and Shuttleworth, W. J. (1991). Stomatal and surface conductance of tropical rainforest. *Agricultural and Forest Meteorol*, **54**, 303–18.

Ebert, T. A. (1987). Estimating growth and mortality parameters by non-linear regression using average sizes in catches. In Pauly, D. and Morgan, G. R. (eds) *Length-based Methods in Fisheries Research*. International Center for Living Aquatic Resources Management, Kuwait Institute for Scientific Research, 35–44.

Hillel, D. (1971). *Soil and Water: Physical Principles and Processes*. Academic Press, New York.

Horton, R. E. (1933). The role of infiltration in the hydrologic cycle. *EOS Trans. Amer. Geophys. Union*, **14**, 446–60.

Lynn, B. H. and Carlson, T. N. (1990). A stomatal resistance model illustrating plant vs. external control of transpiration. *Agricultural and Forest Meteorol.*, **52**, 5–43.

Ricker, W. E. (1958). *Handbook of Computations for Biological Statistics of Fish Populations*. Fisheries Research Board of Canada.

Seber, G. A. F. and Wild, C. J. (1989). *Nonlinear Regression*. Wiley, New York.

Shuttleworth, W. J. (1989). Micrometeorology of temperate and tropical forest. *Phil. Trans. R. Soc. Lond. B*, **324**, 299–334.

Stewart, J. B. (1989). On the use of the Penman–Monteith equation for determining areal evaporatranspiration. In *Estimation of Areal Evapotranspiration* (Proceedings of workshop at Vancouver B.C., Canada, August 1987). IAHS Pub. No. 177, 3–11.

Wright, I., Gash, J. H. C., da Rocha, H. R., Shuttleworth, W. J., Nobre, C. A., Maitelli, G. T., Zamparoni, C. A. G. P. and Carvalho, P. R. A. (1992). Dry season micrometeorology of central Amazonian ranchland. *Q. J. Roy. Meteorol. Soc.*, **118**, 1083–99.

CHAPTER 6

Brown, D. (1992). A graphical analysis of deviance. *Appl. Statist.* **41**, 55–62.

Cordeiro, G. M. and McCullagh, P. (1991). Bias correction in generalized linear models. *J. Roy. Statist. Soc. B*, **53**, 629–43.

Hosmer, D. W. and Lemeshow, S. (1989). *Applied Logistic Regression*. Wiley, New York.

Jorgensen, B. (1987). Exponential dispersion models. *J. Roy. Statist. Soc., B*, 127–61.

Landwehr, J. M., Pregibon, D. and Shoemaker, A. C. (1984). Graphical methods for assessing logistic regression models. *J. Amer. Statist. Ass.*, **79** (385), 61–71.

McCullagh, P. (1980). Regression models for ordinal data. *J. Roy. Statist. Soc. B*, **42**, 109–42.

Nelder, J. A. (1977). A reformulation of linear models. *J. Roy. Statist. Soc., A*, **140**, 48–76.

Pierce, D. A. and Schafer, D. W. (1986). Residuals in generalized linear models. *J. Amer. Statist. Ass.*, **81** (396), 977–86.

Pregibon, D. (1980). Goodness of link tests for generalized linear models. *Appl. Statist.*, **29**, 15–24.

Pregibon, D. (1981). Logistic regression diagnostics. *Annals of Statistics*, **9**, 705–24.

Smyth, G. K. (1989). Generalized linear models with varying dispersion. *J. Roy. Statist. Soc. B*, **51**, 47–60.

Thompson, R. and Baker, R. J. (1981). Composite link functions in generalized linear models. *Appl. Statist.*, **30**, 125–31.

Williams, D. A. (1987). Generalized linear model diagnostics using the deviance and single case deletions. *Appl. Statistics*, **36**, 181–91.

Young, D. H. and Bakir, S. T. (1987). Bias correction for a generalized log-gamma regression model. *Technometrics*, **29**, 183–91.

CHAPTER 7

Atkinson, A. C. and Pearce, M. C. (1976). The computer generation of beta, gamma and normal random variables. *J. Roy. Statist. Soc. A*, **139**, 431–60.

Beale, E. M. L. and Little, R. J. A. (1975). Missing values in multivariate analysis. *J. Roy. Statist. Soc. B*, 129–45.

Benson, M. A. (1950). Use of historical data in flood frequency analysis. *Trans. Amer. Geophys. Union*, **31**, 419–24.

Brown, P. J. (1982). Multivariate calibration. *J. Roy. Statist. Soc. B*, **44**, 287–321.

Buxton, J. R. (1991). Some comments on the use of response variable transformations in empirical modelling. *Appl. Statist.*, **40**, 391–400.

Cohn, T. A. (1988). *Adjusted Maximum Likelihood Estimation of the Moments of Lognormal Populations from Type I Censored Samples*. US Geological Survey Open File Report 88–350.

Cohn, T. A. and Stedinger, J. R. (1987). Use of historical information in a maximum likelihood framework. *J. Hydrology*, **96**, 215–23.

Condie, R. and Lee, K. (1982). Flood frequency analysis with historical information. *J. Hydrology*, **58**, 47–61.

Dempster, A. P., Laird, N. M. and Rubin, D. B. (1977). Maximum likelihood from incomplete data via the EM algorithm. *J. Roy. Statist. Soc. B*, **39**, 1–22.

Ferguson, R. I. (1986). River loads underestimated by rating curves. *Water Resour. Res.*, **22**, 74–6.

Fiering, M. B. (1966). Synthetic hydrology: an assessment. In Kneese, A. V. and Smith, S. C. (eds), *Water Research*. Resources for the Future, Washington, DC, 331–41.

Fiering, M. B. (1967). *Streamflow Synthesis*. Harvard University Press, Cambridge, MA.

Fiering, M. B. and Jackson, B. B. (1971). *Synthetic Streamflows*. Water Resources Monograph 1, Amer. Geophys. Union, Washington, DC, 1–98.

Fishman, G. S. (1973). *Concepts and Methods in Discrete Event Digital Simulation*. Wiley, New York.

Grygier, J. C., Stedinger, J. R. and Yin, H. (1989). A generalised maintenance of variance extension procedure for extending correlated series. *Water Resour. Res.*, **25**, 345–9.

Hirsch, R. M. (1982). A comparison of four record extension techniques. *Water Resour. Res.*, **18**, 1081–8.

Hufschmidt, M. M. and Fiering, M. B. (1966). *Simulation Techniques for Design of Water Resource Systems*. Harvard University Press, Cambridge, MA.

Johnson, R. A. and Wichern, D. W. (1992). *Applied Multivariate Statistical Analysis* (3rd edition). Prentice-Hall, Englewood Cliffs, NJ.

Krzanowski, W. J. (1990). *Principles of Multivariate Analysis: A User's Perspective*. Oxford University Press, Oxford.

Matalas, N. C. and Jacobs, B. (1964). A correlation procedure for augmenting hydrologic data. *US Geological Survey Professional Paper*, **434–E**, E1–E7.

Morrison, D. F. (1976). *Multivariate Statistical Methods*. McGraw-Hill, New York.

Salas, J. D., Delleur, J. W., Yevjevich, V. and Lane, W. L. (1980). *Applied Modelling of Hydrologic Time Series*. Water Resources Publications, Fort Collins, CO.

Vogel, R. M. and Stedinger, J. R. (1985). Minimum variance streamflow record augmentation procedures. *Water Resour. Res.*, **21**, 715–23.

CHAPTER 8

Beven, K. J. (1989). Changing ideas in hydrology: the case of physically-based models. *J. Hydrology*, **105**, 157–72.

Beven, K. J. and Binley, A. (1992). The future of distributed models: model calibration and uncertainty prediction. *Hydrological Processes*, Special Issue on the Future of Distributed Models, 1–20.

Beven, K. J. and Kirkby, M. J. (1979). A physically-based variable contributing area model of basin hydrology. *Hydrol. Sci. Bull.*, **24**, 43–69.

Beven, K. J., Kirkby, M. J. Schofield, N. and Tagg, A. F. (1984). Testing a physically-based flood forecasting model (TOPMODEL) for three UK catchments. *J. Hydrology*, **69**, 119–43.

Beven, K. J. and O'Connell, P. E. (1982). *On the Role of Distributed Models in Hydrology*. Report No. 81. Institute of Hydrology, Wallingford.

Beven, K. J., Warren, R. and Zaoui, J. (1980). SHE: towards a methodology for physically-based forecasting in hydrology. In *Hydrological Forecasting*. IAHS Publication No. 129, 133–7.

Binley, A. M. and Beven, K. J. (1991). Physically-based modelling of catchment hydrology: a likelihood approach to reducing predictive uncertainty. In Farmer, D. G. and Rycroft, M. J. (eds), *Computer Modelling in the Environmental Sciences*. IMA Series No. 28. Clarendon Press, Oxford, 76–87.

Binley, A. M., Beven, K. J., Calver, A. and Watts, L. G. (1991). Changing responses in hydrology: assessing the uncertainty in physically based model predictions. *Water Resour Res.*, 27, 1253–61.

Calver, A. (1988). Calibration, sensitivity and validation of a physically-based rainfall-runoff model. *J. Hydrology*, 103, 102–15.

Crawford, N. H. and Linsley, R. K. (1966). *Digital Simulation in Hydrology: Stanford Watershed Model IV*. Tech. Report No. 39. Dept. Civil Engineering, Stanford University.

Dawdy, D. R. and O'Donnell, T. (1965). Mathematical models of catchment behaviour. *J. Hydraulics Division Proceedings ASCE*, 91 (HY4), 123.

Dooge, J. C. I. (1977). Problems and methods of rainfall-runoff modelling. In Ciriani, T. A., Maione, U. and Wallis, J. R. (eds), *Mathematical Models for Surface Water Hydrology*. Wiley, New York.

Dunne, T. and Black, R. D. (1970). Partial area contributions to storm runoff in a small New England watershed. *Water Resour. Res.*, 6, 1296–1311.

Eeles, C. W. O. (1978). *A Conceptual Model for the Estimation of Historic Flows*. Report No. 55. Institute of Hydrology, Wallingford.

Freeze, R. A. (1980). A stochastic-conceptual analysis of rainfall-runoff processes. *Water Resour. Res.*, 16, 391–408.

Gupta, V. K. and Soorooashian, S. (1985). The relationship between data and the precision of parameter estimates of hydrologic models. *J. Hydrology*, 81, 57–77.

Ibbitt, R. P. (1974). Effects of random data errors on the parameter values for a conceptual model. *Water Resour. Res.*, 8, 70–8.

Ibbitt, R. P. and O'Donnell, T. E. (1971). Fitting methods for conceptual catchment models. *Jour. ASCE Hydraulics Div.*, 97 (HY9), 1331–42.

Kuczera, G. (1983). Improved parameter inference in catchment models. 2: Combining different kinds of hydrologic data and testing their compatibility. *Water Resour. Res.*, 19, 1163–72.

Kuczera, G. (1990). Assessing hydrologic model nonlinearity using response surface plots. *J. Hydrology*, 118, 143–62.

Linsley, R. K. (1982). Rainfall-runoff models: a review. In Singh, V. P. (ed.), *Statistical Analysis of Rainfall and Runoff*. Water Resources Publications, Fort Collins, CO.

Linsley, R. K. and Crawford, N. H. (1960). Computation of a synthetic streamflow record on a digital computer. *Hydrol. Sci. Bulletin*. IAHS Publication No. 51, 526–38.

Mandeville, A. N., O'Connell,, P. E., Sutcliffe, J. V. and Nash, J. E. (1970). River flow forecasting through conceptual models. III: The Ray catchment at Grendon Underwood. *J. Hydrology*, 11, 109–28.

Moore, R. J. and Clarke, R. T. (1981). A distribution-function approach to rainfall-runoff modelling. *Water Resour. Res.*, 17, 1367–82.

Moore, R. J. and Clarke, R. T. (1982). A distribution function approach to modelling basin soil moisture deficit and streamflow. In Singh, V. P. (ed.), *Statistical Analysis of Rainfall and Runoff*. Water Resources Publications, Fort Collins, CO.

Moore, R. J. and Clarke, R. T. (1983). A distribution function approach to modelling basin sediment yield. *J. Hydrology*, 65, 239–57.

Nash, J. E. and Sutcliffe, J. V. (1970). River flow forecasting through conceptual models. I: A discussion of principles. *J. Hydrology*, 10, 282–90.

Nelder, J. A. and Mead, R. (1965). A simplex method for function minimisation. *Computer J.*, 7, 308–13.

Pickup, G. (1977). Testing the efficiency of algorithms and strategies for automatic calibration of rainfall runoff models. *Hydrol. Sci. Bull.*, 22, 257–74.

Pilgrim, D. H. (1975). Model evaluation, testing and parameter estimation in hydrology. In Chapman, T. G. and Dunin, F. X. (eds), *Prediction in Catchment Hydrology*. Australian Academy of Science, 305–33, Canberra.

Rosenbrock, H. H. (1960). An automatic method of finding the greatest or least value of a function. *Computer J.*, 3, 175–84.

Sorooshian, S. and Dracup, J. (1980). Stochastic parameter estimation procedures for hydrologic rainfall-runoff models: correlated and heteroscedastic error cases. *Water Resour. Res.*, **16**, 430–42.
Sorooshian, S., Gupta, V. K. and Fulton, J. L. (1983). Evaluation of maximum likelihood parameter estimation techniques for conceptual rainfall-runoff models: influence of calibration data variability and length on model credibility. *Water Resour. Res.*, **19**, 251–9.
Tucci, C. E. M., Sanchez, J. and Lopes, M. S. (1981). Modelo IPHII de simulação precipitação-vazao na bacia: alguns resultados. *Proc. IV Symposio brasileiro de hidrologia e recursos hidricos*, Fortaleaza, 15–19 November 1981, Vol. 2, 83–103.

CHAPTER 9

Bras, R. L. and Rodriguez-Iturbe, I. (1985). *Random Functions and Hydrology*. Addison-Wesley, Reading, Massachusetts.
Chua, S. H. and Bras, R. L. (1980). *Estimation of Stationary and Non-stationary Fields: Kriging in the Analysis of Arographic Precipitation*. MIT Technical Report 255. Ralph M. Parsons Laboratory for Water Resources and Hydrodynamics, Cambridge, MA.
Chua, S. H. and Bras, R. L. (1982). Optimal estimators of mean areal precipitation in regions of orographic influence. *J Hydrology*, **112**, 23–48.
Cressie, N. A. C. (1993). *Statistics for Spatial Data* (revised edition). Wiley, New York.
Cressie, N. A. C. and Hawkins, D. M. (1980). Robust estimation of the variogram: I. *J. Math. Geology*, **12**, 115–25.
Delfiner, P. and Delhomme, J. P. (1973). Opimum interpolation by kriging. In Davis, J. C. and McCullagh, M. J. (eds), *Display and Analysis of Spatial Data*. Wiley, New York.
Delhomme, J. P. (1978). Kriging in the hydrosciences. *Advances in Water Resour.*, **1**, 251–6.
Delhomme, J. P. (1979). Spatial variability and uncertainty in groundwater flow parameters: a geostatistical approach. *Water Resour. Res.*, **15**, 269–80.
Doctor, P. G. and Nelson, R. W. (1981). Geostatistical estimation of parameters for hydrological transport modelling. *J. Math. Geology*, **13**, 415–28.
Gambolati, G. and Volpi, G. (1979). Groundwater contour mapping in Venice by stochastic interpolators I. Theory. *Water Resour. Res.*, **15**, 281–90.
Gambolati, G. and Volpi, G. (1979). A conceptual deterministic analysis of the kriging technique in hydrology. *Water Resour. Res.*, **15**, 625–9.
Hughes, J. P. and Lettenmaier, D. P. (1981). Data requirements for kriging estimation and network design. *Water Resour. Res.*, **17**, 1641–50.
Huijbregts, C. J. and Matheron, G. (1970). Universal kriging: an optimum approach to trend surface analysis. In *Decision-making in the Mining Industry*. CIM International Symposium, Vol. 12, 159–60.
Journel, A. G. and Huijbregts, C. J. (1978). *Mining Geostatistics*. Academic Press, New York.
Kafritsas, J. and Bras, R. L. (1981). *The Practice of Kriging*. MIT Technical Report 263. Ralph M. Parsons Laboratory for Water Resources and Hydrodynamics, Cambridge, MA.
Kitanidis, P. K. (1983). Statistical estimation of polynomial generalized covariance functions and hydrologic applications. *Water Resour. Res.*, **19**, 909–21.
Matern, B. (1960). Spatial variation. *Comm. Swed. Forestry Inst.*, **49**, 1–144.
Matheron, G. (1973). The intrinsic random functions and their applications. *Adv. Appl. Probability*, **5**, 439–68.
Nielsen, D. R., Biggar, J. W. and Ehr, K. T. (1973). Spatial variability of field-measured soil-water properties. *Hilgardia*, **42**, 215–60.
Villeneuve, J. P., Morin, G., Bobee, B. and Leblanc, D. (1979). Kriging in the design of streamflow sampling networks. *Water Resour. Res.*, **15**, 1833–40.
Volpi, G., and Gambolati, G. (1978). On the use of a main trend for the kriging technique in hydrology. *Adv. Water Resources*, **1**, 345–9.
Volpi, G., Gambolati, G., Cargognin, L., Gatto, P. and Mozzi, G. (1979). Groundwater contour mapping in Venice by stochastic interpolators 2: Results. *Water Resour. Res.*, **15**, 291–7.
Wood, E. F., Sivapalan, M., Beven, K. J. and Band, L. (1988). Effects of spatial variability and scale with implications to hydrologic modelling. *J. Hydrology*, **102**, 29–47.

CHAPTER 10

Box, G. E. P. and Jenkins, G. M. (1970). *Time Series Analysis: Forecasting and Control*. Holden Day, San Francisco.

Chapman, T. G. (1975). Trends in catchment modelling. In Chapman, T. G. and Dunin, F. X. (eds), *Prediction in Catchment Hydrology*. Australian Academy of Science, Canberra.

Klemes, V. (1985). *Sensitivity of Water Resource Systems to Climate Variations*. World Climate Programme 98: World Meteorological Organisation, Geneva.

Index